John the Physician's *Therapeutics*

Studies in Ancient Medicine

Edited by

John Scarborough

Philip J. van der Eijk
Ann Ellis Hanson

VOLUME 37

John the Physician's *Therapeutics*

A Medical Handbook in Vernacular Greek

By

Barbara Zipser

BRILL

LEIDEN • BOSTON
2009

On the cover: Medical Society London 14, p. 295–296, Wellcome Library London, containing a medical compilation.

This book is printed on acid-free paper.

Library of Congress Cataloging-in-Publication Data

Zipser, Barbara.
John the Physician's *Therapeutics* : a medical handbook in vernacular Greek / by Barbara Zipser.
p. cm. — (Studies in ancient medicine ISSN 0925-1421 ; v.37)
Includes bibliographical references and index.
ISBN 978-90-04-17723-9 (hardback : alk. paper)
1. John the Physician, 13th cent. Therapeutics. 2. Medicine, Medieval. 3. Medicine, Greek and Roman. I. Title. II. Title: Therapeutics : a medical handbook in vernacular Greek. III. Series.
[DNLM: 1. John the Physician, 13th cent. Therapeutics. 2. Therapeutics. 3. History, Medieval. 4. Manuscripts, Medical. W1 ST918K v.27 2009 / WZ 294 Z79j 2009]

R128.Z57 2009
610.938—dc22

2009017890

ISSN 0925-1421
ISBN 978 9004 17723 9

PRINTED IN THE NETHERLANDS

Opus exegi.

TABLE OF CONTENTS

LIST OF ILLUSTRATIONS

ACKNOWLEDGEMENTS

After the completion of my PhD at the department of Classical philology at Heidelberg University, I took over a three year fellowship at the Wellcome Trust Centre at UCL, embarking on a new project, an edition of a presumably late Medieval text. As there was virtually no secondary literature on the topic, I had to start from scratch, transcribing an obviously damaged manuscript and deciphering the late dialect of the text it contained. The first weeks of my project resembled a giant jigsaw puzzle. Now, that the edition is completed, I finally have the opportunity to thank all those who have supported me throughout the project.

My UCL post was funded by the Wellcome Trust, without whose generous support the present volume would not be there (072287). The book was sent to press during a subsequent Wellcome Trust Project Grant at Royal Holloway University of London (082296).

Vivian Nutton (London) first pointed the text out to me. Later on, he became the sponsor of my project, and always supported me, also beyond the duration of my fellowship. Vivian also put me in touch with David Bennett (Independent Scholar), who became my main contact for manuscript matters. Klaus-Dietrich Fischer (Mainz) acted as my mentor ever since I was a young doctoral student.

The editorial board of SAM and Brill accepted the book for publication; in particular, I would like to extend my thanks to Philip van der Eijk (Newcastle) for his suggestions on the final version of the book and to Caroline van Erp for editorial assistance.

A number of colleagues contributed in various ways to this project: David Holton (Cambridge) identified a number of obscure words and checked the language of the text for dialectal elements. Guido Mensching (Berlin) looked into a Lombard matter. Chrysanthe Tsitsiou-Chelidone (Komotini) and Dionysios Stathakopoulos (London) helped with lexicographical and grammatical queries. Laurance Totelin (Cambridge) identified a gynecological procedure. Lara Artemis (Wellcome Library), Brigitte Mondrain (Paris) and Agamemnon Tselikas (Athens) looked at a rather enigmatic manuscript.

Dagmar Schweinfurth (Heidelberg) proof read the entire book and Nikolai Serikoff (Wellcome Library) took over the task of double-checking ω. Both worked their way through the text with competence and accuracy; remaining errors are of course my own. I also profited from comments by Sally Bragg and Karen Buckle (both London), Peregrine Horden (London/Oxford), Christos Papadopoulos and Lois Reynolds (both London), Anita von Poser (Heidelberg) and Alain Touwaide (Smithsonian).

Last not least, I would like to acknowledge a group of people who made it possible to actually process the enormous amount of data involved. The core of the project was a stable and versatile IT infrastructure, LaTeX on a Linux platform, at first Ubuntu, and later on SuSE. The edition was written in ledmac, komascript was used for the layout of the book. My thanks go to the open source community who created these excellent programs and made them freely available. Qualified help was provided by Dominik Wujastyk (London), one of the authors of ledmac, Andreas Dafferner (Heidelberg), who wrote the lgreek font, Dieter Irmer (Hamburg), Hannes Pahlke (Darmstadt) and Arthur Zimek (Munich). Helga Satzinger (London) advised on photographic matters.

Barbara Zipser (Royal Holloway University of London)

PART ONE

INTRODUCTION

CHAPTER ONE

AN UNUSUAL TOPIC

This monograph contains the first edition of a Greek medical text from the High Middle Ages. The text is a rare witness from a world we still have little knowledge of, the world of ordinary medical practice — this book is telling us about medicine as a craft, not as a scholarly concept. The text often seems crude and goes into more detail than one would expect in a more learned specimen. It is also partially composed in dialect.

As with most Byzantine medical writings, its theoretical framework is indebted to Galen and other authorities from antiquity. It does not question established medical knowledge, just as modern clinical handbooks would not reflect upon the historical background of physiological concepts such as the vascular system or the discovery of antibiotics, but briefly offer the information one might require in a hospital emergency or when confronted with an unfamiliar case in everyday practice. Thus, the approach is pragmatic and down to earth, the author aims to answer questions such as 'how do I treat a chronic cough' or 'how can I induce labour if the child dies inside the womb'.

It is a very long text, and the work involved in compiling and revising the material involved time-consuming work and philological skill together with first hand medical experience. One might ask who initiated such a project. Was the book commissioned by an authority who also provided the necessary infrastructure for production and made sure the book was passed on to others? Or did a single person prepare it for his own use initially, for instance a practitioner with higher education, who brought one copy on the market which then became a huge success? Was there money to be made, or were philanthropic motives behind what could be seen as an attempt to bring state-of-the-art medical care to ordinary general practice?

The text itself does not give much away about its history, but there are hints from which we can gather that its roots lay in the thirteenth century, maybe on the Greek islands or the surrounding area, in the times after the sack of Constantinople in the fourth crusade, when new states and settlements of the Frankish invaders led to a blossoming of scribal activities and successively to the foundation of monasteries that in turn provided a setting in which book collections could be compiled and preserved for future generations. This is a very likely context for the text to come into existence and flourish, in particular since it was revised several times.[1]

For a brief period of not much more than two hundred years, the mid-thirteenth to the late fifteenth century, the text was widely read and copied and found its way into many medical manuscripts which are now held by libraries all over Europe.[2] After that, and up to the present, the text all but disappeared, and only a small and exclusive number of researchers even mention it.[3] It was probably dismissed because of its dialect and often rough appearance, and never saw the light of day in printed form. The common prejudice that Byzantine medicine was nothing but a period of collecting, compiling and re-arranging medical knowledge — and thus bound to be dull and of no particular interest — did not help. Consequently, the fact this text represents a key to the new developments and innovations of Byzantine medicine remained unnoticed.

A large part of this monograph is devoted to an edition of the two main versions of the text; one of these versions also has a translation and some notes, where it appeared necessary. Other interesting material is presented in analytic indices. The introduction aims at giving some background information on the history of the text, its development, and its sources. Technical issues, such as editorial methods and manuscripts are discussed as well. Finally, two chapters are devoted to the world in which the text was used. This includes information on the language of the two ver-

---

[1] See page 37, 7 and 15.
[2] See page 15.
[3] See page 13.

sions, the way the text talks about language groups and authorities and last but not least the medical practice in which the text was rooted. Additionally, some selected paragraphs of the commentary version of the text are translated and discussed, mainly to illustrate how the commentator worked, and what information he considered worth explaining in more detail.

CHAPTER TWO

THE TEXT

The *Therapeutics* of John the Physician is a working manual designed to serve the basic needs of a practitioner away from the cultural centres of the Byzantine world. In over two hundred chapters, the text contains descriptions of common diseases and provides guidelines on how to prepare medication.

The textual history of the *Therapeutics* is complex. At times it was passed on in more or less exact handwritten copies. But at least twice the text underwent revisions which had a considerable impact. The most dramatic change was the adaptation from learned Byzantine Greek (version ℵ) into the vernacular (version ω). The aim of this redaction was not only to translate the text into the local idiom; the reviser also added a significant amount of material, so that the text was extended by about one quarter of its original length. Some chapters remain of the same length, others are much longer, and the added material included new alternative recipes as well as commentaries on, and explanations of, the original text. It was this commentary which contributed most to the popularity of the text,[4] and ironically, in the course of the second major revision, ω was even translated back into learned Greek.[5]

Despite its popularity, there is no evidence regarding the authorship other than from the book itself. According to some manuscripts, the *Therapeutics* were written by an unknown John who was a physician, but it is unclear whether he was the person who first compiled the text or a later reviser, or whether the name is fictional altogether.[6]

[4] The majority of manuscript copies which have come down to us transmit this later version, compared to just one transmitting the original.

[5] See page 24.

[6] The name 'John' appears in a number of manuscripts of the later version

A likely setting for the development of the *Therapeutics* may have been Cyprus or on the Greek islands[7] in the mid thirteenth to early fourteenth century, but not much later, as the earliest manuscript copy dates from around 1320,[8] and not much earlier, as both main versions mention paper[9] as a medical ingredient (paper became widely available from the early thirteenth century on, and we have to allow for some time for it to be commonly known to the layman.)[10] A castle mentioned in ω was built in early thirteenth century.[11]

ω, but not in the only manuscript transmitting the original version ℵ where the text is attributed to Galen. This could be caused by an error, since the name 'Galen' occurs in the first lines of the preface. On medical authors called 'John' see Temkin 1932: 51-57.

[7] An early manuscript transmitting the later ω version of the *Therapeutics*, MSL 14, was most likely copied in Cyprus, see page 15. A passage in ω also suggests that this version originated from outside the Peloponnese and the Black Sea region, see page 37. Southern Italy can be excluded as a setting because the dialect spoken in this region differs from the idiom of ω.

[8] See page 15.

[9] It appears in a chapter on infected mouth ulcers, 170 in the original and 233 in the later version of the text, and in chapter 73 resp. 89 which is on the treatment of scurf, a condition of the scalp.

[10] For the chronology see for instance Maniaci 2002.

[11] See page 37.

CHAPTER THREE

OTHER TEXTS

A number of practical medical texts similar to the *Therapeutics* survive from the Byzantine period. Yet just a fraction is accessible in printed form,[12] and we have only a very limited amount of indexes, dictionaries, or secondary literature as such on this genre.[13] One of the main reasons is that vernacular Greek was, and often still is regarded as inferior – but there is also a specific methodological issue involved, namely the rather flexible form in which they were transmitted.

Other than the works of the prominent authors of antiquity, practical texts from the Medieval period were often not regarded as rigid entities; scribes did not aim to preserve them in an exact copy but shortened, adjusted and expanded them wherever it appeared necessary. The scribal interventions could range from adding a synonym of a plant name to an entirely new creation based on several manuals. Chapters appear in near identical form in books by different authors, and at least in one instance even a preface is recycled.[14] As a consequence, it becomes increasingly difficult to catalogue manuscripts transmitting working manuals. Similar problems arise in the process of editing. If it is not entirely clear by which routes material has travelled or whether a passage in a practical text is a citation or not, an edition in the conventional manner becomes impossible.

This applies for instance to the so called iatrosophia, a very

---

[12] For more information see Hunger 1978: 2.304.

[13] The only comparable medical texts that were published in a prominent series are on veterinary medicine; the *Hippiatrica* have a language and textual history that is very close to the *Therapeutics*. For more information see McCabe 2006.

[14] This is the case in Theophanes Chrysobalantes and Leo Medicus. See page 9.

common type of book in the Byzantine period.[15] These are excerpts of works by different authors (mainly from the Classical times) which usually but not always fill an entire tome. The length of the excerpts can vary between chapters and single sentences, and mostly but once again not always they are arranged in a thematical order. Often it is hard to distinguish whether the term 'iatrosophion' designates a literary genre or is used as a very flexible form of title for a medical tome, and the terminology can become even more confused. We are aware of medical volumes or groups of texts under this title, but there are also medical codices which would fit the description of an iatrosophion but do not have this name.[16]

Both versions of the *Therapeutics* claim to be an iatrosophion, and one is left wondering about their relationship to other texts with this title, since the *Therapeutics* are too short to fill an entire volume and mostly go back to medieval rather than Classical sources. They partly coincide with Theophanes Chrysobalantes,[17] a comprehensive lexicon on medical practice,[18] and one of the so called Xenon texts, a cluster of medical manuals linked to Byzantine hospitals.[19] The parallels between the three are undeniable and strongly suggest that these texts are related.[20]

But as will be shown in the following, these similarities do not necessarily indicate a linear development from one text to another; on the contrary, it appears as if there must have been a pool of

[15] For an introduction to the genre 'iatrosophion' see Garzya 2003 and Touwaide 2007.

[16] When and where the term 'iatrosophion' was first used is unknown. The problem in tracing the origins of the word are partly of practical nature. If the title of a volume was written on the book cover it might have been lost when a book was rebound. Moreover, titles as such were frequently altered. There is certainly a massive amount of evidence from the fifteenth century onwards. See for instance the manuscripts catalogued in Karas 1994.

[17] This work is partly identical with Leo medicus. For more information see Zipser 2005: 110.

[18] For a detailed study in structure and transmission of Theophanes see Sonderkamp 1987.

[19] For more information and an edition of several xenon texts see Bennett 2003.

[20] For a discussion see Bennett 2003: 259-263.

medical knowledge which is now lost and which was used as the basis for various manuals. These then developed into separate works, and in the end this led to what we have today: a number of partially identical texts of unclear relationship to each other.

The *Therapeutics* consists, roughly speaking, of two cycles going from head to toe, *a capite ad calcem*: the first cycle ends with chapter 65 in the original and with 77 in the later version. The second runs from 66/78 to about 135/164 and is followed by a number of chapters on random topics. The first cycle appears to be a rather clumsy compilation of at least two different sources that were joined together at chapter 13/14.[21] The first part coincides with a Xenon text[22] and the second, more extensive, part is similar to Theophanes Chrysobalantes. The second cycle is in general very similar to Theophanes.[23]

Surprisingly, these similarities are not limited to the original version of the *Therapeutics* as one might expect. Some of the chapters which are specific to the later version correspond to Theophanes as well.[24] And to add to the confusion, some chapters

---

[21] In ℵ (and with slight differences also in ω), the *Therapeutics* start with the diseases of the head (1-3), then followed by the eyes (4-5), runny nose (6), spitting blood (7), pain in the tongue (8), hardening of the bowels (9), liver pain (10), hardening of the spleen (11), kidney pain (12), anasarca (13) and then again with ear-problems (14) and other diseases located in the head, but no general head-problems as for instance headache. In what follows, the sequence *a capite ad calcem* is kept until the end of the cycle. Obviously chapters 9-13 were not meant to be there and should have either been left out or transferred somewhere else.

[22] Θ, chapters Rx1-Rx32 Bennett 2003: 408-420.

[23] Chapters that correspond roughly are (given in the sequence they have in the *Therapeutics* citing the chapter numbering of the later version/the chapter numbering of Theophanes) 32/116, 48/180, 56/140, 62/201, 65/225, 66/226, 67/227, 68/228, 69/229, 70/230, 71/231, 83/1, 86/2, 87/3, 88/6, 89/7, 90/8, 91/10, 92/10, 93/10, 94/12, 95/13, 96/14, 97/17, 98/18, 99/19, 100/20, 102/21, 103/22, 104/22, 105/41, 106/42, 107/43, 108/44, 109/46, 110/48, 111/50, 112/50, 113/51, 114/52, 115/53, 116/54, 117/62, 119/71, 120/71, 122/73, 124/75, 126/77, 129/111, 131/119, 135/127, 136/128, 137/129, 138/130, 139/131, 142/133, 144/139, 145/140, 146/141, 148/151, 149/154, 150/155, 154/158, 155/161, 156/162, 157/164, 159/175, 160/179, 161/196, 162/280, 163/281, 164/282.

[24] This is the case in chapters 48, 71 and 87.

coincide in title but not in their content.[25]

It appears as if both the author or authors of the *Therapeutics* and Theophanes had access to a certain corpus of data on diseases they used as a grid for their work, which led to similar chapter headings in a similar order in both manuals.[26] And to actually compose the chapters they often used similar material. In the case of the *Therapeutics* frequent redactions of the text contributed to (what looks from a modern point of view) a progressive assimilation to Theophanes.[27]

What the form of this postulated corpus might have been is uncertain. From earlier times, we are aware of lists of illnesses which were incorporated into medical manuals.[28] Again, we do not know the purpose of these lists, but it is certainly possible that they were meant to be memorized in teaching. Such a list would explain the similarities between the *Therapeutics* and Theophanes as far as sequence and topic of chapters are concerned. An alternative explanation is that the author of the *Therapeutics* used the index of Theophanes as a guideline for his work, also consulting the main text as a source, whenever suitable.[29] And, of course, it is not impossible that the reason for this striking similarity between both texts is even more complex.

Neither the author nor the commentator of the *Therapeutics* tell us about their background, education and familiarity with medical literature. The only source that is explicitly mentioned is Galen. The original version ℵ claims in the first lines of the preface to the work, that the *Therapeutics* is an epitome of Galen's *De locis affectis*, and the commentator later on added the words

[25] This is the case in 198/254, 216/269 and 243/201.

[26] Most notably, both Theophanes and the *Therapeutics* present fevers in the position in the cycle where one would expect to find the diseases of the heart. In the *Therapeutics* this is the case with chapter ω143 to 147.

[27] Another explanation, that the chronology of the *Therapeutics* version is reversed and that several chapters were left out in the course of a redaction seems less likely; see page 15.

[28] E.g. Aëtius Amidenus VII 2 (= II 254 f. Olivieri) or the pseudo-Galenic *Introductio* 16 (= XIV 767 Kühn). These can also be found in Arabic texts, e.g. Hunain's *questions sur l'* III 163 (= 117 f. Sbath/Meyerhof).

[29] This happened in manuscript **D**, when an unknown redactor used the *Therapeutics* as a source. See page **??**.

ἑρμηνία ... καὶ ἐξήγησις 'translation ... and explanation'. Paradoxically though, the *Therapeutics* is not based on the Galenic treatise *on the affected parts*, and any overlapping content is purely coincidental.[30] It appears as if the author of the original version had access to an epitome or a compilation of medical texts under the title Galen *de locis affectis* he is summarizing in his preface, and which he may have used as a source, but this was not the genuine Galenic work known to us.

[30] See Daremberg 1853: 27.

CHAPTER FOUR

THE MANUSCRIPT TRADITION

The *Therapeutics* are rarely mentioned in secondary literature; a very early reference can be found in Fabricius' *Bibliotheca Graeca*, where it is listed among *inedita*.[31] Daremberg describes the text, presenting transcriptions of some brief extracts and analysing the relationship between Ioannes and the Xenon manuals.[32] Diels lists the text in his comprehensive catalogue of medical manuscripts.[33] Kouzis discusses a number of manuscripts listed in Diels.[34] Sonderkamp's[35] book on the transmission of Theophanes and Bennett's[36] work on Byzantine hospitals and their manuscripts offer an innovative analysis of the transmission. In recent years, Capone Ciollaro published the preface of the *Therapeutics* in an article.[37] Bennett's survey of manuscript witnesses formed the basis for the present edition.

[31] Fabricius 1724: 12.781.

[32] His work is based on manuscripts **ABM**. Daremberg 1853: 23-30.

[33] The entry on Ioannes Archiatrus can be found on II 52; however, some manuscripts containing the *Therapeutics* are listed under the name Ioannes Antiochenus Archiatrus (**D**), 51, Ioannes Iatrosophista (**E**), 54, and Ioannes Spensatus (**C**), 55. Altogether he lists manuscripts **FMRBA**. Diels 1906.

[34] Kouzis 1929: 379-382.

[35] Sonderkamp 1987 describes manuscripts transmitting the medical works of Theophanes, some of which also contain Ioannes.

[36] Bennett 2003. A discussion of the *Therapeutics* can be found on 242-247; see also the descriptions of manuscripts indicated in following chapters of the present edition. I also owe the discovery of a number of manuscripts to David Bennett, especially **G**, which was overlooked by previous research.

[37] This article describes three of the manuscripts transmitting the *Therapeutics* and provides an edition of the preface of the later version based on manuscripts **ABD**. Capone Ciollaro 2006.

1. *The original version* ℵ

The original version has come down to us in a single manuscript, the

- **N** Monacensis *graecus* 551 (Bayerische Staatsbibliothek, Munich), ff. 334r-353v, 15th c.
  The manuscript contains a title 'iatrosophion of Galen', a table of contents (ff. 334r-335v) and a preface (ff. 336r-v), then (ff. 336v-353v) the main part of the *Therapeutics* in 216 chapters. In the table of contents, the headings of two texts following the *Therapeutics* have falsely been added to the list of chapters.
  Hardt 1812: 5.403. Bennett 2003: 455-456.

The codex contains mostly theological works, but contains, towards the end, three medical texts. The *Therapeutics* are in an excellent state of preservation, and the manuscript has very few minor errors and a small *lacuna*.[38]

There are various indications for this being the earlier of the two main versions of the *Therapeutics*. One of the main principles of textual critism is the *lectio difficilior*: it is assumed that when consciously altering a text, scribes or revisors tended to make it more accessible to the readership, e.g. by simplifying it or by adding explanations; applied to the *Therapeutics*, the version without the commentary, i.e. the one transmitted in **N**, is likely to be the earlier one.[39] Moreover, the version transmitted in **N** is written in learned medieval Greek, whereas ω uses a simplified dialectal style; again it is far more likely that a text is made easier to follow.[40]

---

[38] ℵ 133.1.

[39] For instance chapter one first paragraph where the commentator specifies that ivy should be dried in the sun (not just dried), ground as fine as flour (not just ground) and left to soak in olive oil overnight (not just to soak in olive oil). This is additional information that also illustrates the procedures to someone less familiar with medical procedures by using everyday vocabulary. Additions can be found on almost any page in ω.

[40] For instance chapter ℵ 4 and ω 5 where participles are dissolved to form independent clauses linked by the word 'and'. Again, similar simplifications can be found on every page of the text.

The most solid evidence to pin down the chronology of the transmission is the preface of the *Therapeutics*. All the manuscripts other than **N** transmit a text that is difficult or impossible to understand, and the reason for this appear to be marginalia invading the text along with parts of sentences that are obviously corrupt. A comparison between both versions shows that some words which are extant in **N** (and are certainly the correct reading) are likely to have dropped out in the source of ω.[41]

2. *The later version and its descendants*

The manuscripts transmitting the later version fall neatly into separate groups, some of which are further redactions of the text.

2.1. *The later version* ω

This version has come down to us in a large number of manuscripts, but most of them are, for a variety of reasons, in a poor state, and only one contains the entire text in the correct sequence of chapters. The manuscripts can be divided into two main groups. The first is transmitted in a single manucript,

- **L** Medical Society London 14 (Wellcome Library, London), pp. 82 f., 108-271, 14th c.[42]
  The manuscript contains a part of the index and the main text of the *Therapeutics*. The pages were rebound and numbered at a late date, and according to the numbering the correct sequence of the pages would have been 122, 123, 120, 121, 126, 127, 124, 125, 128-271, 82, 83; by mistake, the page numbers 190, 233, 258 are missing whereas 131 has been used twice. Water damage has rendered parts of the text, mostly on the top lines, illegible. The margins have

[41] See page 31.

[42] Dawson 1932: 24 dates the manuscript to around 1500, which appears to be too late. Rather, one would expect **L** to be written in the 14th century. This view is also supported by Brigitte Mondrain and Vivian Nutton. Agamemnon Tselikas dates the handwriting to the 1320s Cyprus.

> been cut, probably when the book was rebound, and some portions of the text on the outer margins are lost. The top and outer margins on some of the pages are torn. In the gutter, the text is often almost illegible because the pages have been bound together too tightly[43]. Since the manuscript does not show any quiremarks it is impossible to determine how many pages of the original codex are missing. A second set of paper **l** was added to the codex. Both **L** and **l** show the same pattern of damage in the margins, which indicate that this happened at an early point of time when they were still used together.[44]
>
> Dawson 1932: 24.

Unlike all the other witnesses to the *Therapeutics*, this manuscript was apparently used by a physician in his medical practice: it is a very small volume, measuring only ~14,4 x ~10,0 cm. The size of the codex makes it inconvenient to read but it had the advantage of portability. The book shows signs of being perused a lot, and an unusually large number of hands have annotated the text, mostly correcting the chapter numbering[45] which indicates that someone used the book. Moreover, several pages are missing,

---

[43] The codex was restored several times: the sides of the binding date to the 18th century, then the back was replaced in the 19th century. This, and more data on the paper and the watermarks of the manuscript were established by Lara Artemis, conservator at the Wellcome Library, London.

[44] The handwriting of **l** is extremely unusual and distinctive. Agamemnon Tzelikas dates it slightly earlier than **L**, and originating from Cyprus. This hand shows, despite obvious differences, some similarities to ms. 10 from St. Neophytos monastery in Cyprus, which, according to Jakovljevic 2002: 86 in turn bears similarities with mss. 55 to 58 of the Esphigmenou monastery on Mount Athos, written by Mark the Hieromonk. The last of these manuscripts is dated to 1316. Moreover, the hand is very similar indeed to Theodoulos Philagres in manuscript 21 of the Archbishopric collection Nicosia, published as manuscript 30 (plate 44) in Constantinides, Browning 1993. According to 151 f. it was copied in Cyprus in 1319.

[45] A preliminary count in both **L** and **l** yielded a total of twenty hands; as stated above, most of the later annotations were corrections of chapter numbering and additions of titles which were missed out when the scribe rubricated the volume. But there are also smaller corrections of the text itself and a number of magical charms.

amongst them a part of the index of the *Therapeutics*, and both the beginning and the end of the text. This pattern of damage is not unusual, for frequently used or otherwise exposed pages are easily torn when a book is handled. The *Therapeutics* seems to have been the most popular text in the volume.

All the other manuscripts form a rather amorphous group Σ.

- **A** Parisinus *graecus* 2236 (Bibliothèque Nationale, Paris), ff. 1ra-42rb, 15th c.
  The manuscript contains a pinax (ff. 1ra-4ra, the first page is torn and some parts are missing), a preface (ff. 4rb-5ra), a title and then the main text of the *Therapeutics* (55a-42rb). There are some *fenestrae*[46] in the main text, but overall all the chapters are in the correct sequence.
  Omont 1888: 219. Bennett 2003: 460-462.

- **B** Parisinus *graecus* 2224 (Bibliothèque Nationale, Paris), ff. 80r-102r, 104v-105r, 15th c.
  The manuscript transmits a form of the text which is obviously not in the correct sequence. It contains the first part of the pinax (f. 80r-v), a title (f. 81r), the main text (ff. 81r-102r), the preface (f. 104v) and then the second part of the pinax (f. 105r). Each part of the text, as indicated above, is written in a different hand which is specific to this section. Moreover, the main part of the text is not in the correct sequence of chapters either and has subdivisions which cannot be found in any other witness. It contains the text from chapters 1 to 7, then 68 to 73, 8 to 22, then 73 with distinctively different readings from its first appearance, then the chapters consecutively up to 98. After that follows a subdivision indicating the next 100 chapters (f. 85r) and all the consecutive chapters up to the end of the text. On f. 97v there is another subdivision. Overall, the source of the manuscript, direct or indirect, must have been damaged

[46] The two larger chunks of text missing are middle of chapter 154 to the end of 156 (*fenestra* on f. 29vb). The next gap is at the end of chapter 174 and reaches to the end of chapter 176 (*fenestra* in f. 33rb.)

and most likely been put together from two different books.
Omont 1888: 216. Bennett 2003: 459-460.

- **C** Parisinus *graecus* 2336 (Bibliothèque Nationale, Paris), ff. 1r-192r, 16th c.
  This manuscript goes back to the same source as **M**; it contains a large part of the *Therapeutics* and has far more mistakes than **M** but no significantly better readings. It contains a title, the preface (pp. 1-5), another title and then the main text (pp. 5-192). Some paragraphs are missing or misplaced.
  Omont 1888: 243

- **G** Iviron Athous 214 (4334) (Holy Mountain), ff. 145r-225r, 15th c.
  The manuscript contains a pinax, the preface, the title and then the main text. The *Therapeutics* is followed, with continuous chapter numbering and without any further indication, by a large number of recipes and other medical material which are also included in the pinax. The exemplar from which this manuscript was copied was obviously damaged; the pinax lists the first 15 chapters, and has then, on the verso of the page, a continuous medical text. This page may have been torn out and put in the wrong way round in the master copy, so that the medical text belonged to the preceding compilation. Next, the pinax lists the following chapters, missing out the chapters from 23 to 73. The main text is in a similar condition. It agrees with **B** in an extensive *lacuna* from chapter 23 to 67 and, in addition, several chapters from the beginning of the book can be found later in the text. Both manuscripts were most likely copied from the same master and they reflect the damage to this manuscript in a different way.
  Karas 1994: 3.161.

- **M** Monacensis *graecus* 288 (Bayrische Staatsbibliothek, Munich), ff. 2r-92r, 14th c.
  The manuscript contains a pinax (ff. 2r-7r), a preface (ff. 7r-8v), a title (ff. 8v), and then the main text of the *Thera-*

peutics. The text is complete and in the correct sequence of chapters. The number of orthographical mistakes is higher than usual, but most are very basic misspellings that can be corrected easily.

Hardt 1806: 3.211-212. Bennett 2003: 454-455.

L preserves the text of the *Therapeutics* in a very good state, and usually has better readings than the other manuscripts. Mostly, the errors in **L** are orthographical and do not bear on the meaning of the text[47]. Other errors occur in pagebreaks at the end of the *verso* where sometimes half a word seems to have dropped out[48]. Since the volume has been rebound, it might well be the case that the missing part of the word was originally there and was just cut off later on by mistake. However, it can not be excluded that these are errors by the scribe.

If one looks at the relationship between **L** and Σ, it is often impossible to rationalize the way in which readings developed because scribes were likely to add or delete words or to change the register of a passage; this makes it impossible to apply stemmatics in the conventional way. I list here a number of possible errors between groups and manuscripts in order to highlight certain tendencies in the transmission. They should not be confused with a proof of the derivation of any witness from another.

In the bipartite stemma of **L** and Σ originating from ℵ, the following are errors of **L** against Σ:

- 89.4: This paragraph is transmitted in Σ and ℵ but is not extant in **L**. According to the stemma, the omission of the paragraph is a mistake in **L**. However, since this is an entire paragraph missing, it might have been left out on purpose because the recipe was not considered useful.

- 120.6: **L** has the incorrect reading κραμπὴν τοῦ γουλίου whereas Σ and ℵ have the correct γουλὶν τῆς κράμβης.

[47] Quite often it is not easy to determine whether these are in fact errors or merely variant spellings of words that did not have a standardized lexical form. Rather distinctive spelling patterns of **L** against the rest of the witnesses are φύλα-φύλλα and τρυμένα-τριμμένα.

[48] For instance chapter ω 238.2, where the page ends with the word ὀδυ.

- 131.8: **L** has the phrase εἴπερ ἔνι μέγας where ℵ and Σ specify the exact age of the patient. In stemmatic terms, **L** is wrong.

Overall, Σ has, compared to **L**, some additional material, but it is impossible to tell whether it is genuine or not.[49] Moreover, Σ has more mistakes than **L**:

- chapter 1.1: ἢ καὶ versus ἢ : Both ℵ and **L** give the instruction to dry ivy or alternatively grind green, i.e. fresh, ivy and soak it in olive oil, then to sieve it and to rub it on the forehead. The words translated as 'or alternatively', ἢ καὶ, appear in this form in ℵ and in **L**. The rest of ω leaves καὶ out, another subgroup of these in turn altering ξηράνας to ξηρὸν assimilating it to χλωρόν which changes the meaning of 'dry (imperative) ivy or alternatively grind green/fresh ivy' to 'grind dry or green/fresh ivy'. The commentator explains the procedure as 'take ivy and dry it in the sun, then grind it and make it like flour, then add green/fresh ivy in olive oil over night' which does not agree with the lemma: the alternatives 'dry ivy' and 'ground fresh ivy' are seen as two ingredients for the same recipe. This misunderstanding by the commentator probably caused the omission of καὶ by Σ.

- 30.1 to 3: **MA** lack some text which is extant in **L**; since a part of this text can be found in ℵ and as the missing part of the chapter is enclosed by the words εἰς τό, this is likely to be a scribal error. Unfortunately, the chapter is missing in **B**, so that it cannot be used as a valid argument.

[49] Σ has additional material against **L** and ℵ in 83.3. The amount of text that can be used for collation is smaller than one would expect. The manuscripts **LAB** have extensive *lacunae* in various places that do not coincide; in addition all manuscripts of Σ tend to leave text out, so that there are a large number of additions and omissions by just a part of the Σ group. Additional material in all of Σ against **L** can be found in 83.3 and 4, chapters 89.4, 194.8, 225.7, 227.4. All of these samples are entire paragraphs and a very large number of shorter additions can be found throughout the edition.

- 89.2: Σ has the reading τρίχας instead of the correct τρύπας which is transmitted in **L**. ℵ has a classical synonym for the correct reading, ὠπαί; *Theophanes* 7 (= I 38, 7 Bernard) has another synonym, τρήματα.

- 129.2: Σ has the reading μυελὸν ἀπὸ τὸν ἐγκέφαλον, ℵ has μυελὸν τῆς κεφαλῆς and **L** has μυελὸν ἀπὸ τὸ κεφάλιν του. Since the expression 'marrow of the head' as found in **L** and ℵ is an obvious synonym for 'brain', Σ's reading 'marrow of the brain' does not seem to be correct. *Theophanes* 111 (= I 340, 7 Bernard) has here simply the word ἐγκέφαλον.

In quite a number of instances, Σ and **L** have variant readings which are equally possible:

- 10: **L** has ἂς βράσουν μετὰ νεροῦ whereas Σ has βράσον με τὸ νερόν. A large number of variants like this can be found throughout the edition.

- 89.10: **L** has χρηστέλαιον whereas Σ has ἔλαιον τὸ χρηστόν.

- 115.1: **L** has φλεβοτόμησέ τον whereas Σ has φλεβοτόμησον.

Within Σ, there is no apparent stemmatic structure. None of the manuscripts was the master copy of another. All witnesses have a number of mistakes they do not share with any other manuscript. Major mistakes are:

- **A** has extensive *lacunae* in 154-156 and 175-176.

- **B** has an extensive *lacuna* (23-67) it shares with **G** but not with any other manuscript.

- **C** was copied from a damaged master; chapters 22, 23 and 24 were pasted in the middle of chapter 4 (p. 7 and 8 of the codex). Moreover, a large number of smaller passages dropped out, e.g. 29.3.

- **G** shares an extensive *lacuna* (23-67) with **B** but with no other manuscript.

- **M** is very close to **C**, e.g. these manuscripts agree in the mistake δορόσταγμα for ῥοδόσταγμα in 4.2; but unlike **C** and the other witnesses, it has an error in 33.3 to 4 where text was left out by mistake.

The only two related manuscripts are **MC** and **BG**; they agree in a number of readings[50], but in an extensive collation of **MC**,[51] the latter never had any significant better or interesting readings but a large number of additional trivial mistakes[52]. The situation of **BG** is similar. Both manuscripts agree in having an extensive *lacuna* in the first part of the *Therapeutics* and are therefore certainly related[53]. However, in a collation of selected chapters with crucial textual problems, **G** did not have any better or interesting readings than **B** but several additional mistakes[54]. Thus neither **C** nor **G** were included in the apparatus.

2.2. *Augmented excerpts*

Three new texts based on the *Therapeutics* are:

- **D** Parisinus *graecus* 2315 (Bibliothèque Nationale, Paris), ff. 117 ff., 15th c.
  On ff. 117r-118v the codex has the preface of the later version of the *Therapeutics*, then a title saying: λόγος καὶ ποίημα καὶ προοίμιον ἰωάννου ἀρχιητροῦ περὶ ἔχων συνοπτικῶς πάντων τῶν παθῶν καὶ τῶν ἀδήλων τὰς θεραπείας καὶ πάντων τῶν διδασκάλλων συνοπτηκὸς ἔχων ἐν τῇ βήβλω ταύταύτη τὰς θεραπείας: διοσκουρίδους καὶ ὑποκράτους καὶ γαληνοῦ περὶ ἰατρικῆς ἐπὶ στύμης εἰς θεραπίαν τῶν ἀνθρωπίνων

[50] **MC** are extremely similar. Some distinctive readings are the omission of the commentary in 1.2, the omission of ὀξύνου in 5 and πίτερα for πίτυρα in 13.

[51] Chapters 1 to 72 and 205 to 231.

[52] E.g. φάματα for the correct φάβατα in 6; it also tends to add or leave out articles, e.g. it has ἀνθρώπου in 26.18, where the other witnesses have τοῦ ἀνθρώπου.

[53] Another very good indicator is the first sentence of 4.2, where very diverse and distinctive readings can be found. **G** agrees here with **B**.

[54] Most of these are spelling errors; more serious is for instance a *lacuna* from the heading of 115 to the end of 118.

σωμάτων δόκιμα καὶ ὀφέλημα. ἀβιτζένα καὶ νικολάου θεραπίες. Then follows a compilation of several medical texts, amongst them also the *Therapeutics*. The texts are mixed up and scattered; e.g. f. 226r corresponds roughly to 68 ff. in the *Therapeutics*, but other parts of the work are missing.[55]

Omont 1888: 235-236. Bennett 2003: 465.

- **E** Parisinus *graecus* 2316 (Bibliothèque Nationale, Paris), ff. 51v-135v, 15th c.
  This manuscript has the preface of the *Therapeutics* and then a medical compilation which is mostly based on the *Therapeutics* (the actual text of the *Therapeutics* ends on f. 119r). However, the text of the *Therapeutics* was revised; e.g. the first sentence of 4.2, which is corrupt in the other witnesses of the later version, is αὐγὸν ἐκζέσας κοπάνισον τὸν κρόκον χωρὶς τοῦ ἄσπρου καὶ κρόκον ἀλέων κουπάνισον ἐν τῷ ἅμα in **E**; in 5 the last sentence in **E** is εἶτα ἔπαρον ἄσπρον ὠοῦ καὶ βάλε τὰ τρίματα καὶ ποιήσας ἔμπραστρον, ἐπίθες ἀπὸ τὸν ἕναν μέλιγγα ἕως τὸν ἄλλων. Thus **E** resembles more a later redaction of the *Therapeutics* with some added material. For practical reasons, it was not included in the present edition, because it would make the apparatus even more complex than it is now. It appeared better to leave more detailled research on this version to further publications.
  Omont 1888: 236-237. Bennett 2003: 465.

- **F** Laurentianus *pluteus* 75.10 (Biblioteca Laurenziana, Florence), ff. 31v-38r, 15th c.
  The manuscript has a heading, indicating a work by Aristotle and John, a pinax (f. 31v), and then an excerpt from the *Therapeutics* with some additional material; in style and content these additions are somewhat different from the *Therapeutics*. Since the fragments of the *Therapeutics*

[55] As David Bennett pointed out, this manuscript is discussed in Jeanselme 1930. Excerpts from this manuscript are published under the name Ioannes Staphidas in Bibliothèque grecque vulgaire: 2.1-8.

were reformulated and as these are just very short passages, it is impossible to determine the source of **F** with certainty. However, the additions are very interesting; therefore selected readings from **F** will be presented in a separate layer of the apparatus.

Bandini 1770: 3.156-158. Bennett 2003: 450-451.

2.3. *A classicizing version*

Two manuscripts transmit a version of the *Therapeutics* that is closer to the classical than to the vernacular:

- **V** Vindobonensis *medicus graecus* 33 (Österreichische Nationalbibliothek, Vienna), ff. 3r-97, 15th c.
  The manuscript contains a fragmentary pinax and then, without title or preface, the main text of the *Therapeutics*.
  Hunger 1969: 2.86-87. Bennett 2003: 476.

- **W** Vindobonensis *medicus graecus* 43 (Österreichische Nationalbibliothek, Vienna), ff. 1r-79v, 16th c.
  The manuscript contains the title ἰατροσόφιον καινόν and then the main text of the *Therapeutics*.
  Hunger 1969: 2.94-95. Bennett 2003: 477-478.

This version is part of the ω family[56]; its characteristic is a tidier style and a grammar that is closer to the classical than to the vernacular: 5 (according to **V**) reads as follows: ὅταν κνίθων$^{\tau}$ οἱ ὀφθαλμοί· λαβῶν φλοῦν ῥοιδέου ὀξύνου, φρύξον εἰς τὸν ἥλιον, εἶτα κοπανίσας ποίησον οἷον τὸ ἄλευρον καὶ βάλον οἶνον ὀλίγον. καὶ ταράξας ἄλειφε τοὺς ὀφθαλμοὺς καὶ ἐπάνω θὲς πανίον βεβρεγμένον μετὰ οἴνου καὶ δύσον αὐτὰ ἕως τὸ πρωΐ. καὶ φλεβοτόμισον αὐτὸν καὶ λοῦσον καὶ τὰ δρυμαῖα πάντα ἀπεχέσθω. καὶ τὸ κρασὶν καὶ τὸ καπνόν καὶ τὸν ἄνεμον. Predicates have been turned into participles, a reduplication perfect has been added and the vernacular word for wine was replaced with the classical term. On the other

[56] This can be stated with certainty, because both versions have a commentary.

hand, it still retains vernacular readings; in 7.1 **VW** have ᾠοῦ as opposed to αὐγοῦ in **L** and Σ. However, in par. 2 of the same chapter all witnesses, including **VW**, have the word αὐγοῦ.

This redaction appears to be secondary to the text of **L** and Σ; it looks as if someone went through the text and randomly replaced vernacular expressions with their classical counterparts.

The basis for this redaction was a manuscript that was neither related to Σ nor to **L**, but it is often closer to **L**[57]:

- 1.1 **LVW** and ℵ have the reading ἢ καί as opposed to the incorrect καί in Σ.
- In 12.2 **VW** have ἐκδάρας, **LA** have ἔκδαρε το and **MB** the incorrect ἔκβαλέ (τό).
- As opposed to the rest of the transmission, **VW** leave 16.2 and 3 out.
- **VW** and Σ do not have 201 which is only transmitted in **L**.
- In the crucial readings of 89 and 129 (errors in Σ against **L**, see page 20) **VW** goes together with **L**.

However, this version has a number of mistakes where the rest of the transmission has a correct reading:

- In 10, **VW** have καὶ ὕδωρ where **L** has ἂς βράσουν μετὰ νεροῦ and Σ has βράσον με τὸ νερόν. The text in **L**, Σ and ℵ says to put bread crumbs into a sauce pan and boil them with water; **VW** just gives the instruction to put bread crumbs and water into a sauce pan which leaves out a step in the preparation of the remedy.
- In 16.1 (which is missing in **L** due to physical damage) **VW** give the instruction to eat, τρόη, a remedy where Σ says to drink it. Eating remedies is not found anywhere else in the *Therapeutics* and has to be considered a mistake.

[57] The analysis of this version is based on a full collation of chapters 1-22 of the *Therapeutics* and on a selective collation of chapters which show distinctive readings in **L** and Σ.

- In 22.1 (again, missing in **L**), **VW** have the obviously incorrect ὄγκους as opposed to ἕλκους right at the end of the paragraph.

These mistakes would almost certainly not have been made by a person with a medical background.

Once in a while, **VW** have interesting new readings, e.g. the addition ἤτοι κανέλαν after κινάμωμον in 15.2, but these are usually mere additions. On the whole, the text of **VW** has to be regarded as a separate version of the text that is interesting in its own right, even more as it sometimes goes together with **L** and against Σ. However, because of its obvious differences to the rest of the transmission it did not appear practical to amalgamate it with the edition of the later version as transmitted in **L** and Σ; an edition of the version of **VW** has to be left to later publications.

The origins of this version are obscure. Both manuscripts were acquired in Constantinople in the sixteenth century by Augier Ghislain Busbecq, who then took them to Vienna, where they are still held today.

2.4. *Excerpts*

Excerpts of the later version also survive; the following three manuscripts contain material that can also be found in the *Therapeutics*. The extracts are too brief and fragmentary to allow for a detailled discussion on their location in the stemma. For practical reasons, they were not included in the present edition.

- **P** Parisinus *graecus* 2294 (Bibliothèque Nationale, Paris), ff. 73v-95v, 15th c.
  This manuscript contains a medical compilation which bears some similarities to the *Therapeutics*.
  Omont 1888: 231. Bennett 2003: 465.

- **R** Angelicus 17 (C.5.4) (Biblioteca Angelica, Rome), ff. 1r-8r, 14th c.
  The manuscript has a title attributing the work to Ioannes archiatrus, and then an excerpt from the later version of the *Therapeutics* (1-24). The language of the text is more

vernacular than in Σ or **L**.
Franchi de' Cavalieri, Muccio 1896: 44-49

- **X** Vindobonensis *theologicus graecus* 231 (Österreichische Nationalbibliothek, Vienna), ff. 77r-93v, 15th c.
This manuscript contains a fragmentary excerpt of ω.
Hunger 1992: 3.108. Bennett 2003: 480.

CHAPTER FIVE

EDITORIAL METHODS

1. *The original version*

The original version is transmitted in just one undamaged manuscript with no serious scribal errors; therefore no issues arise that could complicate a critical edition of the text. However, for the sake of clarity, it appeared best to apply the same rules as in editing the later version, and therefore frequent and purely orthographical errors like iotacism in verb endings or single versus double consonants have been silently corrected.

2. *The later version*

The situation for the later version cannot be more different, and an edition in the conventional way has proved impossible. Even if one leaves out all excerpts and later redactions of the text, there are two main issues which cannot be resolved satisfactorily if one applied the standard editorial methods.[58]

The first and most obvious one is the vernacular idiom of the text. It is consistent neither with modern nor classical Greek and largely not even consistent in itself so that it is impossible to adjust spelling and grammar to fit either language, or even to create a new grammar based on the text. In addition, the number of variant readings is immense, and a complete apparatus listing all the manuscripts would not only be impenetrable but also fill most of the page, nor would an electronic edition of the entire manuscript material render the matter any more intelligible. Thus

[58] Even an electronic edition would not have the necessary flexibility — it might be possible to have a complete input of all relevant manuscript readings and switch between branches of the transmission but it would not resolve linguistic issues or present an immediate overview of the transmission.

it has been necessary to limit the entries in the apparatus and to standardize the language mildly.[59]

In some cases it was clear how to proceed. If, for instance, a classical Greek word was frequently misspelt (and there was no doubt that it was a plain orthographical error), it has been corrected and included in a list of frequent mistakes.[60] This was, for example, the case for κοπάνησον versus κοπάνισον or φύλα versus φύλλα, and the same applies to words which are still in use in modern Greek.[61] Far more difficult to decide is how to proceed if a word is unique to the text, especially when the manuscripts did not agree on its spelling. Any decision on these problems must be deliberate for it creates a new word which would then be cited in the glossary and the index. Overall, it appeared best not to standardize these words and to leave them in different spellings in the main text.

In handling accents, iota subscript and verb endings, I have kept some features of the idiom of the text; e.g. a typical construction which can be found everywhere and that is perfectly understandable is μάλαξέ τα καλὰ where the article (in an enclitic form) is used instead of the classical αὐτά. In these cases the phrase has not been readapted to the classical μάλαξε αὐτὰ καλῶς or καλά. Iota subscript has been added where one would expect to find it in classical Greek. Verb endings after ὅταν, μὴ and similar conjunctions are given in the subjunctive if they were spelt in a way suggesting that this might have been the original reading. For instance ὅταν πίνει or πίνοι is always converted into ὅταν πίνῃ without mentioning the variants in the apparatus. However, μή τα πίνουν is kept in this form.

Thus, any variants with the same pronunciation as a classical or modern Greek word are kept. But if the majority of the

---

[59] A collation of a sample chapter without any standardization can be found in appendix ...

[60] The list can be found in appendix ...

[61] For classical Greek, Liddell Scott Jones 1996. was the obvious choice as a standard dictionary. For modern Greek I consulted a number of lexica, Contopoulos 1880-1882, Kyriakides 1909, Sophocles 1887, Trapp 1994-2001 and in particular Kriaras 1968-1997. Any word in one of these dictionaries was defined as attested. I also used the online epitome of Kriaras.

manuscripts transmit an unknown word or a word with a pronunciation that cannot be mistaken for another word, I aim to preserve this reading and to place significant variants in the apparatus. Minutiae like final ν or punctuation have been standardized without any further mention in the apparatus.

The second major issue in editing the later version is that it is transmitted in two branches which in some places almost appear to be separate versions. The **L** branch is in general shorter and better, but entire pages are missing. The Σ branch is complete, longer but often corrupt. The dilemma an editor faces here is that **L** is needed to correct Σ and Σ to fill the gaps in **L**. Moreover, in the event that **L** and Σ have alternative readings, it is impossible to decide which one is genuine.

Again, I have opted for a practical solution. Since it is impossible to decide whether additional material in Σ is genuine or not, i.e. whether **L** is a shortened or Σ an augmented version, all the text has to be presented. **L**, being the best witness, forms the main text, and where **L** is damaged or does not have some material, **M** steps in. All sources are clearly labelled in the main text. The result is a complete and readable text which does not aim to be a reconstruction of the original but gives the reader all the information that might be required.

The disadvantage of this method is, of course, that it creates a form of the text which is not attested in any witness. Therefore, the chapter numbering by which the text is cited has been established by the editor. Subnumbering between the recipes is added for the sake of clarity.[62]

Other editorial issues are handled in a similar pragmatic manner. Abbreviations are resolved. This is usually straightforward in cases, e.g. ἀνοῦ stands for ἀνθρώπου. If an abbreviation was ambiguous without bearing on the actual meaning of the word, it is assumed that it stands for the most common form of the word.[63]

---

[62] In the manuscripts, the subsections are divided by ἄλλο, ἄλλως, ἕτερον, no indicator at all or by a diacritical sign. In the present edition, all of these were replaced with ἄλλο.

[63] A difficult decision was how to proceed with the case of measurements which are usually, but not always abbreviated: the construction of ἐξαγίου τὸ ἥμισυ was adopted throughout the edition.

For example, κλυστ^ρ becomes κλυστήριν rather than κλυστήριον or other forms of the word. Likewise, τρώ^γ can, depending on the context, either be τρώγει or τρώγῃ.

Even so, a large amount of data is still left in the apparatus. Sometimes scribes tended to reverse the sequence of two word for purely aesthetic reasons. If these transpositions are trivial and just occur in one manuscript, they are omitted in the apparatus. The same applies if one manuscript, usually **M**, added an additional preposition to the heading; e.g. if a heading originally started with ὅταν and one witness has πρὸς ὅταν which is clearly secondary, this is not mentioned.

Omissions of single characters are handled similarly. If a manuscript is damaged and one or two letters are illegible or missing and the transmission in all others is unanimous, this is not mentioned in the apparatus. Of course, this was only the case if the letters in question were trivial and any relevant variant readings were highly unlikely. E.g. this is the case in κ.ι/καί, κοπά..σον/κοπάνισον or .ίξον/μίξον. The only instance in which entire words have been replaced without further mention is the ἄλλο dividing the subsections of the text. In **L**, an important witness, these are written in the outer margins of the book which have been partly torn or are even missing. In other witnesses, the word is often left out or was simply overlooked by the rubricator.

The preface to the later version is, from an editorial point of view, the most difficult to edit. On its own, it is mostly unintelligible and seems damaged beyond restoration. But if it is compared with the original version as transmitted in **N**, parts of sentences suddenly become clearer. Several words and sometimes lines from **N** seem to have dropped out[64] and someone has then filled in text.

---

[64] The crucial word in the first sentence is ὥστε which is extant in ℵ but absent from ω. As it stands, ω makes little sense, because there is no treatment for untreatable diseases. Since there is no reason why someone should change the text of ℵ to the form it has in ω, it must be caused by a mistake or by damage to the manuscript. In the next sentence of ω the absence of a verb in the first half of the sentence makes the text difficult to follow. Again, ℵ has the word διδάξαντος, and there is no obvious reason why someone should have left it out. At the end of the first paragraph of ℵ, ω has the word ἀνδρός where ℵ has ἰατροῦ, and the only reason why someone should replace one

Moreover, marginal annotations seem to have invaded the text.[65]

The logical consequence to restore the preface would be to print it as transmitted in $\aleph$; however, this would also mean deleting content. And as all manuscripts of ω that have a preface agree in their readings except for minor variants, it is highly likely that the attempts to emend the text were in fact made by the commentator. Thus, it appeared more sensible to leave the preface of the later version in the distorted form it has.[66]

The pinax or table of contents is in a similar state of preservation as the preface; in the original version, the pinax lacks any serious mistakes and mostly co-incides with the main text. In the later version, however, all manuscripts except **M** have physical damage in the pages transmitting the pinax. To add to this problem, the surviving text is infested with mistakes; entries are overlooked or merged with the following heading, the chapter numbering is not consistent and headings are rephrased to make them clearer. Most of these changes are, from an editorial point of view, irrelevant, since they reflect very basic scribal mistakes. Moreover, the result of a critical edition of the pinax would have to be adjusted to the main text of the *Therapeutics* to be of any use for the modern reader of the book, which would make the issue even more complex. I therefore decided to leave the pinax of ω out and present a table of contents in English language together with a synopsis of manuscript witnesses and chapters.

---

with another is because there was a gap in the source he used. The transition between the first and the second paragraph in ω makes little sense; the point of the statement attributed to Hippocrates is that one should not only know the general physiology of all living creatures, but also the individual one, i.e. personal disposition or pathological changes, which manifest in the balance of humors. Since the term κρᾶσις is absent from ω, the argument becomes obscure. If one has a look at the words that presumably dropped out in the source the author of ω used, the pattern suggests that at the end of each line a word was deleted by water damage or the like or torn off.

[65] In particular two sentences appear to be interpolated later on and are enclosed in [. . .].

[66] In ω, diacritical signs were used in the same way as throughout the main part of the *Therapeutics*, i.e. citations from $\aleph$ are enclosed in superscript angles ⌜. . . ⌝. Words which are extant in the preface of $\aleph$ but presumably dropped out in ω are enclosed in $<\ldots>$.

CHAPTER SIX

LANGUAGE, PLACE AND DATE

The *Therapeutics* has spread widely within a very brief period of time, undergoing at least two radical redactions. As it does not give us any direct statements on the whereabouts of these redactions, its origins or the places where it was later on copied, all we can rely on is indirect evidence, small pieces of information that were left unintentionally.

The later version ω seems to offer most of these clues,[67] mainly because it is imbued with rare local vocabulary. For two reasons this is only of limited help in identifying its local origins. Firstly, there is not sufficient secondary literature to trace most of these words. The dictionaries we have are not primarily focused on medical vocabulary, not least because very few of these vernacular medical texts have actually been edited.[68] Secondly, even if it were possible, only a limited number of words would actually qualify for such an analysis. The reason for this is the exceptionally large number of variant readings in the manuscript copies of the text. If there are four copies of the same text, it is highly likely that each copy would differ slightly from the other. Mostly, these differences manifest in the length of vowels, an important dialectal characteristic, or a fricative versus a plosive form of a consonant.

It was probably the linguistic makeup of ω that led to the development of these variants. Roughly speaking, its language falls into two different idioms, learned Medieval Greek and its vernacular counterpart. Quotations from $\aleph$, the earliest stage of the text, are written in the learned idiom. When the commentator went through the text he adapted most of $\aleph$ into the Greek spoken at his time and added a number of comments. The result of

[67] The original version $\aleph$ is more polished and lacks a number of distinctive characteristics which are present in ω.

[68] The most detailed information provides Kriaras 1968-1997.

his efforts was a mixture of two idioms, learned and vernacular, contained in one text where the borders between the two are not always clear. In addition, the vernacular reflects a stage in the transition from ancient to modern Greek: as it lacks a standardized grammar and spelling it resembles a continuous development rather than a static system, and it was thus more likely to undergo changes in the process of transmission.

Therefore some manuscript copies have distinctive preferences. **L**, for instance, has the form ὑψηλόν for ψιλόν (fine) throughout the text. Other manuscripts alternate between the learned and vernacular form of a word, e.g. αὐγόν for ᾠον (egg).[69] This flexible attitude towards the exact wording makes it impossible to reconstruct the *Urtext* or autograph of the later version as it was written down by the commentator. As a consequence, it appears questionable whether it is possible to date ω from the language or to identify any dialectal group the commentator might have been part of.

Thus, the only bits of text that can be used as evidence to pin down the place and date at which ω came into existence are passages transmitted unanimously in which the commentator explicitly mentions place names or language groups.[70]

Only two language groups appear in all witnesses of ω, and also, to a limited extent, in ℵ. These are χωρικοί (locals) versus ἰατροί (doctors), describing synonyms in the vernacular versus the learned form of Greek: 'the locals call this plant x, while the doctors call it y.' Whether the commentator refers indeed to the language spoken by doctors can be doubted and would additionally raise the question of the audience of his work, if not physicians. Most likely, he simply means 'medical writers', the medical literature written in learned Greek.[71]

However, these two terms do not allow us to date the text or

---

69 These inconsistencies can be found throughout the text.

70 The methodological issue is of course, that these passages could be later interpolations to the archetype of all existing manuscript copies. Although it is in the end impossible to prove their authenticity, it is highly likely that they were written by the commentator, because they show the same distinctive and highly personalized style.

71 A list of these synonyms is given in appendix ...

to determine its local origins, they only demonstrate the divide between both idioms. Two other language groups are only mentioned in some of the copies of the text. A cluster of manuscripts, Σ, but not manuscript **L**, refers to people called 'the Westerners'. These manuscripts explain a standard Greek word (for hedgehog) with a vernacular term that has no known local roots, and with a description of the animal. Who these mysterious 'Westerners' were remains unclear. It could well be a local reference to a larger vernacular speaking community to the West of where the scribe lived. On the other hand, calling somebody 'Western' might have had more significance; the word was used for groups of people under Western influence, both in terms of government and their religion; 'Westerners' could therefore mean Western Christians as opposed to followers of the Orthodox faith.[72] With a view to the cultural tensions at the time, we may well imagine that someone used the term in a derogatory fashion for speakers of vernacular Greek.

Similarly, another reference to a language group is only transmitted in Σ. But other than in the passage discussed above, **L** is missing a few pages, so that we are unable to tell whether it once contained the text or not. In a recipe, a synonym for a plant name is given (it is a type of garlic), and this synonym is described as 'as the Lombards would call it' (chapter 135).[73] It is an appealing conclusion to understand 'Lombard' here as a name for the Western invaders of the time when the *Therapeutics* was composed but the evidence is not solid enough to support this assumption.

These two rather enigmatic language groups raise a number of important questions. For instance, why are they not attested anywhere else in medical literature? It might be, that the *Therapeutics* transmit a form of dialect which was not used in writing

[72] See Kriaras 1968-1997 s.v.

[73] Unfortunately this chapter is only extant in Σ because some pages are missing in **L**; **V** has the variant reading ἀσκαίλα for the plant in question, **W** has ἀσκέλλα. None of these transmitted words is attested in Greek and they are also absent from any Romance language of the time, at least as far as it is possible to tell from edited text. I owe the analysis of the Romance material to Guido Mensching.

because it was not considered literate. Alternatively, there might be other texts using this vocabulary which are still to be discovered. Or maybe these expressions were only used in a very brief period of time and became obsolete after the political situation changed. Moreover, how do these language groups relate to the two idioms that can be found in the *Therapeutics*?

It is an intriguing issue, and one is left wondering how this vernacular idiom might have been perceived at the time. In medical literature, the *Therapeutics* are the only known text to undergo a revision and translation of this scale. This raises the question by whom this project was commissioned, and what the motives were. If it was a move towards replacing the *lingua franca* of medical literature with the vernacular, it is very likely that this also involved defining a canon of texts or medical knowledge that was considered 'state of the art' reference material. The revision and translation of the *Therapeutics* was certainly a very radical step, and the large number of manuscript copies of ω shows that it was indeed very successful.

As for the date and place of ω, another passage gives us more precise and reliable information. In one of the commentary entries, two place names mentioned[74] are Ποντικόν, commonly used to describe the Black Sea, and Μωρέα, a medieval name of the Peloponnese. The commentator cites a recipe containing Pontic nuts, a well-known ingredient from the Black Sea region,[75] but is uncertain about their exact origin. He gives an explanation that seems bizarre from a modern point of view, namely that these nuts come from 'the Pontikon which is close to Morea'. As David Holton points out, this is probably based on a misunderstanding.[76] A castle called Pontikon on the Peloponnese is mentioned

[74] This passage is transmitted in both subgroups of ω in chapter 106; **V** has, like **L**, the reading Ποντικόν which is certainly the correct form. **W** has Ποντικός.

[75] See for instance Galen *De simpl. med. temp. ac fac.* 7.12 (12.14 K).

[76] If one assumes *Pontikon* is in fact the Black Sea, the author must be to the extreme West of Greece, because from this point of view the Peloponnese is in fact towards the Black Sea. But this would be a very unlikely understanding of geography.

in the *Chronicles of Morea*,[77] and the wording of the passage, Ποντικὸν...ὅπου ἔνι γὰρ εἰς τὸν Μορέαν, κάστρον λαμπρὸν ὑπάρχει suggests that this is indeed the place the commentator is referring to; he obviously heard of Pontic nuts and associated them with a place name that sounded familiar to him although it was not particularly known for nut production. And since he was not sure about the geographical situation, he most likely lived outside the Peloponnese and the Black Sea region.

This passage also allows us to date ω. Pontikon, or Pondikokastro as it is known today, was built by the Franks and is first mentioned in the accounts of the years 1205 to 1209 in the *Chronicles*,[78] which gives a *terminus post quem* for the composition of the commentary in ω. The *Chronicles* themselves are traditionally dated to shortly after 1300.[79] If the commentator is in fact citing the *Chronicles* and not just mentioning the castle, which could very well have had the name 'Pontikon in Morea', he must have lived around the turn of the century or later.[80]

Altogether, the evidence mentioned above points towards the early thirteenth to the fourteenth century, which agrees very well with the approximate dates established before, the mention of paper in a medical recipe indicating a *terminus post quem* of the early to mid thirteenth century for both versions, and the earliest manuscript of ω from the third decade of the fourteenth century.[81]

---

[77] *The Chronicle of Morea* 166.80ff. Rec. H and 167.80ff. Rec. P.

[78] See 112.61 Rec. H and 113.61 Rec. P.

[79] This is the date of the Greek version of the Text. See Schmitt p. XXXVI f. The last year described in the *Chronicles* is 1293.

[80] An alternative explanation is that the *Chronicles*, an extensive epic, were in fact composed successively rather than in one go, using phrases and blocks of texts from earlier sources. This could imply an earlier date.

[81] See page 7 and page 15.

CHAPTER SEVEN

MEDICAL PRACTICE

The *Therapeutics* is, as the name itself indicates, a therapeutic text, and thus almost entirely lacks descriptions of diseases and background information on materia medica, humoral pathology, anatomy and other medical fields.[82] Their aim is to provide basic information to treat a patient, based on a simplified Galenic methodology.

Neither of the main versions of *Therapeutics* (ℵ and ω) contains any direct statements on the context in which the text was used or commissioned. This is not unusual for a medical text of the time. Only very few medical manuals from the Middle Ages would even mention more than the name of the author or the addressee of the text. One of the few exceptions are hospital texts, the so called Xenonika, which are sometimes linked to a specific place.[83] Thus, all conclusions on the context are based on indirect evidence extracted from various parts of the *Therapeutics*.

The original version ℵ contains very little material one could use for this purpose; ℵ is polished and concise, and all one can say is that it omits invasive surgery[84] and covers all areas of general practice. The later version ω on the other hand sometimes gives itself away. Of particular importance are additions and explanations formulated in what looks very much like a first hand account.

---

[82] ℵ contains descriptions of diseases which are absent from ω. Whether these are interpolations to a manuscript transmitting the version or whether these passages were left out during a redaction cannot be determined. The only part of the text providing extensive background information is the preface.

[83] See for more information Bennett 2003: 129 ff.

[84] The *Therapeutics* mentions very few surgical instruments, and all are intended for superficial procedures. Some of these instruments can also be found in Leo Medicus. For more information on surgery in Greek medical texts see Bliquez 1999 and Bliquez 1984.

Not only does the commentator tend to describe tools, ingredients and procedures in more detail than one would expect,[85] he also specifies, at least in a few instances, what a disease looked like without using the standard terminology that was passed on in medical literature over the centuries. Rather, he describes them in his own words.

These additions leads us to a number of important conclusions. Since the commentator mentions only a few ready-made tablets[86] or potions, it is unlikely that the user of the book had access to a pharmacy or a specialized market; on the contrary, he aims at providing recipes as basic as possible. Upon a closer look, goods commonly available such as honey, water or salt are mentioned more often than in other comparable texts. This rather minimalistic and pragmatic approach towards medication of patients also points to limited financial resources.

We have no evidence on the personnel involved. All procedures could, at least from our modern perspective, have been handled by a single person. Preparing medication mostly involves chopping, pounding and boiling ingredients or leaving them to steep, and the result is then applied to the ailing part of the body. Just in one instance – an invasive gynecological procedure[87] – a second person is mentioned; in this case the commentator advises letting 'another woman' help with the treatment.[88] This supports the hypothesis that the text was not aimed at major metropolitan hospitals with better facilities, since these institutions usually had female physicians and specialized gynaecologists.[89]

---

[85] These descriptions can be found throughout the text. See for instance ω 87, where precise instructions are given on how to prepare a dye for blonde hair. The text even specifies stirring the mixture of various ingredients every other day.

[86] For a list see index.

[87] See ω 241.

[88] For obvious reasons, it is not unusual for these procedures to be delegated to women. E.g. Leo medicus *Synopsis* VI 16 (199.8 Ermerins) and VI 21 (203.4 Ermerins) mentions a female physician performing gynecological procedures. See B. Zipser 2005: 113.

[89] The past decades saw a very controversial discussion about Byzantine hospitals. Representative publications are Codellas 1942, Miller 1984, Miller 1985, Nutton 1986, Bennett 2003: 8-315, Horden 2005a.

Reconstructing who the patients were is not always possible. For instance it would be speculative to conclude that they included all age groups, just because the text mentions the elderly and infants.[90] In a few cases, however, we can catch a glimpse of the world of the *Therapeutics*. One passage suggests that some of the patients were subject to physical labour. This can be concluded from the successive additions to the text. Both versions of the *Therapeutics* have instructions how to treat water-blisters and callouses,[91] along with a recipe for ingrown shoe laces. In the latter passage, the commentator specifies the text of the original version saying that this happens in patients who have to work day and night and do not have the time to take their sandals off, which indicates that he actually saw these cases in practice, rather than just relying on textual sources.[92] That at least some of his patients belonged to the lower classes is also supported by a passage in ω 143 where it is specified that a person suffering from a certain type of fever should return to work after one day. Whether this was merely a therapeutic measure or whether it involved something comparable to a modern doctor's certificate is unclear.

Yet it does not appear that all the patients were expected to come from a deprived background, since another passage suggests that his patients survived for a prolonged time, even if they were bedridden. A chapter on the treatment of bedsores, a disease of the chronically ill, contains in ω a detailed description of the places in the body where bedsores occur – both ℵ and Theophanes lack this material.[93] These patients probably relied on support by their families.[94]

---

[90] In ancient medicine, age was correlated with a certain humoral disposition. Thus, mentioning the age in which a specific disease was likely to occur gave the physician vital information on the humours causing the disease.

[91] See chapters ω 68 and ℵ 60.

[92] See chapter ω 67. Both ℵ 59 and Theophanes 227 (= II 210 Bernard) have a more general title πρὸς πληγὰς ποδῶν ἐκ τῶν ὑποδημάτων γινομένας and πρὸς τὰ ἐξ ὑποδημάτων παρατρίμματα.

[93] See ω 66, and ℵ 58 and Theophanes 226 (= 208 f. Bernard) for the more general description: πρὸς ἄρρωστον πληγωθέντα ἐκ τοῦ πολλοῦ κατακεῖσθαι and πρὸς ἐτρίμματα ἐκ μακρᾶς κατακλήσεως.

[94] On the healthcare system of the time see Horden 2005b. Miller 1997.

It would be rather speculative to make more statements on the diseases the physician might have seen in practice. Yet it is striking that if one compares both versions, it becomes apparent that some chapters were left more or less in the form they had before the process of the revision, whereas others were heavily expanded by the commentator. The chapters which grew most are those concerning common diseases such as cough and fever,[95] whereas rare ophthalmological disorders, which would have required specialist treatment, remained almost unchanged.[96]

Overall, the additions in ω are diverse and one is often left wondering why the commentator believed it necessary to explain a certain passage, in particular if his explanations are purely based on common sense. This is for instance the case when he gives instructions on how to remove foreign bodies from the ear.[97] The original version has a clear and brief paragraph reading 'If something gets into the ear: If a flea gets in the ear put warm table-oil in the ear. If something else (gets in the ear) put the tips of the headhair in the ear as far as possible.' The commentator rephrased this passage as follows: 'If something gets in the ear, or a flea: If a flea gets (in) put olive oil in it, as long as it is warm (some manuscripts add the words 'table oil'). If something else gets (in), put the hair from his head, the tips of it, but twisted hair, as far as it gets inside of the ear.'

Everybody might have had this idea, and it does seem a bit odd to include a simple procedure as this in a medical book. But on the other hand, we also hear about the use of sophisticated instruments: In chapter 18 of א an ear-dropper is mentioned: 'If pus flows from the ear: Mix warm wine and honey and put it in the ear, with an ear-dropper. But do not insert the nozzle inside but hold it opposite of the opening, compress the bag and pour it into the ear, but do not be afraid because the wine-honey-mixture will flow out again. Just rinse the pus out of the ear.'

The commentator explained the passage as follows (chapter

Constantelos 1999: 348 ff. Constantelos 1968: 152 ff.

[95] For instance chapters ω 145 (a type of fever), 133 (cough), 55 (jaundice), 64 (sciatica).

[96] See for instance ω 110-117.

[97] This is chapter ω 25 and chapter א 21.

22): 'On pus flowing from the ears: Put warm wine and honey together and mix them. Then take a clyster, but not the one that they administer via the anus, but the one the doctors have which is called ear-(clyster). But do not put the nozzle inside but hold it opposite the hole and compress the pipette to pour out the liquid. But do not be afraid, because the wine-honey-mixture will flow out again, just make sure you rinse the pus out of the ear.' This means that the commentator was familiar with two words for this instrument, both of which are known to us from other medical texts,[98] and was also confident enough to replace one term with another. He also expected his audience to have access to the instrument in question.

The commentator's familiarity with complex medical procedures[99] and methods is also shown in his description how a pessary[100] is to be made; it was used to induce periods, labour (or an abortion?) and urination: 'Grind indigo and nigella seed, mix it with honey and make a bundle from a cloth and put it inside. Then give it to another woman and let her put it into the womb with her hand. But tie a thread to the bundle, and when you want to remove it, you have to pull the thread to remove it. This is called pessary by the doctors.'

Another passage reflects advanced therapeutical measures: in chapter 144 the commentator specifies that a bath in 'warm freshwater' is not to be confused with hot springs, and gives even a rare[101] Galenic term for a specific type of them (Alboula).[102] However, in other cases the commentator struggles with Classical terminology and is obviously not familiar with certain procedures; for instance in a recipe[103] for the treatment of callouses the com-

[98] The instrument ὠτεγχύτης or ὠτικὸν κλυστῆρ is well attested, cf. for instance Galen *Meth. med.* X 358 3 Kühn and Paulus VI 59 2, 9 Heiberg. See page 38 for literature on surgical instruments.

[99] See chapter ω 241; this chapter does not exist in ℵ.

[100] We are aware of a number of descriptions from earlier times, see for instance Paul of Aegina VII 24. 2 (= II 395, 148 Heiberg).

[101] The word is attested in two Galenic works, the *Meth. med.* X 536 Kühn and *De simpl. med. fac.* XI 393.

[102] The passage corresponds roughly to Theophanes chapter 139 (= I 436 Bernard); however, Theophanes lacks any mention of Alboula.

[103] Chapter ω 68, 3.

mentator illustrates the behaviour of a beetle called κανθαρίς as follows: 'Kantharis are called when one finds in summer in the streets insects like wasps and they sit on dung-heaps and twist the dung around and make it round like nuts and spin it with their feet. Take this Kantharis as it is, and smash a rock nicely, and then mix it together and put it on top of the callouses.' The commentator obviously mistook the beetle in question, a blister beetle and a common medical ingredient, for a κάνθαρος or dung-beetle.[104]

An important characteristic of our text is still to be discussed, namely the complete absence of religious charms or magic. The only instance a magical procedure is mentioned is solely specific to **L** and absent from ℵ, Σ and also from the classicizing version transmitted in **VW**. It contains instructions for the treatment of female infertility, namely to 'rip a hen (or a lizard?)[105] to pieces and hold it with your left hand when there is neither sun nor moon and wrap it in a linen cloth and wear it and you will be surprised'.

Oddly, both magic and religion invaded manuscripts transmitting the *Therapeutics*. **L**, a book that was used in practice, contains numerous scholia containing bits from the Greek liturgy along with a magical spell[106]. **M**, a manuscript from the mainstream production, transmits an interpolated Christian charm[107]. It remains a puzzle why both versions of the *Therapeutics* lack any magical or religious content, although it was present in the context in which they were used.

---

[104] For the translation see Liddell Scott Jones 1996. For the use of κανθαρίς in dermatological treatment see Paul of Aegina 7.3.10.

[105] The word γαλεκοτήν is not attested. Klaus-Dietrich Fischer suggested it could be derived from γαλεώτης 'lizard'. My translation 'hen' is based on the following: the second part of the word could be κόττα, a common word for hen, see e.g. Contopoulos and Kyriakides s.v. The first part of the word could be γάλος resp. γαλίνα, according to Kyriakides a Cypriot expression for 'turkey', obviously derived from Latin. Turkeys are not indigenous European and Asian birds and cannot be meant here. It might have been the case that the word was used for different types of fowl in common speech and was than narrowed down later on to a specific type.

[106] These are pages 275 and 278-279 in **l**.

[107] Chapter ω 218.1.

On the whole, we were able to gather a fair amount of information on the world from which our text stems, which must, to a certain degree, reflect the knowledge the commentator gathered in his lifetime. Of particular interest is the way the commentator works through the text and explains things he considers either interesting or ambiguous. His style is distinctive and far from the polished idiom of other, more academic texts which are more familiar to us. And it is this personal and direct way that makes ω special and firmly distinguishes it from other medical texts of the time.

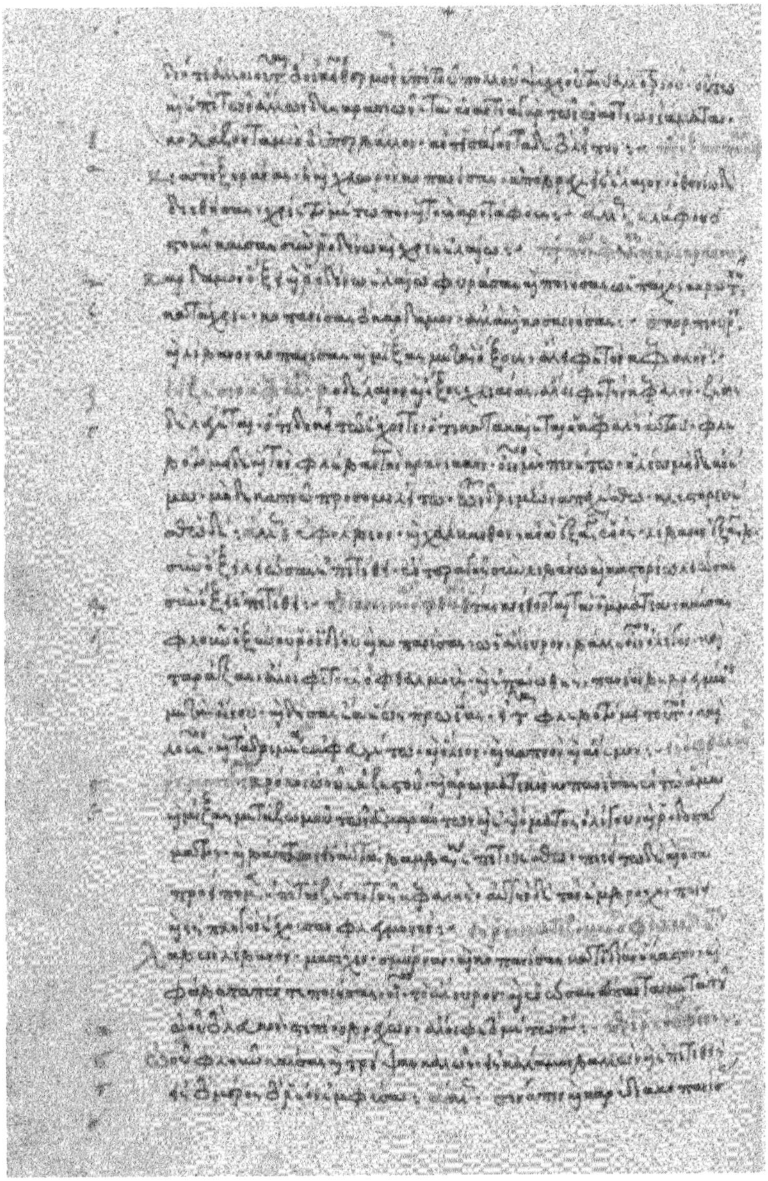

Figure 1: Monacensis *graecus* 551 f. 336v, Bayerische Staatsbibliothek München.

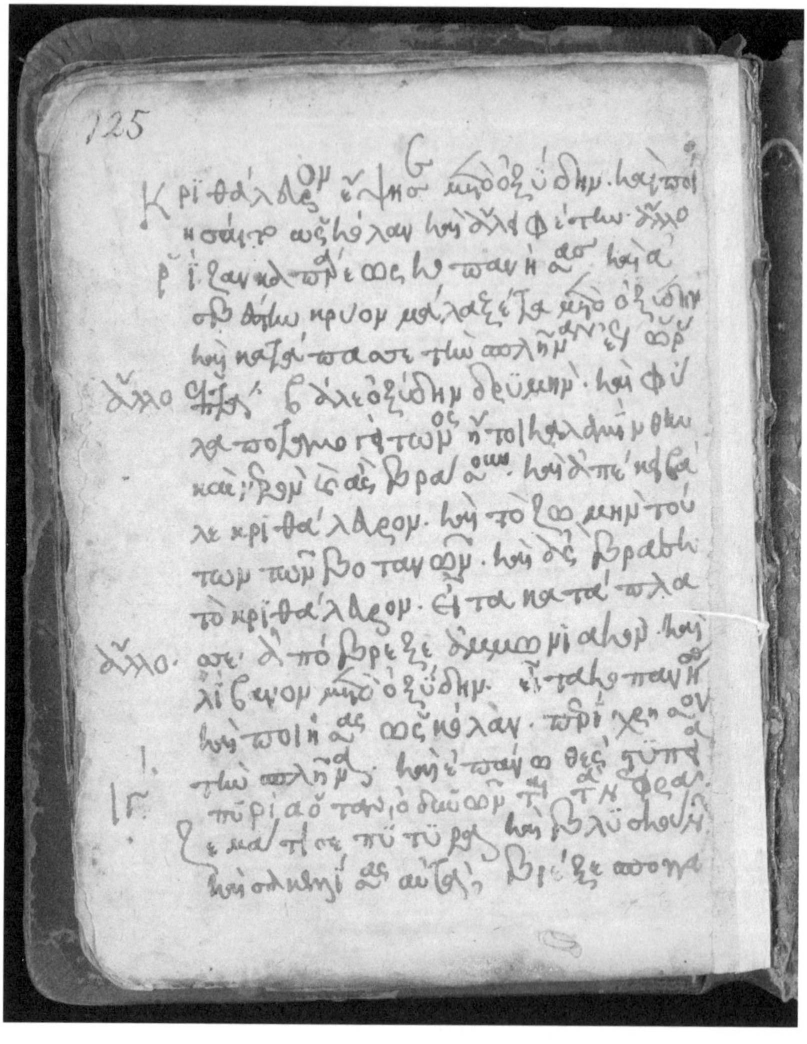

Figure 2: Medical Society London 14 p. 125, Wellcome Library London.

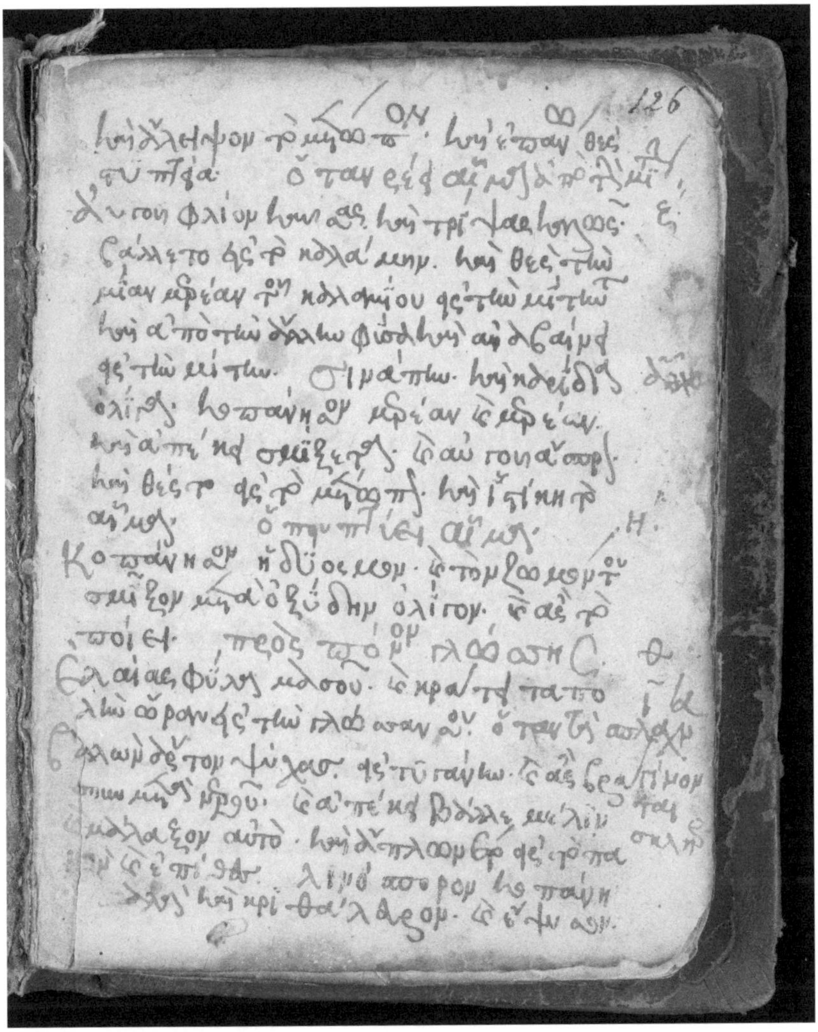

Figure 3: Medical Society London 14 p. 126, Wellcome Library London.

PART TWO

EDITION AND TRANSLATION OF א

THE EDITION

As outlined above, this volume contains the edition of two main versions of the *Therapeutics*, ℵ (aleph) and ω (omega). The first version, ℵ is written in learned Greek, and edited following one manuscript, N, the only witness. Some frequent and trivial mistakes are tacitly corrected; a list of these can be found in the appendix to the edition.

The second version, ω, is edited according to L, which constitutes the main text. Wherever the text in L is physically damaged beyond legibility, I have used M. Variant readings of other manuscripts can be found in the apparatus. If one or more manuscripts have additional text or are missing material compared to the others, this is indicated with upper case arrows and the sigla of the specific manuscript. Again, frequent mistakes are silently corrected. The pinax of ω is left out as it contributes no new material; moreover, this part of the text is infested with trivial mistakes. A transcription of the pinax of L is printed in the appendix.

Since one of the versions is a commentary to the other, most of the text of ℵ can also be found in ω. If ℵ has additional text, this is signified by upper case brackets. Citations from ℵ in ω are enclosed in upper case angles. As for the term citation, it is used in a very broad sense, including also rephrased or translated passages. The chapter numbers of ℵ and ω are, if applicable, followed by the number of the corresponding chapter in the opposite version.

All editorial principles are discussed in the introduction. The appendix holds additional material, for instance a synoptic list of chapter headings along with an elenchus of the manuscripts in which they are transmitted. This list can also be understood as a synopsis of the content of the manuscripts. Moreover, the appendix contains a full collation of a sample chapter and the list of frequent mistakes mentioned above.

CONSPECTUS SIGLORUM

1. *Original version*

N Monac. gr. 551

2. *Commentary version*

A Par. gr. 2236

B Par. gr. 2224

F Laur. gr. 75, 10, sometimes quoted

L MSL 14

M Monac. gr. 288

Σ consensus of MAB

3. *Signa*

ℵ original version as edited

ω commentary version as edited

$^{\triangleright \mathrm{M}}\ldots^{\mathrm{M}\triangleleft}$ text transmitted only in manuscript M

$^{\triangleright \mathrm{om.\ M}}\ldots^{\mathrm{om.\ M}\triangleleft}$ text transmitted in all manuscripts except M

$^{\ulcorner}\ldots^{\urcorner}$ citations from ℵ in ω

$^{\vdash}\ldots^{\dashv}$ text in ℵ that is absent from ω

$^{<}\ldots^{>}$ text in the preface of ℵ that presumably dropped out

13 [14] chapter 13 corresponding to chapter 14 of other version

ἀρχὴ σὺν θεῷ πίναξ συνοπτικοῦ ἰατροσοφίου τοῦ σοφωτάτου Γαληνοῦ

1 πρὸς ὀξὺν πόνον κεφαλῆς
2 πρὸς πόνον κεφαλῆς καὶ ἡμικράνου
3 εἰς ζέσιν κεφαλῆς
4 περὶ κνησμονῆς ὀφθαλμῶν
5 εἰς ὀφθαλμοὺς ῥευματιζομένους
6 περὶ ῥεούσης ῥινός
7 περὶ πτύσεως αἵματος
8 πρὸς πόνον γλώττης
9 πρὸς σκληρότητα σπλάγχνων
10 πρὸς πόνον ἥπατος
11 πρὸς σκληρότητα σπληνός
12 εἰς ὀδύνην νεφρῶν
13 εἰς τὸ ἀναβάσαι σάρκα
14 πρὸς ἔμφραξιν ὤτων
15 πρὸς πιτυρίασιν κεφαλῆς
16 εἰς ὀδύνην ὀφθαλμῶν
17 περὶ ὑποσφαγμάτων
18 πρὸς ὠτία ῥέοντα ἕλκος
19 πρὸς ὠτία ῥέοντα αἷμα
20 πρὸς ὠτία ἔχοντα σκώληκας
21 πρὸς ὅταν πέσῃ τι εἰς ὠτίον
22 πρὸς ὦτα ἔχοντα ἦχον
23 πρὸς ὅταν δυσακούῃ τις
24 πρὸς παρωτίδας τὰ λεγόμενα χελωνικά
25 πρὸς αἷμα ῥέον ἐξ τῆς ῥινός
26 εἰς τὸ γενέσθαι τοὺς ὀδόντας λευκούς
27 πρὸς αἷμα ῥέον ἐκ τῶν οὔλῶν
28 πρὸς σεσηπότα οὔλη στόματος
29 πρὸς κατεσχισμένα χείλη
30 πρὸς πρήσμα γλώττης

---

15 ἀναβάσαι] scripsi cf. ℵ 13 ἀναβά^{ς} N

The beginning with the help of God. The table of contents of the Synoptic Iatrosophion of the Most Wise Galen.

1. For an acute headache
2. For a headache and migraine
3. For a hot head
4. On itching of the eyes
5. For discharge from the eyes
6. For a running nose
7. On spiting of blood
8. On an aching tongue
9. On hardening of the bowel
10. On pain in the liver
11. On hardening of the spleen
12. On kidney pain
13. For anasarca
14. For blocked ears
15. On pityriasis of the head
16. On pain in the eyes
17. On hyposphagma
18. On pus flowing from the ear
19. On blood flowing from the ears
20. On worms in the ears
21. If something gets into the ear
22. If there is a noise in the ear
23. When someone is hard of hearing
24. On the parotids which are called tortoise shells
25. On blood flowing from the nose
26. To make the teeth white
27. If lesions are bleeding
28. On rotting mouth lesions
29. On soreness of the lips
30. On inflammation of the tongue

31 πρὸς δυσωδίαν στόματος
32 περὶ συνάγχην ἤτοι γουργουριακοῦ
33 πρὸς πτύσιν αἵματος
34 πρὸς φλεγμονὴν βυζίων
35 περὶ ἐμέτου
36 περὶ δυσεντερίας
37 περὶ κωλικῆς διαθέσεως
38 πρὸς ἑλμίνθων ἤτοι σκωλήκων ἐν κοιλίᾳ
39 πρὸς λιθίασιν νεφρῶν
40 πρὸς οὐροῦντα ἕλκος
41 πρὸς τὸν οὐροῦντα αἷμα
42 περὶ στραγγουρίας
43 πρὸς φλεγμονὴν ἥπατος
44 πρὸς ἀπόστημα ἥπατος
45 πρὸς ὑδερικούς
46 πρὸς σκληρότητα σπληνός
47 πρὸς πρήσμα σπληνός
48 περὶ ἰκτέρου
49 πρὸς πρήσμα ὀμφαλοῦ
50 πρὸς φλεγμονὴν ὄρχεως
51 περὶ γονορρίας ἤτοι ὅταν ῥέῃ ὁ σπόρος αὐτῶν
52 περὶ πρήσματος ἤτοι ἄλγους ἐντάσεως
53 πρὸς φλεγομονὴν δακτυλίου
54 περὶ ἐξοχάδας
55 περὶ ἐπικραττούντων τῶν καταμηνίων ἃ λέγονται γυναικεῖα
56 πρὸς ἰσχίον
57 περὶ κατασχίσεως τῶν δακτύλων
58 πρὸς ἄρρωστον πληγωθέντα ἐκ τοῦ πολλοῦ κατακεῖσθαι
59 πρὸς πληγὰς ποδῶν ἐκ τῶν ὑποδημάτων γινομένας
60 πρὸς ἥλους γινομένους ἐν ταῖς χερσίν

---

2 ἤτοι] τοῦ ℵ 32   17 πρήσμα] scripsi cf. ℵ 47 πρης$^{\mu}$ N   19 ὀμφαλοῦ] scripsi ὀφθαλμῶν N, cf. pinax to L and note to ω 56   22 πρήσματος] fortasse πριαπισμοῦ   24 ἐξοχάδας] scripsiἐξωχάδας N

31. When the mouth smells
32. On swelling of the throat
33. If someone spits blood
34. On breast inflammation
35. On vomiting
36. On dysentery
37. On colics
38. On helmiths or worms in the stomach
39. On kidney stones
40. On pus in the urine
41. On blood in the urine
42. On strangury
43. On hepatitis
44. On liver abscesses
45. On dropsy
46. On hardening of the spleen
47. On inflammation of the spleen
48. On jaundice
49. On inflammation of the navel
50. On inflammation of the testicles
51. On gonorrhea or when the semen flows
52. On inflammation or pain of the tumescence
53. On the inflammation of the anus
54. On external hemorrhoids
55. If the menses, also called *gynaikeia*, stop
56. On sciatica
57. On soreness of the fingers
58. For patients who get sore from being bed-ridden for a long time
59. On wounds on the feet caused by sandals
60. On calouses on the hands

61 πρὸς παρωνυχίδας
62 πρὸς πτερύγιον
63 πρὸς ὄνυχας ἔχοντας αἷμα
64 πρὸς ψώρας
65 πρὸς πόνον νεύρων
66 πρὸς ὀφθαλμοὺς μὴ βλέποντας
67 πρὸς ἀπορυφάς
68 πρὸς οὐλὴν πληγῆς ὅταν γένηται μαύρη
69 πρὸς ὅταν αἱ τρίχες πίπτουσιν
70 περὶ αὐξήσεως τριχῶν
71 περὶ τοῦ μέλανας τρίχας ποιῆσαι
72 περὶ φθειρῶν τῶν γινομένων ἐν τῇ κεφαλῇ
73 πρὸς τὰ γλυκέα
74 περὶ ἀλωπεκίας καὶ ὀφιάσεως
75 περὶ κεφαλαλγίας
76 περὶ ἐκ ψύξεως κεφαλαλγίας
77 πρὸς τοὺς ἀπὸ στομάχου κεφαλαλγοῦντας
78 περὶ τῆς ἐξ οἴνου κεφαλαλγίας
79 περὶ τοῦ μὴ μεθύειν
80 ἀποφλεγματισμοί
81 ἀποφλεγματισμοὶ καθαρεύοντες τὴν κεφαλήν
82 εἰ θέλης πτέρνεσθαι
83 πρὸς τὸ παῦσαι πταρμούς
84 περὶ τῆς ἀπὸ ψύξεως κεφαλαλγίας
85 περὶ κατάρρου
86 πρὸς ὅταν πέσωσιν αἱ τρίχες τῶν ὀφρύων
87 πρὸς τὸ ποιῆσαι μέλανας τρίχας
88 πρὸς διπλὰς τρίχας ὀφρύων
89 περὶ κριθῆς τῆς ἐν ὀφθαλμῷ
90 πρὸς ὅταν τὸ λευκὸν τοῦ ὀφθαλμοῦ γένηται κόκκινον

---

17 πρὸς] scripsi περὶ N

61. On paronychia
62. On pterygium
63. If the nails have blood
64. On psoriasis
65. On neuralgia
66. On eyes, when they do not see
67. On *aporufai*
68. If a wound gets black
69. When the hair falls out
70. On augmenting hair
71. To make the hair black
72. On head lice
73. On scurf
74. On alopecia and ophiasis
75. On headache
76. On headache caused by cold
77. On headache caused by the stomach
78. On headache caused by wine
79. On not getting drunk
80. Cleansing of phlegm
81. Cleansing the head from phlegm
82. If you want sneezing
83. To stop sneezing
84. On headache caused by cold
85. On catarrh
86. If the hair of the eyebrows falls out
87. How to make hair black
88. On double hair in the eyebrows
89. On styes of the eye
90. If the white of the eye becomes red

91 περὶ ἐμφυσήματος
92 περὶ σκληροφθαλμίας
93 περὶ ξηροφθαλμίας
94 περὶ ἐκτροπίου
95 περὶ αἰγίλωπος
96 πρὸς ὅταν αἱ τρίχες τῶν ὀφθαλμῶν πέσωσιν
97 περὶ ῥυάδος
98 πρὸς λεύκωμα ὀφθαλμοῦ
99 πρὸς πτερύγιον ὀφθαλμοῦ
100 πρὸς ἀμαύρωσιν ὀφθαλμοῦ
101 περὶ ὑπωπίου
102 πρὸς τὸν ἀπὸ ψύξεως πόνον τοῦ ὠτός
103 πρὸς τὸν ἐκ θερμῆς δυσκρασίας πόνον τοῦ ὠτός
104 πρὸς τὰ διὰ παχεῖς χυμοὺς ὀδυνώμενα ὠτία
105 περὶ φλεγμονῆς ὠτίου
106 περὶ αὐξήσεως ὀδόντων παιδίου
107 πρὸς ὄγκωμα ὑποκάτω τῆς γλώττης
108 περὶ ἄφθας
109 περὶ φλεγμονῆς σιαγόνων
110 πρὸς βῆχα
111 πρὸς τοὺς ἀπὸ φωνῆς μέγα βραχνιάσαντες
112 πρὸς δυσπνοίαν νέων μέσων καὶ γερόντων
113 περὶ πνευμονίας ἤγουν δυσπνοίας
114 περὶ ἐμπυήματος ἤγουν πλευρίτου
115 πρὸς φθίσιν
116 πρὸς τὸν ἀπὸ θυμοῦ πυρέττοντα
117 περὶ τριταίου
118 περὶ τεταρταίου
119 περὶ τοῦ ἀφημερινοῦ
120 πρὸς ὅταν γένηται λοιμῶδες, ὃ πυρετικὴν λέγουσιν

14 τὰ] scripsi τοὺς N 14 ὀδυνώμενα ὠτία] scripsi ὀδυνωμένους ὀφθαλμούς N 24 ἐμπυήματος] scripsi ἐμπτυήματος N 26 θυμοῦ] scripsi θ..οῦ N, cf. *ω* 116

91. On emphysema
92. On hardening of the eye
93. On drying of the eye
94. On ectropion
95. On dacryocystitis
97. On flowing
98. On a white spot in the eye
99. On pterygium of the eye
100. On blindness
101. On hypopion
102. On pain in the ears that is caused by cold
103. On pain in the ear that is caused by warm imbalance
104. On pain in the ears caused by thick humours
105. On the inflammation of the ear
106. To induce the growth of teeth in a child
107. On swelling under the tongue
108. On aphthous mouth ulcers
109. For the inflammation of the jaws
110. For cough
111. For those who have a very hoarse voice
112. For dyspnea of young, middle aged and old people
113. On pneumonia and difficulty breathing
114. On empuema or pleuritis
115. On phthisis
116. On someone who is feverish as a result of being irascible
117. On tertian fever
118. On quartan fever
119. On quotidian fever
120. For when the plague strikes, which is called hectic fever

121 πρὸς βυζία ἔχοντα καρκίνου πάθος
122 πρὸς δυσωδίαν μασχαλῶν
123 πρὸς πτύσιν αἵματος πολλὴν καὶ εἰς μίαν
124 πρὸς ὅταν ῥεύγηται ὡς ὠὰ τηγάνου
125 πρὸς τοὺς διὰ ψύξιν ἀνορεκτοῦντας
126 πρὸς τοὺς ἀνορεκτοῦντας διὰ ὕλην
127 πρὸς κυνώδη ὄρεξιν
128 πρὸς κλόξον
129 πρὸς κένωσιν χολῆς ἄνω καὶ κάτω
130 πρὸς φλεγμονὴν ἐντέρων
131 περὶ φλεγμονῆς νεφρῶν
132 πρὸς πάθος τὸ λεγόμενον διαβήτην
133 πρὸς ῥαγάδας ὀρχιδίων
134 πρὸς πνιγμονὴν τὴν ἐξ ἀμαντίων
135 πρὸς τύρωσιν γάλακτος ἐν τῇ κοιλίᾳ
136 πρὸς ἀλφοὺς ὅταν εἰς τὸ πρόσωπον τοῦ ἀνθρώπου ἐξέρχονται ὥσπερ φακαί
137 πρὸς πόνους ἀρμῶν
138 πρὸς ἀπόστημα
139 πρὸς ἀπορυφὰς καὶ ἐκβατά
140 πρὸς δῆξιν ἀσπίδος
141 πρὸς τὸ καταβάσαι γάλα
142 πρὸς γυναίκα μὴ δυναμένην γεννῆσαι
143 πρὸς γλῶσσαν ἁδρᾶν
144 πρὸς τὸ ἐὰν συμβῇ τεθνάναι τὸ παιδίον ἐντὸς τῆς κοιλίας
145 πρὸς δῆξιν ὄφεως καὶ ἐχίδνης
146 περὶ χρυσιασμοῦ
147 πρὸς κοιλίαν ῥέουσαν πολλά
148 πρὸς δῆξιν κυνός
149 πρὸς δῆξιν λυσσῶντος κυνός
150 πρὸς δῆξιν μελισσῶν ἢ καὶ σφηκῶν

121. For breasts suffering from cancer
122. For smelling armpits
123. On spitting a mass of blood once
124. When he vomits something resembling fried eggs
125. For those who have no appetite because of cold
126. For those who have no appetite because of matter
127. On dog-like appetite
128. On hiccup
129. On excreting bile above and below
130. On inflammation of the intestines
131. On inflammation of the kidneys
132. On the disease which is called diabetes
133. On fissures on the testicles
134. On choking because of mushrooms
135. If milk turns to cheese in the stomach
136. On alphoi, if the appear on a person's face like lentils
137. On pain of the joints
138. On ulcers
139. On *aporufai* and swelling
140. For the bite of an asp
141. To induce lactation
142. For a woman if she is unable to conceive
143. For a hardened tongue
144. If the child dies in the womb
145. For the bite of a snake or a viper
146. On jaundice
147. If there is much flow from the abdomen
148. For dog bites
149. If someone is bitten by a rabid dog
150. If someone is bitten by a bee or a wasp

151 πρὸς δυσωδίαν ῥινός
152 πρὸς τὰς ἀπὸ δαρμοῦ πληγάς
153 πρὸς νεφέλην ὀφθαλμῶν
154 πρὸς πόνον ὀδόντων
155 πρὸς σειομένους ὀδόντας
156 πρὸς καῦσιν μέλους τινός
157 πρὸς ποδαλγικούς
158 πρὸς ὅταν οἱ ἁρμοὶ τῶν δακτύλων γίνωνται σκληροί
159 πρὸς τὸ στῆσαι αἷμα
160 πρὸς ἀδυναμίαν στομάχου
161 πρὸς στρόφον
162 περὶ δήξεως σκορπίου
163 πρὸς σκώληκας τοὺς ἐν τῇ πληγῇ γινομένους
164 πρὸς δῆξιν τρυγόνος θαλασσίας
165 πρὸς σπληναρίους καὶ ὀγκωμένους
166 πρὸς φωνὴν ἀποκοπεῖσαν
167 περὶ τῶν λεγομένων χοιράδων
168 πρὸς τὸ παῦσαι δίψαν
169 πρὸς δυσωδίαν στόματος
170 πρὸς σῆψιν οὐλῶν
171 πρὸς βουβῶνας ἤτοι ἀπορυφᾶς
172 πρὸς βιασμούς
173 πρὸς ψύξιν γονάτων
174 εἰς ἐσοχάδας καὶ ἐξοχάδας
175 πρὸς ὅταν ἐξέρχεται ὁ κῶλος τοῦ ἀνθρώπου
176 πρὸς τὰ ἐκβατά
177 πρὸς δῆξιν ἐχίδνης
178 πρὸς τὸ διῶξαι θηρία ἐκ τῆς οἰκείας
179 πρὸς ἡμίκρανον
180 πρὸς ἰσχίον

---

21 ἤτοι] τῆς ℵ 171

151. On a malodorous nose
152. On wounds caused by abrasion
153. On a cloud-like spot in the eye
154. On tooth ache
155. If the teeth are loose
156. If a part of the body is burned
157. On podagra
158. If the joints of the fingers harden
159. To stop bleeding
160. For weakness of the stomach
161. On colics
162. On scorpion bites
163. If worms appear in wounds
164. On bites of a sea roach
165. On those suffering from a spleen disease and a swelling
166. On the loss of voice
167. On the so called scrofula
168. To stop thirst
169. On halitosis
170. On rotting lesions
171. On swollen glands or *aporufai*
172. On stomach cramps
173. On cold knees
174. On internal and external hemorrhoids
175. On a prolapse of the colon
176. On swelling
177. On viper bites
178. To drive pests away from the house
179. On migraine
180. On sciatica

181 περὶ λέπρας
182 πρὸς ψώραν
183 πρὸς λειχῆνας
184 πρὸς τεταρταῖον πυρετόν
185 ἐπίθεμα πρὸς τὸ διῶξαι πυρετὸν καὶ γαστέρα προτρέψαι
186 πρὸς ῥαγάδας χειρῶν καὶ χειλῶν καὶ ποδῶν
187 ἡ σκευασία
188 περὶ ἐκβατῶν καὶ ἀνθράκων
189 πρὸς ἄνθρακας τὰ λεγόμενα μεθύ^{στρ}
190 ἄλειμμα τριχοφυές
191 πρὸς κεφαλαλγίαν καὶ ἑτεροκρανίαν
192 πάσμα σταλτικὸν τῶν ἀπὸ κεφαλῆς ῥευμάτων
193 ἄλλο χρήσιμον πάσμα
194 πρὸς πόνον ἡμικράνου
195 πρὸς καυσίδας καὶ φλεγμονάς
196 πρὸς ὠταλγίαν ὃ καὶ πρὸς πόνον ὀδόντων
197 πρὸς τοὺς ἐν τοῖς ὠσὶ σκώληκας
198 πρὸς τοὺς ἔχοντας σκώληκας καὶ πύον
199 πρὸς πόνους ὤτων δόκιμον
200 πρὸς πόνον ὀδόντων
201 ἕτερον πρὸς πόνον ὀδόντων διδόμενον ἐν παντὶ καιρῷ
202 πρὸς πόνον λαιμοῦ
203 πρὸς συνάγχην καὶ ῥευματιζομένους καὶ ῥέγχοντας
204 ὄσφραντὸν πρὸς κατάρρουν καὶ βῆχας
205 κατάπλασμα εἰς ῥέγχοντας
206 πρὸς χρόνιον βῆχα διὰ πεῖρας
207 περὶ ἀποφλεγματικῶν καὶ πταρμικῶν
208 εἰς ἐπιληπτικοὺς βοήθημα
209 πρὸς πόνον καὶ στρόφον στομάχου

10 ἄλειμμα] scripsi ἄλειμα N

181. On leprosy
182. On psoriasis
183. On lichen
184. On quartan fever
185. Salve to decrease fever and induce defecation
186. For split hands, lips and feet
187. The recipe
188. On swelling and anthrax
189. On anthrax which is also called *methustra*
190. Tincture to induce hair growth
191. On headache and migraine
192. Astringent salve for flux from the head
193. Another astringent salve
194. For headache
195. For burning, inflammation, dim-sightedness, rheumatism and conditions related to itching and blood
196. For ear ache which is also famous for tooth ache
197. For worms in the ears
198. For those with worms and pus
199. Famous for ear ache
200. For tooth ache
201. For tooth ache at all times
202. For a sore throat
203. For tonsillitis, flux from the chest, catarrh, coughing and wheezing
204. Inhalation for catarrh and cough
205. Salve for wheezing and flux from the chest
206. Approved remedy for chronic cough
207. To cleanse from phlegm and to treat sneezing
208. Medication for epileptics
209. For stomach pain and cramps

210 βοήθημα στομαχικόν
211 κοκκία στομαχικά
212 κοκκία ποδαλγικά
213 κατάπλασμα εἰς τοὺς ὀδυνωμένους τὸ ἰσχίον
214 πρὸς πόνον ἰσχίου
215 ἐμβροχὴ εἰς ῥευματιζομένους καὶ ὀδυνωμένους τὸ ἰσχίον
216 ἄλλο εἰς ὀδύνην

210. Medication for the stomach given to those who suffer from indigestion, do not digest or who have a bloated stomach
211. Pills for the stomach
212. Pills for gout
213. Salve for those suffering from sciatica pain
214. For sciatica pain
215. Infusion for rheumatism and sciatica pain
216. Another one for sciatica pain

ἀρχὴ σὺν θεῷ <συνοπτικοῦ ἰατροσοφίου τοῦ σοφωτάτου Γαληνοῦ>

<σύντομος> διδασκαλία τοῦ θαυμασιωτάτου Γαληνοῦ τοῦ σπεύσαντος γράψαι εἰς τὰ τῶν μορίων πάθη ἀνίατα καὶ δυσίατα <ὥστε> εὐίατα γενέσθαι, <καὶ> πρῶτον μὲν καὶ κοινῷ λόγῳ καὶ κεφαλαιωδῶς εἰπεῖν, <διδάξαντος> περὶ τὴν <τοῦ> παντὸς σώματος ἐπιμέλειαν εἶτα καὶ <τὴν> κατὰ τὸν πεπονθότα τόπον γινωμένην ῥαστώνην, ἥτις παρὰ τῶν ἰατρῶν καλεῖται ὠφέλεια καὶ θεραπεία. καὶ πρῶτον διδάσκει περὶ τῶν ἐκτὸς παθῶν ὅτι εἰσὶν εὐίατα, <εἶτα> μέλλει καὶ περὶ τῶν ἐντὸς εἰπεῖν δεομένων τὴν θεραπείαν πολυχρόνιον καὶ ἐπιτήδειον πρὸς τὸ μῆκος τῆς ὁδοῦ τοῦ πάσχοντος τόπου. οἱ ἐντὸς <γὰρ> πεπονθότες τόποι γεγυμνασμένου δέονται <ἰατροῦ> εἰς <τὸ καλῶς διαγινώσκειν αὐτοὺς καὶ μέντοι καὶ πῶς ἔχει κράσεως ἕκαστον.

1 ἰατροσοφίου] scripsi ἰατροσ... N

With the help of God, the beginning of the Synoptic Iatrosophion[1], of the Most Wise Galen:

This is an epitome of the instruction of the Most Wonderful Galen who aimed at writing about the affected parts[2] which are impossible or difficult to treat, in order to make them easy to treat. At the same time he is speaking in a tone intelligible to all,[3] in an abbreviated form, and instructing on the care of the entire body as well as how to soothe the affected parts, which is called help and therapy by the doctors.[4] And in the beginning he talks about the external ailments saying that they are easy to treat, then he speaks about the internal ones which need to be treated for a long time, and suitably, to cover the entire extent of the ailing part.[5] For the correct diagnosis of the internally affected parts requires an experienced physician, and in particular the decision about their humoural mixture.[6].

[1] An Iatrosophion is a common genre in Medieval medicine. It can either refer to a compilation of medical writings or to a manuscript containing a collection of medical texts. Synopsis usually refers to concise and clearly structured medical compendia.

[2] This is a reference to Galen's work *On the Affected Parts*, also known as *De Locis Affectis*; however, the *Therapeutics* are not an epitome of any known Galenic work. Most likely, the author of ℵ used a source that was falsely entitled as Galenic, and transferred the error to his own work.

[3] The Greek term *koinē*, which is used here, usually refers to a simplified, standardized and non-dialectal form of Greek that has its origin in Classical times. The language of ℵ is a Medieval form of this idiom.

[4] The term 'doctor' refers to medical authors, and most importantly their idiom. It is used throughout the *Therapeutics*, see for instance paragraph 146.

[5] This phrase is somewhat obscure. It is unclear whether the author is saying that treatment has to travel a longer way to reach the inner parts of the body, or whether these internal diseases spread over a wider area.

[6] The four humours are, as outlined below, blood, yellow bile, black bile, and phlegm. The qualities are hot, cold, moist and dry, and they can also occur in combination. This yields a total of nine possible mixtures, the ideal one, too hot, cold, dry or moist and too hot/dry, hot/moist, cold/dry or cold/moist.

οὐ μόνον γὰρ> τὴν κοινὴν ἁπάντων φύσιν γινώσκειν <χρή, ὦ σὺ τῶν ἀρετῶν ἱερὸν τέμενος,> ἀλλὰ καὶ τὴν ἰδέαν ἑκάστου, ἥτις ἀπὸ <τῆς κράσεως> τῶν στοιχείων συνίσταται, θερμοῦ <λέγω> καὶ ψυχροῦ καὶ ὑγροῦ καὶ ξηροῦ, ἤγουν πυρὸς ὕδατος ἀέρος καὶ γῆς ἃ γεννῶσι τοὺς χυμοὺς ἐξ ὧν αἷμα καὶ φλέγμα καὶ χολὴ ξανθὴ καὶ μέλαινα.

ἔστιν δὲ τὸ μὲν αἷμα θερμὸν καὶ ὑγρόν, ψυχρὸν δὲ καὶ ὑγρὸν τὸ φλέγμα, θερμὴ καὶ ξηρὰ ἡ ξανθὴ χολή, ψυχρὰ δὲ καὶ ξηρὰ ἡ μέλαινα. ἡ μὲν οὖν ξανθὴ χολὴ θερμὴ καὶ ξηρὰ οὖσα ὡς ἔφημεν, ποιεῖ τὰ τριταϊκὰ νοσήματα ἐν θερμοτέροις σώμασι καὶ ξηροτέροις, ἡ δὲ μέλαινα τὰ τεταρταϊκὰ καὶ τὰ ἄλλα τὰ μελαγχολικὰ νοσήματα οἷον ψώρας, λέπρας, κνησμονὰς καὶ τὰ τούτοις ὅμοια. γίνεται δὲ ἐν σώμασι ψυχροτέροις καὶ ξηροτέροις καὶ ὑγροτέροις. τὸ δὲ φλέγμα τοὺς ἀφημερινοὺς πυρετούς. λέγονται δὲ οὕτως διὰ τὸ συνίστασθαι ἅμα τῷ παρακμάσαι τὸν πρότερον παροξυσμόν. γίνονται δὲ ἐν σώμασι ψυχροτέροις καὶ ὑγροτέροις. τὸ αἷμα ποιεῖ τοὺς συνόχους πυρετούς, οἵτινες συνίστανται ἐν θερμοῖς καὶ ὑγροῖς σώμασι καὶ πολυαίμοις καὶ μὴ ἔχουσι ἔθος κενοῦσθαι.

For it is not only necessary to recognize the common nature of all,[7] o Thou Holy Shrine of Virtues, but also the specific nature of any single individual,[8] which arises from the mixture of the elements, I mean of warm, cold, moist and dry, that is fire, water, air and soil which produce the humours blood, phlegm, yellow bile and black bile.

The blood is warm and moist, cold and moist is phlegm, warm and dry the yellow bile, cold and dry the black bile. Since the yellow bile is warm and dry as we have said, it causes three-day diseases[9] in bodies which are rather warm and dry, the black one causes four-day diseases[10] and other melancholic[11] diseases as psoriasis, leprosy, itching and similar. Such happens in bodies which are rather cold, dry and moist. The phlegm causes two-day fevers.[12] These are called thus, as they arise at the time at which the previous fit of the disease has passed its point of culmination. They develop in bodies which are rather cold and wet. The blood causes continuous fevers which happen in bodies that are rather warm, moist and full of blood and do not have an habit of emptying.

---

[7] This is the ideal model discussed in ancient anatomical and physiological works.

[8] This also includes pathological changes to the ideal body and changes due to diet, circumstances of living, age and the like.

[9] Tertian fever.

[10] Quartan fever.

[11] The term 'melancholic' refers to black bile, *melan* (black) and *cholē* (bile).

[12] Quotidian fever.

φησὶ ῾Ιπποκράτης ὁ ἄριστα διαγνοὺς ἄριστα καὶ θεραπεύει. ὁ γὰρ γινώσκων τὸ σῶμα τοῦδε τοῦ ἀσθενοῦντος ὅτι ὑπάρχον θερμὸν νῦν ἐκ θερμῆς δυσκρασίας νοσεῖ, τῇ ψυχρούσῃ διαίτῃ χρήσεται οὐ πολλῇ ἵνα μὴ καταψύξει λίαν τὸν ἄνθρωπον καὶ οὐ μόνον οὐδὲν ὠφελήσει ἀλλὰ καὶ τοῦ ζῆν ταχέως ἀπαλλάξει αὐτὸν διότι ἀλλοιοῦται τὸ οἰκεῖον θερμὸν ὑπὸ τοῦ πολλοῦ ψυχροῦ τοῦ ἀλλοτρίου. οὕτω καὶ ἐπὶ τῶν ἄλλων δυσκρασιῶν. τὰ ἐναντία γὰρ τῶν ἐναντίων ἰάματα, κολάζοντα μὲν τὸ ὑπερβάλλον, ἀντισάγοντα δὲ τὸ ὀλίγον.

1 φησὶ] cf. Alex. Trall. *Therap.* IX, 3; Stephanus *Schol. in Hipp. Progn.* I,1.

Hippocrates says 'He who makes the best diagnoses also has the best treatment'.[13] For if someone now recognizes that the body of a certain patient is warm and that it is ailing due to a hot imbalance of the humours, he will employ a cool diet, not much however, in order to avoid cooling down the patient excessively, for that would not only be useless, but also it would swiftly rob him of his life, for his own warmth is changed by an excess of alien cold.

The same applies to the other imbalance of the humours. For the opposite is the remedy of the opposite, quelling the excessive, and replenishing whatever is lacking.

[13] This sentence is just attested as a quotation in other works, but not in it original context. For exact references see the apparatus to the edition.

1 [1] α΄ πρὸς ὀξὺν πόνον κεφαλῆς· κισσὸν ξηράνας ἢ καὶ χλωρὸν κοπανίσας ἀπόβρεχε εἰς ἔλαιον, ὀθονίῳ δὲ διηθήσας χρίε τὸ μέτωπον καὶ τοὺς κροτάφους. 2 ἄλλο. ἐλάφου ὀστοῦν καύσας σὺν ῥοδίνῳ καὶ χρίε ἐλαίῳ.

2 [2] β΄ πρὸς πόνον κεφαλῆς καὶ ἡμικράνου· κάρδαμον ὄξει καὶ ῥοδίνῳ ἐλαίῳ φυράσας καὶ ποιήσας ὡς πάχος κηρωτῆς κατάχριε, κοπανίσας τὸ κάρδαμον ἀλλὰ καὶ κοσκινίσας. σκορπίουρον καὶ λίβανον κοπανίσας καὶ μίξας μετὰ ὄξους ἄλειφε τὴν κεφαλήν.

3 [3] γ΄ εἰς ζέσιν κεφαλῆς· ῥοδέλαιον καὶ ὄξος χλιάνας ἄλειφε τὴν κεφαλήν. ζέσις δὲ λέγεται ὅτι δοκεῖ τῷ ἔχοντι ὅτι κατακαίεται ἡ κεφαλὴ αὐτοῦ. φλεβοτόμει δὲ καὶ τὴν φλέβαν τὴν κρανιακήν. οἶνον μὴ πινέτω, ἡλίῳ μηδὲ ἀνέμῳ μηδὲ καπνῷ προσομιλείτω. τῶν δριμέων ἀπεχέσθω, κλυστηρευέσθω δέ. 2 ἄλλο. ⌐εὐφόλβιον καὶ χάλκανθον ἀνὰ ἐξαγίου ἑνός, λίβανον ἐξάγια β’ σὺν ὄξει λειώσας ἐπιτίθει. ἔντερα γῆς σὺν λιβάνῳ καὶ καστορίῳ λειώσας σὺν ὄξει ἐπιτίθει.¬

4 [5] δ΄ περὶ κνησμονῆς ὀφθαλμῶν· ὅταν κνήθωνται τα ὀμμάτια, καῦσας φλοῦν ὀξύνου ῥοϊδίου καὶ κοπανίσας ὡς ἄλευρον βάλε οἶνον ὀλίγον. καὶ ταράξας ἄλειφε τοὺς ὀφθαλμοὺς καὶ ἐπάνω θὲς πανίον βεβρεγμένον μετὰ οἴνου. καὶ δήσας ἔα ἕως πρωΐας. εἶτα φλεβοτόμει τοῦτον καὶ λοῦσον. καὶ τὰ δριμέα ἐκφευγέτω καὶ ἥλιον καὶ καπνὸν καὶ ἄνεμον.

2 διηθήσας] scripsi διιθήσας N

1. For an acute headache: Dry ivy, or take green ivy, soak it in olive oil, sieve it and anoint the forehead and the temples. 2 Burn the bone of a deer with rose oil and apply it with olive oil.[14]

2. For a headache and migraine: Knead cardamom with vinegar and rose-oil until it has the consistency of wax and apply it. But grind the cardamom and sieve it. Grind scorpion wort and frankincense, mix it with vinegar and anoint the head with it.

3. For a hot head: Heat rose-oil and vinegar and anoint the head with it. Scalding is when the patient thinks that his head burns. And also let blood from the cephalic vein.[15] He must not drink wine, and must not be exposed to sun, wind or smoke. He should avoid acrid things, and should be purged.[16] 2 Euphorbium and vitriol one hexagion[17] each, frankincense two hexagia, make them smooth with vinegar and apply. Earthworms with frankincense and castoreum, make smooth with vinegar and apply.

4. On itching of the eyes: When the eyes are itching, burn the peel of a sour pomegranate, pound it like flour and add some wine. And mix it and anoint the eyes and cover with a cloth soaked in wine. And bandage it and leave it until the next morning. Then bleed him and give him a wash. He should avoid acrid things and sun, smoke and wind.

---

[14] The *Therapeutics* consists roughly speaking of two cycles in head to foot order. Quite a lot of the chapters go back to Theophanes Chrysobalantes, a medieval author of uncertain date. Chapters 1 to 13 coincide with a hospital text. For more detailed information on the textual history of the see the chapter 'Other Texts' in the introduction.

[15] The cephalic vein runs across the arm.

[16] This was done with a clyster.

[17] A hexagion equals one and a half drachmae.

5 [4,6] ε′ εἰς ὀφθαλμοὺς ῥευματιζομένους· κρόκον ᾠοῦ ἐκζεστοῦ καὶ ἀρωματικὸν κοπανίσας ἐν τῷ ἅμα καὶ μίξας μετὰ ζωμοῦ τῶν ἀμαράντων καὶ ἑψήματος ὀλίγου καὶ ῥοδοστάγματος καὶ βάπτων εἰς αὐτὰ βαμβάκιν ἐπιτιθέσθω. ποίει τόδε καὶ ὅσα προείπομεν ἐπὶ τὴν ζέσιν τῆς κεφαλῆς. αὐτὴν δὲ τὴν ἐμβροχὴν ποίει καὶ εἰς πληγὴν ἔχουσαν φλεγμονήν. 2 εἰς ῥευματιζομένους ὀφθαλμοὺς ἕτερον· λαβὼν λίβανον μαστίχην σμύρναν καὶ κοπανίσας κατ' ἰδίαν ἕκαστον καὶ φάβατα πέντε ποιήσας οἷον τὸ ἄλευρον καὶ ἑνώσας ἅπαντα μετὰ τοῦ ᾠοῦ τὸ λευκόν, στυππεῖον βρέχων ἄλειφε τὸ μέτωπον.

6 [7] ϛ′ περὶ ῥεούσης ῥινός· ᾠοῦ φλοῦν καύσας καὶ τρίψας καλῶς εἰς κάλαμον βαλὼν καὶ ἐπιτιθεὶς εἰς τὸ μέρος τὸ ῥέον ἐμφύσα. 2 ἄλλο. σινάπιν καὶ καρύδια κοπανίσας κατ' ἰδίαν καὶ ἑνώσας μετὰ τοῦ ᾠοῦ τοῦ λευκοῦ ἐπιτίθει εἰς τὸ μέτωπον καὶ ἵσταται τὸ αἷμα.

7 [8] ζ′ περὶ πτύσεως αἵματος· ἡδυόσμου ζωμὸν μίξας μετὰ ὄξους ὀλίγου πινέτω.

8 [9] η′ πρὸς πόνον γλώσσης· ἐλαίας φύλλα μασῶν ἐπικράτει εἰς τὴν γλῶσσαν.

9 [10] ϑ′ πρὸς σκληρότητα σπλάγχνων· ὅταν τὰ σπλάγχνα γίνονται σκληρά, βαλὼν ψίχας ἄρτου εἰς τήγανον, βράζε μετὰ ὕδατος, εἶτα ἐπίβαλε μέλι καὶ μάλαξον καὶ ἁπλώσας εἰς πανίον ἐπιτίθει. 2 ἄλλο. λινόσπορον κοπανίσας καὶ κριθάλευρον λείαν ἕψησον μετὰ ὕδατος. ὕστερον δὲ βάλε ἕψημα καὶ μάλαξον καὶ ἁπλώσας εἰς πανίον ἐπιτίθει. οὕτως δὲ ποίει καὶ ἔνθα εὕρεις σκληρίαν ὄγκων.

10 [11] ι′ πρὸς πόνον ἥπατος· σελίνου ῥίζας τρεῖς καὶ σκόροδον κοπανίσας ἅμα βράσων αὐτὰ μετὰ οἴνου καλοῦ καὶ πινέτω αὐτὸ κατὰ πρωΐ. τὸ αὐτὸ ποίει ἐπὶ τοὺς δυσουριῶντας. εἰ δὲ πυρέσσει, βράζε αὐτὰ μετὰ ὕδατος.

---

3 ἀμαράντων] scripsi εὐμαράντων N 4 τόδε] scripsi τῷδε N 16 μασῶν] scripsi μασσῶν N

5. For discharge from the eyes: Grind the yolk of a boiled egg and *aromatikon*[18] together and mix this with a decoction of amaranth and some condensed wine and rose extract and dip cotton in it and apply it. Do this and what we prescribed above for a hot head. Prepare the same lotion also for an inflamed wound. 2 Another one for discharge from the eyes: Take frankincense, mastic, myrrh and grind them separately and make five beans like flour and mix everything together with egg white, dip hemp in it and anoint it on the forehead.

6. For a running nose: Burn eggshell, grind it well and put it in a reed and locate it in the part which is running and blow it in. 2 Grind mustard and walnut separately and mix it with egg white and put it on the forehead and the bleeding stops.

7. On spiting of blood: Mix the juice of peppermint with some vinegar and let him drink.

8. On an aching tongue: Chew olive tree leaves and keep them on the tongue.

9. On hardening of the bowel: When the bowels get hard, put bread crumbs in a frying pan, boil them with water, then add honey and smoothe it and put it on a cloth and apply it. 2 Grind linseed and fine barley flour and cook it in water. Later on, add condensed wine and knead it, spread it on a cloth and apply it. Do the same also if you find any hardening in a swelling.

10. On pain in the liver: Grind three celeriacs and garlic and cook them together with good wine and let him drink it in the morning. Do the same with people who have difficulties urinating. If he is feverish, boil them with water.

[18] This adjective means 'aromatic' and can be applied to a variety of words. It is also attested as a name of medications, usually in combination with the specific aromatic ingredient.

11 [12] ια′ πρὸς σκληρότητα σπληνός· λαβὼν κράμβης φύλλα καὶ καθαρίσας αὐτὰ ἀπὸ τῶν νεύρων αὐτῶν, κοπάνισον καὶ βράσας εἰς τηγάνον μετὰ ὕδατος. αὐτὰ εἶτα ἔχων τετριμμένα ἀρσενίκιν καὶ λίβανον πρόσβαλε. καὶ μαλάξας ἐπιτίθει εἰς πανίον. καὶ τίθει εἰς τὸν τόπον μέχρι τριῶν ἡμερῶν. εἶτα ἔκβαλε τὸ πλάσμα, δὸς δὲ καὶ ὄξος αὐτὸν ὀλίγον ἐπιρροφῆσαι. 2 ἄλλο. τὸν φλοῦν τῆς ἰτέας πλησίον τῆς ῥίζης ἐκδάρας, ἕψει μετὰ οἴνου καλοῦ ἐπιβάλλων τὸν οἶνον ἕως τρίτου καὶ βράσον εἶτα δὸς αὐτῷ πίνειν ἕως ἡμέρας ἑπτά. 3 ἄλλο. ἀμμωνιακὸν κοπανίσας καλῶς ἑξάγιον α'. ⌐ἑψήσας μετὰ ὄξους καὶ ποιήσας κολλῶδες, ἄλειφε τὸν σπλῆνα.¬ 4 ἄλλο. ῥίζαν καππάρεως κοπανίσας καὶ ἄσβεστον κρύον καὶ μαλάξας μετὰ ὄξους κατάπλασσε τὸν σπλῆνα ἕως ὥρας ἑπτά.

12 [13] ιβ′ εἰς ὀδύνην νεφρῶν· ζεματίσας πίτυρα καὶ φλησκούνιν καὶ κοπανίσας αὐτὰ βρέξον σπόγγον καὶ πυρίαζε τὸν τόπον. πρῶτον δὲ κλυστερευέσθω.

13 [14] ιγ′ εἰς τὸ ἀναβάσαι σάρκα· ῥητίνην ξηρὰν κοπανίσας ὡς τὸ ἄλευρον πάσε τὴν πληγήν.

14 [15] ιδ′ πρὸς ἔμφραξιν ὠτῶν· βαλὼν εἰς φλοῦν ᾠοῦ ἔψημα καὶ χλιάνας βάλε εἰς τὸ ὠτίον. 2 ἄλλο. στυππεῖα ἁπλώσας εἰς τὴν παλάμην βρέξων αὐτὰ μετὰ τοῦ ᾠοῦ τοῦ λευκοῦ καὶ λεπτοῦ καὶ ἐπάνω ἐπίβαλε κινάμωμον ἔχων τετριμμένον εἶτα ἐπιτίθει εἰς τὸ οὖς.

15 [16] ιε′ πρὸς πιτυρίασιν κεφαλῆς· πρῶτον κένωσον τὸ πᾶν σῶμα ἢ διὰ φλεβοτομίας ἢ διὰ βοηθήματος εἶτα κοπανίσας σεύτλου φύλλα τὸν ζωμὸν ἄλειφε τὴν κεφαλήν. 2 ἄλλο. κοπανίσας λιθάργυρον καὶ τεάφην καὶ λίβανον κατ' ἰδίαν ἕκαστον καὶ μίξας μετὰ ἐλαίου ἄλειφε τὴν κεφαλήν. 3 ἄλλο. βράσας ὕδωρ καὶ ἀγριαγγουρίαν κοπανισμένην καὶ λουπηνάρια καὶ αὐτὰ κοπανισμένα τὸν ζωμὸν ἐκείνων λοῦε τὴν κεφαλήν.

16 [18] ις′ εἰς ὀδύνην ὀφθαλμῶν· πρῶτον τούτους φλεβοτόμει, εἶθ' οὕτως βάλον εἰς τὸν ὀφθαλμὸν αὐτοῦ ἄσπρον ᾠοῦ. 2 ἄλλο. χαμαίμηλα τήλην ῥόδα βράζε μετὰ ὕδατος καὶ βρέχων πανίον πυρίαζε τοὺς ὀφθαλμοὺς χλιαρόν.

23 πιτυρίασιν] scripsi πυτιρίασιν N

11. On hardening of the spleen: Take cabbage leaves and remove the stem, grind them and cook them in a frying pan with water. And then when they are ground, add arsenic and frankincense. And knead it and put it on a cloth and leave on the place for three days. Then remove the paste, and let him have a small sip of vinegar afterwards. 2 Remove the bark of a willow tree somewhere close to the root, boil it with good wine and add the wine up to three times and cook it, then let him drink it for seven days. 3 Grind one hexagion of ammoniac well, cook it with vinegar and make it like a paste and anoint the spleen. 4 Grind the root of a caper bush and cold asbestos and smoothe it with vinegar

12. On kidney pain: Boil bran and penniroyal and grind it and soak a sponge with it and apply a vapour bath to the body part. But purge him before.

13. For anasarca: Grind dry resin like flour and apply on the affected body part.

14. For blocked ears: Pour condensed wine into an egg shell, warm it up and pour it into the ear. 2 Spread hemp onto the palm of your hand, soak it with the white and the thin parts of the egg and, and having ground cinnamon earlier, put it on top and then put it on the ear.

15. On pityriasis of the head: First empty the body, either through blood letting or with the help of remedies, then grind beet root leaves and anoint the head with the juice. 2 Grind litharge, sulphur and frankincense separately, mix it with olive oil and apply it on the head. 3 Boil water and ground wild cucumber and lupins (also ground) and wash the head with the broth.

16. On pain in the eyes: First let their blood, and then put egg white in his eye. 2 Boil chamomile, fenugreek and roses with water, soak a cloth in it and apply a warm vapour bath to the eyes.

17 [21] ιζ΄ περὶ ὑποσφαγμάτων· ὑπόσφαγμα ἐστὶν ὅταν τὸ ἄσπρον τοῦ ὀφθαλμοῦ ἔνι ὡσὰν φακὴ κόκκινον. στάζε οὖν εἰς τὸν ὀφθαλμὸν περιστερᾶς αἷμα ἢ ἅλας καππαδοκικὸν τετριμμένον.

18 [22] ιη΄ πρὸς ὠτία ῥέοντα ἕλκος· οἶνον χλίον καὶ μέλι ἑνώσας ἔμβαλε εἰς τὸ ὠτίον, διὰ ὠτεχύτου. ἀλλὰ μὴ βάλε τὸν αὐλὸν ἐντὸς ἀλλὰ ἔχε αὐτὸν ἐναντίον τῆς ὀπῆς καὶ σφίξας τὸν ἀσκὸν ἔγχεε αὐτὸ εἰς τὸ οὖς πλὴν μὴ φοβηθεὶς πάλιν γὰρ ἐξέρχεται τὸ οἰνόμελι. μόνον ἀποπλύνον τὸ ὠτίον ἀπὸ τοῦ ἕλκους. 2 ἄλλο. ἀλόην τρίψας καλὰ βάλε εἰς τὸ ὠτίον ὀλίγην χωρὶς ὑγροῦ. 3 ἄλλο. ἐλαίας φυλλῶν ζωμὸν μίξας μετὰ μέλιτος στάζε εἰς τὸ ὠτίον. 4 ἄλλο. ῥόδα ἑψήσας μετὰ ὕδατος τὸν ζωμὸν αὐτῶν στάζε εἰς τὸ ὠτίον.

19 [23] ιθ΄ πρὸς ὠτία ῥέοντα αἷμα· πράσου ζωμὸν ἔνσταζε εἰς τὸ ὠτίον. τρυγέαν οἴνου τρίψας καὶ μίξας μετὰ οἴνου χλιαροῦ στάζον εἰς τὸ ὠτίον.

20 [24] κ΄ πρὸς ὠτία ἔχοντα σκώληκας· καλαμίνθιν τρίψας τὸν ζωμὸν αὐτοῦ στάξον εἰς τὸ ὠτίον. 2 ἄλλο. τῆς ῥοδακινέας τὰ φύλλα χλωρὰ κοπανίσας τὸν ζωμὸν αὐτῶν στάξον. τὰ φύλλα τῆς καππάρεως κοπανίσας τὸν ζωμὸν αὐτῶν στάζον εἰς τὸ ὠτίον.

21 [25] κα΄ πρὸς ὅταν πέσῃ τι εἰς ὠτίον· ψύλλος μὲν ἐὰν πέσῃ βάλε χρηστέλαιον χλιαρὸν εἰς τὸ οὖς, εἰ δὲ ἄλλο τι τὰ μαλία τῆς κεφαλῆς ἐκείνου τὰ ἄκρα βάλε εἰς τὸ οὖς ὅσον ἔνι δυνατόν.

22 [26] κβ΄ πρὸς ὦτα ἔχοντα ἦχον· βούτυρον μίξας μετὰ μέλιτος καὶ χλιάνας στάξον εἰς τὸ ὠτίον. 2 ἄλλο. χήνειον ἀξούγγιν λυτὸν ἔνσταζε εἰς τὸ οὖς χλιαρόν. 3 ἄλλο. ἑψήσας κύμινον καὶ ὄξος καὶ ῥοδέλαιον στάξον χλιαρὸν εἰς τὸ οὖς.

23 [26] κγ΄ πρὸς ὅταν δυσακούῃ τις· οἱ τοιούτοι πινέτωσαν βοήθημα, ὕστερον δὲ καὶ τὴν ἱεράν. 2 ἄλλο. αἰγὸς οὖρον ἢ χολὴν στάξον. 3 ἄλλο. καστόριον τρίψας καὶ μίξας μετὰ ἀνηθελαίου στάξον εἰς τὸ ὠτίον χλιαρόν. 4 ἄλλο. κυναρίου χολὴν καὶ χρηστελαίου ἴσα χλιάνας στάξον. 5 ἄλλο. κοκκία δάφνης κοπανίσας καὶ μίξας μετὰ οἴνου παλαιοῦ χλίου στάξον.

---

1 ὑπόσφαγμα] scripsi ὑπόσφιγμα N   19 ψύλλος] scripsi ψίλλος N

17. On hyposphagma: Hyposphagma is when the white of the eye gets red as a lentil. Drip pigeon blood or ground Cappadocian salt into the eyes.

18. On pus flowing from the ear: Mix warm wine and honey and put it into the ear, with an ear-dropper. But do not insert the nozzle inside but hold it opposite of the opening, compress the bag and pour it into the ear, do not be afraid because the wine-honey-mixture will flow out again. Just rinse the pus out of the ear. 2 Grind aloe and put a little into the ear without liquid. 3 Mix the juice of olive tree leaves with honey and drip it into the ear. 4 Boil roses with water and drip the juice into the ear.

19. On blood flowing from the ears: Drip leek juice into the ear. Grind wine sediment, mix it with warm wine and drip it into the ear.

20. On worms in the ears: Grind peppermint and drip the juice into the ear. 2 Grind the fresh leaves of a peach tree and drip the juice. Grind the leaves of a caper tree and drip the juice into the ear.

21. If something gets into the ear: If a flea gets in, put warm table oil into the ear, if it is something else that gets in the ear, put the tips of the head hair in the ear as far as possible.

22. If there is a noise in the ear: Mix butter with honey, warm it up and drip it into the ear. 2 Drip warm melted goose fat into the ear. 3 Boil cumin, vinegar and rose oil and drip it into the ear while it is warm.

23. When someone is hard of hearing: Let them drink a remedy, and afterwards the Holy Remedy.[19] 2 Drops of goat's urine or bile. 3 Grind castoreum, mix it with aniseed oil and drip it into the ear while it is warm. 4 Heat dog's bile and table oil in equal parts and use as drops. 5 Grind laurel berries, mix them with warm old wine and use as drops.

[19] The Holy Remedy or *hiera Galēnou* is one of the few ready made medicines mentioned in the text. Its main ingredients are pepper, hellebore and squill. It was used as a cleansing remedy that also had a toning effect.

24 [27] κδʹ πρὸς παρωτίδας τὰ λεγόμενα χελωνικά· κριθάλευρον καὶ λινόσπορον καὶ τηλάλευρον ἕψησον μετὰ μέλιτος καὶ οἴνου καὶ τίθει. 2 ἄλλο. φύλλα λαπάθου κοπανίσας σὺν ἀξουγγίου ἀναλάτῳ. καὶ ἅλας καὶ πράσα κοπανίσας καὶ ἑνώσας τὰ πάντα ἐπιτίθει. 3 ἄλλο. λουπηναρίων ἄλευρον καὶ ἕψημα καὶ βολβὸν ἑψημένον κοπανιστὸν κατάπλασσε.

25 [28] κεʹ πρὸς αἷμα ῥέον ἐκ τῆς ῥινός· καλαμινθίου φύλλα ξηράνας φύσα διὰ καλάμου εἰς τὴν ῥίνα. ἐὰν δὲ ὁ δέξιος μυκτὴρ ῥέῃ ἀμέτρως ἡμέραν τε καὶ νύκτα, δῆσον τὰς χεῖρας καὶ τοὺς πόδας τὰ δεξιὰ καὶ συκιαστέον εἰς τὸ δεξιὸν μέρος χωρὶς τοῦ κόψαι τὸν τόπον. εἰ δὲ ὁ ἀριστερὸς ποίησον οὕτως ὥσπερ καὶ εἰς τὸ δεξιόν.

26 [29] κϛʹ εἰς τὸ γενέσθαι τὰ ὀδόντια λευκά· κέρατον ἐλάφου καύσας ἕως ἂν γένηται λευκόν. τρίψον αὐτὸ καλά, εἶτα ἄλειφε τοὺς ὀδόντας καὶ τρίβε πλὴν νῆστις. 2 ἄλλο. κριθάριν τρίψας καὶ ἅλας τρίβε τοὺς ὀδόντας.

27 [30] κζʹ πρὸς αἷμα ῥέον ἐκ τῶν οὐλῶν· τὰ φύλλα τῆς ἐλαίας βράσας μετὰ ὕδατος τὸν ζωμὸν ἐκεῖνον κράτει εἰς τὸ στόμα πολλὴν ὥραν. 2 ἄλλο. τῆς μυρσίνης τὰς κορυφὰς καὶ τριακονταφύλλων ἄνθος βράσας μετὰ ὄξους καὶ οἴνου κράτει εἰς τὸ μέρος τῶν οὐλῶν ἔνθα πονεῖ πολλὴν ὥραν.

28 [31] κηʹ πρὸς σεσηπότα οὖλη στόματος· κηκίδιν καὶ σμύρναν κοπανίσας καλῶς ὥσπερ ἄλευρον πλάττε τὰ οὖλη. κρόκον ξηρὸν κοπανίσας καὶ ἅλας καὶ ποιήσας ὥσπερ ἀλεύριν, μίξας μετὰ μέλιτος ἄλειφε τὰ οὖλη.

29 [32] κθʹ πρὸς κατεσχισμένα χείλη· τῶν φοινικίων τὸ ὀστοῦν ἔχει ὡσὰν τζίπαν. ταύτην λαβὼν κόλλησον εἰς τὸ ἐσχισμένον χεῖλος. 2 ἄλλο. χήνειον στέαρ ἄλειφε τὰ χείλη. 3 ἢ μαστίχην τρίψας καὶ μίξας μετὰ τοῦ οἴνου ἄλειφε.

30 [33] λʹ πρὸς πρήσμα γλώττης· γλιστρίδαν μασάτω ἢ φύλλα ἐλαίας τρυφερὰ ἢ ἕψημα μίξας μετὰ ῥοδοσταγματος κράτει εἰς τὴν γλῶτταν πολλὴν ὥραν καὶ κράτει καὶ δροσάτον εἰς τὴν γλῶσσαν.

2 τηλάλευρον] τιλάλευρον N 3 ἀξουγγίου] scripsi ἀξιγγίῳ N 4 πάντα] scripsi πέντε N 5 λουπηναρίων] scripsi λουπηνάριον N 26 κόλλησον] scripsi κώλυσον N

24. On the parotids which are called tortoise shells: Boil barley flour, linseed and fenugreek flour with honey and wine and apply it. 2 Grind sorrel leaves with unsalted fat. Grind salt and leek, mix it and apply it all. 3 Apply lupin flour, condensed wine and boiled purse tassels, all ground.

25. On blood flowing from the nose: Dry peppermint leaves and blow them into the nose through a reed. If the is right nostril flows excessively day and night, bind the right hand and foot and cup him, but do not injure the place. If it is the left one do the same as for the right.

26. To make the teeth white: Burn the horn of a deer until it gets white. Grind it well, then anoint it on the teeth, but let him be fasted. 2 Grind barley and salt and apply it on the teeth.

27. If lesions are bleeding: Boil the leaves of an olive tree with water and keep the broth in the mouth for many hours. 2 Boil myrtle shoots and roses with vinegar and wine and keep it on the part where the lesions are for several hours.

28. On rotting mouth lesions: Grind gall nut and myrrh well like flour and apply it on the lesions. Grind dry saffron and salt and make it like flour, mix it with honey and apply it on the lesions.

29. On soreness of the lips: Use the stone of a peach as a salve. Take it and stick it on the sore lip. 2 Apply goose fat on the lip. 3 Grind mastic, mix it with wine and apply.

30. On inflammation of the tongue: Let him chew puslane or succulent leaves of an olive tree. Mix condensed wine with rose water and keep it on the tongue for several hours, also keep dew on the tongue.

31 [34] λα΄ πρὸς δυσωδίαν στόματος· ῥόδα ξηρὰ καῦσας καὶ ποιήσας οἷον ἀλεύριν τρίβε τοὺς ὀδόντας. 2 ἔτι στάχος βράσας μετὰ οἴνου κράτει ἐν τῷ στόματι.

32 [35] λβ΄ περὶ συνάγχης τοῦ γουργουριακοῦ· φλεβοτόμησον αὐτὸν μὴ ἐκβάλης δὲ ἐκ πρώτης πολλὴν ἵνα μὴ λυποθυμήσῃ καὶ ἀποθάνῃ. καὶ παρελθούσης ἡμέρας φλεβοτόμησον πάλιν τοῦτον, κἂν πυρέσσῃ κἂν οὐ πυρέσσῃ. οἶνον μὴ πινέτω. ἄντλησον τοὺς πόδας αὐτοῦ. ἔστω δὲ ζεστός. ἀνέμῳ μὴ προσομιλήτω. ἐὰν δὲ οὐδὲν ὠφεληθῇ, φλεβοτόμησον αὐτὸν ὑποκάτω τῆς γλώττης ἤτοι τὴν γλῶτταν κατάκοψον μετὰ ξηραφίου ὀλίγον κατασχίσας, ἔτι μελάνθιον ἑψήσας μετὰ μέλιτος καὶ οἴνου, ἄλειφε τὴν ῥίζαν τῆς γλώττης καὶ τὴν κοιλίαν ἰάτρευε. καθ' ἑκάστην κλυστήρευε. εἰς δὲ τὸν τράχηλον ἐπιτίθει μαλία βεβρεγμένα μετὰ ἐλαίου χλιαροῦ, πιτύρων ζωμὸν κράτει εἰς τὸ στόμα. 2 ἄλλο. γάλα χλίον γλυκὺ κράτει εἰς τὸ στόμα. 3 ἄλλο. πηγάνου φύλλα χλωρὰ κοπανίσας τὸν ζωμὸν τούτων μίξας μετὰ γάλακτος ἄλειφε τὴν ῥίζαν τῆς γλώττης.

33 [36] λγ΄ πρὸς πτύσιν αἵματος· φλεβοτόμησον, ἄλειφε τοὺς πόδας καὶ τὰς χεῖρας ἐλαίῳ παλαιῷ. τροφὴ δὲ αὐτῷ αὐστηρὰ ἔστω. οἶνον ῥοΐδια ὀξύνα μῆλα ἀκράγουρα καὶ ἀπίδια. 2 ἄλλο. βάτου φύλλων ζωμὸν μίξας μετὰ βώλου τετριμμένου πινέτω. 3 ἄλλο. μαλία ἄπλυτα βρέξαι εἰς ὄξος χλίον καὶ ῥοδέλαιον πυρίαζε τὸ στῆθος.

34 [37] λδ΄ πρὸς φλεγμονὴν βυζίων· βράσας ὄξος καὶ οἶνον καὶ βρέξας ἐν αὐτῷ σπόγγον καινὸν καὶ ἀποπιέσας ἐπιτίθει εἰς τὰ βυζία καὶ δήσας ἃς ἕνι μίαν ἡμέραν καὶ πάλιν ποίει οὕτως. 2 ἄλλο. ψιχίας ἄρτου βράσας μετὰ τοῦ οἴνου καὶ ὄξους καὶ μαλάξας μετὰ γλιστρίδος κοπανισμένης ἐπιτίθει. 3 ἄλλο. κριθάλευρον καὶ τήλης ἄλευρον ἕψησον μετὰ ὕδατος καὶ ὕστερον βάλε μέλι. καὶ μάλαξον τοῦτο καὶ ἐπιτίθει ἐπάνω.

---

19 ἀκράγουρα] scripsi ἀκράγουρ. N

31. When the mouth smells: Burn dry roses, make them like flour and apply on the teeth. 2 Also boil vermilion with wine and keep it in the mouth.

32. On swelling of the throat: Bleed him, but do not let a large amount of blood the first time, so that he does not faint or die. And bleed him again after a day, whether he is feverish or not. He must not drink wine. Rinse his feet. But he must be hot. He must avoid wind. If it does not help, bleed him from underneath his tongue, that is: cut underneath his tongue with a razor blade a small incision, then boil nigella with honey and wine, apply it to the root of the tongue and attend to the stomach. Purge him every day. Put hair soaked in warm oil on the neck, keep bran juice in the mouth. 2 Keep warm sweet milk in the mouth. 3 Grind fresh rue leaves, mix the juice with milk and apply it to the root of the tongue.

33. If someone spits blood: Bleed him, anoint his feet and hands with old oil. His food must be tart. Wine, sour pomegranates, apples, cucumbers and pears. 2 Mix the juice of raspberry leaves with ground earth and let him drink. 3 Soak unwashed hair in warm vinegar and rose oil and apply a vapour bath for the chest.

34. On breast inflammation: Boil vinegar and wine and soak a new sponge in it, put it on the breast and squeeze it and then bandage it and leave it for one day and then repeat. 2 Boil bread crumbs with wine and vinegar, knead it with ground purslane and apply. 3 Boil barley flour and fenugreek flour with water and then add honey. Knead it and apply.

35 [38] λε´ περὶ ἐμέτου· ἀλόην μαστίχην στύρακα λάδανον λίβανον ἀψινθέας σπόρον κοπανιστὸν ἕψημα ὀλίγον ἄλευρον σίτινον καὶ οἰνάνθην καὶ οἶνον ὀλίγον ποίησον ἔμπλαστρον καὶ ἐπιτίθει εἰς τὸ στῆθος. 2 ἄλλο. τοῦ ἡδυόσμου τὸν ζωμὸν καὶ τοῦ ῥοϊδίου μίξας μετὰ μέλιτος βράξε ἕως τοῦ γενέσθαι παχὺν ὀλίγον καὶ τρωγέτω τὴν ἑσπέραν ὁπόταν μέλλῃ κοιμηθῆναι. μὴ δειπνείτω δέ. χρήσθω δὲ καὶ κλυστηρίοις.

36 [39,40] λϛ´ περὶ δυσεντερίας· ⊢δυσεντερία λέγεται ἡ ἕλκωσις τῶν ἐντέρων. γίνεται δὲ ποτὲ μὲν ἐκ μεταπτώσεως ποτὲ δὲ ἀπὸ χολωδῶν καὶ δριμέων ἐκκρίσεων. σημεῖον δὲ ἔχει τὸ αἱματῶδες καὶ ξύσματα καταφέρειν ἔθ' ὅτε καὶ ὑμένας ἐκ τῶν ἐντέρων καταφέρεσθαι. ἀλλ' εἰ μὲν ἐν τοῖς λεπτοῖς ἐντέροις ἐστὶ τὸ ἕλκος, συναναμεμιγμένα εἰσὶ τὰ καταφερόμενα τῇ κόπρῳ. εἰ δὲ ἐν τοῖς παχέσιν ἐστὶν ἡ ἕλκωσις κεχωρισμένα εἰσὶ τῆς κόπρου καὶ μᾶλλον δὲ ἐμέτων τὴν θεραπείαν χρήζει ὥσπερ τῶν διὰ τοῦ στόματος προσφερομένων βοηθημάτων τοῖς ἄνωθεν λεπτοῖς.⊣ λημνίαν σφραγίδα τρίψας, πινέτω μετὰ ὕδατος χλίου. 2 ἄλλο. ὕδωρ χλιαρὸν μετὰ λημνίας σφραγίδος κλυστηρευέσθω ἄνευ ἅλατος καὶ ἐλαίου καὶ μέλιτος. 3 ἄλλο. βάτος καὶ πολυκόμπιον ἤτοι ἵππουριν βράσας μετὰ οἴνου πινέτω. 4 ἄλλο. συκαμινέας καρπὸς ἄγουρος ξηρανθεὶς καὶ κοπανισθεὶς ἄς τον πίνῃ μετὰ ὕδατος χλίου. ὁμοίως καὶ τὸν βάτου καρπὸν χρῶ κλυστηρίῳ τοιούτῳ. ὀρύζην μετὰ ὕδατος βράσας τὸν χυλὸν αὐτοῦ ἑνώσας μετὰ αἰγείου στέατος ἴατρευε. 5 ἄλλο. πινέτω τὴν θηριακὴν καὶ μετὰ ὕδατος ἰατρευέσθω αὐτήν. 6 ἄλλο. ἴατρευε αὐτοὺς οὕτως· χλίον οἶνον καὶ μέλι καὶ πηγανέλαιον καὶ βῶλον τετριμμένον. 7 ἄλλο. βάτου φύλλα καὶ μυρσίνης ῥοδίου φύλλα καὶ σχίνου καὶ δάφνης καὶ κηκίδιν τετριμμένον ἕψησον ταῦτα πάντα μετὰ ὕδατος καὶ πυρίαζε τὴν ἕδραν βρέχων σπόγγον καινόν. κρύον μὴ πινέτω. ἀνέμων καὶ φλεβοτομίας ἀπεχέσθω.

35. On vomiting: Make a plaster out of aloe, mastic, storax, ladanum, frankincense, ground wormwood seed, some condensed wine, wheat flour, vine flower and some wine and put it on the chest. 2 Mix peppermint and pomegranate juice with honey, boil it until it gets a little thick, and let him eat it in the evening before retiring. He must not have supper. He must also purge.

36. On dysentery: ⊢Dysentery is an ulceration of the intestines. Sometimes it happens because of digestion, sometimes because of exsudates containing bilious and acrid elements. It it is indicated by a blood-like appearance, with shreds being discharged, and sometimes part of the lining of the intestines. If the ulceration is located in the small intestine, the discharge is mixed with feces. If the ulceration is in the large intestine, it is free from feces and requires rather a therapy through vomiting, and likewise if it is in the upper part of the small intestine it requires the medication administered by mouth.⊣[20] Grind Lemnian earth and let him drink with warm water. 2 Purge him with warm water, and Lemnian earth without salt, oil and honey. 3 Boil raspberries and mare's tail with wine and let him drink. 4 Let him drink dried unripe mulberries with warm water. Similarly use mulberries for purging. Boil rice with water and mix the broth with goat's fat and use it. 5 Let him drink theriac and use it with water. 6 Treat them like this: warm wine, honey, rue oil and ground earth. 7 Boil raspberry and myrtle leaves, the leaves of pomegranates and mastic, bay leaves, ground gall nut together in water, soak it in a new sponge and apply a vapour bath to the anus. He must not drink cold things. He must keep away from wind and bloodletting.

[20] This passage cannot be found in ω. It largely coincides with Theophanes Chrysobalantes 167 (= II 34 Bernard). Notably, it concerns diagnostics, which are largely absent from ω.

37 [41] λζ΄ περὶ κωλικῆς διαθέσεως· ἰάτρευε τοῦτον οὕτως διὰ πηγανελαίου καὶ καστορίου τετριμμένου. 2 ἄλλο. ἀψίνθιον βράσας μετὰ ὕδατος τὸν ζωμὸν αὐτοῦ πότιζε καθ' ἑκάστην ἡμέραν πρωΐ χλίον καὶ περιτρίψας ὀλίγον μίξας μετὰ ὄξους καὶ μέλιτος ὀλίγον πότιζε αὐτόν. 3 ἄλλο. πότισον αὐτὸν τὴν πικρὰν ὡσεὶ ἐξαγίου τὸ ἥμισυ μετὰ οἴνου χλιαροῦ. 4 ἄλλο. τήλης φύλλα ἢ τὰ κοκκία βράσας μετὰ χρηστελαίου ἰάτρευε αὐτὸν χλίον.

38 [42] λη΄ πρὸς ἑλμίνθας ἤτοι σκώληκας ἐν κοιλίᾳ· πότισον τούτου τῶν χαμαιμήλων τὸν ζωμόν. 2 ἄλλο. ῥοδέλαιον μετὰ ἑψήματος πότισον. 3 ἄλλο. ἡδύοσμον μετὰ ὕδατος βράσας πότισον χλίον. 4 ἄλλο. ἀλόην τρίψας πότισον μετὰ ὕδατος χλιαροῦ. 5 ἄλλο. μελάνθιν κοπανίσας πότισον μετὰ ὕδατος χλιαροῦ. 6 ἄλλο. μυελὸν ἐλάφου ἄλειψον τὴν ῥάχιν καὶ τὴν κοιλίαν.

39 [43] λθ΄ πρὸς λιθίασιν νεφρῶν· ἀγρίαν καὶ ἀδίαντον βράσας μετὰ ὕδατος πότισον χλίου νηστικόν. 2 ἄλλο. ἐρεβίνθιν μαῦρον πότισον χλιαρὸν βράσας μετὰ ὕδατος. 3 ἄλλο. λοῦε αὐτὸν συχνῶς καὶ ἡνίκα ἐξέρχεται ἀπὸ τοῦ λοετροῦ πότιζε αὐτὸν ἀγρίας ζωμὸν καὶ ἀδίαντον ἢ τοῦ ἐρεβινθίου τοῦ μαύρου τὸν ζωμὸν. 4 ἄλλο. ἀνδράχνης σπόρον κοπανίσας πότισον μετὰ οἴνου χλιαροῦ. 5 ἄλλο. φλεβοτόμησον τοῦτον. ἔστω ζεστά. ἀνέμῳ μὴ προσομιλείτω.

40 [45] μ΄ πρὸς οὐροῦντα ἕλκος· καύσας ναρθήκην τὴν στάκτην πότισον μετὰ ἑψήματος χλίου. βράσας μέλι μετρίως ἐπίβαλε λινόκοκκον καὶ ἀγριαγγουρίας σπόρον καὶ κουκουνάρια καὶ στάχος πάντα κοπανίσας οἷον τὸ ἀλεύριν. ἐσθιέτω δὲ ἐξ αὐτοῦ ὅσον καρύδιν ὅταν μέλλῃ κοιμηθῆναι. 2 ἄλλο. ἀμύγδαλα πικρὰ ἐσθιέτω τὴν νύκτα καὶ μὴ δειπνείτω. 3 ἄλλο. βράσας μέλι ἐπίβαλε ἐντὸς ἀμύγδαλα πικρὰ καὶ κρόκον κοπανιστὸν ἐσθιέτω καὶ μὴ δειπνείτω.

41 [46] μα΄ πρὸς τὸν οὐροῦντα αἷμα· φλεβοτόμει τούτους ἀπ' ἀγκῶνος τὴν καθόλου φλέβαν. ῥαφανίου σπόρον πότισον μετὰ ὄξους καὶ μέλιτος ὀλίγον.

37. On colics: Treat him in the same way with rue oil and ground castoreum. 2 Boil wormwood with water and let him drink the warm broth on every day in the morning. Stir it a little and add vinegar and some honey and let him drink it. 3 Let him drink Bitter Medication[21] about half a hexagion with warm wine. 4 Boil fenugreek leaves or fruit with table oil and use it warm.

38. On helmiths or worms in the stomach: Give him chamomile juice to drink. 2 Give him rose oil with condensed wine to drink. 3 Boil peppermint with water and give him to drink while it is warm. 5 Grind nigella and give to drink with warm water. 6 Anoint the spine and the stomach with deer marrow.

39. On kidney stones: Boil grass and maidenhair with water and give to drink when he is fasted. 2 Give him black chickpeas to drink while they are warm after you have boiled them in water. 3 Bathe him frequently and when he leaves the bath give him the juice of grass or maidenhair to drink or the juice of black chickpeas. 4 Grind purslane seed and give him to drink with warm wine. 5 Bleed him. He must be warm. He must avoid wind.

40. On pus in the urine: Burn ferula and give the ash to drink with warm condensed wine. Boil honey and add some linseed and the seed of wild cucumbers, after you have ground it all like flour. He should eat the amount of a walnut of it at bedtime. 2 He should eat bitter almonds at night and must not have supper. 3 Boil honey and throw bitter almonds inside and ground saffron. Give him to eat, and he must not have supper.

41. On blood in the urine: Let their blood from the usual vein on the bend of the arm.[22] Give radish seed to drink with vinegar and some honey.

---

[21] This term usually refers to a medicine based on aloe.

[22] The term 'usual vein' is not attested. It could either describe 'the vein commonly used for blood letting under these circumstances'. Alternatively, it could be a term for a specific blood vessel. This might have been the median cubital vein, the vein in the bend of the elbow which is still used for phlebotomy.

42 [47] μβ′ περὶ στραγγουρίας· στραγγουρία ἐστὶν ὅταν κατουρῇ ὀλίγον μετὰ βίας πόνου καὶ ἀνάγκης. φλεβοτόμει αὐτοὺς τὴν καθόλου φλέβαν. λοῦε συνεχέστερον. τροφὴ αὐτῷ εὔχυμος οἷον κωβιὸς καλαμάρια κάβουροι ποτάμιοι καὶ πέστρουβαι καὶ λυθρινάρια. 2 ἄλλο. δενδρολίβανον βράσας μετὰ οἴνου πότιζε αὐτὸν καθ' ἑκάστην ἡμέραν χλίον ἕως ἡμέρας δέκα.

43 [49] μγ′ πρὸς φλεγμονὴν ἥπατος· φλεγμαίνοντος τοῦ ἥπατος γίνεται πόνος εἰς τὸ δεξιὸν μέρος καὶ κρατεῖ ἕως τῶν ὤμων καὶ εἰς τὸ τέλος τῶν πλευρῶν τῶν κοντῶν καὶ ξηρὸς καὶ δίψα καὶ ἀνορεξία καὶ ἔμετος χολῆς. φλεβοτόμησον αὐτὸν εὐθὺς ἀπ' ἀγκῶνος δεξιοῦ τὴν καθόλου. εἰς τὸ ἧπαρ τίθει κατάπλασμα κριθῆς καὶ τήλης ἄλευρον. καὶ φοινίκια κοπανισμένα βράσας μετὰ ὕδατος καὶ ἁπλώσας εἰς πανίον ἐπιτίθει. κλυστήρευε αὐτὸν μετὰ θερμοῦ ὕδατος καὶ ἐλαίου καὶ μέλιτος καὶ ἅλατος.

44 [50,51] μδ′ πρὸς ἀπόστημα ἥπατος· κάπνεον βαλὼν καὶ βράσας μετὰ ὕδατος πότισον αὐτὸν νῆστιν. 2 πρὸς σκληρότητα ἥπατος· ἀμμωνιακὸν καὶ ἀξούγγιν ἐλάφου καὶ ταύρου ποιήσας ἔμπλαστρον ἐπιτίθει εἰς τὸ ἧπαρ. 3 ἄλλο. ἀγρίας καὶ ἀδιάντου ζωμὸν πότιζε.

45 [52] με′ πρὸς ὑδερικούς· τοὺς τοιούτους τζυκνίδας σπόρον πότισον κοπανίσας μετὰ ὕδατος χλίου. 2 ἄλλο. σέλινον πολυκόμπιν καὶ ἄσαρ βράσας μετὰ ὕδατος πότιζε καθ' ἑκάστην ἡμέραν πρωῒ νῆστιν. πότιζε καὶ τὴν πικρὰν κατὰ δύο ἡμέρας μετὰ χλίου οἴνου ἀπὸ ἡμίσεος ἑξαγίου. 3 ἄλλο. πότισον καὶ τὴν θηριακὴν μετὰ χλίου οἴνου ἀπὸ ἡμίσεος ἑξαγίου. 4 ἄλλο. πότισον καὶ τὴν θηριακὴν μετὰ χλίου οἴνου. ἀπεχέτω ὕδατος καὶ μαγειρίας καὶ δείπνου καὶ ἀναπαύσεως. μηδὲ ἐκ πρωΐας ἐσθιέτω. οἶνον θερμὸν μὴ πινέτω. ἡ τροφὴ αὐτοῦ ἔστω τοιαύτη· ἰχθύες πετραῖοι, ᾠὸν ῥοφητὸν πλὴν μόνον τὸν κρόκον ἰχθὺν χλωρὸν ὠπτόν, φακῆς δὲ καὶ ἐρεβινθίου καὶ φασουλίου καὶ τῶν ἄλλων ἑσπερίων καὶ ὀπωρῶν πάντων ἀπέχου. 5 ἄλλο. ἀψίνθιον βράσας μετὰ ὕδατος τὸν ζωμὸν αὐτοῦ πότιζε χλιαρὸν καθέκαστον πρωΐ.

---

4 λυθρινάρια] scripsi λιθρινάρια N 9 ἀνορεξία] scripsi ἀνωρεξία N

42. On strangury: Strangury is when he urinates little with force, pain and urgency. Let their blood from the usual vein. Bathe him often. His food must be juicy like gudgeon, octopus, sea crabs, trout and pagellus. 2 Boil rosemary with wine and let him drink it warm every day for ten days.

43. On hepatitis: If the liver is inflamed, there is pain in the right side and it reaches the shoulders and the end of the short ribs, and there is dryness, thirst, anorexia and vomiting of bile. Bleed him from the usual vein directly at the bend of the right arm. Put barley salve and fenugreek flour on the liver. And boil ground dates with water and apply them on a cloth and cover. Purge him with warm water, oil, honey and salt.

44. On liver abscesses: Take smoked grey vines, boil with water and give him to drink while he is fasted. 2 On hardening of the liver: make a plaster with ammoniac and deer and bull fat and put it on the liver. 3 Let him drink the juice of grass and maidenhair.

45. On dropsy: Give them ground nettle seed to drink with warm water. 2 Boil celery, mare's tail and hazelwort with water and let him drink every day in the morning when he is fasted. Let him also drink the Bitter Medication every other day with warm wine, around half a hexagion. 3 Also, give him theriac to drink with warm wine around half a hexagion. 4 Also give him also theriac to drink with warm wine. He must keep away from water, pastries, supper and rest. And he must not eat in the morning. He must not drink warm wine. He diet should be: stone fish, raw eggs except for the yolk, cooked pale fish. He should abstain from lentils, chickpeas, beans and all other supper food and from fruit. 5 Boil wormwood with water and let him drink the broth warm every day in the morning.

46 [54] μϛʹ πρὸς σκληρότητα σπληνός· φλεβοτόμησον αὐτὸν ἀπὸ τῆς ἀριστερᾶς χειρός, ἢ τὸ ἄκρον αὐτῆς ἢ τὸν μικρὸν δάκτυλον. 2 ἄλλο. κριθάλευρον βράσας μετὰ ὕδατος καὶ εἰς πανίον ἁπλώσας τίθει εἰς τὴν σπλῆνα. ἐχέτω δὲ καὶ χοίρειον ἀξούγγιν καὶ ὄρνιθος καὶ χήνειον, ὅλα δὲ ἄπαστα. τήλης ἄλευρον κάρδαμον κοπανιστὸν ἰσχάδων ζωμὸν βράσας καὶ ἁπλώσας εἰς πανίον ἐπιτίθει.

47 [53] μζʹ πρὸς πρήσμα σπληνός· ἀμμωνιακὸν κοπανίσας βρέχε τὴν νύκτα ὅλην εἰς ὄξος δριμύ. καὶ τότε ἄλειφε τὴν σπλῆνα πολλὴν ὥραν καθεκάστην ἡμέραν πρωῒ χλίον. 2 ἄλλο. κενταύριον βράσας μετὰ ὕδατος πότιζε αὐτὸν πρωῒ νῆστιν. κοπιαζέτω ὅσον δύναται καὶ τῶν ψυχρῶν ἀπεχέσθω. χρῶ δὲ καὶ τῇ ἐμπλάστρῳ τῇ λεγομένῃ ἡ οὐσία μέλαινα.

48 [55] μηʹ περὶ ἰκτέρου· ἴκτερον ἐστὶν ὅταν τὸ ὅλον σῶμα καὶ οἱ ὀφθαλμοὶ γένωνται κίτρινοι. φλεβοτόμει οὖν τοὺς τοιούτους. πότισον καὶ βοήθημα. 2 ἄλλο. ῥίζαν λαπάθου καὶ σελίνου καὶ ἀδιάντου βράσας πινέτω ἀντὶ ὕδατος ἄλλου. 3 ἄλλο. ἀψίνθιον βράσας μετὰ ὕδατος πότιζε αὐτὸν πρωῒ νῆστιν. πότισον καὶ τὴν πικρὰν πρωῒ μετὰ οἴνου. 4 ἄλλο. κριθάλευρον καὶ τήλης ἄλευρον καὶ λινόσπορον κοπανιστὸν βράσας μετὰ οἴνου καὶ ὄξους καὶ ἁπλώσας εἰς πανίον ἐπὶ τὸ ἧπαρ τίθει. οἴνου καὶ ἡλίου ἀπεχέσθω.

49 [56] μθʹ πρὸς πρήσμα ὀμφαλοῦ· τρυγέαν οἴνου τρίψας καὶ ῥόδα καὶ κηκίδιν μίξας μετὰ οἴνου τίθει ἐπάνω. ἀφ' οὗ δὲ θέσεις τοῦτο, θὲς ἐπάνω σπόγγον νέον βεβρεγμένον μετὰ ὄξους καὶ οἴνου καὶ δῆσον.

---

7 πρήσμα] scripsi πρίσμα N 21 ὀμφαλοῦ] scripsi ὀφθαλμῶν N, cf. pinax to L

---

21 ὀμφαλοῦ] Cf. the note to chapter ω 56.

46. On hardening of the spleen: Bleed him from the left hand, either from the middle or from the little finger. 2 Boil barley flour with water, apply it on a cloth and put it on the spleen. It should have also pig, bird and goose fat, all unsalted. Boil fenugreek flour, ground cardamom and the broth of dried figs, apply it on a cloth and cover.

47. On inflammation of the spleen: Grind ammoniac and soak it in acrid vinegar over the entire night. And then anoint it warm on the spleen for several hours every day in the morning. 2 Boil centaury with water and give him to drink in the morning when he is fasted. He should work as much as he can and should stay away from cold. Use also the plaster called the Black Matter.[23]

48. On jaundice: Jaundice is when then entire body and the eyes get yellow. Let their blood. Give remedies to drink. 2 Boil sorrel root, celeriac and root of maidenhair and he should drink it instead of other water. 3 Boil wormwood with water and let him drink in the morning when he is fasted. Give him also the Bitter Medication to drink in the morning with wine. 4 Boil barley flour and fenugreek flour and ground linseed with wine and vinegar, apply it on a cloth and put it on the liver. He should keep away from wine and sunlight.

49. On inflammation of the navel: Grind wine sediment and mix it with roses and gall nut with wine and apply. When you apply it, put a new sponge that has been soaked with vinegar and wine on top and bandage.

[23] This medication was mainly based on minerals and metals. It was said to have a softening effect.

50 [57] ν´ πρὸς φλεγμονὴν ὄρχεως· φλεβοτόμει τοὺς τοιούτους εἰς τοὺς ἀστραγάλους τῶν δύο ποδῶν. 2 ἄλλο. σταφίδας καθαρίσας ἐκ τῶν κοκκίων αὐτῶν καὶ κοπανίσας καὶ κύμινον κοπανιστὸν καὶ φαβίτιον ἄλευρον ἑψήσας μετὰ ὕδατος καὶ ἁπλώσας εἰς πανίον ἐπιτίθει εἰς τοὺς ὄρχεις. 3 ἄλλο. λινόσπορον κοπανίσας καλῶς ἕψησον μετὰ ἑψήματος ἢ τήλης ἄλευρον ἑψήσας μετὰ μέλιτος καὶ οἴνου. καὶ κρίνου ῥίζαν ἑψήσας μετὰ κριθαλεύρου καὶ ἑψήματος τίθει ἐπάνω. ὠφελεῖ δὲ καὶ ἡ παρὰ τῶν ἰατρῶν λεγομένη ἀνετὴ καὶ ἡ διὰ χυλοῦ.

51 να´ [29] περὶ γονορρίας ἤτοι ὅταν ῥέῃ ὁ σπόρος αὐτῶν· κοπιαζέτω πολλά, ἀνάπαυσιν μηδεμίαν ἐχέτω. τροφὴ αὐτῷ ἔστω ξηρὰ οἷον ἄρτος ὕδωρ ἰχθὺς ὀπτός. λιπαρῶν δὲ καὶ τῆς λεγομένης πηκτῆς καὶ τοῦ λευκοῦ τοῦ ᾠοῦ ἀπεχέσθω.

52 [59] νβ´ περὶ πρήσματος ἤτοι ἄλγους ἐντάσεως· ὠμὸν πήγανον καὶ τὸν πήγανον ἐσθιέτω. ὠφελεῖ δὲ καὶ ἔμετος πρὸ τροφῆς καὶ τρέψις ποδῶν καὶ χειρῶν καὶ κόπος.

53 [60] νγ´ πρὸς φλεγομονὴν δακτυλίου· ἰντύβου χυλὸν καὶ ῥοδέλαιον καὶ ψιμμίθιον καὶ λίβανον τετριμμένον ποιήσας ὡς ἂν ζυμάριν, ἐπιτίθει μετὰ πανίου. 2 ἄλλο. ψίχας ἄρτου καὶ κρόκον κοπανιστὸν μαλάξας ἀκριβῶς ἐπιτίθει.

---

2 σταφίδας] scripsi ταφίδας N 14 πρήσματος] fortasse πριαπισμοῦ 14 ἐντάσεως] ἐντάσε N ἐντάσεως index N 14 πήγανον] scripsi πηγάνου N

50. On inflammation of the testicles: Bleed these patients from the ankles of both feet. 2 Grind raisins after you have removed the seed and boil them with ground cumin and bean flour with water and apply it on a cloth and put it on the testicles. 3 Grind linseed well and boil it with condensed wine; alternatively boil fenugreek flour with honey and wine. And boil lily root with barley flour and condensed wine and apply. The medication which is called by the doctors Relaxing[24] and the one called Through Juices.[25]

51. On gonorrhea or when the semen flows: He should work a lot and should not have a break. His diet should be dry, like bread, water and cooked fish. He should abstain from fat, jelly and egg white.

52. On inflammation or pain of the tumescence[26]: He should eat unripe rue and rue. Vomiting before meals, rubbing of hand and feet and work are also beneficial.

53. On the inflammation of the anus: Make chicory juice, rose oil, lead white and ground frankincense like yeast and apply it with a cloth. 2 Mix bread crumbs and ground saffron thoroughly and apply.

[24] This term is not attested elsewhere, except for the corresponding chapter in Theophanes Chrysobalantes 192 (= II 118).

[25] This term could be used for a variety of medications, for instance one based on apple juice. Theophanes uses it, but does not mention its ingredients.

[26] This heading appears obscure at first sight; the term translated as 'tumescence', *entasis*, is known to the Classicist mainly in its archaeological meaning, where it describes the bulging of a column. It is used in medical writers to designate tension or trauma which would make little sense in this context. As ω reveals, it means 'erection', and the disease described is priapism.

54 [61,62] νδ΄ περὶ ἐξοχάδα· ἐὰν ὦσιν πρησμέναι ἄλειφε αὐτὰς ῥοδέλαιον εἶτα πᾶττε ψιμμίθιον τετριμμένον ἢ λιθάργυρον ἢ μαστίχην. ἐὰν δὲ ὦσι πρησμέναι, πάλιν φλεβοτόμησον αὐτόν. 2 ἄλλο. μελάνθιν κοπανίσας ἀποπλύνας πρῶτον αὐτὰς μετὰ οἴνου, πάττε τὸ μελάνθιον. 3 ἄλλο. στάκτην ἐλαίας καὶ κολοκύνθης ξηρᾶς πάττε ἀποπλύνας πρῶτον μετὰ οἴνου. ἐὰν κρέμωνται ἔξω ὀγκωμέναι ἀπὸ τοῦ αἵματος, ἄλειψον ταῦτα κυκλαμίνου χυλόν. ⊢ἐπιτίθει βδέλλαν μίαν ἢ καὶ δύο ἐπάνω. 4 ἄλλο. ἀξούγγιν καὶ πήγανον κοπανίσας καὶ ἑνώσας ἄλειφε τὸν τόπον. 5 ἄλλο. λινέλαιον καὶ τετριμμένην τεάφην ἑνώσας καὶ χλιάνας ἐπίχριε ὅσον ἐνδέχεται μετὰ βαμβακίου ἢ σπόγγου τὰς ἐξοχάδας. 6 ἄλλο. βερονίκιν, μᾶλλον τὸ ὑλωδέστερον τούτου, καὶ τεάφην καὶ ὀρίγανον καὶ ἑνώσας ὑποκάπνιζε οὕτως. χύτραν ἢ ἕτερον ἀγγεῖον τρύπησον ἵνα ἐνέρχηται στενούμενος ὁ καπνός. κάθισε ἐπάνω τοῦ σκεύους ἐκείνου καὶ ὑποκαπνίζου. ἔστι δὲ λίαν ὠφέλιμον. φλεγμαινουσῶν δὲ χηνάξουγγον χλιάνας ἐπιτίθει. οἶνον δὲ πυρίαζε συνεχῶς.⊣ ἐὰν δὲ πολὺ αἷμα ῥέῃ ἀπ' αὐτάς, μυρσίνης σχίνου βάτου φύλλα ῥούδιν κοπανιστὸν ῥοϊδίου φλοῦν κηκίδιν κοπανιστὸν τριακοντάφυλλα ξηρὰ κοπανισμένα βράσας μετὰ ὄξους ὀλίγου καὶ οἴνου βρέχε σπόγγον νέον καὶ πυρίαζε αὐτὰς συνεχῶς. 7 ἄλλο. πρῶτον μὲν ἀπόπλυνε αὐτὴν μετὰ οἴνου. εἶτα ἔχων ψιμμίθιν καὶ κηκίδιν βῶλον τετριμμένα καλῶς πάττε ἐπάνω.

55 [63] νε΄ περὶ ἐπικρατούντων τῶν καταμηνίων ἃ λέγονται γυναικεῖα· φλεβοτόμησον ταύτην εἰς τοὺς ἀστραγάλους τῶν δύο ποδῶν. πότιζε αὐτὴν καὶ τὴν λεγομένην ἱερὰν καὶ καστόριον τετριμμένον μετὰ οἴνου χλιαροῦ, ἔστω καὶ ζεστὰ καὶ λουέτω συχνάκις.

11–12 ὑλωδέστερον] scripsi ὑλοδεστερον N

54. On external hemorrhoids: If they are inflamed, anoint them with rose oil, and then apply ground lead white, litharge or mastic. If they are inflamed, bleed him again. 2 Grind nigella, wash them first with wine and then apply the nigella. 3 Apply the ash of an olive tree and a dry pumpkin after you have washed them with wine. If they are hanging on the outside because they are swollen from the blood, anoint them with violet juice. ⊢Put one or two leeches on top. 4 Grind fat and rue, mix it and apply on the place. 5 Mix linseed oil and ground sulphur, warm it up and apply to the hemorrhoids with a cotton cloth or a sponge as far as it is possible. 6 Mix speedwell, rather the more substantial bits of it, sulphur and oregano and apply a smoke bath from below like this: make a hole in a jug or another vessel so that the smoke comes out in a stream. Sit on top of this equipment and apply a smoke bath from below. It is very beneficial. If they are inflamed apply warm goose fat. Apply vapour baths with wine often.⊣ If there is a lot of blood flowing out from them, boil the leaves of myrtles, mastic and raspberries and a ground rose, the peel of a pomegranate, ground gall nut, ground dried roses with some vinegar and wine, soak a new sponge and apply frequent vapour baths. 7 First wash it with wine. Then prepare lead white, gall nut and earth, all ground well, and then apply.

55. If the menses, also called *gynaikeia*,[27] stop: Let her blood from the ankles of both her feet. Let her drink the Holy Medication and ground castoreum with warm wine. It should be warm and she should bathe frequently.

[27] This word is Greek for 'women's matters'.

56 [64] νϛ´ πρὸς ἰσχίον· ⊢ἰσχιὰς γίνεται φλέγματος μυξώδους ἢ χολῆς ἐπιρρευσάσης ἐν τοῖς συνδέσμοις τῶν ἄρθρων τῆς κοτύλης. ἔρχεται δὲ ἡ ὀδύνη ἀπὸ τὸ κάτω σφαίρωμα καὶ τὸν βουβῶνα τόπον ἄχρι γόνατος.⊣ φλεβοτόμει οὖν αὐτοὺς τὴν καθόλου φλέβαν. 2 ἄλλο. καστόριον τρίψας πότιζε αὐτούς μετὰ ὄξους ὀλίγου καὶ μέλιτος. 3 ἄλλο. ἰάτρευε αὐτοὺς τὸν ζωμὸν τοῦ λουπηναρίου. 4 ἄλλο. καστόριον τρίψας καὶ πεπέρι καὶ εὐφόλβιον καὶ πύρεθρον καὶ κηρὸν καὶ χαλβάνιν καὶ ἀναλύσας μετὰ πηγανελαίου ἄλειφε τὸν τόπον χλιαρόν. ὠφελεῖ καὶ ἡ πηγανερὰ καὶ ἰβεριάδα λεγομένη ἐπιτιθέμενα ἔνθα ὁ πόνος.

57 [65] νζ´ περὶ κατασχίσεως τῶν δακτύλων· ⊢τὴν ἀνωτέρω ἰβερίδα ἔμπλαστρον μετὰ χοιρείου ἀξουγγίου ἄλειφε τοὺς κατεσχισμένους τόπους.⊣ 2 ἄλλο. ψιμμίθιον λιθάργυρον κολοφωνίαν τρίψας καλὰ καὶ ἑνώσας μετὰ χρηστελαίου καὶ χοιρείου ἀξουγγίου ἐν ἡλίῳ ἄλειφε τοὺς τόπους.

58 [66] νη´ πρὸς ἄρρωστον πληγωθέντα ἐκ τοῦ πολλοῦ κατακεῖσθαι· κηρίον καὶ ἔλαιον καὶ λιθάργυρον μίξας καὶ ἀναλύσας ἄλειφε τοὺς τόπους. 2 ἄλλο. τρυγέαν μετὰ ὕδατος τηγανίσας ἄλειφε τοὺς τόπους.

59 [67] νθ´ πρὸς πληγὰς ποδῶν ἐκ τῶν ὑποδημάτων γινομένας· πετζὶν καύσας τὴν στάκτην ἐκείνου πάττε τὰς πληγάς. ἀφ᾽ ὅτου δὲ ἀποπρησθῶσιν αἱ πληγαὶ καύσας κηκίδιν πάττε τὰς πληγάς.

60 [68] ξ´ πρὸς ἥλους γινομένους ἐν ταῖς χερσίν· μελάνθιον κοπανίσας ἄλειφε τοὺς τόπους.

61 [69] ξα´ πρὸς παρωνυχίδας· παρωνυχὶς ἀπόστημα ἐστὶν περὶ τὴν ῥίζαν τοῦ ὄνυχος συνισταμένη. κηκίδιν κοπανίσας καὶ ἑνώσας μετὰ μέλιτος ἄλειφε τὸν τόπον. φοινίκια κοπανίσας καὶ μαλάξας μετὰ οἴνου ἐπιτίθει. 2 ἄλλο. ⊢λουπηναρίων ἄλευρον μετὰ μέλιτος μαλάξας ἐπιτίθει.⊣

62 [70] ξβ´ πρὸς πτερύγιον· πτερύγιον ἔνι ὁπόταν εἰς τὸ πλάγιον τοῦ ὄνυχος αὐξηθῇ σὰρξ μακρά. ἀρσενίκιν ψιλὸν πάττε ἐπάνω. εἶτα ἐπιτίθει ἐπάνω σπόγγον καινόν, βρέξας μετὰ οἴνου. 2 ἄλλο. ῥοϊδίου φλοῦν ξηράνας καὶ κοπανίσας καὶ ἑνώσας μετὰ μέλιτος ἐπιτίθει.

---

5 πότιζε αὐτούς] Bennett πότιζε αὐτούς. πότιζε αὐτοὺς N 6 ἰάτρευε] scripsi λάτρευε N 6 λουπηναρίου] scripsi λουμπιναρίου N

56. On sciatica: ⊢Sciatica happens when mucous phlegm or bile flow into the ligaments of the hip joint. The pain reaches from the lower end of the buttocks and the groin to the knees.⊣[28] Let their blood from the usual vein. 2 Grind castoreum and give it to drink with a little vinegar and honey. 3 Administer the the juice of lupins. 4 Grind castoreum, pepper, euphorbium, pellitory, wax and ferula and dissolve it with rue oil and anoint the place when it is warm. The medication called Rue and the one called Pepperwort is also beneficial when applied to the place.

57. On soreness of the fingers: ⊢Apply the Pepperwort Plaster mentioned above with pig fat on the sore places.⊣ 2 Grind lead white, litharge and colophonium well, mix it with table oil and pig fat and apply on the places in sunlight.

58. For patients who get sore from being bed-ridden for a long time: Mix wax, oil and litharge and dissolve it and apply it to the places. 2 Cook wine sediment with water and apply to the places.

59. On wounds on the feet caused by sandals: Burn leather and apply the ash on the wounds. When the wounds swell, burn gall nut and apply in the wounds.

60. On calouses on the hands: Grind nigella and apply on the places.

61. On paronychia: Paronychia is an ulcer arising from the root of the nail. Grind gall nut, mix it with honey and anoint on the place. Grind dates, mix them with wine and apply. 2 ⊢Mix lupin flour with honey and apply.⊣

62. On pterygium: Pterygium is when a lot of flesh grows next to the nail. Put fine arsenic on it. Then put a new sponge soaked with wine on top. 2 Dry the peel of a pomegranate, grind it, mix it with honey and apply.

---

[28] This passage roughly corresponds to Theophanes Chrysobalantes 215 (= II 180 Bernard).

63 [72,73] ξγ΄ πρὸς ὄνυχας ἔχοντας αἷμα· ⊢ἄλευρον ψιλὸν ἑνώσας μετὰ πίσσης ἐπιτίθει.⊣ 2 ἄλλο. ἰξὸν δρύνου ἐπιτίθει εἰς τὸν τόπον. 3 ἄλλο. φλεβοτόμει τοὺς τοιούτους κατὰ τὸν τοῦ ἔαρος καιρόν. πότιζε αὐτὸν ἱεράν. 4 ἄλλο. φαγὼν πρῶτον ῥεφάνια, ἀνάγκαζε αὐτὸν ἐμεῖν. 5 ἄλλο. πότιζε ὄξους μετὰ κεδρέαν. 6 ἄλλο. πότιζε αὐτοὺς καὶ τὴν θηριακὴν ὡσεὶ κοκκία ἕξ, μετὰ οἴνου λούσας αὐτὸν πρῶτον. καθ' ὃν μέλλει καιρὸν ἐξέρχεσθαι ἀπὸ τοῦ λουτροῦ, ἄλειφε αὐτὸν μυρσινέλαιον ἢ οἰνάνθην ἢ σχίνου ἔλαιον. 7 ἄλλο. κηκίδιν κοπανίσας καὶ φλοῦν σηπέας μετὰ ὄξους μίξας ἄλειφε τὴν κοιλίαν καὶ τὸ ὅλον σῶμα.

64 [74,75] ξδ΄ πρὸς ψώρας· τοὺς τοιούτους φλεβοτόμει. πότισον καὶ βοήθημα ἔχοντα πλὴν τὸ μέλαν. 2 ἄλλο. λιθάργυρον καὶ ἀγριοσταφίδα καὶ τεάφην τρίψας καὶ ἑνώσας μετὰ ἐλαίου καὶ ὄξους καὶ ὑδραργύρου καὶ δαφνελαίου ἄλειφε ἐν ἡλίῳ τὴν ψώραν. 3 ἄλλο. λιθάργυρον καὶ ῥούδιν καὶ ψιμμίθιον καὶ συρικὸν καὶ ὑγροπίσσην τρίψας καὶ ἑνώσας μετὰ ὄξους ἄλειφε τὸν τόπον ἐν ἡλίῳ. 4 ἄλλο. ἀρσενίκιν τρίψας καὶ ἑνώσας μετὰ οἴνου ἄλειφε ἐν ἡλίῳ. 5 ἄλλο. †κύραν ἄλευρον ἑνώσας μετὰ ὄξους ἄλειφε τὸν τόπον ἐν ἡλίῳ. ἂς ἔχῃ δὲ καὶ κάρδαμον τετριμμένον 6. ἄλλο. τεάφην τρίψας καὶ ἑνώσας μετὰ ὄξους ἄλειφε ἐν ἡλίῳ.

65 [76,77] ξε΄ πρὸς πόνον νεύρων· ἀσβέστιν μίξας μετὰ τρυγέαν ἀλείφου. πεντάνευρον κοπανίσας καὶ μίξας μετὰ μέλιτος ἐπιτίθει.

66 [78,80] ξϛ΄ πρὸς ὀφθαλμοὺς μὴ βλέποντας· πράσα κοπανίσας τὸν ζωμὸν αὐτῶν μίξας μετὰ μέλιτος στάζε εἰς τοὺς ὀφθαλμούς. 2 ἄλλο. ⊢λαπάθου τῆς ῥίζης τὸν φλοῦν καθαρίσας πρῶτον καὶ καύσας καὶ ἑνώσας μετὰ ὄξους ἄλειφε. 3 ἄλλο. καρπὶν τρίψας καὶ ἑνώσας μετὰ ὑγροπίσσης ἄλειφε. 4 ἄλλο. καππάρεως φλοῦν τρίψας καὶ τεάφην καὶ λίβανον καὶ ἑνώσας μετὰ ὄξους ἄλειφε.⊣

8 οἰνάνθην] scripsi οἰνάνθιν N 15 συρικὸν] scripsi cf. ω 74 συρνικὸν N

63. If the nails have blood: ⊢Mix fine flour with pitch and apply.⊣ 2 Put a mistletoe from an oak on top. 3 Let their blood during spring. Give him the Holy Medication to drink. 4 First let him eat radish, then make him vomit. 5 Let him drink some vinegar with cedar oil. 6 Let him drink about six kokkia[29] of theriac, but first let him wash with wine. When he is about to leave the bath, anoint him with myrtle oil or vine flowers or mastic oil. 7 Grind gall nut and the shell of a cuttle fish, mix it with vinegar and anoint the stomach and the whole body.

64. On psoriasis: Bleed them. And let them drink a medication that has rather black properties. 2 Grind litharge, stavesacre and sulphur and mix them with oil, vinegar, quicksilver and laurel oil and anoint on the psoriasis in the sunlight. 3 Grind litharge, a rose, lead white, lead oxide and liquid pitch, mix it with vinegar and apply it on the place in sunlight. 4 Grind arsenic, mix it with wine and apply it in sunlight. 5 Mix flour of .....[30] with vinegar and apply to the place in sunlight. Let it also contain some ground cardamom. 6 Grind sulphur, mix it with vinegar and apply in sunlight.

65. On neuralgia: Mix asbestos with wine sediment and apply. Grind plantain, mix with honey and apply.

66. On eyes, when they do not see: Grind leek, mix the juice with honey and drip into the eyes. 2 ⊢Clean the bark of sorrel root, burn it, mix it with vinegar and apply. 3 Grind hellebore, mix it with liquid pitch and apply. 4 Grind the bark of capers, sulphur and frankincense, mix it with vinegar and apply.⊣

[29] A kokkion was a measurement for an amount about the size of a pepper corn or grain.

[30] This word is not attested.

67 [81] ξζʹ πρὸς ἀπορυφάς· φλεβοτόμει τοὺς τοιούτους καὶ μίαν καὶ δευτέραν. ὅταν δὲ ἔνι ἄγουρος, βρέξας σπόγγον καινὸν εἰς ὄξος τίθει ἐπάνω, εἶτα δῆσον. ποίει δὲ καὶ τοῦτο μίαν καὶ δὶς καὶ τρίς. 2 ἄλλο. μαλία προβάτου βρέξας εἰς χλίον ῥοδέλαιον τίθει ἐπάνω. ἀφ' ὅτου δὲ ὡριμάσῃ, ἀμμωνιακὸν ἑνώσας μετὰ μέλιτος ἐπιτίθει. 3 ἄλλο. ῥητίνην κοπανίσας καὶ μίξας μετὰ ὄξους καὶ μέλιτος μαλάξας ἐπιτίθει. 4 ἄλλο. προζύμιν μετὰ μέλιτος μαλάξας, ἐπιτίθει. 5 ἄλλο. σῦκα κοπανίσας καὶ μίξας μετὰ μέλιτος ἐπιτίθει.

68 [82] ξηʹ πρὸς οὐλὴν πληγῆς ὅταν γένηται μαύρη· λιθάργυρον καὶ λευκοπέπερι τρίψας καὶ μίξας μετὰ ῥοδελαίου ἄλειφε τὴν πληγήν. 2 ἄλλο. εὐζώμου σπόρον κοπανίσας καὶ μίξας μετὰ χολῆς βοός καὶ λιθαργύρου ἄλειφε τὴν πληγήν. 3 ἄλλο. τεάφην τρίψας καὶ ἑνώσας μετὰ ὄξους ἄσπρου, ἄλειφε τὴν πληγὴν συχνῶς.

69 [83] ξθʹ πρὸς ὅταν αἱ τρίχες πίπτουσιν· ἀλόην τρίψας καὶ μίξας μετὰ οἴνου παλαιοῦ αὐστηροῦ ἄλειφε τὸν τόπον. 2 ἄλλο. λάδανον μαλάξας μετὰ οἴνου ἄλειφε τὴν κεφαλήν.

70 [85] οʹ περὶ αὐξήσεως τριχῶν· λάδανον ἀναλύσας μετὰ οἴνου ἄλειφε τὰς τρίχας. 2 ἄλλο. πολύτριχον καὶ λάδανον καὶ κολοκύνθης κομμάτια βράσας μετὰ βροχίνου ὕδατος βάλε αὐτὸ εἰς ἀγγεῖον ὑέλινον καὶ κρέμασον ἐν ἡλίῳ ἡμέραις κ'. ταράσσων αὐτὸ καθ' ἑκάστην ἡμέραν καὶ πληρωθεισῶν τῶν ἡμερῶν βάπτε τὸ κτένιον σου καὶ κτενίζου.

71 [86] οαʹ περὶ τοῦ μέλανας τρίχας ποιῆσαι· κηκίδιν τρίψας καὶ ῥινίσματα σιδήρου βράσας μετὰ ὄξους καὶ οἴνου αὐστηροῦ βάλε εἰς ἀγγεῖον ὑέλινον καὶ κρέμασον ἐν ἡλίῳ ἡμέρας δέκα. εἶτα ἄλειφε τὰς τρίχας τῆς κεφαλῆς τὸ πρωΐ. καὶ ἂς ἔνι οὕτως ἕως τὸ ἄλλον πρωΐ καὶ τότε κτενίζου.

---

5 ἀμμωνιακὸν] scripsi ἀμμωνικὸν N

67. On *aporufai*[31]: Let their blood on the first and the second day. If he is young, soak a new sponge in vinegar, put it on top and bandage it. Do this on the first day two or three times. 2 Soak goat's hair in warm rose oil and cover. After some time mix ammoniac with honey and cover. 3 Grind resin, mix it with vinegar and honey, knead it and apply. 4 Mix yeast with honey and apply. 5 Grind figs, mix them with honey and apply.

68. If a wound gets black: Grind litharge and white pepper, mix it with rose oil and anoint it on the wound. 2 Grind rocket seed, mix it with cow's bile and litharge and anoint it on the wound. 3 Grind sulphur, mix it with white vinegar and anoint it on the wound repeatedly.

69. When the hair falls out: Grind aloe, mix it with old dry wine and anoint it on the place. 2 Mix ladanum with wine and anoint it on the head.[32]

70. On augmenting hair: Dissolve ladanum with wine and anoint it on the hair. 2 Boil maidenhair fern, ladanum and pieces of pumpkin with rain water, put it into a glass vessel and leave it in the sun for twenty days. Stir it every day, and when the time is over, soak your comb in it and comb yourself.

71. To make the hair black: Grind gall nut and boil iron file dust with vinegar and dry wine, put it into a glass vessel and leave it in the sun for ten days. Then anoint it into the head hair in the morning, and leave it there until the next morning, and then comb.

[31] This word is not attested anywhere else; it seems to describe swollen lymph glands, see ℵ 171 and related passages.

[32] This is where the second cycle of diseases begins.

72 [88] οβ′ περὶ φθειρῶν τῶν γινομένων ἐν τῇ κεφαλῇ· λουπηνάρια ὠμὰ βρέχων εἰς ὕδωρ νυχθήμερον ἄλειφε τὰς τρίχας καὶ κτενίζου. 2 ἄλλο. λουπηνάρια ὠμὰ βράσας μετὰ ὕδατος ἀντὶ ἄλλου θερμοῦ λούου αὐτό. 3 ἄλλο. ἀγριοσταφίδα κοπανίσας καὶ μίξας μετὰ χρηστελαίου ἄλειφε τὴν κεφαλήν. 4 ἄλλο. ὑδράργυρον μετὰ ὄξους καὶ ἐλαίου μίξας ἄλειφε τὴν κεφαλήν. 5 ἄλλο. ὑδράργυρον μίξας μετὰ χοιρίου ἀξουγγίου ἀναλυτοῦ ἄλειφε τὴν κεφαλήν. τὰ αὐτὰ δὲ ποίει καὶ ἐπὶ τοὺς ἔχοντας κονίδας.

73 [89] ογ′ πρὸς τὰ γλυκέα· πότισον πρότερον βοήθημα. φλεβοτόμησον δὲ καὶ κρανιακήν. τοὺς τόπους δὲ τῆς κεφαλῆς ὅπου εἰσὶν ὥσπερ ὠπαί, πυρίαζε μετὰ ὕδατος, λουπηναρίων βράσας αὐτά. πρῶτον δὲ ξήρισον καλὰ τὴν κεφαλήν. 2 ἄλλο. μυρσίνην βράσας μετὰ οἴνου καὶ ὕδατος πυρίαζε. 3 ἄλλο. καύσας χαρτὶν βαμβύκινον τὴν στάκτην ἐκείνου μίξας μετὰ οἴνου ἀυστηροῦ καὶ μαλάξας ἐπιτίθει. 4 ἄλλο. λιθάργυρον τρίψας καὶ ἅλας ἑνώσας μετὰ οἴνου καὶ ἐλαίου καὶ ὄξους ἄλειφε τὴν κεφαλήν.

74 [90] οδ′ περὶ ἀλωπεκίας καὶ ὀφιάσεως· ἀλωπεκία ἐστὶν ὅταν αἱ τρίχες τῆς κεφαλῆς πέσωσιν. ὀφίασις δὲ ὅταν γένηται ἡ κεφαλὴ λεῖα μὴ ἔχουσα τὸ σύνολον τρίχα. ἐὰν ἴδης τὸ δέρμα τῆς κεφαλῆς ὅτι ἔστι μέλαν, πότισον βοήθημα ἔχον τὸ μέλαν πλέον τῶν ἄλλων εἰδῶν. ἐὰν δὲ ἴδης αὐτὸ μήτε πολλὰ μέλαν μητὲ λευκὸν ἀλλὰ μέσον, ἀλόην τρίψας πότισον μετὰ οἴνου καὶ τὴν λεγομένην πικρὰν ὡσεὶ κοκκία δώδεκα μετὰ οἴνου καθ' ἑκάστην ἡμέραν. 2 ἄλλο. βλησκούνιν καὶ ὕσσωπον καὶ σινήπιν καὶ ὀρίγανον βράσας μετὰ μέλιτος ὀλίγου καὶ ὄξους. τὸ ὄξος αὐτῶν βάλε εἰς τὸ στόμα σου καὶ κράτει αὐτὸ πολλὴν ὥραν καὶ τοῦτο ποίει καθ' ἑκάστην ἡμέραν. 3 ἄλλο. πανίον χονδρὸν βρέχων εἰς τὸ ὄξος τὸ δριμύ, τρίβε τὸ δέρμα τῆς κεφαλῆς ἕως ἂν γένηται κόκκινον.

13–14 βαμβύκινον] scripsi βαμβίκινον N

72. On head lice: Boil raw lupins in water for one night and one day, anoint it into the hair and comb. 2 Boil raw lupins with water and use it to wash instead of other warm water. 3 Grind stavesacre, mix with table oil and anoint it onto the head. 4 Mix quicksilver with vinegar and olive oil and anoint it onto the head. 5 Mix quicksilver with dissolved pig fat and anoint it onto the head. Do the same for those who have ticks.

73. On scurf: First give him medication to drink. Bleed him from the cephalic vein. The parts of the head that have holes treat with vapour baths of water and lupins you have boilt in it. But first shave the head well. 2 Boil myrtles with wine and water and apply as a vapour bath. 3 Burn cotton paper, mix the ash with dry wine, knead it and apply. 4 Grind litharge, mix salt with wine, oil and vinegar and anoint it onto the head.

74. On alopecia and ophiasis: Alopecia is when the hair of the head falls out. Ophiasis is when the head gets smooth and does not have any hair at all. When you see that the skin of the head is black, let him drink a medication that has more black properties than others. If you see that it is neither black nor white but intermediate, grind aloe, give him to drink with wine and the so called Bitter Medication, around twelve kokkia with wine every day. 2 Boil penniroyal, hyssop, mustard and oregano and boil it with some honey and vinegar. Put this vinegar in your mouth and leave it for several hours and do this every day. 3 Soak a coarse cloth in bitter vinegar, and anoint the skin of the head until it gets red.

75 [91,92,93] οε΄ περὶ κεφαλαλγίας· ἐὰν ὁ ἄρρωστος αἰσθάνηται ὅτι ἔχει βάρος ἡ κεφαλὴ αὐτοῦ, σημεῖον πλήθους ἐστίν. πότισον αὐτὸν βοήθημα, ἀλλὰ δὴ καὶ τὴν ἱεράν, καὶ φλεβοτόμησον. τὴν ἱερὰν δὲ τοῦ Λογαδίου. εἰ δὲ φαίνεται αὐτὸν ὅτι μετὰ βελόνης κατακεντᾶται τὸ δέρμα τῆς κεφαλῆς αὐτοῦ καὶ ἔχει καὶ καύσωνα ἡ κεφαλὴ αὐτοῦ πολλὴν πότιζε αὐτὸν βοήθημα ἔχοντα ξανθὰ πλέον τῶν ἄλλων εἰδῶν. λοῦε δὲ αὐτὸν πυκνὰ μετὰ γλυκέων ὑδάτων. εἰ δὲ φαίνεται ὅτι ἀποταυρίζεται ἡ κεφαλὴ αὐτοῦ σημεῖον ἔνι ὅτι ἀπὸ πνευμάτων ἐνοχλεῖται. καὶ πότισον αὐτὸν πικράν. ὄσφραινε δὲ πέπερι τετριμμένον καὶ κυκλάμινον. ἀλειφέσθω δὲ καὶ ῥοδέλαιον τὴν κεφαλήν. 2 ἄλλο. ὠμοτριβὲς ἔλαιον μίξας μετ' ὄξους ἄλειφε τὴν κεφαλήν.

76 [94] ο϶΄ περὶ ἐκ ψύξεως κεφαλαλγίας· ἐὰν δὲ φαίνηται αὐτὸν ὅτι τὸ δέρμα τῆς κεφαλῆς αὐτοῦ ἔνι ψυχρὸν παγωμένον καὶ αἱμωδιασμένον, ἄλειφε τὸν τοιούτον ἀνηθέλαιον ἢ πηγανέλαιον χλίον μετὰ καστορίου ἢ ἐφολβίου ὀλίγου.

77 [95] οζ΄ πρὸς τοὺς ἀπὸ στομάχου κεφαλαλγοῦντας· νοήσεις δὲ ὅτι ἀπὸ στομάχου κεφαλαλγεῖ, ὅτι ὁ πόνος τῆς κεφαλῆς φαίνεται τοιοῦτος ὥσπερ κατακεντώμενος ὑπό τινος ὥρας καὶ ὥρας. πηδῶσι δὲ καὶ αἱ φλέβες τῆς κεφαλῆς καὶ οἱ μήλιγγοι αὐτοῦ καὶ τὰ ὠτία ἔχουσιν ἦχον. κλυστήρευε δὲ αὐτὸν καθ' ἡμέραν μετὰ μέλιτος θερμοῦ καὶ ἅλατος καὶ ἐλαίου. τὴν δὲ κεφαλὴν ἄλειφε μετὰ ἀνηθελαίου μίξας καὶ ὄξος ὀλίγον. κέλευε δὲ αὐτὸν καὶ ἐμεῖν. πότιζε δὲ αὐτὸν τὴν πικρὰν. εἰς δὲ τὸν στόμαχον αὐτοῦ τίθει ἔμπλαστρον τὴν διὰ σπερμάτων. πυρίαζε δὲ μετὰ οἴνου παλαιοῦ καὶ ἀψινθίου καὶ τριψιδίου στάχους καὶ μαστίχης. βράσας πυρίαζε ὡσεὶ πανία κ' ἢ ι' ἢ ιε'. εἶτα λουέσθω. ἔστω δὲ ἡ τροφὴ αὐτοῦ τοιαύτη· ἀπεχέτω τῶν παστῶν καὶ τῶν ὀσπρίων. μονοφαγεῖτο δὲ καὶ τὸν οἶνον ὀλίγον πινέτω.

14–15 αἱμωδιασμένον] scripsi cf. ω 94 μοδιασμένον N 17 πρὸς] scripsi περὶ N

75. On headache: If the patient feels that his head is heavy, it is the sign of some kind of excess. Let him drink a medication, and certainly the Holy one, and bleed him. The Holy Medication according to Logadios.[33] If he feels as if the skin of his head is being pricked with a needle and his head has a strong burning sensation, let him drink a medication that has more white properties than others. Bathe him frequently with fresh water. If he feels that his head is stretched, it is a sign that he is suffering from pneuma. Let him drink the Bitter Medication. Let him inhale ground pepper and violets. Let him anoint his head with rose oil. 2 Mix cold pressed olive oil with vinegar and anoint onto the head.

76. On headache caused by cold: When he feels that the skin of his head is cold as ice and swollen, anoint him with aniseed oil, warm rue oil with castoreum or some euphorbium.

77. On headache caused by the stomach: You will recognize that he has headaches caused by the stomach, because the pain in the head appears as if he was pricked by something for hours. The veins in the head are pulsing, and his temples and his ears have a noise. Purge him every day with warm honey, salt and olive oil. Anoint his head with aniseed oil mixed with a little vinegar. Instruct him to vomit. Let him drink the Bitter Medication. Put a plaster on his stomach, the one called Through Seeds.[34] Apply vapour baths with old wine, wormwood, ground vermilion and mastic. Boil it and apply a vapour bath for about twenty, ten or eleven measures. Then let him bathe. His diet should be: he should abstain from pastry and pulse. He should only eat once a day and should drink little wine.

---

[33] Logadios is mentioned in other, earlier sources, but we do not not have any evidence on who he was.

[34] This medication was used to ease pain; it consisted of a variety of plant seeds, for instance celery seed.

78 [96] οη΄ περὶ τῆς ἐξ οἴνου κεφαλαλγίας· εἰ μὲν οὐκ ἔπεψε τὸν οἶνον ἐμεῖν ἀνάγκαζε αὐτὸν μετὰ χλίου ὕδατος. εἰ δὲ πέψας ἔτι κεφαλαλγεῖ, ἀλειφέσθω ῥοδέλαιον ἢ ὄξος μετὰ ῥοδελαίου. 2 ἄλλο. κράμβης φύλλων ζωμὸν ἄλειφε τὴν κεφαλήν. 3 ἄλλο. ἐσθιέτω τὴν κράμβην ἑψημένην ἕως ἂν ἡ νόσος λυθῇ. ἐσθιέτω δὲ μηλία καὶ ἀπίδια καὶ ῥοΐδια.

79 [96,97] οθ΄ περὶ τοῦ μὴ μεθύειν· γουλὶν κράμβης ἐσθίων οὐ μὴ μεθύεις. 2 ἄλλο. ὕσσωπὸν καὶ βλησκοῦνιν καὶ ὀρίγανον καὶ σάμψυχον βράσας μετὰ ὄξους καὶ ἑψήματος, τὸν ζωμὸν τούτων κράτει εἰς τὸ στόμα καὶ ἐξέρχονται πολλὰ ὑγρά.

80 [97] π΄ ἀποφλεγματισμοί· ἀγριοσταφίδα καὶ πέπερι καὶ μαστίχην κοπανίσας καὶ ἑνώσας μάσα. 2 ἄλλο. ζίγγιβερ τρίψας καὶ ἑνώσας μάσα.

81 [97,98] πα΄ ἀποφλεγματισμοὶ καθαρεύοντες τὴν κεφαλήν· σεύτλου φύλλων ζωμὸν βάλε εἰς τὴν ῥίνα. 2 ἄλλο. κυκλαμίνου ζωμὸν βάλε εἰς τὴν ῥίνα. 3 ἄλλο. μελάνθιν κοπανίσας καὶ μίξας μετὰ χρηστελαίου βάλε εἰς τὴν ῥίνα.

82 [99] πβ΄ εἰ θέλης πτέρνεσθαι· καρπὶν λευκὸν καὶ καστόριον καὶ πέπερι τρίψας καὶ ἑνώσας δῆσον εἰς πανίον καὶ ὀσφραινέσας. 2 ἄλλο. κηκίδιν τρίψας καὶ δήσας εἰς πανίον ὀσφραίνου.

83 [100] πγ΄ πρὸς παῦσαι πταρμούς· μυρίζου βασιλικοῦ σπόρον ἢ φύλλεον συνεχῶς.

84 [101,102] πδ΄ περὶ τῆς ἀπὸ ψύξεως κεφαλαλγίας· δάφνης κοκκία καὶ πηγάνου σπόρον καὶ δαδίου ῥίζαν κοπανίσας βράζε μετὰ ὄξους. εἶτα βάλε κηρίον ὀλίγον εἰς τὸν ζωμὸν αὐτῶν καὶ ἀλείφου. 2 ἄλλο. ῥοδακίνου ὀστοῦν κοπανίσας τὸ ἐντὸς αὐτοῦ τρίψας καὶ μίξας καὶ μαλάξας μετὰ ὄξους ἄλειφε τὸ μέτωπον. μετὰ πανίου τιθεὶς αὐτὸ ὡς ἔμπλαστρον εἰς τὸ μέτωπον. 3 ἄλλο. ⊢εὐζώμου σπόρον τρίψας καὶ λίβανον καὶ σὺν ὄξει μαλάξας ἐπιτίθει μετὰ πανίου εἰς τὸ μέτωπον.⊣ 4 ἄλλο. εὐφόλβιον καὶ λίβανον τρίψας μετὰ ὄξους ἑνώσας τίθει εἰς τὸ μέτωπον μετὰ πανίου.

21–22 ἢ φύλλεον] scripsi ἠψίλεον N

78. On headache caused by wine: If he did not digest the wine, make him vomit with warm water. If he digested it and is still suffering from headaches, anoint him with rose oil or vinegar with rose oil. 2 Anoint the juice of cabbage leaves on the head. 3 Let him eat the cabbage cooked until the disease is cured. Let him eat apples, pears and pomegranates.

79. On not getting drunk: If you eat the stem of a cabbage, you will never get drunk. 2 Boil hyssop, pennyroyal, oregano and marjoram with vinegar and condensed wine, keep the broth in your mouth and all the liquids will come out.

80. Cleansing of phlegm: Grind stavesacre, pepper and mastic and chew it together. 2 Grind ginger and chew it together.

81. Cleansing the head from phlegm: Put the juice of beet root leaves into the nose. 2 Put violet juice into the nose. 3 Grind nigella, mix it with table oil and put it into the nose.

82. If you want sneezing: Grind white hellebore, castoreum and pepper, mix it and bind it into a cloth and let him inhale. 2 Grind gall nut, bind it into a cloth and let him inhale.

83. To stop sneezing: Inhale basil seed or flea wort frequently.

84. On headache caused by cold: Grind laurel berries, rue seed and pine root and boil it with vinegar. Then add some wax to the broth and apply. 2 Grind the stone of a peach, the inner bits, mix and knead it with vinegar and anoint onto the forehead. Put it with a cloth on the forehead as a plaster. 3 ⊢Grind rocket seed and frankincense and knead it with vinegar and apply onto the forehead with a cloth.⊣ 4 Grind euphorbium and frankincense, mix it with vinegar and apply to the forehead with a cloth.

85 [103,104] πε΄ περὶ κατάρρου· ὅταν ῥέῃ τὸ ὑγρὸν ἀπὸ τοῦ οὐρανίσκου λέγεται κατάρρους. ὅταν δὲ ἀπὸ τῆς ῥινὸς λέγεται κόρυζα. καὶ εἰ μὲν ἐκ θερμῆς ἀσθενεῖ ἔνι τὸ ὑγρὸν λεπτὸν καὶ δριμύ. ἄλειφε τὴν κεφαλὴν αὐτοῦ ῥοδέλαιον ἀπὸ ἀγουρῶν ἐλαιῶν ἐσκευασμένον. λοῦε αὐτὸν πυκνά. ὀσφραίνου δὲ μελάνθιν καὶ κύμινον δήσας εἰς τὸ πανίον. ἐὰν ἀπὸ ψύξεως ἐγένετο ἐκεῖνο τὸ ὑγρὸν παχὺ καὶ ψυχρόν, ἐσθιέτωσαν οὖν ὀλίγον. τῶν ψυχρῶν ἀπεχέσθωσαν. ἀνέμῳ καὶ καπνῷ μὴ προσομιλείτωσαν. ἄλειφε τὴν κεφαλὴν ἀνηθέλαιον ἢ πηγανέλαιον ἢ ναρδέλαιον. τὴν δὲ ῥίνα αὐτοῦ ἐντὸς ἄλειφε λίβανον τρίψας καὶ ἑνώσας μετὰ χρηστελαίου. τὸ στῆθος ἄλειφε βαρσαμέλαιον ἢ πηγανέλαιον ἢ ἀνηθέλαιον ἢ ναρδέλαιον ἢ σάμψυχον.

86 [105] πϛ΄ πρὸς ὅταν πέσωσιν αἱ τρίχες τῶν ὀφρύων· φοινικίου ὀστοῦν καύσας καὶ τρίψας καὶ μετὰ ῥοδελαίου μίξας ἄλειφε τὴν κεφαλήν.

87 [106] πζ΄ πρὸς τὸ ποιῆσαι μέλανας τρίχας· λάδανον μαλάξας μετὰ οἴνου καὶ ἐλαίου ῥαφανίου χρίε. 2 ἄλλο. ἀξούγγιν αἰγιδίου ἀναλύσας χρίε. 3 ἄλλο. κέρατον τράγου καύσας καὶ μετὰ ναρδελαίου ἑνώσας χρίε.

88 [107] πη΄ πρὸς διπλὰς τρίχας ὀφρύων· νυκτερίδα σφάξας μετὰ χρυσαφίου τὸ αἷμα ἄλειψον τὸν τόπον. πλὴν πρῶτον ἀνασπάσας τὰς διπλὰς τρίχας. 2 ἄλλο. σκολόπενδραν θαλασσίαν κοπανίσας καὶ ἑνώσας μετὰ τοῦ χυλοῦ τοῦ κωνείου, ἄλειφε τὸν τόπον.

89 [108] πθ΄ περὶ κριθῆς τῆς ἐν ὀφθαλμῷ· τεάφην τρίψας καὶ ἑνώσας μετὰ ὄξους ἄλειφε.

90 [109] ϙ΄ πρὸς ὅταν τὸ λευκὸν τοῦ ὀφθαλμοῦ γένηται κόκκινον· ὅταν τὸ ἄσπρον τοῦ ὀφθαλμοῦ γένηται κόκκινον καὶ δοκεῖ τὸν ἄνθρωπον ὅτι ἐστὶ μεγαλότερον τὸ ἄσπρον τοῦ ὀφθαλμοῦ παρὰ τὴν κόρην, καλεῖται παρὰ τῶν ἰατρῶν τὸ πάθος χήμωσις. φλεβοτόμησον τοὺς τοιούτους φλέβαν κρανιακήν. ἐντὸς δὲ τοῦ ὀφθαλμοῦ βάλε ᾠοῦ τὸ λευκόν. οἴνου ἡλίου ἀνέμου κονιορτοῦ λουτροῦ πάμπαν ἀπεχέσθω.

2 κόρυζα] scripsi κόριζα N 28 χήμωσις] scripsi χήλωσις N

85. On catarrh: If liquid flows from the roof of the mouth it is called catarrh. If from the nose, it is called runny nose. And when he is suffering from a hot disease, the liquid is thin and bitter. Anoint his head with rose oil that was made of unripe olives. Bathe him frequently. Let him inhale nigella and cumin you have bound into a cloth. If this liquid arises from a cold disease, it is thick and cold, and they should eat little. They should abstain from all cold. They should not be in contact with wind and smoke. Anoint the head with aniseed oil, rue oil or nard oil. Anoint his nose inside with frankincense you have ground and mixed with table oil. Anoint the chest with balsam oil, rue oil, aniseed oil, nard oil or marjoram.

86. If the hair of the eyebrows falls out: Burn the stone of a date, grind it, mix it with rose oil and anoint it on the head.

87. How to make hair black: Knead ladanum with wine and radish oil and apply. 2 Dissolve goat's fat and apply. 3 Burn the horn of a goat, mix it with nard oil and apply.

88. On double hair in the eyebrows: Slaughter a bat with a golden knife and apply the blood on the place. But remove the double hair beforehand. 2 Grind a sea centipede mix it with the juice of hemlock and apply on the place.

89. On styes of the eye: Ground sulphur, mix it with vinegar and apply.

90. If the white of the eye becomes red: If the white of the eye becomes red and if the person appears as if the white of the eye is higher than the pupil, the doctors call it chemosis. Let their blood from the cephalic vein. Put egg white into the eye. He should abstain completely from wine, sunlight, wind, dust and baths.

91 [110] ϙα´ περὶ ἐμφυσήματος· τὸ ἐμφύσημα ὄγκος ἐστὶ του βλεφάρου. ὅταν δὲ τὸ βλέφαρον ἔχῃ κνησμὸν χλιάνας ὄξος καὶ οἶνον πυρίαζε. 2 ἄλλο. φακὴν βράσας μετὰ ὕδατος πυρίαζε τοὺς ὀφθαλμούς. 3 ἄλλο. κρόκον βράσας μετὰ οἴνου γλυκέου πυρίαζε τοὺς ὀφθαλμούς.

92 [111] ϙβ´ περὶ σκληροφθαλμίας· σκληροφθαλμία ἐστὶν ὅταν σαλεύῃ ὁ ὀφθαλμὸς μετὰ βίας καὶ μετὰ πόνου. εἰσὶ καὶ κόκκινοι καὶ δάκρυον οὐδὲν ῥέουσιν. πυρίαζε αὐτοὺς μετὰ ὕδατος ποταμίου πλὴν χλίου μετὰ σπόγγου καινοῦ. χήνειον ἀξούγγιν ἢ ὀρνίθειον τίθει εἰς τοὺς ὀφθαλμούς, ὅταν μέλλῃς κοιμηθῆναι. τὴν δὲ κεφαλὴν ἄλειφε χρηστέλαιον.

93 [112] ϙγ´ περὶ ξηροφθαλμίας· ξηροφθαλμία ἐστὶν ὅταν κνήθωνται τὰ βλέφαρα αὐτοῦ. οὐδὲν δακρύουσιν. λοῦε αὐτοὺς πυκνῶς.

94 [113] ϙδ´ περὶ ἐκτροπίου· ἐκτρόπιον ἐστὶν ὅταν τὸ βλέφαρον κάτω ἐξηστραφῇ. μολύβδιν καύσας μετὰ τεάφην μίξας ἄλειφε τὸ βλέφαρον.

95 [114] ϙε´ περὶ αἰγίλωπος· αἰγίλωψ ἐστὶν ὅταν πλησίον τῆς ῥινὸς εἰς τὸ ἐντὸς μέρος εἰς τὴν ἄκραν τοῦ ὀφθαλμοῦ γένηται ἀπόστημα. 2 ἄλλο. κρίνου ῥίζαν κοπανίσας καὶ μίξας μετὰ μέλιτος ἐπιτίθει. 3 ἄλλο. λουπηνάρια κοπανίσας πάττε τὴν πληγήν.

96 [115] ϙϛ´ πρὸς ὅταν αἱ τρίχες τῶν ὀφθαλμῶν πέσωσιν· πρῶτον φλεβοτόμησον τοὺς τοιούτους. εἶτα πότιζε βοήθημα ἔχων τὰ ξανθὰ πλείονα τῶν ἄλλων διότι ἀπὸ τῆς ξανθῆς χολῆς γέγονεν. 2 ἄλλο. τῶν φοινίκων τὰ ὀστᾶ κοπανίσας καὶ καστόριον καὶ κρόκον καὶ ἀρκούδας ἀξούγγιν λυτὸν μίξας τὰ πάντα ἄλειφε τοὺς τόπους τῶν βλεφάρων. πλὴν πρόσεχε μὴ ἐντὸς εἰσέλθῃ τοῦτο τοῦ ὀφθαλμοῦ καὶ ποιήσῃ βλάβην. πράσα καὶ κρόμμυα καὶ σκόροδα μὴ ἐσθιέτω. οἶνον πολὺν μὴ πινέτω. πεπέρεως καὶ ὀσπρίων καὶ πολλῶν λουτρῶν ἀπεχέσθω.

91. On emphysema: Emphysema is a swelling of the eyelid. If the eyelid is itching, warm vinegar and wine and apply a vapour bath. 2 Boil lentils with water and apply a vapour bath to the eyes. 3 Boil saffron with sweet wine and apply a vapour bath to the eyes.

92. On hardening of the eye: Hardening of the eye is when the eye moves with force and pain. Also, they are red and have no tears. Apply vapour baths with water from a river with a new sponge as long as it is warm. Put goose or bird fat on the eyes before bedtime. Anoint table oil on the head.

93. On drying of the eye: Drying of the eyes is when the eyelids are itching. They do not have tears. Bathe them frequently.

94. On ectropion: Ectropion is when the lower eyelid is turned outside. Burn lead, mix it with sulphur and apply to the eyelid.

95. On dacryocystitis: Dacryocystitis is when an ulcer arises close to the nose in the inner corner of the eye. 2 Grind lily root, mix it with honey and apply. 3 Grind lupins and apply on the wound.

96. When the hair of the eyes falls out: First of all let their blood. Then let them drink medication that has more bright properties than others because it arises from yellow bile. 2 Grind the stones of dates, castoreum, saffron and dissolved bear fat, mix it all and apply it to the places on the eyelids. But beware that this does not make its way into the eye and causes harm. He should not eat leek, onions and garlic. He should not drink much wine. He should abstain from pepper, pulse and excessive baths.

97 [116] ϙζ′ περὶ ῥυάδος· ῥυὰς λέγεται ὅταν ἀπὸ τὴν ἄκραν τοῦ ὀφθαλμοῦ πλησίον τῆς ῥινὸς εἰς τὸν μέγαν κανθὸν καὶ εἰς τὸν μικρὸν ῥέῃ ἀκουσίως τὸ δάκρυον. σπόγγον νέον εἰς κεδρέαν βρέξας καὶ καῦσας τὴν στάκτην αὐτοῦ πάττε ἐκεῖ ἔνθα ῥέει τὸ δάκρυον. 2 ἢ τὴν στάκτην ἑνώσας μετὰ αἵματος δρακοντίου ἄλειφε τὸν τόπον. οἶνον καὶ κρόμμυα καὶ πράσα φάγε. ὁμοίως ἥλιον καὶ ἄνεμον καὶ καπνὸν καὶ τὸ λουτρόν.

98 [117] ϙη′ πρὸς λεύκωμα ὀφθαλμοῦ· σηπίας ὄστρακον καύσας καὶ μετὰ μέλιτος ἑνώσας βάλε εἰς τὸν ὀφθαλμόν. 2 ἄλλο. ⊢χρηστέλαιον παλαιὸν βάλε εἰς τὸν ὀφθαλμόν.⊣ τροφὴ δὲ ἔστω αὐτῷ χλωρά. τὰ παστὰ δὲ ἐκφευγέτω καὶ τὸ ἄσπρον τοῦ ᾠοῦ εἰ μὴ τὸν κρόκον μόνον.

99 [118] ϙθ′ πρὸς πτερύγιον ὀφθαλμοῦ· πτερύγιον καλεῖται τὸ πάθος ὅταν ἀπὸ τὸν μέγαν κανθὸν τοῦ ὀφθαλμοῦ αὐξάνῃ σάρκα στενὴ καὶ μικρὰ καὶ ἔρχηται ἕως τὴν κόρην. 2 ἄλλο. μαγνήτιν τρίψας καὶ κρόκον καὶ μετὰ μέλιτος ἑνώσας ἄλειφε τὸν τόπον ἐκεῖνον.

100 [119,120] ρ′ πρὸς ἀμαύρωσιν ὀφθαλμῶν· ἀμαύρωσις λέγεται ὅταν χωρίς τινος πάθους οὐ βλέπει ὁ ἄνθρωπος. φλεβοτόμησον οὖν τοὺς τοιούτους καὶ πινέτω βοήθημα. ἄνεμον δὲ καὶ κονίορτον καὶ ἥλιον καὶ καπνὸν καὶ ὄσπρια καὶ παστὰ καὶ λάχανα ὅση σοι δύναμις φεῦγε, εἰ μὴ χλωρὴν μαγειρίαν καὶ κρόκον μόνον ᾠοῦ καὶ οἶνον ὀλίγον ζεστόν. 2 ἀμβλυωπία ἐστὶν ὅταν ποτὲ μὲν βλέπει ὀλίγον ποτὲ δὲ οὐδ᾽ ὅλως. φλεβοτόμει οὖν αὐτὸν εἰς τὴν ἄκραν τοῦ ὀφθαλμοῦ, ὅπου ἐστὶ πλησίον τῆς ῥινός ὃ λέγουσι μέγαν κανθόν. ἄλειφε τοὺς ὀφθαλμοὺς ἔλαιον πρῶτον, εἶτα ἀποσπόγγισον τοὺς ὀφθαλμοὺς μετὰ ὕδατος χλίου. 3 ἄλλο. γουλὶν τῆς κράμβης ἐσθιέτω ὠμόν. ἄνηθον δὲ καὶ ἐλαίαν μαύραν καὶ μαρούλια καὶ πράσα καὶ λάχανα καὶ σέλινα ἔκφευγε. 4 ἄλλο. μελάνθιν κοπανίσας καὶ μετὰ ὕδατος μίξας καὶ μαλάξας ἄλειφε τοὺς ὀφθαλμούς.

---

15 μαγνήτιν] scripsi μαγνῆτην N

97. On flowing: Flowing is called when tears are flowing involuntarily from the top of the eye into the big corner of the eye close to the nose and also into the small corner. 2 Mix ash with dragon blood and apply to the place. He should eat wine, onions and leek. Similarly sunlight, wind, smoke and baths.

98. On a white spot in the eye: Burn the shell of a cuttle fish, mix it with honey and put it into the eye. 2 ⊢Put old table oil in the eye.⊣ He should eat pale things. He should abstain from pastry and egg white, just the yolk.

99. On pterygium of the eye: Pterygium is called if the small and narrow flesh from the inner corner of the eye extends to the pupil. 2 Grind magnet and saffron and mix it with honey and apply to the place.

100. On blindness: Blindness is called when the person does not see without having a disease. Let their blood and he should drink medication. Abstain from wind, dust, sunlight, smoke, pulse, pastry and vegetables as far as possible, except for white cooked meat, just the yolk of the egg and some warm wine. 2 Dimsightedness is when he sees sometimes little, sometimes not at all. Bleed him from the edge of his eye which is close to the nose which is called inner corner. Anoint the eyes first with olive oil, then sponge them with warm water. 3 Let him eat the raw stem of a cabbage. He should avoid aniseed, black olives, lettuce, leek, vegetables and celery. 4 Grind nigella, mix it with water, knead and apply to the eyes.

101 [122,123] ρα′ περὶ ὑπωπίου· ὅταν ὑποκάτω τοῦ βλεφάρου γένηται μέλαν, τοῦτο δὲ τὸ πάθος γίνεται εἰς τὰς γυναίκας καὶ εἰς τὰ παιδία. καλεῖται παρὰ τῶν ἰατρῶν ὑπώπιον. ῥεφάνιν κοπανίσας τίθει ἐπάνω. 2 ἄλλο. τυρὸν τρυφερὸν κοπανίσας τίθει. 3 ἄλλο. λιθάργυρον τρίψας καὶ μετὰ μέλιτος μίξας τίθει εἰς τὴν πληγήν. 4 ἄλλο. κύμινον τρίψας καὶ δήσας εἰς πανίον πυρίαζε τὸν τόπον βρέχων εἰς ὄξος θερμόν. 5 ἄλλο. κηρίον καὶ ῥοδέλαιον καὶ ἀψίνθιον τρίψας καὶ λίβανον καὶ ἡδύοσμον καὶ λιθάργυρον μίξας καὶ ἀναλύσας ἐπίχριε. ὅταν τὸ κάτω βλέφαρον γένηται κόκκινον ἢ ὥσπερ ξανθὸν κοπάνισον φύλλα ὑσσώπου καὶ σταφίδας ἐκτὸς τῶν κοκκίων καὶ τίθει εἰς τὴν πληγήν. 6 ἄλλο. ἀψίνθιον κοπανίσας ἐπιτίθει. 7 ἄλλο. κυκλάμινον κοπανίσας τίθει εἰς τὴν πληγήν. οἶνον μὴ πινέτω, μηδὲ ἡλίῳ μηδὲ ἀνέμῳ προσομιλείτω. φλεβοτόμει τούτους κρανιακήν.

102 [124] ρβ′ πρὸς τὸν ἀπὸ ψύξεως πόνον τοῦ ὠτός· ὅταν ἀπὸ ψύξεως γένηται πόνος εἰς τὸ ὠτίον, καστόριον τρίψας καὶ μετὰ πηγανελαίου μίξας στάζε εἰς τὸ ὠτίον, ἢ μετὰ ναρδελαίου ἢ μετὰ δαφνελαίου. 2 ἄλλο. πέπερι τρίψας μετὰ ναρδελαίου στάζε. 3 ἄλλο. σκόροδον κοπανίσας καλὰ καὶ βράσας μετὰ χρηστελαίου στάζε. ἀνέμον καὶ ὄσπρια καὶ παστὰ φεῦγε.

103 [125,126] ργ′ πρὸς τὸν ἐκ θερμῆς δυσκρασίας πόνον τοῦ ὠτός· ὅταν γένηται πόνος εἰς τὸ ὠτίον διὰ θερμὴν δυσκρασίαν, ἄσπρον ὠοῦ μετὰ ῥοδελαίου μίξας στάζε εἰς τὸ ὠτίον. 2 ἄλλο. ἄσπρον ὠοῦ μίξας μετὰ τοῦ χυλοῦ τοῦ στρύχνου καὶ χλιάνας στάζε. ἥλιον δὲ καὶ ἄνεμον ἐκφευγέτω.

104 [127] ρδ′ πρὸς τὰ διὰ παχεῖς χυμοὺς ὀδυνώμενα ὠτία· νοήσεις ὅτι διὰ παχεὶς χυμοὺς ὀδυνᾶται τὸ ὠτίον οὕτως· δοκεῖ ὅτι βάρος ὑπάρχει ἐν τῇ κεφαλῇ καὶ τὰ ὠτία ἔχουσιν ἦχον. ἐὰν δὲ βάρος ἡ κεφαλὴ οὐκ ἔχει, δοκεῖ δὲ αὐτὸν ὅτι ἀποταυρίζεται τὸ δέρμα τῆς κεφαλῆς σημεῖον ἐστὶ ὅτι τοῦ ὠτίου τὸ πάθος ὑπάρχει καὶ οὐχὶ τῆς κεφαλῆς. ὄξος καὶ μέλι χλιάνας στάζε εἰς τὸ ὠτίον. 2 ἄλλο. προβάτου χολὴν μίξας μετὰ ῥοδελαίου καὶ χλιάνας στάζε εἰς τὸ ὠτίον. 3 ἄλλο. πράσου ζωμὸν μίξας μετὰ μέλιτος καλοῦ καὶ χλιάνας στάζε εἰς τὸ ὠτίον.

20 πόνον] scripsi πόνου N 25 τὰ] scripsi τοὺς N 25 ὀδυνώμενα ὠτία] scripsi ὀδυνωμένους ὀφθαλμούς N

101. On hypopion: If dark colour manifests under the eyes. This disease happens mainly in women and children. It is called hypopion by the doctors. Grind a radish and apply. 2 Grind fat cheese and apply it. 3 Grind litharge, mix it with honey and apply onto the wound. 4 Grind cumin, bind it into a cloth, soak it in warm vinegar and apply a vapour bath on the place. 5 Grind wax, rose oil and wormwood and mix it with frankincense, peppermint and litharge and apply after having dissolved it. If the lower eyelid gets red or like yellow, grind the leaves of hyssop and stavesacre without the fruits and apply to the wound. 6 Grind wormwood and apply. 7 Grind violets and apply to the wound. He should not drink wine, nor should he encounter sunlight or wind. Let their blood from the cephalic vein.

102. On pain in the ears that is caused by cold: If pain arises in the ear from cold, grind castoreum, mix it with rue oil and drip it into the ear, or with nard oil or laurel oil. 2 Grind pepper with nard oil and use as drops. 3 Grind garlic well, boil with table oil and use as drops. He should avoid wind, pulse and pastry.

103. On pain in the ear that is caused by warm imbalance: If pain arises in the ear because of warm imbalance, mix egg white with rose oil and drip it into the ear. 2 Mix egg white with the juice of nightshade, warm it up and use as drops. He should avoid sunlight and wind.

104. On pain in the ears caused by thick humours: That is how you will recognize that the ear aches because of thick humours: it appears as if there is heaviness in the head and the ears are ringing. If the head does not have heaviness, and it appears as it the skin of the head is stretched, this is a sign that it is a disease of the ear and not of the head. Heat vinegar and honey and drip it into the ear. 2 Mix goat's bile with rose oil, warm it up and drip it into the ear.

105 [128] ρε´ περὶ φλεγμονῆς ὠτίου· φλεβοτόμησον πρῶτον αὐτοὺς πυρίαζε δὲ μετὰ χρηστελαίου. 2 ἄλλο. ἀξούγγιν χλιάνας χήνειον στάζε. 3 ἄλλο. ἕψημα χλιάνας στάζε εἰς τὸ ὠτίον. οἶνον ὀλίγον πινέτω. ἀνέμῳ δὲ καὶ ἡλίῳ μὴ προσωμιλείτω καὶ καπνῷ.

106 [129] ρϛ´ περὶ αὐξήσεως ὀδόντων παιδίου· λαγωοῦ μυελὸν τῆς κεφαλῆς ἄλειφε τὰ οὖλη.

107 [130] ρζ´ πρὸς ὄγκωμα ὑποκάτω τῆς γλώττης· ὅταν γίνηται ὄγκος εἰς τὴν γλῶσσαν, ὑποκάτω δὲ γίνεται μᾶλλον ὑπὸ τῆς φλέβας. κηκίδιν τρίψας μετὰ ἑψήματος ἄλειφε τὸν ὄγκον. 2 ἄλλο. ἐλαίας φυλλῶν ζωμὸν ἄλειφε τὴν γλῶσσαν, ἔνθα ἔνι τὸ πάθος.

108 [131] ρη´ περὶ ἄφθας· ἄφθα ἐστὶν ὅταν τὰ χείλη τοῦ παιδίου σήπωνται διὰ τὴν δριμύτητα τοῦ γάλακτος. γίνεται δὲ τὸ πάθος καὶ εἰς τελείους ἀνθρώπους. μυελὸν ἐλάφου ἄλειφε τὰ χείλη. ἀπιδίου ζωμὸν ἄλειφε τὰ χείλη. 2 ἄλλο. τὴν ῥίζαν τοῦ μαλάθρου καύσας τὴν στάκτην πάττε τὰ χείλη. οἶνον πολὺν μὴ πινέτω. φλεβοτόμησον αὐτὸν εἰ ἔστι δωδεκαχρόνων ἢ δεκαπέντε ἢ εἴκοσι. ⊢τὴν φλέβαν δὲ κόπτε τὴν μέσην. ἐσθιέτωσαν οὖν ὀλίγα.⊣

109 [132] ρθ´ περὶ φλεγμονῆς σιαγόνων· γάλα γλυκὺ χλίον πρόβειον κράτει ἐν τῷ στόματι. 2 ἄλλο. ὕδωρ βρόχιον βράσας πυρίαζε τὸν τόπον μετὰ σπόγγου καινοῦ ἱκανὴν ὥραν. 3 ἄλλο. κριθάλευρον καὶ λινόσπορον κοπανίσας καλὰ καὶ βράσας μετὰ ὕδατος καὶ οἴνου ἐπιτίθει εἰς τὸν τόπον μετὰ πανίου. τοῦτο δὲ ποίει ἡμέρας τρεῖς, πρωῒ καὶ ἑσπέρας. 4 ἄλλο. βράσας γλυκύριζον καὶ κρόκον κράτει εἰς τὸ στόμα.

---

23 γλυκύριζον] scripsi γλυκύρριζον N

105. On the inflammation of the ear: First let their blood, apply vapour baths with table oil. 2 Heat goose fat and use as drops. 3 Warm condensed wine and drip it into the ear. He should drink little wine. He should avoid wind, sunlight and smoke.

106. To induce the growth of teeth in a child: Anoint the brain of a hare on the lesions.

107. On swelling under the tongue: If there is a swelling on the tongue, it usually occurs underneath, under the veins. Grind gall nut with condensed wine and apply to the swelling. 2 Anoint the tongue with the juice of olive leaves where the disease is.

108. On aphthous mouth ulcers: Aphtha is when the lips of a child rot because of the acrid nature of the milk. This disease also affects adults. Anoint the lips with the marrow of a deer. Anoint the lips with pear juice. 2 Burn fennel root and apply the ash to the lips. He should not drink a lot of wine. Bleed him if he is twelve, fifteen or twenty years old. ⌐Cut the middle[35] vein. They should eat little.¬

109. For the inflammation of the jaws: Put warm sweet goat's milk into the mouth. 2 Boil rain water and apply a vapour bath to the place for the appropriate time. 3 Grind barley flour and linseed well, boil it with water and wine and apply it to the place with a cloth. Do this for three hours, in the morning and the evening. 4 Boil liquorice and saffron and keep it in the mouth.

[35] This might be the median cubital vein, which is situated in between the cephalic and basilic vein in the bend of the elbow.

110 [133] ρι΄ πρὸς βῆχα· τὸν λεγομένον στύρακα ἴσα ἐρεβινθίῳ κατάπινε. μὴ δειπνείτω δέ. ὠφελεῖ δὲ ὁ στύραξ καὶ εἰς τὸν κατάρρουν. 2 ἄλλο. τὸ πρωῒ καταλύσας σῦκα μικρὰ κομμάτια καὶ εἰς οἶνον βρέχων ἐσθιέτω. ἀφ' ὅτου δὲ φάγῃ αὐτὰ ἂς πίνῃ οἶνον γλυκὺν ὀλίγον. πινέτω δὲ καὶ ἐσθιέτω καὶ τὴν λεγομένην γροῦταν. ὅταν μέλλῃ κοιμηθῆναι ἐσθιέτω ἀμύγδαλα πικρὰ πέντε. εἰ δὲ μή, γλυκέα δέκα. καὶ ἐπάνω οἶνον θερμόν. οἶνον πινέτω. εἰ δὲ οὐκ ἔχει γλυκέα ἐσθιέτω φοινίκια τρία. 3 ἄλλο. μέλι βράσας μετρίως βάλε ἀμύγδαλα γλυκέα κοπανισμένα ἑκατόν. κουκουνάρια καὶ καταστατὸν κοπανίσας ἀμύγδαλα πικρὰ κεκοπανισμένα εἴκοσι καὶ ἐὰν οὐ πυρέσσῃ, βάλε καὶ πέπερι. κόψας αὐτὸ χονδρὸν καὶ γλυκύριζον κόψας αὐτὸ καλά. καὶ λινόσπορον κοπανιστὸν καὶ φοινίκια κοπανιστὰ λιπαρὰ καὶ φύλλα ξηρὰ ὑσσώπου κοπανιστὰ καὶ μαστίχην καὶ στάχος καὶ τριψίδιν καὶ κύμινον ὀλίγον καὶ λευκὸν πέπερι καὶ μακροπέπερι καὶ τζουκνίδας σπόρον τρίψας τὰ πάντα καλὰ ἕνωσον καὶ βάλε εἰς τὴν χύτραν τοῦ μέλιτος καὶ ἀπὸ τότε μηδὲν βράσῃς αὐτά. 4 ἄλλο. ἄλειφε τὸ στῆθος αὐτοῦ βούτυρον προβάτου, μὴ τρίψας δὲ αὐτὸ ἢ χήνειον στέαρ ἢ ὀρνίθειον. τίθει δὲ καὶ τὴν διὰ βουτύρου εἰς τὸ στῆθος αὐτοῦ ἢ τὴν ἀνηθίνην.

111 [134] ρια΄ πρὸς τοὺς ἀπὸ φωνῆς μέγα βραχνιάσαντες· λοῦε αὐτὸν συνεχῶς, ἡ τροφὴ δὲ αὐτῷ ἔσται καταστατὸς καὶ ὠὰ ῥοφητὰ καὶ γλυκὺ γάλα πρόβειον χλίον.

1 στύρακα] scripsi στήρακα N 2 στύραξ] scripsi στήραξ N 11 γλυκύριζον] scripsi γλυκύρριζον N 17 προβάτου] scripsi cf. ω 110 πρόσφατον N

110. For cough: Drink the so called storax, the amount of a chickpea. He should not have supper. Storax is also useful for catarrh. 2 Dissolve figs into small pieces, soak in wine and let him eat in the morning. And when he eats them, let him drink some sweet wine. Let him drink and eat the so called groats. When he is about to go to bed, he should eat five bitter almonds. If not, ten sweet almonds and then warm wine. He should drink wine. If he does not have sweet almonds, he should eat three dates. 3 Boil honey sensibly and add one hundred ground almonds. Grind pine kernels and starch, and add twenty ground bitter almonds and if he is not feverish add also pepper. Grind it coarsely, grind liquorice nicely as well. And ground linseed and ground juicy dates and ground dry leaves of hyssop, mastic, vermilion, mixed spice, some cumin, white pepper, long pepper, nettle seed, grind everything well, mix it and put it into a jug full of honey, and after that do not boil it. 4 Anoint his chest with goat's butter, but do not grind it, or goose or bird fat. But put also the Butter Remedy on his chest or the Aniseed one.

111. For those who have a very hoarse voice: Bathe him frequently, his diet should be starch, raw eggs and warm sweet goat's milk.

112 [135,137] ριβʹ πρὸς δυσπνοίαν νέων μέσων καὶ γερόντων· δύσπνοια ἐστὶν ὅταν ἀγανακτῶσιν ἀναπνέοντες καὶ μάλιστα ψύχους ὄντος κακῶς διάκεινται οἱ δυσπνοϊκοί. τὴν σκύλλαν ἐψήσας εἰς τοὺς ἄνθρακας ἐπάνω λάμβανε τὰ φλούδια αὐτῆς καὶ βρέχων εἰς μέλι καὶ ὄξος ὀλίγον ἔσθιε. 2 ἄλλο. πότισον αὐτὸν και τὴν λεγομένην ἱεράν. ἐσθιέτω δὲ ῥεφάνια. ἀναγκαζέτω ἐμεῖν χυμούς. εἰ καιρὸν ἔχει πολὺν ἐξότου οὐκ ἐφλεβοτομήθῃ, φλεβοτομησάτω τότε. αἴρας ἄλευρον καὶ σῦκα κοπανιστὰ μαλάξας μετὰ ὕδατος, τίθει εἰς τὸ στῆθος μετὰ πανίου. 3 ἄλλο. ἄλειφε τὸ στῆθος πηγανέλαιον. ἀμμωνιακὸν τρίψας μετὰ οἴνου θερμοῦ πινέτω. ἀπεχέσθω δὲ καὶ ἀνέμου καὶ πάντων τῶν ψυχρῶν. 4 ἄλλο. ἐὰν ὁ πόνος κρατεῖ εἰς τὸν ὠμὸν φλεβοτόμησον αὐτόν. εἰ δὲ εἰς τὴν κοιλίαν ἢ εἰς τὰ σπλάχνα πότισον βοήθημα. κενταύριον βράσας καὶ πίτυρα τὸν ζωμον ἐκείνων μετὰ μέλιτος καὶ ἅλατος ἰάτρευε καὶ ἐλαίου. 5 ἄλλο. κριθάριν φρύξας πυρίαζε τὸν τόπον. 6 ἄλλο. χαμαίμηλα καὶ ἀψίνθιον καὶ ἰσχάδας βράσας μετὰ ὕδατος καὶ κριθάλευρον μετ' αὐτῶν καὶ λινοσπόρου τὸ ἄλευρον καὶ ἁπλώσας εἰς πανίον ἐπιτίθει ἔνθα πονεῖ. παρελθούσης τῆς ἑβδόμης ἡμέρας ποίησον ἐκλεκτόν μετὰ ἀμυγδάλων γλυκέων καὶ τῆς τζουκνίδης τὸν σπόρον. 7 ἄλλο. ἔμπλαστρον τὸ λεγόμενον πηγανερὰ ἐπιτίθει εἰς τὸν τόπον. 8 ἄλλο. συκίαζε δὲ αὐτὸν ἐπάνω εἰς τὸν τόπον, ἐὰν οὐδὲν παύῃ ἡ ὀδύνη. 9 ἄλλο. κηρὸν καὶ τεάφην καὶ τερεβινθίνην κοπανίσας καὶ ἀναλύσας μετὰ χοιρείου ἀξουγγίου καὶ κυπρινελαίου καὶ βουτύρου καὶ ποιήσας ὡς κηρωτὴν τίθει εἰς τὸν ὀδυνώμενον τόπον. 10 ἄλλο. πήγανον καὶ ὕσσωπον καὶ καλαμίνθιν καὶ ἶριν καὶ χαμαιπίτυν βράσας μετὰ μελικράτου εἰς τὸν ζωμὸν τῶν βοτάνων ἐμβαλὼν ῥέον τετριμμένον κοκκία ἕξ. πότιζε καθ' ἑκάστην ἡμέραν, ὀλίγον μὲν τὸ ῥέον πλείων δὲ ὁ ζωμὸς ἔστω.

3 σκύλλαν] scripsi σκύλαν N 20 πηγανερὰ] scripsi cf. ω 137 om. N 22 τερεβινθίνην] scripsi τερεβινθύνην N

112. For dyspnea of young, middle aged and old people: Dyspnea is when they are in discomfort when they breathe, and they are in a particularly bad condition when it is cold. Boil squill over charcoal, take its bark, soak it in honey and some vinegar and eat it. 2 Let him drink the so called Holy Medication. He should eat radish. Force him to vomit the humours. If it has been a long time since his blood has been let, then bleed him. Knead darnel flour and ground figs with water, put on the chest with a cloth. 3 Anoint the chest with rue oil. Grind ammoniac with warm wine and let him drink. He should abstain from wind and all cold things. 4 If the pain reaches to the shoulder, bleed him. If the pain reaches to the stomach or the intestines, let him drink medication. Boil centaury and bran, and administer the broth with honey, salt and oil. 5 Roast barley and apply a vapour bath to the place. 6 Boil chamomile, wormwood and dried figs with water, and with it barley flour and linseed flour, spread it onto a cloth and put it where the pain is. After seven days make a lozenge with sweet almonds and nettle seed. 7 Put the so called Rue Plaster on the place. 8 Cup him on top of the place, if the pain does not stop. 9 Grind wax, sulphur and turpentine, dissolve it with pig fat, henna oil and butter and make it a waxlike consistency and put it on the place where the pain is. 10 Boil rue, hyssop, peppermint, iris and ground pine with mead and add six hexagia of ground rhubarb to the vegetable broth. Give him to drink every day, whereas there should be little rhubarb and more broth.

113 [138] ριγ΄ περὶ πνευμονίας ἢ δυσπνοίας· πνευμονία ἐστὶν ὅταν πτύῃ μετὰ βηχὸς πτύελον παχὺν κίτρινον δυσώδη. φλεβοτόμησον αὐτὸν μὴ μέχρι λειποθυμίας. 2 ἄλλο. πινέτωσαν χυλὸν γυμνοκρίθου καὶ σεμίδαλιν καὶ θίασον ἀμυγδάλων γλυκέων καὶ δροσάτον. 3 ἄλλο. τίθει εἰς τὸ στῆθος τὴν πηγανερὰν ἥτις καὶ ἔχει οὕτως· κηρίον καὶ χήνειον ἀξούγγιν καὶ μυελὸν ἐλάφειον ἀναλύσας τίθει.

114 [139] ριδ΄ περὶ ἐμπυήματος ἢ πλευρίτου· ἐμπύημα λέγεται ὅταν χρονίσαντος τοῦ πλευρίτου βήχῃ καὶ πτύῃ παχέα, δυσώδη τε καὶ ἑλκώδη. ὕσσωπον καὶ θυμὸν μετὰ ὕδατος βράσας πίνε τὸν ζωμόν. 2 ἄλλο. γλυκύριζον καὶ σησάμιν μετὰ μέλιτος βράσας τρῶγε ἐξ αὐτοῦ καθ' ἑσπέραν ὀλίγον. καὶ ἐπάνω πίνε θερμὸν οἶνον.

115 [139,140,141,142] ριε΄ πρὸς φθίσιν· φθίσις γίνεται ἀπὸ ἐμπυήματος πρῶτον ἣν λέγουσι οἱ ἰατροὶ ἑκτικὴν νόσον. πότιζε τοῦ Ἔσδρα τὴν ἀντίδοτον. εἰς δὲ τὸ στῆθος τίθει τὴν διὰ βουτύρου καὶ κυπρίνου καὶ τὴν Μνάσεον ἔμπλαστρον. 2 ἄλλο. πότιζε καὶ τὴν Μιθριδάτην καὶ τὴν θηριακήν. εἰ δὲ πυρέσσῃ, πότισον αὐτὸν χυλὸν γυμνοκρίθου ἔχοντα καὶ σταφίδας καθαριστὰς ἀπὸ τῶν κοκκίων αὐτῶν καὶ ζίνζιφα καὶ δαμάσκηνα καὶ φοινίκια δύο ἢ τρία. ὕστερον δὲ ἀφ' ὅτου βράσουν, βάλε καὶ μαστίχην καὶ σάχαρ, εἰ δὲ μὴ μέλι ὀλίγον. 3 εἰς φθισιν· φθίσις γίνεται ὅταν ὁ πνεύμων τοῦ ἀνθρώπου ποιήσῃ ἕλκος καὶ πτύῃ ἕλκη καὶ μαραίνεται. θὲς ἐπάνω εἰς τὴν θώρακα κηρωτὰς τὰς διὰ βουτύρου. 4 ἄλλο. ἄλειφε καὶ πηγανέλαιον ἢ ἀνηθέλαιον τὸ στῆθος αὐτοῦ. ὕστερον καὶ τὴν Μνάσεον θὲς ἔμπλαστρον. 5 ἄλλο. πότιζε αὐτὸν τὸ λεγόμενον μελίκρατον. 6 ἄλλο. καταστατὸν καὶ γλυκύριζον βράσας μετὰ ὕδατος πίνε. πότιζε αὐτοὺς καὶ τὴν ἰσόθεον καὶ τὴν δι' ὀρόβων καὶ τοῦ Ἔσδρα καὶ τὴν θηριακήν. ὡσεὶ κοκκία τρία ἢ τέσσαρα τὸ πρωῒ μετὰ οἴνου ὀλίγου.

1 πνευμονία] scripsi περιπνευμονία N 10 γλυκύριζον] scripsi γλυκύρριζον N 12–13 ἐμπυήματος] scripsi ἐμπτυήματος N 16 Μιθριδάτην] scripsi ἡμιθτριδάτην N 25 γλυκύριζον] scripsi γλυκύρριζον N

113. On pneumonia and difficulty breathing: Pneumonia is when he coughs up a thick, yellow and stinking sputum. Bleed him, but not until he faints. 2 They should drink the juice of barley, fine flour, orgeat of sweet almonds and dew. 3 Put the Wax Remedy on the chest which is like this: dissolve wax, goose fat and deer marrow and apply.

114. On empuema or pleuritis: Empyema is called when he coughs up something thick, stinking and puslike during a chronic pleuritis. Boil hyssop and thyme with water and drink the broth. 2 Boil liquorice and sesame with honey and some of it in the evening. And drink warm wine with it.

115. On phthisis: Phthisis arises from primary empyema, which the doctors call consumptive disease. Let him drink the Antidote of Esdra.[36] Put the Butter Remedy on the chest and the Henna Remedy and the Mnaseos Plaster.[37] 2 Let him drink the Mithridates Remedy[38] and theriac. If he is feverish, let him drink juice of barley which also has raisins which have been cleaned from their seed, and red dates, damsons and two or three dates. Then, when it boils, add mastic and sugar, and maybe a little honey. 3 On phthisis: Phthisis arises when the lung of a person produces pus and he spits pus and wastes away. Put wax like remedies with butter on the chest. 4 Anoint his chest also with rue oil or aniseed oil. Finally also apply the Mnaseos Plaster. 5 Let him drink the so called mead. 6 Boil starch and liquorice with water and drink it. Let them also drink the Godlike Remedy[39] and the one Through Vetch and the Esdra Remedy and theriac. Approximately three or four kokkia in the morning with some wine.

---

[36] This medication has drying properties. It consists mainly of cardamom and saffron.

[37] This remedy has softening properties. It consists of oil, litharge and a variety of other ingredients.

[38] This medication was seen as a panacea; it consists, amongst other ingredients, of cinnamon and roses.

[39] This medication consists of castoreum, euphorbium and a variety of other ingredients.

116 [143] ρις′ πρὸς τὸν ἀπὸ θυμοῦ πυρέττοντα· ὅταν ἀπὸ θυμοῦ ὁ ἄνθρωπος πυρέξῃ, ἐκ πρώτης ἡμέρας λοῦε αὐτὸν καὶ τὸ ὅλον σῶμα ἄλειφε χρηστέλαιον, εἶτα καὶ θρέψον αὐτόν. ταῦτα δὲ ποίει καὶ τοὺς ἀπὸ κόπου καὶ λύπης καὶ φροντίδος πυρέξαντας.

117 [144] ριζ′ περὶ τριταίου· τριταῖος πυρετὸς ἔνι, ὅταν μίαν ἡμέραν ἔχει ἄνεσιν καὶ τὴν ἄλλην πυρέσσῃ. ἐσθιέτω μαιούλια ἴντυβα μετὰ ὄξους. ἰάτρευε αὐτὸν οὕτως· μολόχια καὶ πίτυρα βράσας τὸν ζωμὸν ἐκεῖνον μετὰ ἐλαίου καὶ ἅλατος καὶ μέλιτος. οἶνον καὶ παστὰ ἐκφευγέτω. 2 ἄλλο. σελίνου ἀνήθου ἀγρίας τὰς ῥίζας κιχώριον εὐπατόριον πολυκόμπιν δενδρολίβανον βράσας μετὰ ὕδατος, ὕστερον δὲ καὶ σάχαρ ἀντὶ μέλιτος βαλὼν πίνε. 3 ἄλλο. ἀψίνθιον βράσας μετὰ ὕδατος πότιζε τοὺς τοιούτους κατὰ πρωΐ. 4 ἄλλο. ὅταν δὲ καιρὸς τοῦ λουθῆναι γλυκέα ὕδατα καὶ εἰ ἀγαπᾷ τὸ λοετρόν καὶ δὶς τῆς ἡμέρας λοῦε αὐτόν. 5 ἄλλο. πότιζε τοὺς τοιούτους δροσάτον καὶ ζουλάπιν.

118 [145] ριη′ περὶ τεταρταίου· τεταρταῖον ἐστίν, ὅταν δύο ἡμέρας ἔχῃ ἄνεσιν καὶ τὴν ἄλλην πυρέσσῃ. γίνεται ἐπὶ ψυχρῷ καὶ ξηρῷ χυμῷ τῷ μελαγκολικῷ. τὴν πρώτην ἡμέραν μήτε φλεβοτομήσῃς μήτε καθάρῃς αὐτόν. εἰ δὲ ἔχει πολλὰς ἡμέρας ὅπου οὐκ ἐφλεβοτόμησεν, ἂς ἐκβάλῃ ὀλίγον αἷμα, καὶ ἕως ἐξέρχεται μαῦρον ἂς ῥέῃ, πλὴν δὲ πρόσεχε μὴ λειποθυμήσῃ. θρέψον δὲ αὐτὸν ὀψάριν χλωρόν. 2 ἄλλο. πότιζε αὐτὸν καὶ τὸ διὰ τριῶν πεπέρεων, τὸ λεγόμενον διοσπολιτικόν. 3 ἄλλο. πέπερι τρίψας μετὰ χλίου ὕδατος πότιζε. ⊢καὶ τὸν ζωμὸν τοῦ σελινορρίζου καὶ ἀνηθορρίζου καὶ τῶν ὁμοίων ὡς ἐπὶ τοῦ τριταίου εἴπομεν.⊣ εἰς δὲ τὸν σπλῆνα τίθει τὰ λεγόμενα Πολυάρχια. 4 ἄλλο. πρῶτον κένωσον τὸν μελαγχολικὸν χυμὸν διὰ γαστρός, εἶτα πότιζε τὴν θηριακήν καὶ τὸ δι' ὀποῦ τοῦ Κυραναϊκοῦ φάρμακον. ἐσθιέτω τὰ παστὰ καὶ τὰ θερμαντικά οἷον πέπερι καὶ τὰ σκόροδα.

14 ἡμέρας] scripsi ἡμέραν N

116. On someone who is feverish as a result of being irascible: If someone is feverish because of irascibility, bathe him from the first day on and anoint the entire body with table oil, and then feed him. Do the same for those who are feverish from exhaustion, sadness and thinking.[40]

117. On tertian fever: Tertian fever is when he has a remission for one day and a fever for the other. He should eat lettuce and chicory with vinegar. Treat him like this: Boil mallow and bran and their broth with olive oil, salt and honey. He should abstain from wine and pastry. 2 Boil the root of celery and wild aniseed, chicory, agrimony, mare's tail and rosemary with water, in the end add sugar instead of honey and drink it. 3 Boil wormwood instead of water and let them drink in the morning. 4 When it is time to bathe, fresh water, and if he likes bathing, bathe him also twice a day. 5 Let them drink dew and zulapin.

118. On quartan fever: Quartan fever is when he has a remission for two days and then a fever. It arises from the melancholic humour which is cold and dry. On the first day do not bleed him or purge. If it is several days since his blood has been let, he may lose a little blood, and let it run, as long as it is black, but be aware that he does not faint. Feed him with white fish. 2 Let him drink the Three Pepper Remedy which is called Diospolitikon. 3 Grind pepper and let him drink with warm water. ⊢And the juice of celeriac, fennel and the like as we said in the chapter on the tertian fever.⊣ Put the so called Polyarchia[41] on the spleen. 4 First purge the melancholic humour through the stomach, then let him drink theriac and the Opium Remedy of Kyrenaikos. He should eat pastry and warming things like pepper and garlic.

---

[40] Fevers do not have any specific point of manifestation. Therefore they tended to be discussed in separate books rather than in a head-to-foot collection. Here, fevers can be found between the diseases of the lung and the breasts, which corresponds with the location of the heart.

[41] This medication consists of henna, cadamom, wax and other ingredients.

119 [146] ριθ´ περὶ τοῦ ἀφημερινοῦ· ἐπὶ δὲ τοῦ ἀφημερινοῦ πυρετοῦ εἰς τὰς ἀρχὰς μὲν πότισον τὸ ὀξύμελι καὶ τὰ διουρητικά φάρμακα. ἐθιέτω καὶ ῥεφάνια, εἶτα ἐμεῖτω. ἐσθιέτω καὶ κύμινον. ἔστω καὶ ζεστά.

120 [147] ρκ´ πρὸς ὅταν γένηται λοιμώδες, ὃ πυρετικὴν λέγουσιν· ὅταν γένηται νόσημα λοιμώδες, ὃ λέγουσι πυρετικὴν οἱ ἰδιῶται, θεράπευσον οὕτως· πρῶτον κλυστήρευε καὶ εἰ πολλὰς ἡμέρας ἔχει, ἐξ ὅτου οὐκ ἐφλεβοτόμησεν, φλεβοτόμει ὀλίγον αἷμα. πότιζε καὶ τὰ διουρητικά. εἰ δὲ καὶ ἔσται ἐντός, πότισον καὶ ὕδωρ, ὅσον θέλης ψυχρόν. 2 ἄλλο. τρίψας ἀρμένιον βῶλον πότιζε αὐτὸν μετὰ ὕδατος χλίου. 3 ἄλλο. πότιζε καὶ τὴν λεγομένην θηριακήν, καὶ τὴν ἀθανασίαν μετὰ ὀλίγου οἴνου τὸ πρωΐ νῆστιν.

121 [148,149] ρκα´ πρὸς βυζία ἔχοντα καρκίνου πάθος· λιθάργυρον τρίψας καὶ μετὰ χρηστελαίου καὶ ὄξους ἑνώσας ἄλειφε τὸν πόδα. 2 ἄλλο. ἀγριοσταφίδα καὶ ἀγριοκάρδαμον καὶ ψιμμίθιν τρίψας καὶ ἑνώσας μετὰ μυρσινελαίου ἄλειφε τὸν τόπον. ἀφ' ὅτου δὲ θεραπευθῇ ὁ κάρκινος εἰ θέλης να μηδὲν τὸ πάθη ποτὲ τὸ βυζίον τοῦτο τὸ πάθος, ποίησον οὕτως· κύμινον τρίψας καὶ μετὰ ὕδατος μαλάξας τίθει εἰς τὰ βυζία, ἐπάνω δὲ τίθει σπόγγον καινὸν βεβρεγμένον μετὰ οἴνου καὶ ὄξους. εἶτα δῆσον αὐτόν. παρελθουσῶν δὲ ἡμερῶν τριῶν ἔπαρον κύμινον. βολβὸν δὲ κοπανίσας καὶ μαλάξας μετὰ μέλιτος τίθει. εἶτα δῆσον. ποίει δὲ τοῦτο ἐκ τρίτου τὸν μῆνα.

122 [150] ρκβ´ πρὸς δυσωδίαν μασχαλῶν· στυπτηρίαν ὑγρὰν καὶ σμύρνην τρίψας καὶ ἑνώσας μετὰ οἴνου ἄλειφε τὰς μασχάλας. ὅπου δὲ ἔνι βρῶμος τοῦτο ποίει. 2 ἄλλο. λιθάργυρον καὶ σμύρναν καὶ ἄμωμον τρίψας καὶ μίξας μετὰ οἴνου καλοῦ, ἄλειφε τὰς μασχάλας.

123 [136] ρκγ´ πρὸς πτύσιν αἵματος πολλὴν καὶ εἰς μίαν· ὅταν πτύῃ τις αἷμα πολὺ καὶ εἰς μίαν, σημεῖον ἐστὶν ὅτι ἐκόπει ἡ φλέβα, ὥσπερ ὅταν ἀφ' ὑψηλοῦ πέσῃ καὶ ἐμέσῃ αἷμα. πότισον αὐτὸν βῶλον. εἰ δὲ κατάρρους ὢν πτύῃ αἷμα φλεβοτόμησον αὐτόν. εἰ δὲ πτύῃ αἷμα ἀφρῶδες, σημεῖον ὅτι ἐκ τῶν ἀναπνευστικῶν ἐστίν. μυρσίνης φύλλα βράσας μετὰ ὄξους πυρίαζε τὸ στῆθος. τίθει δὲ εἰς τὸ στῆθος τὸ ἔμπλαστρον τὸ δι' ἰτεῶν. 2 ἄλλο. λημνίαν σφραγίδα πότιζε μετα οἴνου χλίου.

33 λημνίαν] scripsi λυμνίαν N

119. On quotidian fever: For quotidian fever in the beginning let him drink oxymel and diuretic remedies. He should eat radish and then he should vomit. He should also eat cumin. He should keep warm.

120. For when the plague strikes, which is called hectic fever: If there is the plague which the common people call hectic fever, treat it like this: First bleed him, and if it is several days since his blood has been let, let some blood. Let him also drink diuretic remedies. If it is also inside, let him drink cold water as much as you like. 2 Grind Armenian earth and let him drink with warm water. 3 Let him drink the so called theriac, and the Immortal Remedy[42] with some wine in the morning when he is fasted.

121. For breasts suffering from cancer: Grind litharge, mingle it with table oil and vinegar and apply it to the basis. 2 Grind stavesacre, wild cardamom and lead, mix it with myrtle oil and apply to the place. And if the cancer is treated and you do not want that the breast will be affected again, do this: Grind cumin, mix it with water and apply on the breast, on top put a new sponge soaked in wine and vinegar. Then bandage it. After three days, take cumin. Grind purse tassels, mix it with honey and apply. Then bandage it. Do this on the third day of the month.

122. For smelling armpits: Grind a liquid astringent with myrrh, mix it with wine and apply to the armpits. Do this when there is smell. 2 Grind litharge, myrrh and cardamom, mix it with good wine and apply to the armpits.

123. On spitting a mass of blood once: If someone spits a lot of blood at one time, it is a sign that a vein is ruptured, like when someone falls from a height and vomits blood. Let him drink earth. If he spits blood when he has a catarrh, bleed him. If he spits bloody froth, it is a sign that it comes from the airways. Boil myrtle leaves with vinegar and apply a vapour bath to the chest. Put the Willow Plaster on the chest. 2 Let him drink Lemnian earth with warm wine.

---

[42] This medication has an anodyne effect. It consists mainly of nard, saffron and opium.

124 [151] ρκδʹ πρὸς ὅταν ῥεύγηται ὡς ὠὰ τηγάνου· ὅταν ὁ ἄνθρωπος ῥεύγηται ὡς ὠὰ τηγάνου, σημεῖον ἐστὶν ὅτι θερμὴ δυσκρασία ἐν τῇ κοιλίᾳ. ψωμὸν βρέχων εἰς ὄξος μεμιγμένον μετὰ ὕδατος ἐσθιέτω καὶ ὀξύγαλα παχύ καὶ μαρούλια καὶ ἴντυβα καὶ πηκτῆς τρέμουσαν. ἀντὶ οἴνου πινέτω ὄξος μετὰ ὕδατος.

125 ρκεʹ πρὸς τοὺς διὰ ψύξιν ἀνορεκτοῦντας· τοὺς διὰ ψύξιν ἀνορεκτοῦντας θεράπευε οὕτως· πινέτω οἶνον θερμὸν ἄκρατον καὶ ὕδωρ καὶ γάρος μικτόν. ἄνηθον καὶ πήγανον καὶ μακεδονησίου ῥίζαν βράσας μετὰ ὕδατος πίνε. ὕστερον βάλε καὶ πέπερι κοπανιστόν. 2 ἄλλο. μέλι βράσας ὡς ποιεῖς τὸ κυδωνάτον, βάλε πέπερι τετριμμένον καὶ μακροπέπερι κοπανιστά. καὶ ταράξας καλῶς ἔσθιε ἐξ αὐτοῦ τὸ βραδὺ ἴσον καρυδίῳ. εἰ δύναται δὲ ἄς το καταπίνῃ. εἶτα πινέτω ἐπάνω θερμὸν οἶνον. μηδὲν δειπνάτω. ἀφ' ὅτου δὲ πληρώσῃ τοῦτο ἐσθιέτω τὸ διὰ καλαμίνθης. σκευάζεται δὲ οὕτως· μέλι βράσας καὶ τρίψας καλαμινθίου τὰ φύλλα ὁμοίως καὶ γληχῶνα καὶ μακεδονησίου φύλλα καὶ θύμου καὶ πέπερι ὀλίγον. τὰ πάντα τρίψας καλῶς βάλε ἐν τῷ μέλιτι καὶ τάραξον αὐτὰ καὶ τρῶγε ἐξ αὐτοῦ. 3 ἄλλο. πίνε καὶ τὴν διὰ μόσχου ὡσεὶ κοκκία †δ.. καὶ τὴν θηριακὴν καὶ σκόροδα ὠμὰ καὶ ἑψημένα ἔσθιε.

126 [153] ρκϛʹ πρὸς τοὺς ἀνορεκτοῦντας διὰ ὕλην· τοὺς δὲ ἀνορεκτοῦντας ἐκ τοῦ ἔχειν ὑγρὰ ἐν τῷ στομάχῳ πότιζε αὐτοὺς πικρὰν ὡσεὶ κοκκία ὀκτὼ μετὰ οἴνου κατὰ πρωΐ. τὰ παστὰ μηδόλως ἐσθιέτω μηδὲ ζωμὰ πινέτω μηδὲ ὑδροποτείτω μηδὲ δειπνείτω. 2 ἄλλο. ῥόδα ξηρὰ καὶ ἀλόην καὶ κηκίδιν καὶ φοινίκια καὶ οἰνάνθην ὀλίγον καὶ μελίλωτον καὶ μαστίχην καὶ κρόκον καὶ ἀψινθίου σπόρον καὶ πίσσαν ὀλίγην τρίψας τὰ πάντα καλῶς ποίει ἔμπλαστρον εἰς τὸν στόμαχον.

127 [154] ρκζʹ πρὸς κυνώδη ὄρεξιν· κυνώδης ὄρεξις ἔνι ὅταν ἐσθίῃ τις καὶ πάλιν ἐσθίῃ ὥσπερ οἱ κύνες. γίνεται τὸ πάθος ἐκ ψυχρῶν χυμῶν. πινέτωσαν γάρος μετὰ ὕδατος καὶ τὸ διὰ τριῶν πεπερέων.

---

7 ἀνορεκτοῦντας] scripsi ἀνωρεκτοῦντας N 8 μακεδονησίου] scripsi μακεδονίου N 26 πίσσαν] scripsi πίσαν N

124. When he vomits something resembling fried eggs: If a person spits something like fried eggs, it is a sign that he has a warm imbalance in the stomach. Soak bread in vinegar which is mixed with water and he should eat it, and thick sour milk, lettuce, chicory and jelly. Instead of wine he should drink vinegar with water.

125. For those who have no appetite because of cold: Those who have no appetite because of the cold treat like this: He should drink warm undiluted wine and water mixed with fish sauce. Boil aniseed, rue and parsley root in water and drink. Then add ground pepper. 2 Boil honey as you would do for quince jam, add ground pepper and ground long pepper. And mix it well and eat from it about the amount of one walnut. If possible let him drink it as well. Then let him drink warm wine. He should not have supper. If this is done he should eat the Peppermint Remedy. It is made like this: Boil honey and grind peppermint leaves and in the same amount pennyroyal, parsley and thyme leaves and some pepper. Grind it all well, add honey, stir it and eat from it. 3 Drink also the Musk Remedy, about .... kokkia, and theriac and let him eat raw and cooked garlic.

126. For those who have no appetite because of matter: Let those who have no appetite because of liquids in the stomach drink bitter remedies, approximately eight kokkia with wine in the morning. He should eat no pastry at all, not should he drink broth nor water, nor should he have supper. 2 Grind dry roses, aloe, gall nut, dates, vine flower, mélilot, mastic, saffron, wormwood seed and some pitch well altogether and make a plaster for the stomach.

127. On dog-like appetite: Dog-like appetite is when someone eats and eats again like the dogs. This disease is caused by cold humours. Let them drink fish sauce with water and the Three Pepper Remedy.

128 [155] ρκη΄ πρὸς κλόξον· τὸν ἔχοντα κλόξον θεράπευε οὕτως· πήγανον βράσας μετὰ οἴνου πινέτω. 2 ἄλλο. ⊢σέλινον ἑψήσας μετὰ μέλιτος καὶ οἴνου πινέτω.⊣ 3 ἄλλο. καστόριον τρίψας μετὰ ὄξους καὶ οἴνου πινέτω μὴ βράσας δὲ τοῦτο.

129 [157] ρκθ΄ πρὸς κένωσιν χολῆς ἄνω καὶ κάτω· τὸν κενοῦντα χολὴν καὶ ἄνω καὶ κάτω θεράπευε οὕτως· ψαλίδας κοπανίσας τὸν ζωμὸν πίνε. 2 ἄλλο. ῥόδα ξηρὰ βράσας καὶ οἰνάνθην μετὰ οἴνου αὐστηροῦ πίνε. 3 ἄλλο. ῥοΐδιον ὄξυνον κοπανίσας πίνε τὸν ζωμόν. πίνε αὐστηρὸν οἶνον. 4 ἄλλο. ψίχας ἄρτου εἰς αὐστηρὸν οἶνον βρέχων ἔσθιε. ἄντλει τοὺς πόδας μετὰ θερμοῦ ὕδατος καὶ ἂς ἔνι ζεστά. 5 ἄλλο. κοκκία σταφίδων κοπανίσας μετὰ αὐστηροῦ οἴνου πίνε. τὴν αὐτὴν θεραπείαν ποίησον καὶ τὸν ἐξ ἄλλου τινὸς ψυχρὰν θέντα καὶ λυθέντα τὴν κοιλίαν. ζεστὰ ὁ τοιοῦτος ἔστω. ψυχρὸν μὴ πινέτω, καὶ ἐὰν οὐ πυρέσσῃ, λοῦε αὐτόν.

130 [158] ρλ΄ πρὸς φλεγμονὴν ἐντέρων· ὅταν συμβῇ φλεγμονὴ εἰς τὰ ἔντερα ἢ ξηρὰ κόπρος πέσῃ εἰς τοὺς πόρους τῶν ἐντέρων, ξερᾷ τὴν κόπρον ὁ ἄνθρωπος. καὶ καλεῖται τὸ νόσημα παρὰ τῶν ἰατρῶν ἰλεός. καὶ ἔστιν σχεδὸν ἀνίατον πάθος, εἰ διὰ φλεγμονὴν τῶν ἐντέρων ἐστίν. εἰ δὲ δι' ἔμφραξιν σωτήριον. καὶ ἐὰν ὁ πάσχων ᾖ δεκατεσσάρων χρονῶν, ποίησον ἐμβροχὰς καὶ καταπλάσματα καὶ πυρίας. εἰ δὲ ἔνι κ' χρονῶν ὁ ἄρρωστος, φλεβοτόμησον. εἰ δὲ ἀκμάζων ἤτοι μ' χρονῶν, φλεβοτόμησον καὶ αὐτόν. οἶνον μὴ πινέτωσαν. ἔστω δὲ καὶ ζεστά.

131 [159] ρλα΄ περὶ φλεγμονῆς νεφρῶν· φλεγμαινόντων τῶν νεφρῶν πυρέττουσι καὶ ἐμοῦσι χολήν. τοὺς τοιούτους φλεβοτόμησον. 2 ἄλλο. τῆς τήλης τὰ κοκκία κοπανίσας καὶ τοῦ ἀνήθου τὸν σπόρον μαλάξας μετὰ ὕδατος ἐπιτίθει ἐπάνω τῶν νεφρῶν. 3 ἄλλο. τὸν σπόρον τοῦ πέπονος τρίψας μετὰ μέλιτος καὶ οἴνου πίνε. 4 ἄλλο. πυρίαζε τοὺς νεφροὺς μετὰ οἴνου καλοῦ. τὸ διὰ χυλοῦ ἔμπλαστρον καὶ τὸν Μνασέου ἐπιτίθει εἰς τοὺς νεφρούς. 5 ἄλλο. κηρὸν ἀναλύσας μετὰ ῥοδελαίου καὶ κρόκου ὠοῦ ἐπιτίθει.

6 ψαλίδας] scripsi ψαλλίδας N 18 ἰλεός] scripsi ἴλεως N 28 πέπονος] scripsi πέπωνος N

128. On hiccup: Someone suffering from hiccup treat like this: Boil rue with wine and he should drink it. 2 ⌐Boil celery with honey and wine and he should drink it.¬ 3 Grind castoreum and he should drink it with vinegar and wine but do not boil it.

129. On excreting bile above and below: Someone excreting bile above and below treat like this: Grind vine tendrils and drink the juice. 2 Boil dry roses and vine flowers and let him drink with dry wine. 3 Grind a sour pomegranate and drink the juice. Drink dry wine. 4 Eat bread crumbs soaked in dry wine. Rinse the feet with warm water and let him keep warm. 5 Grind the fruit of stavesacre and drink with dry wine. Administer the same therapy for someone who has another form of cold disposition and is suffering from diarrhea. He should keep warm. He must not drink cold things, and if he is not feverish, bathe him.

130. On inflammation of the intestines: If there is an inflammation of the bowels or if dry feces get into the passages of the bowels, the person vomits feces. And this disease is called ileos by the doctors. And it is an almost incurable disease if it happens because of an inflammation of the bowels. If it is because of a constipation there is hope. And if the patient is fourteen years old, use infusions, plasters and vapour baths. If the patient is twenty years old, bleed him. If he is middle aged or forty years old, also bleed him. They should abstain from wine. He should keep warm.

131. On inflammation of the kidneys: If the kidneys are inflamed, they have fever and vomit bile. Let their blood. 2 Grind the fruit of fenugreek, mingle with aniseed, knead with water and apply it over the kidneys. 3 Grind melon seed and drink with honey and wine. 4 Apply a vapour bath of good wine to the place. Put the Juice Plaster and the Remedy of Mnaseos on the kidneys. 5 Dissolve wax and apply with rose oil and egg yolk.

132 [160] ρλβ΄ πρὸς πάθος τὸ λεγόμενον διαβήτην· διαβήτης ἐστὶν ὅταν ἡνίκα πίνῃ, ὀρέγηται κατουρῆσαι. τοὺς τοιούτους πότισον πολυγόνου χυλὸν ἢ τοῦ λεγουμένου ῥάσδου. 2 ἄλλο. φοινίκια βράσας τὸν ζωμὸν αὐτῶν πίνε. 3 ἄλλο. κριθάλευρον βράσας μετὰ ὕδατος, ὕστερον βαλὼν ὄξος καὶ ῥοδέλαιον ἑψήσας καὶ πάλιν ὕστερον ζωμὸν ψαλίδων βαλὼν καὶ φύλλα βαλὼν ἀμπέλου τίθει εἰς τοὺς νεφρούς.

133 [161] ρλγ΄ πρὸς ῥαγάδας ὀρχιδίων· ῥαγὰς λέγεται παρὰ τῶν ἰατρῶν ὅταν οἱ ὄρχεις κατασχισθῶσιν, ἡ ἕδρα ...... ῥητίνην τρίψας καὶ μετὰ ῥοδελαίου μαλάξας ἄλειφε τοὺς τόπους τοὺς κατεσχισμένους, πλὴν τὴν ῥητίνην πρῶτον φρύξας εἰς τὸν ἥλιον. 2 ἄλλο. κρόκον ᾠοῦ ἢ ὄπιον μετὰ τοῦ κρόκου τοῦ ᾠοῦ τίθει εἰς τὰς πληγάς. 3 ἄλλο. κισσοῦ φύλλα χλωρὰ πάττε τὴν πληγήν. 4 ἄλλο. τὴν στάκτην τῶν φύλλων τοῦ κισσοῦ μίξας μετὰ ῥοδελαίου εἰς μολύβδινον ἀγγεῖον καὶ ταράξας ἐπιτίθει. ἀποσπογγίσας πρῶτον μετὰ οἴνου τὰς πληγὰς πάττε ῥόδα κοπανισμένα. εἰ δὲ φλεγμαίνει τὸ αἰδοῖον, κοπάνισον τρυφερὰ φύλλα ἀμπέλου καὶ μίξας μετὰ ψιμμιθίου ἐπιτίθει. ἄντλει δὲ τὸ αἰδοῖον πρῶτον μετὰ ὕδατος θαλασσίου.

134 [162,163] ρλδ΄ πρὸς πνιγμονὴν τὴν ἐξ ἀμανίτων· ἅλμην μετὰ ὕδατος πότιζε. 2 ἄλλο. νίτρον τρίψας καὶ μίξας μετὰ ὕδατος βαλὼν εἰς τὸ στόμα αὐτοῦ ταραττέτω ἔνθεν καὶ ἔνθεν ἱκανὴν ὥραν. ἐσθιέτω δὲ καὶ σκόροδα. 3 ἄλλο. κλήματος στάκτην πότισον μετὰ οἴνου καὶ ὄξους διπλοῦ. 4 ἄλλο. νίτρον τρίψας πότισον ὁμοίως. 5 ἄλλο. ᾠὰ ὀρνιθερὰ ἐκζέσας μετὰ ὄξους καὶ οἴνου τρῶγε. πήγανον βράσας μετὰ χρηστελαίου πίνε. 6 ἄλλο. ὀρίγανον βράσας μετὰ οἴνου καὶ μέλιτος πινέτω. ὅσον δὲ δύναται, ἃς ἐμῇ.

135 [164] ρλε΄ πρὸς τύρωσιν γάλακτος ἐν τῇ κοιλίᾳ· ἐὰν τυρωθῇ τὸ γάλα εἰς τὴν κοιλίαν τοῦ ἀνθρώπου, πότισον αὐτὸν χυλὸν καλαμινθίου. 2 ἄλλο. θυμὸν βράσας μετὰ οἴνου πινέτω.

---

2 ὀρέγηται] scripsi ὠρέγηται N 3 πολυγόνου] scripsi πολυγώνου N 6 ψαλίδων] scripsi ψαλλίδων N 9 ......] lacuna N

132. On the disease which is called diabetes: Diabetes is when he wants to urinate while he is still drinking. Let them drink mare's tail juice, or the juice of the so called nard. 2 Boil dates and drink the broth. 3 Boil barley flour with water, then add vinegar and rose oil and boil and then again add the juice of vine tendrils, add vine leaves and put it on the kidneys.

133. On fissures on the testicles: Fissure is called by the doctors when the testicles are split, the anus .... Grind resin, knead it with rose oil and anoint it onto the split parts, but dry the resin first in sunlight. 2 Put egg yolk or opium with egg yolk on the wounds. 3 Apply fresh leaves of ivy on the wound. 4 Mix the ash of ivy leaves with rose oil in a lead vessel, stir it and apply. Rinse the wounds with a sponge and wine first, then apply ground roses. If the penis is inflamed, grind juicy vine leaves, mix them with lead white and apply. But rinse the penis first with sea water.

134. On choking because of mushrooms: Let him drink salty water with water. 2 Grind saltpetre, mix it with water, put it into his mouth where he should move it to and fro for the appropriate time. Let him also eat garlic. 3 Let him drink the ash of vine branches with wine and the double amount of vinegar. 4 Grind saltpetre and let him drink in the same way. 5 Boil bird eggs thoroughly with vinegar and wine and eat them. Boil rue with table oil and drink it. 6 Boil oregano with wine and honey and he should drink it. Let him vomit as far as this is possible.

135. If milk turns to cheese in the stomach: If the milk in the stomach of a person turns into cheese, let him drink peppermint juice. 2 Boil thyme with wine and let him drink.

136 [167] ρλϛ′ πρὸς ἀλφοὺς ὅταν εἰς τὸ πρόσωπον τοῦ ἀνθρώπου ἐξέρχονται ὥσπερ φακαί· ἀλφοὶ εἰσὶν ὅταν εἰς τὸ πρόσωπον τοῦ ἀνθρώπου ἐξέρχονται ὡσὰν φακῆς κοκκία. δενδρολίβανον κοπανίσας καὶ μετὰ κεδρέας μίξας καὶ χρηστελαίου ἄλειφε τὸ πρόσωπον. 2 ἄλλο. τεάφην τρίψας καὶ λιθάργυρον καὶ ἀγριοσταφίδα καὶ μετὰ χρηστελαίου μίξας ἄλειφε. 3 ἄλλο. δαδὶν λιπαρὸν βράσας μετὰ χρηστελαίου καλῶς καὶ τὸ ἔλαιον αὐτοῦ μίξας μετὰ τετριμμένου λιθαργύρου καὶ ὄξους ἄλειφε, πλὴν ἐν λουτρῷ. 4 ἄλλο. ἀγριοσταφίδα τρίψας καὶ τεάφην καὶ μίξας μετὰ ὄξους ἄλειφε.

137 [170] ρλζ′ πρὸς πόνους ἁρμῶν· τοὺς πόνους τῶν ἁρμῶν καὶ τὰς σκληρότητας μαλακίζει ἀμμωνιακὸν βεβρεγμένον εἰς ὄξος δριμὺ νυχθήμερον μεμαλαγμένον μετὰ μέλιτος καὶ ἐπιτιθέμενον. 2 ἄλλο. πύρεθρον καὶ ζίγγιβερ καὶ πέπερι τρίψας καὶ μετὰ χρηστελαίου ἑνώσας ἄλειφε τοὺς πόνους. 3 ἄλλο. κεδρέαν μίξας μετὰ χρηστελαίου ἄλειφε τὸν τόπον. 4 ἄλλο. τὴν ῥίζαν τοῦ κισσοῦ καύσας καὶ τὴν στάκτην μίξας μετὰ κρόκου ᾠοῦ τίθει εἰς τὸν τόπον. ὑοσκυάμου σπόρον κοπανίσας καὶ νίτρον ὁμοίως μετὰ ὕδατος πινέτω. 5 ἄλλο. ἄλειφε τὸν τόπον ἔνθα ἔνι ὁ πόνος τῶν ἁρμῶν ἐλάφειον ἀξούγγιν ἢ ὄρνιθος ἢ ταύρου.

138 [171] ρλη′ πρὸς ἀπόστημα. ἀπόστημα σκορπίζει καὶ μαλακίζει· ὑοσκυάμου φύλλα κοπανισμένα μετὰ χοιρίου ἀξουγγίου τίθει εἰς τὸ ἀπόστημα. ⊢καλαμίνθης φύλλα χλωρὰ κοπανίσας τίθει ἐπάνω.⊣ 2 ἄλλο. σπαθοβότανον κοπανίσας τίθει ἐπάνω. 3 ἄλλο. ⊢μαλάχην τρίψας καὶ πίσσαν καὶ κριθάλευρον καὶ βούτυρον καὶ ὄξος δριμὺ καὶ ἑνώσας βράσον αὐτὰ ὁμοῦ καὶ βρέχων στυππεία ἄλασσε τὸν τόπον συχνά.⊣ 4 ἄλλο. ἀγριομολόχης τὰ φύλλα βράσας μετὰ ὕδατος γλυκέως καὶ κοπανίσας τίθει. 5 ἄλλο. κριθάλευρον καὶ λινόσπορον βράσας μετὰ ὕδατος βάλε χοίρειον ἀξούγγιν καὶ χήνειον καὶ ὄρνιθος ἐλάφου μυελὸν καὶ ταύρου ἀξούγγιν καὶ ἀρκούδας καὶ ἀλώπεκος τὰ πάντα λυτὰ μαλάξας τίθει μετὰ πανίου.

---

23 σπαθοβότανον] scripsi παθοβότανον N

136. On alphoi, if they appear on a person's face like lentils: Alphoi are when on the face of a person appears something like the seeds of lentils. Grind rosemary, mix it with cedar oil and table oil and anoint the face. 2 Grind sulphur, litharge and stavesacre, mix it with table oil and apply. 3 Boil a juicy piece of pine wood well with table oil, mix the oil with ground litharge and vinegar and apply, but in the bath. 4 Grind stavesacre and sulphur, mix it with vinegar and apply.

137. On pain of the joints: Ammoniac which is soaked for one night and one day in acrid vinegar, mingled with honey and applied softens pain in the joints and hardening. 2 Grind pellitory, ginger and pepper, mix it with table oil and apply on the aching part. 3 Mix cedar oil with table oil and anoint the place. 4 Burn the root of ivy, mix the ash with egg yolk and apply to the place. Grind henbane seed and saltpetre in the way and he should drink it with water. 5 Anoint the place where the joint pain is located with deer, bird or bull marrow.

138. On ulcers. This dissipates and softens an ulcer: Put ground henbane leaves with pig fat on the ulcer. ⸢Grind juicy peppermint leaves and apply them.⸣ 2 Grind sword grass and apply. 3 ⸢Grind mallow, pitch, barley flour, butter and acrid vinegar, mix it and boil it together, soak hemp in it and apply it to the place repeatedly.⸣ 4 Boil the leaves of wild mallow with fresh water, grind and apply. 5 Boil barley flour and linseed with water, add pig, bird or deer fat and bull, bear and fox marrow, dissolve and mix it all and apply with a cloth.

139 [172] ρλθ΄ πρὸς ἀπορυφὰς καὶ ἐκβατά· καρπὶν κοπανίσας καὶ μετὰ προπόλεως μίξας τίθει ἐπάνω. 2 ἄλλο. κριθάλευρον βράσας μετὰ τοῦ χυλοῦ τῶν σευτλοφύλλων καὶ μαλάξας τίθει ἐπάνω. 3 ἄλλο. κρόμμυον ἑψήσας ἐπ' ἀνθράκων καὶ κοπανίσας τίθει. 4 ἄλλο. πίσσαν τρίψας πάσσε ἐπάνω.

140 [173] ρμ΄ πρὸς δῆξιν ἀσπίδος· φύλλα μηλέας λαβὼν τὸν ζωμὸν τούτων πίνε μετὰ οἴνου. τὰ δὲ φύλλα κοπανίσας τίθει ἐπάνω. 2 ἄλλο. μυρσίνης φύλλα κοπανίσας τίθει ἐπάνω. 3 ἄλλο. ἡδύοσμον κοπανίσας πότισον μετὰ θερμοῦ ὕδατος. 4 ἄλλο. κόστου φύλλα κοπανίσας τίθει εἰς τὴν πληγήν. 5 ἄλλο. χαλβάνιν τρίψας καὶ μίξας μετὰ χρηστελαίου ἄλειφε. 6 ἄλλο. ἀξούγγιν ἀρκούδας λύσας ἄλειφε τὴν πληγήν.

141 [176] ρμα΄ πρὸς τὸ καταβάσαι γάλα· λυγαίας σπόρον κοπανίσας πότισον μετὰ οἴνου. 2 ἄλλο. ἀνήθου τὴν ῥίζαν βράσας πίνε. 3 ἄλλο. μαράθρου φύλλα καὶ ῥίζαν καὶ σπόρον βράσας μετὰ οἴνου πίνε. 4 ἄλλο. μελάνθιν κοπανίσας πότιζε μετὰ οἴνου. 5 ἄλλο. ἅλμην θαλασσίαν ἄντλει συνεχῶς. 6 ἄλλο. ψιμμίθιν καὶ λιθάργυρον τρίψας καὶ μίξας μετὰ ὄξους καὶ ὕδατος καὶ ἅλατος ἄλειφε τὰ βυζία.

142 [177] ρμβ΄ πρὸς γυναίκα μὴ δυναμένην γεννῆσαι· ζῶντας καβούρους ὑποκάπνιζε.

143 [178] ρμγ΄ πρὸς γλῶσσαν ἁδράν· γλώσσης ἁδρότητα μαλακίζει. μέλι βράσας κράτει ἐπάνω πανίου ἔχον ἐντὸς λινόκοκκον καὶ ἡνίκα ἱδρώσῃ τὸ πανὶν ἐκ τοῦ ἀτμοῦ τοῦ μέλιτος τρίβε τὴν γλῶσσαν. ἡδύοσμον χλωρὸν κοπανίσας τρίβε τὴν γλῶσσαν. 2 ἄλλο. ἡδυόσμου χλωροῦ ζωμὸν μίξας μετὰ ῥοδελαίου καὶ χαμαιμηλαίου ἄλειφε τὴν γλῶσσαν. ἐλαίας φύλλα μάσα καὶ ἐπικράτει εἰς τὴν γλῶσσαν.

144 [183] ρμδ΄ πρὸς τὸ ἐὰν συμβῇ τεθνάναι τὸ παιδίον ἐντὸς τῆς κοιλίας· ἀψίνθιον βράσας μετὰ οἴνου πότιζε. 2 ἄλλο. καππάρεως φλοῖον ἢ τὴν ῥίζαν βράσας μετὰ οἴνου πότισον. 3 ἄλλο. κινάμωμον τρίψας καὶ σμύρνην πότισον. 4 ἄλλο. μακεδονησίου σπόρον κοπανίσας πότισον μετὰ ὕδατος. 5 ἄλλο. ὀρίγανον βράσας μετὰ ὕδατος πότιζε.

---

4 κρόμμυον] scripsi κρόμυον N

139. On *aporufai* and swelling: Grind hellebore, mix it with propolis and apply. 2 Boil barley flour with the juice of beet root leaves, mix it and apply. 3 Cook an onion on charcoal, grind it and apply. 4 Grind pitch and apply.

140. For the bite of an asp: Take the leaves of an apple tree and drink their juice with wine. Grind the leaves and apply them. 2 Grind myrtle leaves and apply them. 3 Grind peppermint and let him drink with warm water. 4 Grind costus leaves and put them on the wound. 5 Grind ferula, mix it with table oil and apply. 6 Dissolve the bear's fat and apply to the wound.

141. To induce lactation: Grind the seed of a chaste tree and let her drink with wine. 2 Boil anise root and let her drink. 3 Boil the leaves and the root of fennel with wine and drink. 4 Grind nigella and let her drink with wine. 5 Rinse her frequently with sea water. 6 Grind lead white and litharge, mix it with vinegar, water and salt and apply to the breasts.

142. For a woman if she is unable to conceive: Apply a vapour bath with living shrimps.

143. For a hardened tongue: This softens a hardened tongue. Boil honey, put it on a cloth that has linseed inside, and when the cloth starts to sweat from the vapour of the honey, anoint the tongue. Grind fresh peppermint and anoint the tongue. 2 Mix the juice of fresh peppermint with rose oil and chamomile oil and apply to the tongue. Grind willow leaves and keep them on the tongue.

144. If the child dies in the womb: Boil wormwood with wine and let her drink. 2 Boil the bark of capers or their root with wine and let her drink. 3 Grind cinnamon and myrrh and let her drink. 4 Grind parsley seed and let her drink with water. 5 Boil oregano with water and let her drink.

145 [184] ρμε′ πρὸς δῆξιν ὄφεως καὶ ἐχίδνης· καστόριον τρίψας πότιζε μετὰ οἴνου. 2 ἄλλο. κρίνου φύλλα κοπανίσας τίθει εἰς τὴν πληγήν. 3 ἄλλο. ξυλοκέρατα κοπανίσας καλῶς πότισον μετὰ οἴνου.

146 [187] ρμϛ′ περὶ χρυσιασμοῦ· κίτρινον ὅταν γένηται τὸ πρόσωπον ὅπερ καλοῦσιν οἱ ἰδιῶται χρυσιασμὸν οἱ δὲ ἰατροὶ ἴκτερον ποίει οὕτως. ἀδίαντον μετὰ οἴνου βράσας πότιζε. 2 ἄλλο. ἀλόην τρίψας πότιζε μετὰ ὕδατος. 3 ἄλλο. ἀψινθίου κορυφὴν μετὰ ὕδατος βράσας ἢ μετ' οἴνου πότισον. 4 ἄλλο. σκολοπένδριον βράσας μετὰ ὕδατος πίνε. 5 ἄλλο. ῥίζαν λαπάθου βράσας μετὰ οἴνου πίνε. τοῦ χρυσολαχάνου τὸν σπόρον κοπανίσας μετὰ ὕδατος ἢ μετὰ τυρογάλακτος πίνε.

147 [189] ρμζ′ πρὸς κοιλίαν ῥέουσαν πολλά· ῥίζαν βάτου καὶ τὰ βλαστάρια κοπανίσας τὸν ζωμὸν αὐτῶν πίνε. 2 ἄλλο. μουρέας ῥίζαν κοπανίσας καὶ ἑψήσας μετὰ οἴνου πίνε. 3 ἄλλο. παλιουρίας φύλλα καὶ τὴν ῥίζαν βράσας μετὰ οἴνου πίνε. μυρσίνης φύλλα βράσας μετὰ οἴνου ἰάτρευε. 4 ἄλλο. σχίνου φύλλα καὶ τὸν σπόρον ἑψήσας μετὰ οἴνου ἰάτρευε. 5 ἄλλο. κολοκύνθην ἑψήσας μετὰ οἴνου πίνε. 6 ἄλλο. κυπαρίσσου τρυφερὰ βλαστάρια βράσας μετὰ οἴνου πίνε.

148 [190] ρμη′ πρὸς δῆξιν κυνός· μελισσόφυλλα κοπανίσας πότιζε μετὰ οἴνου. 2 ἄλλο. καλαμινθίου φύλλα κοπανίσας καὶ μετὰ τῆς στάκτης μίξας καὶ μαλάξας τίθει εἰς τὴν πληγήν. 3 ἄλλο. σκόροδον κοπανίσας πότιζε μετὰ ὕδατος. 4 ἄλλο. πήγανον χλωρὸν κοπανίσας καὶ μαλάξας μετὰ χρηστελαίου τίθει εἰς τὴν πληγήν. 5 ἄλλο. ἀγριοσυκῆς φύλλα κοπανίσας τρυφερὰ καὶ μετὰ μέλιτος μαλάξας τίθει εἰς τὴν πληγήν. 6 ἄλλο. πολυγόνου χυλὸν πότισον μετὰ οἴνου.

149 [193] ρμθ′ πρὸς δῆξιν λυσσῶντος κυνός· γεντιανῆς τὴν ῥίζαν βράσας μετὰ οἴνου πότισον. 2 ἄλλο. καβούρους ποταμίους καύσας τὴν στάκτην πότισον μετὰ οἴνου. 3 ἄλλο. βδέλλιον σὺν οἴνῳ πίνε. 4 ἄλλο. λίθον αἱματίθην τρίψας εἰς ἀκόνην πότισον μετὰ οἴνου.

150 [198] ρν′ πρὸς δῆξιν μελισσῶν ἢ καὶ σφηκῶν· μολοχίου ἀγρίου τὰ φύλλα κοπανίσας τίθει εἰς τὴν πληγὴν ἢ ἡμέρου. 2 ἄλλο. δάφνης φύλλα κοπανίσας τρυφερὰ τίθει εἰς τὴν πληγήν. 3 ἄλλο. πήγανον χλωρὸν κοπανίσας τίθει εἰς τὴν πληγήν.

145. For the bite of a snake or a viper: Grind castoreum and let her drink with wine. 2 Grind lily leaves and apply to the wound. 3 Grind carob beans well and let him drink with wine.

146. On jaundice: If the face turns yellow, what the common people call jaundice and the doctors call icterus, do the following: Boil maidenhair with wine and let him drink. 2 Grind aloe and let him drink with water. 3 Boil the shoots of wormwood with water or wine and let him drink. 4 Boil Hart's tongue fern with water and drink. 5 Boil sorrel root with wine and drink. Grind the seed of orach with water or with butter milk and drink.

147. If there is much flow from the abdomen: Grind raspberry roots and its shoots and drink the juice. 2 Grind mulberry roots, boil it with wine and drink. 3 Boil the leaves and the root of Christ's thorn with wine and drink. Boil myrtle shoots with wine and administer. 4 Boil mastic leaves and its seeds with wine and administer. 5 Boil pumpkin with wine and drink. 6 Boil the juicy shoots of a cypress with wine and drink.

148. For dog bites: Grind melissa and drink with wine. 2 Grind peppermint leaves, mix them with ash, knead and apply to the place. 3 Grind garlic and let him drink with water. 4 Grind fresh rue, mix it with table oil and apply to the wound. 5 Grind the juicy leaves of a wild fig tree, knead with honey and apply to the wound. 6 Let him drink mare's tail juice with wine.

149. If someone is bitten by a rabid dog: Boil gentian root with wine and let him drink. 2 Burn sea shrimps and let him drink the ash with wine. 3 Drink bdellium with wine. 4 Grind hematite stone with a whetstone and let him drink with wine.

150. If someone is bitten by a bee or a wasp: Grind the leaves of wild mallow and put them on the wound. Alternatively take domesticated mallow. 2 Grind juicy bay leaves and put them on the wound. 3 Grind fresh rue and put it on the wound.

151 [199] ρνα´ πρὸς δυσωδίαν ῥινός· φύλλα κισσοῦ ἢ καλαμίνθης κοπανίσας τὸν χυλὸν αὐτῶν χλιάνας στάξον εἰς τὴν ῥίνα. 2 ἄλλο. ὀριγάνου χλωροῦ τὸν χυλὸν στάξον. 3 ἄλλο. μελάνθιον κοπανίσας καὶ μίξας μετὰ χρηστελαίου ἄλειφε ἐντός τε καὶ ἔξω αὐτήν.

152 [200] ρνβ´ πρὸς τὰς ἀπὸ δαρμοῦ πληγάς· βλαστάρια πλατάνου κοπανίσας καὶ μετὰ χοιρείου ἀξουγγίου μαλάξας τίθει εἰς τὴν πληγήν. 2 ἄλλο. λημνίαν σφραγίδα τρίψας καὶ μετὰ ὕδατος μαλάξας ἄλειφε τὰς πληγάς. βῶλον τρίψας καὶ μίξας μετὰ ὕδατος ἄλειφε. 3 ἄλλο. ἡδύοσμον χλωρὸν κοπανίσας καὶ μαλάξας μετὰ χοιρίου ἀξουγγίου τίθει.

153 [203] ρνγ´ πρὸς νεφέλην ὀφθαλμῶν· δρακοντίας βοτάνης χυλὸν ἄλειφε ἔξω τοὺς ὀφθαλμούς. 2 ἄλλο. τριφύλλιον κοπανίσας τὸν ζωμὸν μίξον μετὰ μέλιτος καὶ ἄλειφε ἔξω τοὺς ὀφθαλμούς. 3 ἄλλο. σμύρναν τρίψας καὶ μετὰ μέλιτος ἑνώσας ἄλειφε τοὺς ὀφθαλμούς.

154 [205] ρνδ´ πρὸς πόνον ὀδόντων· καρπὶν μαῦρον βράσας μετὰ ὄξους ἂς κρατῇ εἰς τὸ μέρος τὸ ὀδυνώμενον. 2 ἄλλο. βλαστάρια πλατάνου τρυφερά βράσας μετὰ ὄξους κράτει. 3 ἄλλο. βάτου φύλλα βράσας μετὰ ὄξους κράτει.

155 [206] ρνε´ πρὸς σειομένους ὀδόντας· ὕσσωπον μετὰ ὄξους βράσας κράτει εἰς τὸ στόμα. 2 ἄλλο. κυκλαμίνου φλοῦν ἑψήσας μετὰ ὄξους κράτει εἰς τὸ στόμα. 3 ἄλλο. καρπὸν καππάρεως βράσας μετὰ ὄξους κράτει εἰς τὸ στόμα. 4 ἄλλο. τὸν φλοῦν τῆς ῥίζης τῆς καππάρεως βράσας μετὰ ὄξους κράτει εἰς τὸ στόμα. 5 ἄλλο. ῥοϊδίων φύλλα βράσας μετὰ ὄξους κράτει.

156 [207] ρνϛ´ πρὸς καῦσιν μέλους τινός· ἀρνόγλωσσον κοπανίσας τίθει ἐπάνω. 2 ἄλλο. συκαμίνου φύλλα ἑψήσας μετὰ ἐλαίου καὶ ὕστερον κατακόψας αὐτὰ μικρὰ τίθει ἐπάνω.

157 [208] ρνζ´ πρὸς ποδαλγικούς· κρίνου ῥίζαν κοπανίσας τίθει ταύτην ἔνθα ἡ ὀδύνη. 2 ἄλλο. τὸν σπόρον τῆς γογγυλίδος βράσας μετὰ ὕδατος τὸ ὕδωρ ἐκεῖνο ἄντλει τοὺς πόδας. 3 ἄλλο. ῥεφανίου ῥίζαν κοπανίσας τίθει εἰς τὸν τόπον. 4 ἄλλο. σπέρμα ῥεφανίου βράσας μετὰ ὕδατος ἄντλει τοὺς πόδας.

---

20 κυκλαμίνου] scripsi κυκαμίνου N

151. On a malodorous nose: Grind the leaves of ivy or peppermint, warm the juice up and drip it into the nose. 2 Drip the juice of fresh oregano. 3 Grind nigella, mix it with table oil and apply it internally and externally.

152. On wounds caused by abrasion: Grind the shoots of plane tree, knead it with pig fat and apply it to the wound. 2 Grind Lemnian earth, knead it with water and apply to the wounds. Grind earth, mix it with water and apply. 3 Grind fresh peppermint, knead it with pig fat and apply.

153. On a cloud-like spot in the eye: Anoint the juice of tarragon plant externally on the eye. 2 Grind clover, mix the juice with honey and apply it outside on the eye. 3 Grind myrrh, mix it with honey and apply to the eyes.

154. On tooth ache: Boil black hellebore with vinegar, and let him keep it in the place that hurts. 2 Boil the juicy shoots of a plane tree with vinegar and keep it. 3 Boil raspberry leaves and keep them.

155. If the teeth are loose: Boil hyssop with vinegar and keep it in the mouth. 2 Boil the bark of violet with vinegar and keep it in the mouth. 3 Boil a caper with vinegar and keep it in the mouth. 4 Boil the bark of the root of a caper plant with vinegar and keep it in the mouth. 5 Boil the leaves of a pomegranate tree with vinegar and keep it.

156. If a part of the body is burned: Grind arnica and apply it. 2 Boil mulberry tree leaves with olive oil, then chop them up into small bits and apply them.

157. On podagra: Grind lily root and put it where it hurts. 2 Boil the seed of turnip with water and use the water to rinse the feet. 3 Grind radish root and put it on the place. 4 Boil radish seed with water and rinse the feet.

158 [209] ρνη΄ πρὸς ὅταν οἱ ἁρμοὶ τῶν δακτύλων γίνωνται σκληροί· βαλὼν ἀμμωνιακὸν εἰς ὄξος βρέχε δύο νυχθήμερα. εἶτα μαλάξας αὐτὸ μετὰ ὑγροπισσίου τίθει ἐπάνω εἰς τὴν σκληρότητα. 2 ἄλλο. θαλάσσιον ὕδωρ ἄντλει τοὺς πόδας σου. 3 ἄλλο. ἅλας καὶ νίτρον τρίψας καὶ μετὰ χρηστελαίου ἑνώσας ἄλειφε συνεχῶς τοὺς ὀδυνωμένους τόπους.

159 [210] ρνθ΄ πρὸς τὸ στῆσαι αἷμα· ἀδιάντον κοπανίσας τὸν χυλὸν τούτου πότιζε μετὰ οἴνου. 2 ἄλλο. τεάφην τρίψας τί ἔνθα ῥέῃ. ἡδύοσμον χλωρὸν κοπανίσας καὶ μετὰ ὄξους καὶ οἴνου τὸν χυλὸν μίξας πίνε. 3 ἄλλο. ἀμυγδαλέας δάκρυον σὺν οἴνῳ πινόμενον. 4 ἄλλο. βάτου φύλλα τρυφερὰ κοπανίσας τίθει ἔνθα ῥέῃ τὸ αἷμα. ἡδύοσμον τρίψας τίθει εἰς τὴν ῥίνα.

160 [211,212,213] ρξ΄ πρὸς ἀδυναμίαν στομάχου· πολυγόνου φύλλα κοπανίσας τίθει εἰς τὸν στόμαχον. 2 ἄλλο. ῥέον Ποντικὸν σὺν ὕδατι πίνε. 3 ἄλλο. γεντιανῆς τὴν ῥίζαν κοπανίσας πότισον μετὰ οἴνου. 4 ἄλλο. γλιστρίδος φύλλα κοπανίσας καὶ μετὰ πιτύρων μίξας τίθει εἰς τὸν στόμαχον. 5 ἄλλο. δάφνης φύλλα κοπανίσας τὸν χυλὸν πότισον. 6 ἄλλο. μακεδονησίου σπόρον κοπανίσας πότισον μετὰ οἴνου. 7 ἄλλο. ζοχὶν ἥμερον ἢ καὶ ἄγριον τρῶγε. 8 ἄλλο. χαμαίμηλα βράσας πότισον μετὰ ὕδατος.

161 [215] ρξα΄ πρὸς στρόφον· τοῦ ἀνήθου τοῦ ξηροῦ τὴν κορυφὴν καὶ τὸ σπέρμα βράσας μετὰ ὕδατος, πότιζε. 2 ἄλλο. πήγανον καὶ ἄνηθον βράσας μετὰ ὕδατος πότιζε. 3 ἄλλο. σάμψυχον κοπανίσας τὸν χυλὸν πότιζε μετὰ οἴνου. 4 ἄλλο. κύμινον βράσας μετὰ ἐλαίου ἰάτρευε. 5 ἄλλο. καστόριον πότισον μετὰ οἴνου. 6 ἄλλο. ναρθηκίου σπόρον.

162 [216] ρξβ΄ περὶ δήξεως σκορπίου· τὸν καρπὸν τοῦ ἀσφοδέλου τρίψας μετὰ οἴνου πίνε. 2 ἄλλο. σάμψυχον κοπανίσας καὶ μετὰ ὄξους καὶ ἅλατος τετριμμένου ἑνώσας τίθει εἰς τὴν πληγήν. 3 ἄλλο. μελισσόφυλλα κοπανίσας πότιζε μετὰ οἴνου, ἢ τὰ φύλλα αὐτοῦ τίθει εἰς τὴν πληγήν.

---

4 νίτρον] scripsi νίτροντρον N

158. If the joints of the fingers harden: Put ammoniac into vinegar for two days and nights. Then knead it with liquid pitch and apply on top of the hardening. 2 Rinse your feet with sea water. 3 Grind salt and saltpetre, mix it with table oil and apply it frequently on the hurting parts.

159. To stop bleeding: Grind maidenhair, and let him drink the juice with wine. 2 Grind some sulphur onto where it flows. Grind fresh peppermint, mix the juice with vinegar and wine and drink. 3 Gum of an almond tree drunk with wine. 4 Grind the fresh leaves of raspberry and put it where the blood flows. Grind peppermint and put it on the nose.

160. For weakness of the stomach: Grind the leaves of mare's tail and put on the stomach. 2 Drink rhubarb with water. 3 Grind gentian root and let him drink with wine. 4 Grind the leaves of purslane, mix with bran and apply to the stomach. 5 Grind bay leaves and let him drink the juice. 6 Grind parsley seed and let him drink with wine. 7 Eat domesticated or wild thistle. 8 Boil chamomile and let him drink with water.

161. On colics: Boil the top of a dried anise plant and aniseed with water and let him drink. 2 Boil rue and anise with water and let him drink. 3 Grind marjoram and let him drink the juice with wine. 4 Boil cumin with olive oil and administer. 5 Let him drink castoreum with wine. 6 Ferula seed.

162. On scorpion bites: Grind the fruit of asphodelos and drink it with wine. 2 Grind marjoram, mix it with vinegar and ground salt and apply on the wound. 3 Grind melissa and let him drink it with wine, or alternatively put its leaves on the wound.

163 [217] ρξγ´ πρὸς σκώληκας τοὺς ἐν τῇ πληγῇ γινομένους· κοπάνισον φύλλα πτελέας καὶ τίθει ἐπάνω. 2 ἄλλο. ἀψίνθιον χλωρὸν κοπανίσας τίθει ἐπάνω. 3 ἄλλο. κεδρέαν ἄλειψον. τὸ αὐτὸ ποίει καὶ ἐπὶ τῶν ὠτίων τῶν ἐχόντων σκωλήκια.

164 [218] ρξδ´ πρὸς δῆξιν τρυγόνος θαλασσίας· ἐλελίσφακον τρίψας μετὰ οἴνου πότισον. 2 ἄλλο. τὰ φύλλα αὐτοῦ κοπανίσας τίθε εἰς τὴν πληγήν.

165 [219,220] ρξε´ πρὸς σπληναρίους καὶ ὀγκωμένους· σάμψυχον βράσας μετὰ ὕδατος, πότισον τὸν ζωμὸν αὐτοῦ μετὰ οἴνου. 2 ἄλλο. καππάρεως φλοῦν κοπανίσας πότιζε μετὰ οἴνου. 3 ἄλλο. κριθάλευρον βράσας μετὰ ὕδατος καὶ ὕστερον κοπανίσας συκῆς ῥίζαν καὶ μίξας ἐπιτίθει. εἰ δὲ ἔχει φλεγμονὴν ἡ σπλῆνα, κοπάνισον γλησχούνιν καὶ μετὰ τῶν πιτύρων μίξας τίθει εἰς τὴν σπλῆνα. 4 ἄλλο. λουπηναρίων πικρῶν ἄλευρον καὶ πίτυρα μαλάξας μετὰ ὕδατος τίθει ἐπάνω. φύλλα κράμβης κοπανίσας καὶ μετὰ πιτύρων μίξας καὶ μαλάξας τίθει ἐπάνω. κολοκύνθης χλωρᾶς τὰ ξύσματα τίθει ἐπάνω. 5 ἄλλο. λινόσπορον κοπανίσας καὶ τήλης σπόρον βράσας μετὰ ὕδατος ὕστερον δὲ βάλε χήνειον ἀξούγγιν καὶ χοίρειον καὶ ἀλώπεκος καὶ ὄρνιθος καὶ ταύρου καὶ μόσχου καὶ ἄρκου καὶ μυελὸν ἐλάφειον καὶ ἑνώσας τὰ πάντα καὶ μαλάξας τίθει ἐπάνω εἰς τὴν φλεγμονήν.

166 [221] ρξϛ´ πρὸς φωνὴν ἀποκοπεῖσαν· σκόροδα ὠμὰ καὶ ἑψημένα ἐσθιέτω. 2 ἄλλο. χαλβάνιν ποιήσας κοκκία ἴσα τῷ ἐρεβινθίῳ ὅταν μέλλῃ κοιμηθῆναι καταπινέτω αὐτά. καὶ ἐπάνω πινέτω θερμὸν οἶνον καὶ ἄλλο μηδὲ ἐσθιέτω.

167 [223] ρξζ´ περὶ τῶν λεγομένων χοιράδων· ἀρνόγλωσσον κοπανίσας τίθει ἐπάνω. 2 ἄλλο. ἀδίαντον κοπανίσας τίθει ἐπάνω. 3 ἄλλο. λουπηναρίων ἄλευρον μίξας μετὰ ὄξους τίθει ἐπάνω. 4 ἄλλο. λάπαθον κοπανίσας καὶ μετὰ ὄξους μίξας τίθει. 5 ἄλλο. ἀρωματοβοτάνου χυλὸν μίξας μετὰ πιτύρων τίθει. 6 ἄλλο. κυκλαμίνου ῥίζαν κοπανίσας τίθει ἐπάνω.

---

2 ἀψίνθιον] scripsi ψίνθιον N

163. If worms appear in wounds: Grind the leaves of an elm tree and apply them. 2 Grind fresh wormwood and apply. 3 Smear with cedar oil. Do the same also when the ears have worms.

164. On bites of a sea roach: Grind sage and let him drink with wine. 2 Grind its leaves and apply onto the wound.

165. On those suffering from a spleen disease and a swelling: Boil marjoram with water and let him drink the juice with wine. 2 Grind the bark of a caper plant and let him drink with wine. 3 Boil barley flour with water and then grind the root of a fig tree, mix it and apply. If the spleen is inflamed, grind pennyroyal, mix it with bran and apply to the spleen. 4 Knead the flour of bitter lupins and bran with water and apply it. Put splinters of green pumpkin on top. 5 Grind linseed and fenugreek seed, boil with water and then add goose fat and pig, fox, bird, bull, calf and bear fat and deer marrow, mix it all, knead it and apply it on the inflammation.

166. On the loss of voice: He should eat raw and cooked garlic. 2 Make chickpea sized pills of ferula and let him swallow when he is about to go to bed. And on top he should drink warm wine and he should not eat anything else.

167. On the so called scrofula: Grind arnica and apply it. 2 Grind maidenhair and apply it. 3 Mix lupin flour with vinegar and apply it. 4 Grind sorrel, mix it with vinegar and apply. 5 Mix spice tree juice with bran and apply. 6 Grind lily root and apply.

168 [226] ρξη΄ πρὸς τὸ παῦσαι δίψαν· ἐλαίας κολυμβάδας κράτει ὑπὸ τῆς γλώσσης σου. 2 ἄλλο. ᾠὸν ὀρνίθειον ὠμὸν ῥόφα κατὰ πρωῒ νῆστις. 3 ἄλλο. μαρουλίου σπόρον βράσας καὶ γλυσκύριζον μετὰ ὕδατος πίνε ψυχράνας πρῶτον.

169 [233] ρξθ΄ πρὸς δυσωδίαν στόματος· πέπερι κοκκία δέκα πέντε τρίψας καὶ ξυλοκέρατα καύσας καὶ τρίψας καὶ αὐτὰ καὶ φλοῦν ῥοιᾶς ἑξάγιον ἓν κοπανίσας καὶ τρίψας καὶ αὐτό. καὶ κηκίδιν ἓν τρυπητὸν καὶ μετὰ μέλιτος μίξας ἄλειφε τὸν τόπον ὅπου ἔνι σαπημένος. 2 ἄλλο. μύρον καλὸν ἄλειφε. 3 ἄλλο. ἀρσενίκιν τρίψας καὶ μετὰ μέλιτος μίξας ἄλειφε στυππεῖα. καὶ τὸ στυππείον βάλε μετὰ τοῦ δακτύλου σου εἰς τὸ στόμα σου.

170 [233] ρο΄ πρὸς σῆψιν οὐλῶν· φλοῦν ἰτέας καύσας καὶ χαρτὶν τὴν στάκτην αὐτῶν μετὰ πεπέρεως τετριμμένου καὶ μέλιτος ἑνώσας τίθει εἰς τὴν σῆψιν. 2 ἄλλο. πέπερι τρίψας καὶ μετὰ ὑδραργύρου καὶ μέλιτος ἑνώσας ἄλειφε τὸν τόπον. 3 ἄλλο. λουπηναρίων ἄλευρον καὶ ἀρκούδας χολὴν μίξας ἄλειφε.

171 [234] ροα΄ πρὸς βουβῶνας ἤτοι ἀπορυφᾶς· ἀσβέστιν ζωντανὸν μετὰ μέλιτος καὶ σαπουνίου μαλάξας θὲς ἐπάνω. 2 ἄλλο. ⊢κρίνου ῥίζαν κοπανίσας τίθει.⊣

172 [235] ροβ΄ πρὸς βιασμούς· ῥεφανίου σπόρον κοπανίσας καὶ κηκίδιν καὶ βράσας μετὰ οἴνου πότισον. 2 ἄλλο. ⊢βάτου κορυφὰς βράσας μετὰ παλαιοῦ οἴνου πότισον τὸ πρωΐ.⊣ 3 ἄλλο. ῥοϊδίων φλούδια χλωρὰ ἢ καὶ ξηρὰ καὶ πίσσαν ξηρὰν κάπνιζε ὑποκάτω. 4 ἄλλο. ὀρίγανον βράσας μετὰ οἴνου πότισον.

173 [238] ρογ΄ πρὸς ψύξιν γονάτων· ἀσβέστιν τρίψας καὶ μετὰ ὑγροπίσσης μαλάξας ἁπλωνέτω εἰς τὸ πανίν καὶ τίθει ἐπάνω. 2 ἄλλο. δάφνης κοκκία κοπανίσας καὶ ἑνώσας μετὰ θαλασσίου ὕδατος ἐπίθες ἔνθα ἡ ψύξις τῶν γονάτων.

---

3 γλυσκύριζον] scripsi γλυκύρριζον N   17 ἤτοι] scripsi τῆς N, cf. ω 234, pinax ℵ

168. To stop thirst: Keep pickled olives under your tongue. 2 Swallow raw egg in the morning when you are fasted. 3 Boil lettuce seed and liquorice with water and drink it after you cooled it down.

169. On halitosis: Grind fifteen pepper corns, burn carob beans, and chop it as well, and grind one hexagion of the peel of a pomegranate and chop it as well. And a ground gall nut and mix it with honey, and apply to the place where it is rotting. 2 Apply good myrrh. 3 Grind arsenic, mix it with honey and apply onto hemp. And put the hemp with your finger into your mouth.

170. On rotting lesions: Burn the bark of a willow tree and paper, mix the ash with ground pepper and honey and apply it to the rotten bit. 2 Grind pepper, mix it with quicksilver and honey and apply to the place. 3 Mix lupin flour and bear bile and apply.

171. On swollen glands or *aporufai*: Knead smouldering asbestos with honey and soap and apply it. 2 ⌐Grind lily root and apply.¬

172. On stomach cramps: Grind radish seed and gall nut, boil it with wine and let him drink. 2 ⌐Boil the tips of raspberry plant with old wine and let him drink in the morning.¬ 3 Apply a vapour bath from below with the fresh or dried peels of pomegranates and dry pitch. 4 Boil oregano with wine and let him drink.

173. On cold knees: Grind asbestos, knead it with liquid pitch apply it on a cloth and put it on top. 2 Grind laurel berries, mix them with sea water and apply where the cold part of the knees is.

174 [242] ροδ′ εἰς ἐσοχάδας καὶ ἐξοχάδας· ὀφιδίου δέρμα εἰς χρηστέλαιον βρέξας διὰ χωνῆς ὑποκαπνίζου. 2 ἄλλο. τὸν φλοῦν τῆς ῥίζης τῆς παλιουρέας βράσας μετὰ οἴνου πότισον ἐντὸς τοῦ λοετροῦ. ἔστω δὲ νῆστις ἄχρις ὡρῶν τεσσάρων. 3 ἄλλο. ⊢ῥόδα ξηρὰ καὶ κρόκον καὶ ἀλόην καὶ ᾠοῦ τὸ λεπτὸν μετὰ ῥοδελαίου ἑνώσας σὺν τῷ λεπτῷ τοῦ ᾠοῦ βαλὼν καὶ χοίρειον ἀξούγγιν κατάπλαττε εἰς πανίον καὶ ἐπιτίθει εἰς τὸν τόπον.⊣ 4 ἄλλο. θύννας κεφαλὴν ἢ παλμίδας καύσας τὴν στάκτην πάττε ἐπάνω. 5 ἄλλο. πολυκόμπιν ξηρὸν τηγανίσας μετὰ χρηστελαίου ἄλειφε. 6 ἄλλο. κρόκον κοπανίσας καὶ συρικὸν καὶ ἑνώσας ἄλειφε. 7 ἄλλο. ῥίζαν κρίνου καὶ κριθάλευρον μίξας καὶ μετὰ ὕδατος βράσας ἅπλωνε εἰς πανίον καὶ ἐπιτίθει. 8 ἄλλο. ἀποσπογγίσας πρῶτον τὸν κῶλον μετὰ ὄξους πάσσε καλάμου φύλλα κεκαυμένα. 9 ἄλλο. ᾠοῦ κρόκον καύσας πάττε τοῦτον. 10 ἄλλο. ἐρεβινθίων πάμπαν ἀπεχέσθω.

175 [243] ροε′ πρὸς ὅταν ἐξέρχεται ὁ κῶλος τοῦ ἀνθρώπου· καταστατὸν τρίψας καὶ ψιμμίθιον καὶ μετὰ οἴνου μαλάξας ἄλειφε. 2 ἄλλο. μελίλωτα καὶ ῥόδα ξηρὰ καὶ ψιμμίθιον μετὰ οἴνου βράσας ἄλειφε τὸν τόπον. 3 ἄλλο. ἀμανιτάριν ἑψήσας μετὰ ὕδατος καὶ ὕστερον κοπανίσας καὶ μετὰ βουτύρου ἑνώσας ἄλειφε μετὰ τοῦ δακτύλου σου τὸν τόπον. 4 ἄλλο. ἄλειφε πρότερον μετὰ οἴνου τὸν τόπον. εἶτα ἔχε τετριμμένον τοῦ κέγχρου τὸ πίτυρον κοπανισμένον καὶ πάσσε ἐπάνω. εἶτα βάλε τὸν κῶλον αὐτοῦ ἐντὸς ἀλειμμένον ὄντα μετὰ τῶν πιτυρῶν.

176 [245] ρος′ πρὸς τὰ ἐκβατά· ἀλεύριν σίτινον μαλάξας μετὰ χυλοῦ τῶν σεύτλων τίθε ἐπάνω. 2 ἄλλο. ὑαλίον τρίψας καὶ μετὰ ὄξους καὶ σαπωνίου καὶ χρηστελαίου μαλάξας τίθει ἐπάνω. 3 ἄλλο. καρπὶν μαῦρον τρίψας καὶ μετὰ προπόλεως μίξας καὶ μαλάξας τίθει ἐπάνω. 4 ἄλλο. μολύβδιν ῥινήσας καὶ μίξας μετὰ τοῦ χυλοῦ τοῦ χλωροῦ καλάμου θὲς ἐπάνω. 5 ἄλλο. ἀγριοπεριστέρου κόπρον τρίψας καὶ μετὰ μέλιτος μαλάξας τίθει ἐπάνω. 6 ἄλλο. ῥίζαν λαπάθου κοπανίσας καὶ ἑνώσας μετὰ χοιρείου ἀξουγγίου θὲς ἐπάνω.

177 [247] ροζ′ πρὸς δῆξιν ἐχίδνης· πήγανον κοπανίσας τὸν χυλὸν αὐτοῦ πότιζε. 2 ἄλλο. ἄμωμον τρίψας πότισον μετὰ τοῦ χυλοῦ τοῦ πηγάνου.

2 χωνῆς] scripsi χόνης N 7 θύννας] scripsi cf. ω 242 θυν.. N 28 μολύβδιν] scripsi μολύμβην N

174. On internal and external hemorrhoids: Soak the skin of a snake in table oil and apply a vapour bath from below through a funnel. 2 Boil the bark of the root of Christ's thorn and let him brink it with wine in the bath. He should then be fasted for four hours. 3 ⌐Dry roses, saffron, aloe and the thin parts of an egg mix with rose oil, and together with the thin parts of the egg also add pig fat, apply it onto a cloth and put it on the place.¬ 4 Burn the head of a tunny fish or of a ......[43] and apply the ash. 5 Fry dry mare's tail with table oil and apply. 6 Grind saffron and lead oxide, mix and apply. 7 Mix the root of lily and barley flour, boil it with water, apply on a cloth and put it on top. 8 First rinse the colon with vinegar and then put burnt reed on top. 9 Burn egg yolk and put it onto him. 10 He should abstain completely from chickpeas.

175. On a prolapse of the colon: Grind starch and lead white, knead it with wine and apply. 2 Boil melilot, dry roses and lead white with wine and apply to the place. 3 Boil a mushroom in water, then grind it and mix it with butter and apply it with your finger on the place. 4 First anoint the place with wine. Then take ground millet bran and apply it. The put the colon inside, after you have anointed it with the bran.

176. On swelling: Knead wheat flour with beet root juice and apply it. 2 Grind glass, knead it with vinegar, soap and table oil and apply it. 3 Grind black hellebore, mix it with propolis, knead and apply. 4 File lead, mix it with the juice of fresh reed and apply. 5 Grind the dung of wild dove, knead it with honey and apply. 6 Grind the root of sorrel, mix it with pig fat and apply.

177. On viper bites: Grind rue and let him drink the juice. 2 Grind cardamom and let him drink with the juice of rue.

[43] This word is not attested, but it appears to describe a type of fish.

178 ροη′ ⊢πρὸς τὸ διῶξαι θηρία ἐκ τῆς οἰκείας· λυγαίας φύλλα τίθει εἰς τὴν οἰκίαν πανταχόθεν. 2 ἄλλο. σπόρον λυγαίας θυμία τὴν οἰκίαν. 3 ἄλλο. ἐλάφου κέρας θυμίου τὴν οἰκίαν. 4 ἄλλο. καρδάμου σπόρον κάπνιζε τὴν οἰκίαν. 5 ἄλλο. καλαμίνθιν κάπνισέ σου τὸν οἶκον.⊣

179 [248] ροθ′ πρὸς ἡμίκρανον· γῆς ἔντερα ἤτοι σκώληκας μεθ' ὧν οἱ ἁλιεῖς δελεάζουσι τοὺς ἰχθῦς τούτους φρύξας μίξας μετὰ κοκκίων πεπέρεως δεκαπέντε τετριμμένων καὶ μετὰ ὄξους, ἄλειφε τὴν κεφαλήν. 2 ἄλλο. ἀγριοπεριστερᾶς κόπρον τρίψας μετὰ ὄξους λειώσας ἄλειφε. 3 ἄλλο. ⊢σινάπιν καὶ καρδάμου σπόρον τρίψας καὶ μίξας μετὰ ὄξους ἄλειφε.⊣ 4 ἄλλο. λυγαίας σπόρον κοπανίσας καὶ μίξας μετὰ χρηστελαίου ἄλειφε τὸν τόπον καὶ τοὺς μήνιγγας. 5 ἄλλο. ἀσβέστιν ζωντανὸν μετὰ μέλιτος μαλάξας ἄλειφε. 6 ἄλλο. ἀμύγδαλα πικρὰ κοπανίσας καὶ μαστίχην καὶ μετὰ ὄξους μίξας ἄλειφε.

180 [249] ρπ′ πρὸς ἰσχίον· ἐρεβίνθια βράσας μετὰ ὕδατος τρῶγε καὶ πίνε τὸν ζωμὸν αὐτῶν. 2 ἄλλο. σινήπι κοπανίσας καὶ πέπερι καὶ μίξας μετὰ μέλιτος ἄλειφε. 3 ἄλλο. ἀλόην τρίψας καὶ ῥητίνην καὶ πέπερι τίθει ἐπάνω προαλείψας μέλι. 4 ἄλλο. ἑψημένα σκόροδα μετὰ χρηστελαίου καὶ γάρους ἐσθιέτω. 5 ἄλλο. βούτυρον μίξας μετὰ ἀνηθελαίου ἄλειφε τὸν τόπον.

181 [251] ρπα′ περὶ λέπρας· ⊢ἡ λέπρα γίνεται ἐπὶ χυμῷ διαβρωτικῷ ὅθεν καὶ λέπιδες ἀποπίπτουσι ἀπὸ τῆς ἐπιφανείας τοῦ δέρματος μετὰ κνησμοῦ, ἀλλ' ἡ μὲν λέπρα διὰ βάθους ἐπινέμεται τὸ δέρμα κυκλοτερῶς, ἡ δὲ ψωρὰ ἐπιπολεωτέρα ἐστὶ καὶ ποικίλως ἐσχηματισμένη. τούτους δεῖ φλεβοτομεῖν καὶ καθαίρειν διὰ τῶν μέλαινα κενούντων χολήν. ὕστερον δὲ χρηστέον τοῖς ὑπογεγραμμένοις.⊣ ἀσφοδέλου ῥίζης ⊢χυλὸν οὐγγίας ιϛ' θείου ἀπύρου μάννης ἀνὰ ἑξάγια ι' νίτρου ἑξάγια η' χρίε δι' ὄξους. ἀνεμώνης ῥίζα σὺν ὄξει λειωθεῖσα.⊣

---

28 μάννης] scripsi μάνης N 29 ἀνεμώνης] Nutton νεμόνης N

178. ⊢To drive pests away from the house: Put the leaves of chaste tree everywhere in the house. 2 Smoke the house with the seed of chaste tree. 3 Smoke the house with the horn of a deer. 4 Smoke the house with cardamom seed. 5 Smoke your house with peppermint.⊣

179. On migraine: Burn the intestines of the earth or worms which the fishermen use to attract the fish, mix with fifteen ground pepper corns and with vinegar, apply to the head. 2 Grind the dung of a wild dove, make it smooth with vinegar and apply. 3 ⊢Grind mustard and cardamom seed, mix it with vinegar and apply.⊣ 4 Grind the seed of chaste tree, mix it with table oil, apply to the place and the temples. 5 Knead smouldering asbestos with honey and apply. 6 Grind bitter almonds and mastic, mix with vinegar and apply.

180. On sciatica: Boil chick peas with water, eat them and drink the broth. 2 Grind mustard and pepper and mix it with honey and apply. 3 Grind aloe, resin and pepper and apply it, but beforehand anoint it with honey. 4 He should eat boiled garlic with table oil and fish sauce. 5 Mix butter with anise oil and apply to the place.

181. On leprosy: ⊢Leprosy arises from a corrosive humour, which is why bits fall off from the surface of the skin with an itching. But the leprosy stays inside the skin in a circular form, whereas psoriasis is rather located at the surface, and it has a varied shape. You have to let their blood and purge them with something that removes black bile. Then you have to use what is listed below:⊣ asphodelos root ⊢juice sixteen ounces, native sulphur and frankincense ten hexagia each, saltpetre eight hexagia, apply with vinegar. Anemone root made smooth with vinegar.⊣

182 [252] ρπβ΄ πρὸς ψώραν· ἀγριοσταφίδα κοπανίσας καὶ καρδάμου σπόρον καὶ μίξας μετὰ ὄξους καὶ μαλάξας ἄλειφε τὴν ψώραν. 2 ἄλλο. κρίνου ῥίζαν κοπανίσας καὶ κόπρον αἰγὸς καὶ μετὰ ὄξους ἄλειφε. 3 ἄλλο. φύλλα ῥοδάφνης ἑψήσας μετὰ ἐλαίου ἔκβαλε τὰ φύλλα καὶ τότε βάλε κηρίον καὶ ῥοδέλαιον καὶ λιθάργυρον καὶ ψιμμίθιν καὶ ὑγροπίσσιν καὶ ὄξος καὶ μίξας πάντα ἄλειφε.

183 [253] ρπγ΄ πρὸς λειχήνας· ὑγροπίσσιν ἄλειφε τὰς λειχῆνας. 2 ἄλλο. ⊢σίελον ἀνθρώπου ἄλειφε τὰς λειχῆνας.⊣ 3 ἄλλο. δαμασκηνέας δάκρυον ἀποβρέξας νυχθήμερον εἰς ὄξος ἄλειφε. 4 ἄλλο. λυγαίας φύλλα κοπανίσας καὶ μίξας μετὰ ὄξους τίθει ἐπάνω. 5 ἄλλο. καππάρεως φύλλα τρίψας καὶ μίξας μετὰ ὄξους ἄλειφε. 6 ἄλλο. τεάφην τρίψας καὶ μίξας μετὰ ὄξους ἄλειφε.

184 ρπδ΄ ⊢πρὸς τεταρταῖον πυρετόν· κενταύριον πετροσελίνου σπέρμα ἴσα λαβὼν κόψας τε καὶ σήσας δίδου πίνειν μετὰ ἀφεψήματος καλαμίνθης ἢ ῥεφανῶν ἀγρίων. ἑψήσας τὰς ῥίζας πότιζε. καὶ ὑάκινθος δὲ λειωθεὶς καὶ ἐν τῷ στόματι κατεχόμενος τὸν πυρετὸν μειοῖ καὶ ἄδιψον ποιεῖ.⊣

185 ρπε΄ ⊢ἐπίθεμα πρὸς τὸ διώξαι πυρετὸν καὶ γαστέρα προτρέψαι· λαβὼν ἠριγέροντα καὶ ἀλθαίαν κόψας ἀμφότερα ἐν τριβλίῳ βαλὼν καὶ ἀναλύσας ταῦτα μετὰ κρόκου ᾠῶν καὶ ῥοδοστάγματος ἐξ αὐτῶν ἄλειφε τὸν πυρέσσοντα πρωῒ καὶ ἑσπέρας.⊣

186 ρπϛ΄ ⊢πρὸς ῥαγάδας χειρῶν καὶ χειλῶν καὶ ποδῶν· πρὸς ῥαγάδας χειρῶν καὶ χειλῶν καὶ ποδῶν καὶ πρὸς ἱδρωτίδας ἄλειμμα καὶ πρὸς τὰς γινομένας ἐν αὐτοῖς ἐγκαύσεις ἀπὸ τοῦ προσομιλῆσαι ἀνέμων χειμερινῶν ἢ χιόνι καὶ πρὸς τὰς ἐπιπολῆς τοῦ δέρματος κεγχροειδεῖς ἑλκώσεις ἅς τινες ἱδρωτήρια καλοῦσιν.⊣

8 σίελον] scripsi ἱέλον N

182. On psoriasis: Grind stavesacre and cardamom seed, mix it with vinegar, knead and anoint on the psoriasis. 2 Grind lily root and goat's dung and apply with vinegar. 3 Boil the leaves of oleander with oil, remove the leaves, add wax, rose oil, litharge, lead white, liquid pitch and vinegar, mix it all and apply.

183. On lichen: Anoint liquid pitch on the lichen. 2 ⊢Anoint human saliva on the lichen.⊣ 3 Soak gum of a damson tree in vinegar for one night and one day and apply. 4 Grind the leaves of chaste tree, mix with vinegar and apply it. 5 Grind caper leaves, mix with vinegar and apply. 6 Grind sulphur, mix it with vinegar and apply.[44]

184. ⊢On quartan fever: Take the same amount of centaury and parsley seed, grind and sieve, and let him drink with a decoction of peppermint or wild radish. Boil the roots and let him drink. And hyacinth puree decreases fever and makes him less thirsty if it is kept in the mouth.⊣

185. ⊢Salve to decrease fever and induce defecation: Take groundsel and marsh mallow, grind both and put in a bowl, dissolve it with egg yolks and rose oil and anoint the feverish person in the morning and the evenings.⊣

186. ⊢For split hands, lips and feet: A salve for split hand, lips and feet and for the treatment of pustules and for the heat arising in them when they are in contact with winter winds or snow, and for superficial millet like skin ulcerations which some call *hidroteria*.[45]⊣

---

[44] The following chapters are not part of ω. Since most of them differ in style and content, it is very likely that they were added later to one of the manuscripts transmitting ℵ. This is not uncommon — compilations tended to be augmented in the process of copying.

[45] *Hidrōs* is the Greek word for 'sweat'; hence this this might be an inflammation of sweat glands, a serious and recurrent form of acne.

187 ρπζʹ ⊢ἡ σκευασία· νάρδου ζάμβακα ῥοδόσταγμα μοσχέλαιον σμύρναν κηρὸν ἄσπρον θυμίαμα ἀνὰ ἑξάγια τρία, μόσχου ἑξάγιον ἕν. τὸ μὲν λάδανον ἅμα τῷ κηρῷ πρῶτον ἐν λοπαδίῳ βαλὼν ἕψει πυρὶ μαλθακῷ μέχρις ἂν κηρωθῶσιν καὶ ἀναμιγῶσιν ἀλλήλοις. ἀναζεσάντων δὲ τούτων ἐπίβαλε τὴν σμύρναν τὸν λίβανον τὴν μαστίχην τὴν ῥητίνην τὸ θυμίαμα καὶ τὰ λοιπά. ἐπειδὰν δὲ ποτὲ ταῦτα ζέσαντα ἀλλήλοις ἀναμιχθῶσιν ἐπίβαλε τούτοις τὰ ὑγρά. ἀρθείσης δὲ τῆς λοπάδος ἀπὸ τοῦ πυρὸς πρότερον ἐμβάλειν τὸ ῥοδόσταγμα, ἔπειτα τὸ μοσχέλαιον τὸν ζάμβακα τὸ νάρδον. ἐνυπαρχούσης δὲ ἐν τῷ πυρὶ ἀνάπτεται σφόδρα καὶ ὑποχεῖται ἡ ἐν αὐτῇ ὕλη καὶ ἐμπίπτει ἔξω.⊣

188 ρπηʹ ⊢περὶ ἐκβατῶν καὶ ἀνθράκων· ἐκβατὰ καὶ ἀπορυφᾶς καὶ θλάσματα καὶ ἄνθρακας ὠφελεῖ λαπάτου ῥίζα μετὰ ἀξουγγίου ἐπιτιθημένη.⊣

189 ρπθʹ ⊢πρὸς ἄνθρακας τὰ λεγόμενα μεθύστρας· ἔλαιον ἀλείψας τὸν τόπον ἐπίπασσε καλαμίνθιν ἢ κόπρον ἀνθρώπου νεάζουσαν. 2 ἄλλο. σκοροδολάζαρος σὺν ὑγροπίσσῃ καὶ ὀλίγῃ ἅλατι ἑνούμενος τῷ τόπῳ ἐπιθέμενος ὠφελεῖ. 3 ἄλλο. ἔλαιον μετὰ ἅλατος τετριμμένον καὶ καπνίας φούρνου ἢ μαγειρίου ἑνώσας ἐπίθες κατὰ δεσμῶν ἐν πανίῳ. καὶ κεδρίαν μετὰ ἅλατος τετριμμένου ἐπίθες ἐπάνω λύσας. 4 ἄλλο. κρίνου ῥίζαν λεάνας ἐπίθες, ἐπάνω δὲ φύλλα λαχάνου.⊣

190 ρϙʹ ⊢ἄλειμμα τριχοφυές· λαγωοῦ κοιλίαν καύσας καὶ καλλίτριχον τὴν βοτάνην φρύξας ἢ ταύτης μὴ οὔσης ἀβροτόνου τὴν κόμην. ἀμφότερα ἐξ ἴσου σταθμίσας καὶ λαδάνου τὸ ἴσον καὶ ἑνώσας αὐτὰ πάντα προσμίξας καὶ μυρσίνῳ ἐλαίῳ ἐπιχέας καὶ ἀναμαλάξας ὡς γλοιῶδες γενέσθαι ἐπίχριε.⊣

191 ρϙαʹ ⊢πρὸς κεφαλαλγίαν καὶ ἑτεροκρανίαν· ἔμπλαστρον ἡ πανάρετος κατὰ τοῦ μετώπου ἐπιτιθεμένη παύει πᾶσαν κεφαλαλγίαν καὶ ἑτεροκρανίαν. 2 ἄλλο. κοχλίοι χερσαῖοι λειοτριβούμενοι μετὰ τῶν σαρκῶν αὐτῶν καὶ σὺν τῷ λευκῷ τοῦ ᾠοῦ ἑνούμενοι καὶ ἀπὸ μήνιγγος εἰς μήνιγγα χριόμενοι μεγάλως ὠφελοῦσιν.⊣

23 ἄλειμμα] scripsi ἄλειμα N 27 γλοιῶδες] scripsi γλιῶδες N

187. ⊢The recipe: Three hexagia each of nard, mixed spice, rose water, musk oil, myrrh, white wax and incense, one hexagion of musk. First put ladanum together with wax into a sauce pan and cook on a low flame until they are wax like and mingled. When they boil, add myrrh, frankincense, mastic, resin, mixed spice and the rest. And after they have mingled when they boil add the liquids to it. When you have placed the sauce pan over the fire, first add the rose water, then the musk oil, the mixed spice and the nard. If it remains in the fire, it burns heavily, the matter in it will come out and flow outside.⊣

188. ⊢On swelling and anthrax: Applying sorrel root with fat is beneficial for swelling, *aporufai*, bruises and anthrax.⊣

189. ⊢On anthrax which is also called *methustra*[46]: Anoint the place with olive oil, and then put on top peppermint or fresh human feces. 2 Mix garlic with pitch and some salt and it is beneficial when applied to the place. 3 Grind olive oil with salt and mix it with ash from the oven or the cooker and apply it bandaging it with a cloth. And dissolve cedar oil with ground salt and apply it. 4 Make lily root smooth and apply, and on top the leaves of vegetables.⊣

190. ⊢Tincture to induce hair growth: Burn the stomach of a hare and dry the plant called maidenhair; if you do not have it, wormwood leaves. Take the same weight of either of them, and also of ladanum, mingle it all and mix it and pour myrtle oil on top of it, knead it so that it becomes like earth and apply.⊣

191. ⊢On headache and migraine: The Virtue Plaster ends all headache and migraine when applied to the forehead. 2 Land snails ground smoothly together with their flesh and mixed with egg white and applied from one temple to the other are a great help.⊣

[46] This word is not attested in a medical sense. Literarily, it is Greek for 'drunk woman'.

192 ρϙβ′ ⊢πάσμα σταλτικὸν τῶν ἀπὸ κεφαλῆς ῥευμάτων· ἅλας λίτραν μίαν νίτρον τὸ ἔρυθρον λίτρας τρεῖς ἅλας Γάγγρηνον λίτραν μίαν χάλκανθον στυπτηρίαν οὐγγίας η’ πάντα ταῦτα λειώσας καλῶς τριχίνῳ σηθὲς κοσκίνῳ. βαλὼν δὲ εἰς τζουκάλιν ἄθικτον δεδεμένα μετὰ λινοῦ σακουλλίου ἔα σκεπάσας αὐτὰ ἀσφαλῶς ἕως ἂν ἑνωθέντα γένηται ξηρίον ἕν. γίνεται δὲ τοῦτο ἐὰν τοῦ λουτροῦ τῆς καμίνου καρβούνια ἐγχώσῃς αὐτοῖς, ἐκεῖνα καὶ ἐάσῃς καίεσθαι νυκτὸς ὁλόχληρον μίαν. ἐκ τούτου τοίνυν τοῦ ξηρίου λαμβάνων τὶ μέρος καὶ διὰ τῆς τρίψεως χνοοποιῶν ἐπειδὰν ἐξέρχεσθαι μέλλῃς ἐκ τοῦ λουτροῦ τότε ἐπιπάττειν δεῖ τὴν κεφαλὴν μετὰ τῶν τῆς χειρὸς δακτύλων στερῶς ἵνα τὸ ξηρὸν διέλθων τὰς τρίχας εἰς αὐτὸν τῆς κεφαλῆς τὸν χρῶτα διέλθῃ.⊣

193 ρϙγ′ ⊢ἄλλο χρήσιμον πάσμα σταλτικὸν τῶν ἀπὸ κεφαλῆς ῥευμάτων· καφουρᾶς οὐγγίαν α’ καρυοφύλλου κιναμώμου στάχους ἄσαρ στυπτηρίας σχιστῆς κέδρου καρποῦ ἀνὰ οὐγγίαν α’ κηκίδων ἀθρύπτων οὐγγίας β’ νίτρου αἰγυπτιακοῦ οὔγγιας γ’ ἀμιάντου οὐγγίας β’ ἀριστολοχίας μακρᾶς και στρογγύλης ἀνὰ οὐγγίαν α’ πολυτρίχου οὐγγίαν α’ γλήχωνος οὐγγίαν α’ ὀριγάνου οὐγγίας β’ εὐζώμου σπέρματος οὐγγίαν α’ νάπυος ἴρεως ἀνὰ οὐγγίαν α’ κόκκου ἀληθινοῦ οὐγγίας β’. ταῦτα τρίψας καὶ σήσας καὶ ἑνώσας μετὰ τὸ λουθῆναι ἐπίπαττε τὴν κεφαλήν.⊣

194 ρϙδ′ ⊢πρὸς πόνον ἡμικράνου· ἀμμωνιακοῦ λιβάνου εὐφορβίου σμύρνης τραγακάνθης κυμίνου ἀνὰ οὐγγίαν α’ ὄξει ἀναλαμβανέσθω εἶτα ἐπὶ τῆς χείρας ὄξει διάλυε καὶ ἐπὶ τοῦ μετώπου ἐπίχριε.⊣

195 ρϙε′ ⊢πρὸς καυσίδας καὶ φλεγμονὰς καὶ ἀμβλυοπίας καὶ ῥευματισμοὺς καὶ κνησμώδεις καὶ αἱματώδεις διαθέσεις· πηγάνου φύλλα χλωρὰ λειοτριβήσας καὶ ἐκπιέσας τὸν ἐξ αὐτῶν ζωμὸν ἀφελόμενος βάλε ἐν χαλκῷ ἰγδίῳ νύκτα μίαν. εἰθούτως ἐν ἀγγείῳ ὑελίνῳ εἰσάγαγε καὶ χρῶ.⊣

196 ρϙϛ′ ⊢πρὸς ὠταλγίαν ὃ καὶ πρὸς πόνον ὀδόντων δεδοκίμασται· σκόρδων σκελίδια ἑπτὰ καθαρίσας καὶ †μετανανου^{δυ}† λεάνας καὶ ἐλαίου βρόχον ἕνα ἐπιστάξας ἐν πανίῳ τε βαλὼν καινῷ ἐκπίεσον. τὸν δὲ ὀπὸν †ἃς πρωμίῳ† βαλὼν καὶ χλιάνας ἔνσταζε τῷ ἀλγοῦντι ὠτίῳ.⊣

---

5 σακουλλίου] scripsi σακού^{λ} N 22 εὐφορβίου] scripsi εὐφολβίου N 33 ὀπὸν] scripsi ὠπὸν N

192. ⊢Astringent salve for flux from the head: One litre of salt, three litres of red saltpetre, one litre of Gangra salt, eight ounces of astringent blue vitriol, make all this smooth well and sieve it with a fine sieve. Bind it into a sack linen and put it into a new vessel, leave it there with a lid on, until it mingles and becomes one dry substance. This will happen when you add the charcoal from the oven used to heat the bath, and when you leave them to burn over one entire night. Take a part of this dry substance and turn it into a powder by rubbing it; and when you are about to get out of the bath you have to apply it to the head with the fingers of your hand firmly, so that the dry substance permeates the hair and permeates to the skin of the head itself.⊣

193. ⊢Another astringent salve for the flux from the head: One ounce of camphour, one ounce each of clove tree, cinnamon, vermilion, hazelwort, ground astringents and juniper berry, two ounces of whole gall nut, three ounces of Egyptian saltpetre, two ounces of asbestos, one ounce each of long and round birthwort, one ounce of maiden hair fern, one ounce of pennyroyal, two ounces of oregano, one ounce of rocket seed, one ounce each of mustard and iris, two ounces of genuine kermes oak. Grind it, sieve, mix and apply onto the head after bathing.⊣

194. ⊢For headache: One ounce each of ammoniac, frankincense, euphorbium, myrrh, tragakanth and cumin, he should mix it with vinegar, then dilute in the hand with vinegar and apply onto the forehead.⊣

195. ⊢For burning, inflammation, dim-sightedness, rheumatism and conditions related to itching and blood: Grind fresh leaves of rue until they are smooth, press the juice out, remove it and put it into a copper mortar over night. Then transfer it into a glass vessel and use it.⊣

196. ⊢For ear ache, something also famous for tooth ache: Clean seven cloves of garlic and make them smooth with ...... and with olive oil and press it with a noose after you have put it into a new cloth. But let the .... make it smooth and drip into the aching ear.⊣

197 ρϙζʹ ⊢πρὸς τοὺς ἐν τοῖς ὠσὶ σκώληκας· βούτυρον ἄναλον ἡλίου θάλποντος ἀναλύσον τοῦτο ἐν αὐτῷ τῷ ἡλίῳ καὶ βαλὼν εἰς τὸ οὖς γέμισον αὐτό. ἐν τούτῳ γὰρ ἀναπλέῃ ἡ σκώληξ ἡνίκα ὑπὸ τοῦ ἡλίου ἀναθερμανθῇ.⊣

198 ρϙηʹ ⊢πρὸς τοὺς ἔχοντας σκώληκας καὶ πύον· φύλλα καππάρεως κόψας καὶ τὸν ἐξ αὐτῆς ζωμὸν χλιάνας ἔνσταζε εἰς τὸ οὖς.⊣

199 ρϙθʹ ⊢πρὸς πόνους ὤτων δόκιμον· σινήπι ἀπόζεσας εἰς λινοῦν ἔμβαλε καὶ ἀποπιέσας τὸ ἐπιστάζον ὑγρὸν ἕνωσον γυναικείῳ γάλακτι καὶ ἐπίσταζον εἰς τὸ οὖς. 2 ἄλλο. ζοχὶν τρίψας δῆσον εἰς λινοῦν καὶ ἐν ἀτμῷ θερμοῦ ὕδατος ἐπικρέμασον ὥστε μεταλαβεῖν αὐτοῦ τε καὶ τῆς αὐτοῦ θερμότητος καὶ τῆς ποιότητος. εἶτα ἀποπίεσον τοῦτο ἐπὶ τοῦ ἀλγοῦντος ὠτίου. τοσαύτην δὲ ἐχέτω θερμότητα ὅσην ὁ κάμνων δύναται φέρειν τοῦτο πεπείραται. 3 ἄλλο. γλήχωνα καὶ κύμινον βάλε εἰς πῦρ, εἶτα τῇ ἀναθυμιάσει τοῦ γλήχωνος καὶ τοῦ κυμίνου στυππεῖον θερμάνας ἐπίθες. 4 ἄλλο. ὀπίου ὅσον κριθῆς τὸ μέγεθος λύσας σὺν γάλακτι θέρμανον καὶ ἐπίσταξον εἰς τὸ οὖς. καὶ παύει τὸν πόνον. 5 ἄλλο. θεῖον ἄπειρον καθαρὸν βάλε εἰς κάρβουνα καὶ ὑποθυμίασον καὶ παύεται ὁ πόνος.⊣

200 ςʹ ⊢πρὸς πόνον ὀδόντων· πεντανεύρου σπέρμα ἐπιτιθέμενον παύει τὸν πόνον. 2 ἄλλο. ῥητίνης ξηρᾶς πευκίνης ἐπιτιθημένης καὶ τὴν ὀδύνην παύει διακρατουμένης καὶ τὸ σεσηπὼς τῶν ὀδόντων ἀποκαθαίρει. 3 ἄλλο. σεῦτλον μετ' ὄξους ἑνούμενον καὶ τριβόμενον ὠφελεῖ ἐπιτιθέμενον.⊣

201 σαʹ ⊢πρὸς πόνον ὀδόντων ἐν παντὶ καιρῷ διδόμενον· ἀναμαλάξας μαστίχην εἰθούτως πάνυ τρίψας λεπτῶς ἀγρίας σταφίδος κόκκους θ' καὶ βαλὼν μέσον καὶ ἀναδιπλώσας καὶ τῇ χειρὶ μαλάξας καὶ μέρος ὀλίγον κηροῦ λευκοῦ πρὸς τὸ γενέσθαι ἁπαλώτερον. εἰθούτως μάσα καὶ πτύε. τοῦτο παραυτίκα κουφίζει τὸν πόνον καὶ ὅτε μέλλει ἐκβάλειν αὐτό. σὺν τῇ ἐκβαλῇ κλύζε τὸ στόμα μετὰ χλίου πρὸς τὰ κινηθέντα ὑγρὰ ἐξέλθειν.⊣

---

11 ποιότητος] scripsi πιότητος N

197. ⊢For worms in the ears: Dissolve unsalted butter in the warm sunlight, put it into the ear and fill it up. For the worm will swim on its surface when it is warmed up by the sunlight.⊣

198. ⊢For those with worms and pus: Grind the leaves of capers, warm their juice up and drip it into the ear.⊣

199. ⊢Famous for ear ache: Boil mustard, put it into a cloth, press the liquids out, mix it with women's milk and drip it into the ear. 2 Grind thistle, bind it into a linen cloth and let it hang in the vapour of warm water, so that it acquires its warmth and consistency. Then compress it over the aching ear. It should be as warm as the patient can stand, this is approved. 3 Put pennyroyal and cumin into a fire, then use the vapours of the pennyroyal and the cumin to warm up hemp and apply it. 4 Dissolve opium, about the amount of one barley corn, warm it in milk and drip onto the ear. And the pain stops. 5 Put native sulphur on charcoal, apply a vapour bath and the pain stops.⊣

200. ⊢For tooth ache: Plantain seed stops the pain when applied. 2 If you apply dried resin of a pine tree and keep it there it will stop the pain, and it also cleanses rotting teeth. 3 Mix beet root with vinegar and grind it and it is beneficial when applied.⊣

201. ⊢For tooth ache at all times: Knead mastic, and then grind nine fruits of stavesacre until they are smooth and put it in the middle and fold it over and knead it by hand with a small bit of white wax so that it gets softer. And then chew it and spit it out. This relieves the pain immediately also when he is going to spit it out. After spitting it out rinse the mouth with warm water so that the liquids that have been moved come out.⊣

202 ϛβʹ ⊢πρὸς πόνον λαιμοῦ· ὠφελεῖ καὶ εἰς συνάγχην καὶ εἰς ῥευματιζόμενον θώρακα, εἰς κυνάγχην καὶ βῆχαν καὶ δύσπνοιαν καὶ ὀρθόπνοιαν, πρὸς στόμαχον φυσώμενον καὶ πνευματούμενον καὶ τοὺς ἀπὸ κεφαλῆς τὸν θώρακα ῥευματιζόμενον. ἔστιν καὶ γυναικειῶν ἀγωγόν. 2 ἔχει δὲ οὕτως· σελίνου σπόρον οὐγγίας δʼ μαράθρου σπέρμα οὐγγίας γʼ πεπέρεως οὐγγίας βʼ εὐζώμου σπέρμα οὐγγίαν αʼ σίνωνος συριακοῦ δραχμὰς βʼ μέλιτος τὸ ἀρκοῦν.⊣

203 ϛγʹ ⊢πρὸς συνάγχην καὶ ῥευματιζομένους τὸν θώρακα καὶ πρὸς κατάρρουν καὶ βήσσοντας καὶ ῥέγχοντας· ἀμμωνιακὸν ὅσον μέγεθος φάβατος μασσώμενον πρὸς τὸ πεπονθὸς μέρος καὶ καταπινόμενον οὐ μόνον ἴαται ἀναστέλλοντα ῥεύματα ἀλλὰ καὶ τὴν κοιλίαν προτρέπει. ἔστι δὲ καὶ πεπτικὸν μετὰ δεῖπνον διδόμενον καὶ ἐμπνευμάτωσιν σπληνὸς παύει.⊣

204 ϛδʹ ⊢ὀσφραντὸν πρὸς κατάρρουν καὶ βῆχας· πυρέθρου κόστου ἴσα τρίψας βαλὼν εἰς λινοῦν ἑσπέρας ὀσφραίνου.⊣

205 ϛεʹ ⊢κατάπλασμα εἰς ῥέγχοντας καὶ ῥευματιζομένους τὸν θώρακα· μέλι χλιάνας ἄλειφε τὸν θώρακα καὶ ἐπίπασσε γλήχωνα τετριμμένον καὶ ἐπάνω ἐπίθες μαλίον.⊣

206 ϛϛʹ ⊢πρὸς χρόνιον βῆχα διὰ πεῖρας· ὑσσώπου Κρητικοῦ γλήχωνος κόμης ἐφίσης βαλὼν τρίψας τε καὶ σήσας καὶ μέλιτι ἀπεφρυσμένῳ ἑνώσας χρῶ. τοῦτο καὶ πρὸς τοὺς τὸν θώρακα ῥευματιζομένους ἐνεργέστατον.⊣

207 ϛζʹ ⊢περὶ ἀποφλεγματικῶν καὶ πταρμικῶν· λαβὼν ὑδροσάτον βάλε ἐν χύτρᾳ καὶ ὀρίγανον καὶ ἀγριοσταφίδα καὶ ἰσχάδας καὶ γλήχωνα καὶ πύρεθρον καὶ μαστίχιν τετριμμένον καὶ ἑνώσας χρῶ. 2 ἄλλο. κυκλάμινον τρίψας καὶ στρούθιον ὁμοίως καὶ μόσχον καὶ μετὰ τὸ λεᾶναι σήσας ὀσφραίνου.⊣

208 ϛηʹ ⊢εἰς ἐπιληπτικοὺς βοήθημα· πύρεθρον κόψας καὶ λεπτῶς κινήσας πότιζε ἐξ αὐτοῦ ἑξάγιον ἓν μετὰ ἁγιάσματος τῶν ἁγίων θεοφανίων ἕως ἡμέρας ιαʼ ἐν ἑκάστῃ ἡμέρᾳ ἑξάγιον ἕν. τοῦτο δόκιμόν ἐστι προφυλακτικόν.⊣

12 πεπτικὸν] scripsi πεμπτικὸν N 20–21 ἀπεφρυσμένῳ] scripsi ἀπεφρισμένῳ N

202. ⊢For a sore throat: It is also beneficial for a sore throat, flux from the chest, angina, cough, difficulty breathing, if breathing is only possible in an upright position, for a bloated stomach with wind and those suffering from a flux from the chest caused by the head. It also induces periods. 2 It is like this: four ounces of celery seed, three ounces of fennel seed, two ounces of pepper, one ounce of rocket seed, two drachmae of Syriac parsley and the appropriate amount of honey.⊣

203. ⊢For tonsillitis, flux from the chest, catarrh, coughing and wheezing: Ammoniac, the amount of a bean, applied onto the ailing part and drunk does not only cure uprising humours but also encourages the stomach. It also supports digestion after a meal and stops the inflation of the spleen.⊣

204. ⊢Inhalation for catarrh and cough: Grind pellitory and costus in equal parts, put it into a linen cloth and inhale in the evening.⊣

205. ⊢Salve for wheezing and flux from the chest: Warm honey and apply it to the chest and put ground pennyroyal on top and on top put hair.⊣

206. ⊢Approved remedy for chronic cough: Put together hyssop from Crete, pennyroyal and hair, grind it, sieve it, add dried honey and use it. This is also beneficial for flux from the chest.⊣

207. ⊢To cleanse from phlegm and to treat sneezing: Take dew, put it into a jug, and ground oregano, stavesacre, dried figs, pennyroyal, pellitory and mastic, mix it and use it. 2 Grind violet and soap wort and also musk, make it smooth, sieve it and inhale.⊣

208. ⊢Medication for epileptics: Grind pellitory and move it until it is smooth and let him drink one hexagion of it with Holy water for eleven days, every day one hexagion. This is a famous prophylactic.⊣

209 ϛθʹ ⊢πρὸς πόνον καὶ στρόφον στομάχου· κυμίνου καὶ πεπέρεως δάφνης φύλλων τρυφερῶν κόψας αὐτὰ καὶ σήσας πότιζε μετὰ οἴνου. 2 ἄλλο. ἀμμωνιακὸν μετὰ τὸ δεῖπνον τρωγόμενον οὐ μόνον πέπτει ἀλλὰ καὶ κοιλίαν ὑπάγει καὶ ἐμπνευματώσεις διαφορεῖ.⊣

210 σιʹ ⊢βοήθημα στομαχικὸν διδόμενον βραδὺς πεπτοῦσιν ἢ καὶ ἀπεπτοῦσι καὶ τοῖς φυσωμένοις τὸν στόμαχον· γομφίτου λίτραν α' κρόκου ἀληθινοῦ οὐγγίαν α' πλατυκυμίνου πεπέρεως κοινοῦ ἀνὰ ἐξάγια θ' πεπέρεως λευκοῦ ἐξάγια β' πεπέρεως μακροῦ ἐξάγια δ' ὑσσώπου Κρητικοῦ ἐξάγια β' μακεδονησίου σπέρματος ἐξάγια β' ἄμεως ἐξάγια β' μαράθρου σπέρματος καὶ ἀνίσου ἀνὰ ἐξαγίου ἑνός. ἀμφότερα λειώσας καὶ μέλιτος τὸ ἀρκοῦν ἀναλαβόμενος χρῶ ἐξάγιον μεθ' ὕδατος.⊣

211 σιαʹ ⊢κοκκία στομαχικά· ἀλόης ἐξάγιον ἓν κιναμώμου ἐξάγιον ἓν μαστίχιν ἐξαγίου τὸ ἥμισυ ῥόδον ἀληθινὸν ὁμοίως μετὰ χυλοῦ ἀψινθίου.⊣

212 σιβʹ ⊢κοκκία ποδαλγικά· μαλάσσουσι τὴν γαστέρα καθαίρουσι χολὴν μέλαιναν καὶ ξανθὴν καὶ φλέγμα, οὐκ ἐῶσί τε ῥευματισθῆναι τοὺς πόδας ἢ τὰς χεῖρας οὐδὲ τὸ λοιπὸν σῶμα λαμβανόμενα ἅπαξ ἢ δὶς τῆς ἑβδομάδος. ἡ δὲ σύνθεσις αὐτῆς ἥδε. 2 ἡ σκευασία· ἀλόης ἡπατικῆς οὐγγίας γ' μαστίχης οὐγγίαν α' ἀγαρικοῦ οὐγγίας ἥμισυ ἐπιθύμου οὐγγίας ἥμισυ ῥέου ναρδοστάχυος φύλλα ἀνὰ δραχμῶν ἕξ. οἴνῳ καλῷ ἀναλάμβανε καὶ ποίει κοκκία ὀροβιαῖα. τὴν γραφὴν αὐτῶν ἔλαβον παρὰ Μαριανοῦ στρατηλάτου.⊣

213 σιγʹ ⊢κατάπλασμα εἰς τοὺς ὀδυνωμένους τὸ ἰσχίον· γλήχωνος σινήπεος ἀμφοτέρους τρίψας ἐν τριβλίῳ μετ' οἴνου ἕψει καὶ κατάπλασσε τὸ ἰσχίον καὶ ἐπιδέσμει. 2 ἄλλο. πύρεθρον τρίψας καὶ νάρδῳ ἑνώσας κατάπλαττε. τοῦτο δὲ καὶ ἐπὶ παντὸς μορίου ναρκωθέντος χρῶ.⊣

214 σιδʹ ⊢πρὸς πόνον ἰσχίου· ἀμπέλου κλήματα καύσας καὶ τὴν τέφραν μετὰ πίσσης ὑγρὰς ἀναζυμώσας ποίει ὡς ἔμπλαστρον ἢ ἐν βαλανείῳ μετὰ τὸ λούσασθαι ἐπιτίθει εἰς τὴν ψώραν καὶ εἰς τὸ ἰσχίον καὶ ἄπονος ἔσῃ.⊣

209. ⊢For stomach pain and cramps: Grind some cumin, pepper and fat bay leaves, sieve it and let him drink with wine. 2 Ammoniac eaten after a meal does not only digest but also empties the stomach and dissolves bloating.⊣

210. ⊢Medication for the stomach given to those who suffer from indigestion, do not digest or who have a bloated stomach: One litre of storax, one ounce of true saffron, nine hexagia each of flat cumin and common pepper, two hexagia of white pepper, four hexagia of long pepper, two hexagia of hyssop from Crete, two hexagia of parsley seed, two hexagia of ajowan seed, one hexagion each of fennel seed and aniseed, both ground smoothly. Add the appropriate amount of honey. Use one hexagion with water.⊣

211. ⊢Pills for the stomach: One hexagion of aloe, one hexagion of cinnamon, half a hexagion of mastic, real rose, make it the same way with wormwood juice.⊣

212. ⊢Pills for gout: They make the stomach smooth and cleanse black and yellow bile and phlegm, and they will not allow for the feet or hands to suffer from a flux, or the rest of the body, if taken once or twice a week. This is the recipe: 2 The recipe: Three ounces of hepatic aloe, one ounce of mastic, half an ounce of agaric, half an ounce of epithymos, six drachmae each of rhubarb and nard leaves. Add good wine and make pills of the size of vetch. I received this instruction from Marianos Stratelates.⊣

213. ⊢Salve for those suffering from sciatica pain: Ground both pennyroyal and mustard in a bowl, boil with wine, apply to the hip and bind it. 2 Grind pellitory, mix it with nard and apply. The same use also for all numb parts.⊣

214. ⊢For sciatica pain: Burn the vine twigs and mingle the ash with liquid pitch and make a plaster or apply it in the bath after washing on the psoriasis or the hip and he will be free of pain.⊣

215 σιε´ ⊢ἐμβροχὴ εἰς ῥευματιζομένους καὶ ὀδυνωμένους τὸ ἰσχίον· κυπρίνου ἐλαίου ὄξους ἀνὰ οὐγγίαν α' κέγχρυος κεκομμένης ἑξάγιον ἓν ἑνώσας πάντα ἐν τριβλίῳ καὶ ἀναζέσας πάντα μετὰ ἐρίων ἢ μαλίων πυρίαζε τὸν ὀδυνώμενον τόπον. 2 κατάπλασμα εἰς τὸ αὐτό· θερμῶν πικρῶν μετὰ μέλιτος καὶ ὄξους τοῦ ἀρκοῦντος μίξας ἕψησον καὶ κατάπλασσε.⊣

216 σιϛ´ ⊢ἄλλο εἰς ὀδύνην ἰσχίου· ῥίζης ἀγριοκαρδάμου τὸν φλοῦν ἀνελόμενος κόψον αὐτὸν λεῖαν καὶ μετὰ στέατος χοιρείου ἀναμίξας ἅπλωσον εἰς πανίον καὶ ἐπιτίθει εἰς τὸ ἰσχίον ἔνθα ἡ ὀδύνη καὶ ἔα φορεῖν αὐτὸ ὥρας τρεῖς καὶ μετὰ ταῦτα πρόσταξον παρευθὺς τὸν κάμνοντα ἐν τῷ βαλανείῳ γινόμενον λουθῆναι καλῶς.⊣

2 κέγχρυος] scripsi κάγχρυος N 8 λεῖαν] scripsi λίαν N

215. ⊢Infusion for rheumatism and sciatica pain: One ounce each of henna oil and vinegar, one ounce of ground millet mix it all in a bowl, let it boil with wool or hair and apply a vapour bath to the aching place. 2 Salve for the same: Mix warm and bitter ingredients with the appropriate amount of honey and vinegar, boil and apply.⊣

216. ⊢Another one for sciatica pain: Remove the bark of the root of wild cardamom, ground it until smooth, mix it with pig fat, spread it on a cloth and put it on the hip where the pain is and let him wear it for three hours and after that send the patient directly to the bath where he can bathe well.⊣

PART THREE

EDITION OF ω

▷MA ἐν τῇδε τῇ βίβλῳ περιέχονται κεφάλαια θεραπευτικὰ τῶν ἀδήλων, φανερῶν ἐπὶ πεπονθότων τόπων σν' MA◁

▷Σ ⌜ἀρχὴ σὺν θεῷ⌝ τῆς παρούσης δέλτου καὶ ἑρμηνία καὶ ⌜διδασκαλία⌝ καὶ ἐξήγησις ⌜τοῦ θαυμασιωτάτου⌝ καὶ λογιωτάτου ⌜ ἀρχιατροῦ τοῦ Γαληνοῦ τοῦ σπεύσαντος εἰς τὰ τῶν μορίων πάθη ἀνίατα καὶ δυσίατα⌝ λέγω ἅμα καὶ ⌜εὐίατα γενέσθαι⌝ ὠφέλειαν καὶ ἴασιν. ⌜πρῶτον μὲν κοινῷ λόγῳ κεφαλαιωδῶς εἰπεῖν, περὶ τὴν παντὸς σώματος ἐπιμέλειαν εἶτα καὶ κατὰ τὸν πεπονθότα τόπον γενομένην ῥαστώνην ἥτις παρὰ τῶν ἰατρῶν καλεῖται ὠφελεία καὶ θεραπεία καὶ πρῶτον μὲν διδάσκει περὶ τῶν ἐκτὸς παθῶν, ὅτι εἰσὶν εὐίατα⌝ κατὰ τοὺς πεπονθότας τόπους· καὶ ἐφ' ἑξῆς μέλλει ⌜καὶ περὶ τῶν ἐντὸς εἰπεῖν,⌝ τῶν ⌜δεομένων τὴν θεραπείαν πολυχρόνιον καὶ ἐπιτήδιον πρὸς τὸ μῆκος τῆς ὁδοῦ τοῦ πάσχοντος τόπου,⌝ ὅτι ⌜οἱ ἐντὸς πεπονθότες τόποι γεγυμνασμένου δεῖνται⌝ ἀνδρὸς ⌜εἰς⌝ τὰ τῆς ἰατρικῆς μέρη φημί, εἶτα καὶ τὰ διαστήματα τοῦ ἀλγοῦντος τόπου· φησὶ γὰρ ὁ πάντων τῶν καλῶν ἡγεμὼν Ἱπποκράτης, ὅτι οὐ δεῖ ⌜τὸν ἰατρὸν γινώσκειν μόνον τὴν κοινὴν φύσιν,⌝ ἁπάντων λογικῶν ζώων φημὶ καὶ ἀλόγων, ⌜ἀλλὰ καὶ τὴν ἰδίαν.⌝

κοινὴ γὰρ φύσις ἐστὶν ἐπὶ πᾶσιν, ὅτι πάντα συνίστανται ἀπὸ τεσσάρων στοιχείων ἤγουν θερμὸν, ψυχρὸν, ξηρὸν, ὑγρόν. θερμὸν μὲν λέγεται τὸ πῦρ, ξηρὸν ἡ γῆ, ψυχρὸν τὸ ὕδωρ, ὑγρὸν τὸ ἔαρ. καὶ πάλιν θερμὸν μέν ἐστιν ἡ ξανθὴ χολή, ἥτις καὶ ξηρὰ ὑπάρχει· ξηρὸν ὁ μελαγχωλικὸς χυμός, ὅστις καὶ ψυχρὸς εἴη· ὑγρὸν τὸ

---

1 ἐν] ... A 1–2 ἀδήλων, φανερῶν] om. A 2 πεπονθότων] ποθότων M 2 σν'] σνβ' A 5 μορίων] μωρίων M 6 καὶ δυσίατα] om. A 6 γενέσθαι] γενέσθω εἰς A 6 ἴασιν] ἴσιν M καὶ add. A 7 μὲν] καὶ add. M εἰς add. B 7 τὴν] τοῦ add. AB 9 ἥτις παρὰ τῶν ἰατρῶν καλεῖται ὠφελεία καὶ θεραπεία] om. A 10–11 κατὰ τοὺς] om. A 11 πεπονθότας] πεπονθότα A 11 τόπους] πάθη B πάθη καὶ A 11 καὶ περὶ] πλὴν B 12 καὶ] om. B 12 ἐπιτήδιον] ἐπὶ τίδιον M ἐπὶ A 13 πρὸς τὸ μῆκος τῆς ὁδοῦ] om. A 13 ὅτι] om. B 14 τόποι] τόπους M 14 δεῖνται] δέονται B 15 τὰ] τοῦ B 16 οὐ δεῖ] οὐδεὶς A 17 τὸν ἰατρὸν] τῶν ἰατρῶν A 17 γινώσκειν] γινώσκει A 17 ἁπάντων] τῶν add. AB 18 ἀλόγων] καὶ τῶν λοιπῶν add. MB 19 ὅτι] ἐν ᾧ οἱ A 19 πάντα] πάντες A τὰ πάντα B 19 ἀπὸ] ὑπὸ A 20 θερμὸν] θερμοῦ B 20 ψυχρὸν] ψυχροῦ B 20 ξηρὸν] ξηροῦ B 20 ὑγρόν] καὶ ὑγροῦ καὶ B καὶ add. A 21 μὲν] om. B 22 ἐστιν] om. B 23 μελαγχωλικὸς] μελαγχονικὸς MB 23 εἴη] ἐστὶ B

αἷμα, ὅπερ καὶ θερμὸν ὀλίγον. ὅσον πρὸς τὴν ξανθὴν χολήν, ⌜ἡ δὲ ξανθὴ χολὴ⌝ καθ' ὃ φύσει κούφη καὶ λεπτὴ καὶ ποθοῦσά τὴν ἄνω φοράν. ⌜εἴωθεν ποιέειν τὰ τριταϊκὰ νοσήματα. ἐν θερμοτέροις σώμασι καὶ ξηροτέροις.⌝ [ἐστὶ δ' ὅτε καὶ ἐν σώμασι θερμοτέροις καὶ ὑγροτέροις.] ⌜ὁ μελαγχολικὸς χυμός εἴωθεν ποιεῖν μὴ μόνον τοὺς τεταρταίους πυρετούς, ἀλλὰ καὶ τὰ μελαγχολικὰ νοσήματα, ἄλφους, λέπρας, ψώρας καὶ κνησμονάς. καὶ τὰ ἄλλα τούτοις ὅμοια. γίνονται δὲ καὶ ἐν σώμασι ψυχροτέροις καὶ ξηροτέροις. τὸ φλέγμα ποιέει τοὺς ἀφημερινοὺς πυρετούς. ὅτι ἅμα τῷ παρακμᾶσαι τὸν πρότερον παροξυσμὸν εἰσβάλλει ὁ ἕτερος παροξυσμός. γίνεται ἐν σώμασι ψυχροτέροις καὶ ὑγροτέροις⌝ ὡς ἡ γεροντικὴ ἡλικία καὶ τὰ τῶν βρεφῶν σώματα. [συμβαίνειν εἴωθε τὸν πυρετὸν τοῦτον τρόπῳ τοιῷδε, ὡς ὁ λογιώτατος ἰατρὸς φησὶ ὁ Γαληνὸς ▷om. A ὅτι τὰ παιδία μηδὲν ἄλλο εἰδότες τῶν κατὰ τὸν βίον πλὴν ἐσθίειν, ἀδήφαγα γὰρ τὰ παιδία. om. A◁] ⌜τὸ αἷμα ποιεῖ τοὺς συνόχους πυρετούς, ὅπερ καὶ θερμὸν ὑπάρχει καὶ ὑγρόν. γίνεται γὰρ ἐν σώμασι πολυαίμοις καὶ μὴ ἐχόντων ἔθος κενοῦν⌝ τὸ λυποῦν καὶ πλεονάζον αἷμα. ἰδία δ'

1 ὀλίγον] om. B 2 καθ' ὃ] καθόλου A 2 κούφη] om. B 2 τὴν] τι A 3 τριταϊκὰ] τρίτα ἰκὰ MB τρίτα ἱκανὰ A 3 θερμοτέροις] θερμωτέροις M 4 σώμασι] καὶ κουφοτέροις add. A 4 ἐστὶ δ'] ἔστιν A ἔστι δὲ B 4–5 θερμοτέροις καὶ] om. B 5 ὁ] om. M 5 ποιεῖν] ποιέειν B 6 τεταρταίους] τετάρτους M 6 μελαγχολικὰ] scripsi μελαγχονικὰ M μελαγχωνικὰ A μελικὰ B 7 καὶ] om. B 7 κνησμονάς] κνιμονὰς M 7 τούτοις] τούτης B 8 γίνονται] γίνεται B 8 ξηροτέροις] ξυροτέροις M 9 ποιέει] ποιεῖν A 9 τῷ] scripsi τὸ MAB 9 παρακμᾶσαι] παρεκμᾶσαι MA 10 εἰσβάλλει] εἰσβάλλη M εἰσβάλει B 10 παροξυσμός] καὶ add. A 10 γίνεται] δὲ add. B 11 ψυχροτέροις] ὑγροτέροις B καὶ ξηροτέροις add. A 11 ὑγροτέροις] ψυχροτέροις B 11 ἡ γεροντικὴ ἡλικία καὶ τὰ] om. B 11 ἡλικία] λικία A 12 τῶν] ἐν τοῖς τῶν B 12 σώματα] σώμα.τ. A σώμασι καὶ ἐν τῇ γεροντικῇ ἡλικίᾳ B 12 συμβαίνειν] συμβαίνει B 12 εἴωθε] γίνεσθαι add. A 12 τὸν] ὁ B 12 πυρετὸν τοῦτον] πυρετὸν add. M πυρετὸς οὗτος B 13 τοιῷδε] τοιούτῳ B 13 ὁ Γαληνὸς] ὁ Γαλινὸς M Γαληνὸς A om. B 13 ὅτι] om. B 14 μηδὲν] οὐδὲν B 14 εἰδότες] εἰσὶν εἰδότα B 14 τῶν] om. B 14 ἀδήφαγα] ἀδδηφάγα MB 15 τὸ] δὲ add. B 15 συνόχους] συνέχους B 15 καὶ] om. AB 16 γὰρ] δὲ B 17 ἐχόντων] ἔχουσιν B 17 κενοῦν] κενεῖν M ἐκκενοῦν B 17 λυποῦν] λοιποῦν B 17 πλεονάζον] πλεονάζων B 17 ἰδία] ἰδίαι A ἰδέαι B 17 δ'] δὲ B ὑπάρχουσι add. A

φύσις, ὅτι ὁ μὲν ἄνθρωπος ὑπάρχει θερμός, ὁ δὲ ψυχρός, ὁ δὲ ξηρός, ὁ δὲ ὑγρὸς ἢ κατὰ συζυγίαν. ὅτι ὁ μὲν πάλιν τῇ κράσει ξηρὸς καὶ θερμός, ὁ δὲ ξηρὸς καὶ ψυχρός, ὁ δὲ θερμὸς καὶ ὑγρός. καὶ πάλιν ὁ αὐτὸς ⌜῾Ιπποκράτης φησὶν ὅτι ὁ γὰρ ἄριστα διαγνοὺς, ἄριστα καὶ θεραπεύει ἤγουν ὁ ἰατρὸς, ὅπου γινώσκει, ὅτι τὸ σῶμα τοῦδε τοῦ νοσοῦντος ὑπάρχει θερμόν. καὶ νῦν ὑπάρχει θερμὴ δυσκρασία. καὶ μὴ μόνον, οὐ ποιεῖ ὄφελος, ἀλλὰ καὶ τὸν ἄνθρωπον ἀπάγει ταχεώς εἰς μόρον ἤτοι θάνατον, διότι⌝ ἐκμοχλεύει ⌜τὸ ἀλλότριον καὶ τὸ οἰκεῖον θερμόν,⌝ εἰς ὃ ἔζη τὸ σῶμα. ⌜οὕτως καὶ ἐπὶ τῶν ἄλλων δυσκρασιῶν, ἀλλὰ τὰ ἐναντία ποιέειν, ὡς εἰκός. τὰ γὰρ ἐναντία τῶν ἐναντίων ἰάματα. κολάζονται μὲν τὸ ὑπερβάλλον, ἀντισάγοντα δὲ τὸ λυποῦν.⌝ ▷om. Bκαὶ πάλιν τὰ ἐναντία τῶν ἐναντίων ἰάματα.om. B◁ ἤγουν τὴν θερμὴν δυσκρασίαν ψύχειν, τὴν δὲ ψυχρὰν θερμαίνειν, τὴν δὲ ξηρὰν ὑγραίνειν, ▷om. Bτὴν δὲ ὑγρὰν ξηραίνειν.om. B◁ κατὰ συζυγίαν ἤγουν τὴν μὲν θερμὴν καὶ ξηρὰν δυσκρασίαν ψύχειν καὶ ὑγραίνειν, τὴν ψυχρὰν καὶ ὑγρὰν θερμαίνειν καὶ ξηραίνειν.Σ ◁

---

1 φύσις] φύσεις A φήσις B 1 ἄνθρωπος] om. A 1 ὁ] ἡ M 2 πάλιν] ἐστὶ A ὑπάρχει add. B 2 κράσει] κράση MB 3 ὁ] ἡ M 4 γὰρ] om. B 4 διαγνοὺς, ἄριστα καὶ] om. A 5 ἤγουν] om. A 5 ἰατρὸς] ἠατρὸς B 5 τὸ] τω M 6 νοσοῦντος] ἤγουν τοῦ ἀρωστοῦντος add. B 6 νῦν] εἰ add. B 6 θερμὴ δυσκρασία] θερμὴν δυσκρασίαν M χρῆται τῇ ψυχοῦσῃ ἀγωγῇ ὡσ εἰκὸς ἤγουν ἐνδεχόμενον ἵνα μὴ κατὰ ψύξει πολλὰ τὴν θερμὴν δυσκρασίαν B 6 καὶ] ἢ A 7 οὐ ποιεῖ] ο᾽ποι εἰ B 7 καὶ] om. M 7 ἄνθρωπον ἀπάγει ταχεώς] ἄν̄...αχέως ἀπάγει A 7 ἀπάγει] εὐπάγει B 8 ἤτοι θάνατον] om. A 8 ἤτοι] εἰς add. B 8 καὶ] om. A 9 οἰκεῖον θερμόν] οἰκεῖ ἀνθερμήν A 10 ἐναντία] ἐναντίαν A 10 ποιέειν] πιέειν M 10 εἰκός] εἰκῶς M 11 κολάζονται] κολάζοντα A, κολάζον᾽ B 11 ἀντισάγοντα] ἀντ᾽ εἰσάγον τὰ B 12 λυποῦν] scripsi λοιπὸν MB λεῖπον A 12 τῶν] ἐναντία add. M 13 δυσκρασίαν] εὐκρασίαν B 13 ψύχειν] ψυχρόν A 13 δὲ] om. B 13 ψυχρὰν] θερμὴν M 13 θερμαίνειν] ψύχειν M θερμένη B 14 δὲ] om. B 14 ὑγραίνειν] ὑγραίνη B 14 ξηραίνειν] ὑγραίνειν M 14 κατὰ] καὶ κατὰ B 15 μὲν] om. B 15 καὶ ξηρὰν] om. B 15 δυσκρασίαν] εὐκρασίαν B 15 ψύχειν] ψύχην B 16 ὑγραίνειν] ὑγραίνην B 16 θερμαίνειν] θερμαίνη B 16 ξηραίνειν] τὴν θερμὴν καὶ ὑγρὰν ψύχειν καὶ ξηραίνειν B

▷om. L λόγος καὶ ποίημα καὶ προοίμιον Ἰωάννου ἀρχιατροῦ περιέχων συνοπτικῶς πάντων τῶν παθῶν καὶ τῶν ἀδήλων τὰς θεραπείας om. L◁

1 [1] α′ ⌜πρὸς ὀξὺν πόνον κεφαλῆς· κισσὸν ξηράνας ἢ καὶ χλωρὸν κοπανίσας ἀπόβρεχε εἰς ἔλαιον. ὀθονίῳ δὲ διηθήσας χρίε τὸ μέτωπον καὶ τοὺς κροτάφους.⌝ τουτέστιν ἔπαρε κισσὸν καὶ ξήρανον αὐτὸν εἰς ἥλιον. εἶτα κοπάνισον καὶ ποίησον αὐτὸν οἷον ἀλεύριν. καὶ πάλιν βάλε κισσὸν χλωρὸν εἰς ἔλαιον ἡμερόνυκτον. καὶ ἂς βρέχηται πλὴν τὰ φύλλα. εἶτα ῥίψας τὰ φύλλα ἔπαρε τοῦ ἄλλου κισσοῦ τοῦ ξηροῦ τὸ ἀλεύριν καὶ ἕνωσε αὐτὰ με τὸ ἔλαιον τοῦ χλωροῦ κισσοῦ. καὶ τὸ ἔλαιον ἄλειφε τὸ μέτωπον καὶ τοὺς κροτάφους ἤτοι τοὺς μήλιγγας. 2 ἄλλο. ⌜ἐλάφου ὀστοῦν καῦσας λέανον σὺν ῥοδίνῳ ἐλαίῳ καὶ χρίε.⌝ ▷om. M ἤτοι ἀφ᾽ ὅτου καύσης τοῦ ἐλάφου τὸ ὀστοῦν, τρίψον καὶ ποίησον αὐτὸ οἷον τὸ ἀλεύριν. εἶτα μίξον αὐτὸ μετὰ τοῦ ῥοδελαίου καὶ ἄλειφε τὸ μέτωπον καὶ om. M◁ τοὺς μήλιγγας.

2 [2] β′ πρὸς πόνον κεφαλῆς καὶ ἡμικράνου· ⌜κάρδαμον ὄξει καὶ ῥοδίνῳ ἐλαίῳ φυρώσας καὶ ποιήσας πάχος κηρωτῆς κατάχριε.⌝ ἤτοι ἐπάρας κάρδαμον καί κοπάνισον αὐτὸ καὶ ποίησον οἷον τὸ ἀλεύριν.

---

1 ποίημα καὶ] πο......... A 1 ἀρχιατροῦ] ἀρχιητροῦ M ἀρχ...τροῦ A 1–2 περιέχων] om. B 2 ἀδήλων] καὶ φανερῶν add. A^2 2–3 θεραπείας] πρὸς ἓν ἕκαστον τὴν τάξιν add. B ἀρχὴ τῶν κεφαλαίων add. A 5 ξηράνας] ξηρὸν M 5 καὶ] om. Σ 6 διηθήσας] δηθήσας M διαθήσας A 7 καὶ τοὺς] om. M 7–8 ξήρανον αὐτὸν] ξήρανέ τον MA ξήρατον B 8 κοπάνισον] κοπάνισέ τον MA αὐτὰ add. B 8 ποίησον] ποίησε A 8 αὐτὸν] om. MB τὸν A 8 οἷον] τὸ add. B 8 ἀλεύριν] ἄλευρον M 9 ἡμερόνυκτον] om. A 9 ἂς] .. A 9–10 βρέχηται] ἡμερονύ.ιον add. A 10 εἶτα ῥίψας τὰ φύλλα] καὶ Σ 10 τοῦ ἄλλου] καὶ τοῦ M 11 τοῦ] om. MA 11 ξηροῦ] om. A 11 ἀλεύριν] ἤγουν τοῦ ξηροῦ add. A 11 αὐτὰ] το M τα AB 11 με τὸ ἔλαιον] μετὰ τοῦ ἐλαίου Σ 13 τοὺς μήλιγγας] τοὺς μήνιγγας MA fenestra .:. B 13 ὀστοῦν] ὀστοῦν add. A 13 λέανον] ἤτοι τρίψον add. M 14 ὅτου] οὗ B 14 καύσης] καύσῃ A 15 ὀστοῦν] ὀστέον A 15 τρίψον] τρίψε το AB 15 αὐτὸ] τοῦτο A τοῦτον B 15 μίξον αὐτὸ] μίξε το A μίξας B 16 μετὰ τοῦ] μετ. A με τὸ B 16 ῥοδελαίου] ῥοδέλαιον AB 16 τὸ μέτωπον καὶ] om. A 16 τοὺς μήλιγγας] τοὺς μύνιγγας καὶ τὸ μέτωπον M καὶ τὸ μέτωπον add. A 18 ῥοδίνῳ] ῥοδένῳ M 18 φυρώσας] λιώσας B 19 ἐπάρας] ἔπαρον Σ 19 κοπάνισον] om. A 19 αὐτὸ καὶ ποίησον οἷον τὸ ἀλεύριν] om. MA

εἶτα μίξον αὐτὸ με τὸ ὄξος καὶ με τὸ ῥοδέλαιον καὶ ποίησον αὐτὸ ὡς ἂν ζυμάριν μὴ πολλὰ σκληρόν. εἶτα ἄλειφε τὴν κεφαλὴν καὶ τὸ ἡμίκρανον. 2 ἄλλο. ⌜σκορπίουρον καὶ λίβανον κοπανίσας ἀναλαβὼν ποίει ὁμοίως.⌝ ἤτοι κοπάνισον τὸ λίβανον πρῶτον καὶ ποίησον ψιλὸν πολλά, εἶτα τὸ σκορπίουρον καὶ μίξον αὐτὰ μετὰ ὄξους καὶ ποίησον ὡς κόλλαν λεπτήν καὶ ἄλειφε τὴν κεφαλὴν καὶ τὸ μέτωπον καὶ τοὺς μήλιγγας.

3 [3] γʹ εἰς ζέσιν κεφαλῆς· ⌜ῥοδέλαιον βάλε καὶ ὀξείδιν καὶ χλίανέ τα. εἶτα ἄλειφε τὸ κεφάλιν τοῦ ὅλον. 2 ζέσις δὲ λέγεται, ὅταν φαίνεται τινὰ ὅτι καίεται ἡ κεφαλὴ αὐτοῦ καὶ ἐξάπτει. πλὴν πρῶτον φλεβοτόμησον αὐτὸν κρανιακὴν φλέβα.⌝ εἶτα ἄλειφέ τον τὸ ὀξείδιν καὶ τὸ ῥοδέλαιον χλίον. ⌜καὶ κρασὶν μὴ πίνῃ μηδὲ ἥλιος τὸν κρούῃ μηδὲ ἄνεμος μηδὲ κάπνος. μηδὲ δριμέα τρώγῃ μηδὲ θερμὰ καὶ ἂς ἰατρεύηται καὶ κλυστήριν.⌝

4 [5] δʹ εἰς ὀφθαλμούς· ▷MAἄλειφε μύρον τρίτον καὶ ἴαται. ἢ ὄνυχας ὄνου ἀρσενικοῦ καῦσον καὶ τρίψον καὶ μίξον γάλα γυναικεῖον ἀρσένου παιδὸς μητρὸς καὶ ἄλειφε.MA◁ 2 ἄλλο. ⌜αὐγὸν ἐκζέσας

---

6 ὡς] κολύκιον εἶτι add. F 12 χλίον] κρίον F 12–13 μηδὲ ἥλιος... καπνὸς] ἂς προσέχηται ἀπὸ τὸν ἥλιον καὶ τὸν ἄνεμον καὶ τὸν καπνὸν F 14 κλυστήριν] ἑρμη^ν τοῦ κυστη^ρ· μολόχη καὶ πίτερον σι^τρ καὶ σεύτλα καὶ ἂς βράσουν με τὸ νερόν. ἔπειτα σακέλλησον αὐτὰ καὶ τότε βάλε ἅλας καὶ μέλι καὶ ἐλάδιν. τὸ μὲν μέλι ἂς ἔνι δύο μοίρας τὸ δὲ ἔλαιον μία καὶ ἅλας ἀρκετὸν add. F

---

1 μίξον αὐτὸ] μίξε το MA ἄμιξαίτα B 1 με τὸ] μετὰ B 1 ὄξος] ὀξείδιν Σ 1 με τὸ] τὸ B 1 αὐτὸ] αὐτὰ B 3 σκορπίουρον] σκορδοπίουρον M σκόρδον πίουρον A 3 λίβανον] λιβάνιν M 3 κοπανίσας] κοπάνισον B 4 ποίει ὁμοίως] om. B 4 ἤτοι] om. B 4 τὸ] τὸν Σ 5 σκορπίουρον] σκορδοπίουρον M 5 μίξον] μίξε MA 5 αὐτὰ] τα MA om. B 5 μετὰ ὄξους] με τὸ ὀξείδιν MB 7 μήλιγγας] μήνιγγας A μίληγκους B 8 ῥοδέλαιον] ῥόδιον ἤτοι ῥοδόλαιον Σ 8 καὶ] εἰς M om. A 8 ὀξείδιν] ὄξος M 9 εἶτα] scripsi εἶ L καὶ MA ἔπητα B 9 τὸ κεφάλιν] τὴν κεφαλὴν AB 9 ὅλον] ὅλην Σ 9 ὅταν] τὸν add. AB 10 τινὰ] τὸν ἄνθρωπον M om. AB 10 αὐτοῦ] τοῦ Σ 11 φλεβοτόμησον] φλογοτόμηον B 11 αὐτὸν] τοῦτον M τον A 11 ἄλειφέ] ἄλειψον ἤγουν τὴν κεφαλήν M 11 τον] την add. A τοῦ M 11 τὸ] om. B 12 μηδὲ] ὁ add. AB 13–14 ἂς ἰατρεύηται] ἰασιατρεύεται B 15 ὀφθαλμοὺς] πόνον ὀφθαλμῶν MA 17 μητρὸς] om. A

κοπάνισον τὸν κρόκον χωρὶς τὸ ἄσπρον καὶ ἕτερον κρόκον ὃν βάνωσιν εἰς φᾶν. κοπάνισον ἐν τῷ ἅμα, εἶτα κοπάνισον ἀμάραντον καὶ τὸν ζωμὸν μίξον με τὸν κρόκον τοῦ ῴοῦ καὶ τοῦ ἄλλου κρόκου. καὶ βάλε καὶ ἕψημα ὀλίγον καὶ ῥοδόσταγμα καὶ μίξε τα ὅλα καὶ βρέξε βαμβάκιον εἰς τὸν ζωμόν. καὶ θές το εἰς τὰ ὀμμάτιά του. καὶ φλεβοτόμησον τοῦτον. καὶ κρασὶν μὴ πίῃ. καὶ ἂς ἀπέχῃ καὶ τὰ ἄλλα τὰ εἶπα ὅταν ἔχῃ ζέσιν ἡ κεφαλή. αὐτὴν δὲ τὴν ἰατρίαν ποίησον καὶ εἰς πληγὴν ὅπου ἔχει φλεγμονὴν⌝ ὡς ἂν ἑστίαν καὶ καίεται ὁ τόπος.

5 [4] ε′ ⌜ὅταν κνήθωνται οἱ ὀφθαλμοί· ἔπαρον φλοῦδιν ῥοϊδίου ὀξύνου καὶ φρύξε το εἰς τὸν ἥλιον καὶ κοπάνισον. καὶ ποίησον οἷον τὸ ἀλεύριν. καὶ βάλε κρασὶν ὀλίγον καὶ τάραξέ το. εἶτα ἄλειφε τοὺς ὀφθαλμοὺς καὶ ἐπάνω θὲς πανὶν βρεμμένον με τὸ κρασὶν καὶ δήσε το ἕως τὸ πρωΐ. καὶ φλεβοτόμησον τοῦτον καὶ λοῦσέ τον καὶ τὰ δριμέα πάντα ἂς ἀπέχηται, καὶ τὸ κρασὶν καὶ τὸν ἥλιον καὶ τὸν καπνὸν καὶ τὸν ἄνεμον.⌝

6 [5] ϛ′ εἰς ῥευματιζομένους ὀφθαλμούς· ⌜ἔπαρον λίβανον μασ-

---

1 κρόκον] κόρκον F 1–3 ἕτερον... κρόκου] με ζωμὸν ἀμαράντου F 4 καὶ] κρόκον τὸν βάνουν εἰς τὸ γλυκόζυνον κοπανισμένον καὶ add. F

---

1 τὸ ἄσπρον] τοῦ ἄσπρου Σ 1 ἕτερον] ἄλλον MB om. A 1 κρόκον] om. A 1 ὃν] οἷον MB om. A 1–2 βάνωσιν] βάνουν M βάλλουν B om. A 2 εἰς] om. Σ 2 φᾶν] γλυκόζυνον καὶ M τὸν γλυκόζυνον B ζαφαρᾶν A 2 ἀμάραντον] ἀμάραντα MB 3 ζωμὸν] τῶν add. M αὐτοῦ add. B 3 μίξον] μίξε τον A μίξε το B 3 ῴοῦ] αὐγοῦ Σ 3 κρόκου] ἤγουν τοῦ ζαφαρᾶ A 4 καὶ] εἰς M 4 ῥοδόσταγμα] δορόσταγμα M 5 βρέξε] βρέξον A 5 βαμβάκιον] βαμβάκιν Σ 5 ζωμόν] τους add. AB 5 το] τα B 5 ὀμμάτιά] ὠτία B 5 του] om. A τὸν B 6 τοῦτον] κρανιακὴν add. A 6 κρασὶν] οἶνον M 6 πίῃ] πίνῃ Σ 6 ἂς ἀπέχῃ] ἀπέχηται MB 6 καὶ] om. AB 7 ἄλλα] ὅμοια A ὅλα B 7 εἶπα] εἶπον M εἴπαμεν A 7 κεφαλή] add. fenestram M 7 αὐτὴν] ταύτην MA 7 ἰατρίαν] ἰατρείαν A 8 ποίησον] ποίει M 8 ὅπου] ὅταν A 8 ἑστίαν] scripsi ἡστίαν L ἑστιά M ἰσθία A πῦρ B 10 ὅταν κνήθωνται οἱ ὀφθαλμοὶ] εἰς κνησμὸν ὀμμάτων M κνήσις λε. ὅταν ξύεται καὶ δακρίζ. in marg. M[2] εἰς κνησμὸν ὀμμάτων, ὅταν κνίθονται οἱ ὀφθαλμοὶ AB 10 ῥοϊδίου] ῥοδίου LMA 11 ὀξύνου] om. M 12 ἀλεύριν] ἄλευρον M 12 το] τὰ M 13–14 κρασὶν καὶ δήσε το ἕως] πανὶν καὶ B 14 το] τα A 14 τὸ] om. M 14 λοῦσέ] λουέ M 15 ἂς ἀπέχηται] ἀσπέχει B 15 καὶ τὸν] om. B 15 ἥλιον] om. Σ 17 ῥευματιζομένους ὀφθαλμούς] transp. A 17 λίβανον] λιβάνιν M 178.17–179.1 μαστίχην] μαστίχιν M

τίχην, σμύρναν καὶ κοπάνισον αὐτὰ ἕνα καὶ ἕνα. καὶ φάβατα πέντε καὶ ποίησον καὶ αὐτὰ ἀλεύριν. εἶτα ἔπαρον ἄσπρον αὐγοῦ καὶ βάλε τὰ τρίμματα καὶ ἕνωσέ τα καὶ ἔπαρον καὶ ἄλειψον τὸ μέτωπον. καὶ ἐπάνω θὲς στυππεῖα.⌝

7 [6] ζʹ ὅταν ῥέῃ αἷμα ἀπὸ τῆς μύτης· ⌜αὐγοῦ φλοῖον καῦσας καὶ τρίψας καλῶς βάλε το εἰς τὸ καλάμιν. καὶ θὲς τὴν μίαν μερέαν τοῦ καλαμίου εἰς τὴν μύτην τοῦ καὶ ἀπὸ τὴν ἄλλην φύσα καὶ ἀναβαίνει εἰς τὴν μύτην. 2 ἄλλο. σινάπιν καὶ καρύδια ὀλίγα κοπάνισον μερέαν καὶ μερέαν. καὶ ἀπέκει μίξε τα. καὶ αὐγοῦ ἄσπρον καὶ θές το εἰς τὸ μέτωπον. καὶ ἱστήκει τὸ αἷμα.⌝ ▷Bψάριν καὶ ἀλόην ἐπίσης τρίψας καὶ πάσας θές το καὶ ἵσταται.B◁

8 [7] ηʹ ὅπου πτύῃ αἷμα· ⌜κοπάνισον ἡδύοσμον καὶ τὸν ζωμόν του μίξον μετὰ ὀξείδιν ὀλίγον. καὶ ἄς το πίῃ.⌝

9 [8] θʹ πρὸς πόνον γλώσσης· ⌜ἐλαίας φύλλα μασοῦ. καὶ κράτει τὰ πολλὴν ὥραν εἰς τὴν γλῶσσαν σοῦ.⌝

10 [9] ιʹ ὅταν τὰ σπλάχνα γίνωνται σκληρά· ⌜βαλὼν ἄρτου ψίχας

8–9 μερέαν... ἀπέκει] χώρια καὶ χώρια εἶτα F 10 ἱστήκει] σταθήσεται F 10 τὸ αἷμα] καλαμίνθου τὰ φύλλα ξηράνας καὶ κοπανίσας φύσα μετὰ καλάμου τὴν ῥίναν. ἄλλο καλόν. στάξον τὰ ἠρήνα εἰς τὸ αἷμα θὲς κάρβουνα ἁπτόμενα εἶτα θὲς λίβανον καὶ μαλὶν πρόβιον εἰς τὰ κάρβουνα ἀπάνου καὶ ἂς δέχηται τὸν ἀτμὸν καὶ σταθήσεται. ἄλλο. καβαλήναν ὄνου βάλε εἰς ὄξους καὶ τὸ ἀποσφηνιαζέτο καὶ στάζε τὸ εἰς τὴν ῥίναν. ἐὰν ἀπὸ τὸ δεξιὸν μέρος τὸ ῥοθού^ν ῥέῃ βάλε βενετούζαν εἰς τὸ ἀριστερὸν μέρος ἀποκά^τ τοῦ πλευροῦ. εἰ δὲ ῥέῃ εἰς τὸ ἀριστερὸν βάλε εἰς τὸ δεξιὸν καὶ βρόχοσέτον εἰς τοὺς μεγάλους δακτούλους ποδῶν καὶ χειρῶν F

---

1 σμύρναν] εἰς μίρναν B 2 ποίησον... ἀλεύριν] τρίψον αὐτὰ M ποίησον αὐτὰ ἀ.... A ποίησον αὐτὰ ἀλεύριν B 2 αὐγοῦ καὶ] αὐγ.... A 3 ἕνωσέ] ἕνωσον B 3 τα] αὐτὰ B 3 ἔπαρον] .επα.. L 3 ἄλειψον] ἄλειφε A 3 τὸ μέτωπον] τὸν τόπον A 4 στυππεῖα] στυπτεῖα L 5 ὅταν] ὁποῦ AB 5 τῆς μύτης] τὴν μύτην Σ 5 φλοῖον] scripsi φλίον L φλούδιον M φλοῦδιν B 6 καλῶς] καλὰ M 6 το] om. A τον B 6 τὸ] om. M 7 τὴν ἄλλην] ἄλλης A τῆς ἄλλης B 7 φύσα] φύσησον A 7 ἀναβαίνει] ἂς ἀναβῇ A 8 μύτην] του add. Σ 8 σινάπιν] σινάπια A 8 καρύδια] καρύδιν B 8 ὀλίγα] om. M ὀλίγον B 8–9 μερέαν... ἀπέκει] om. M 9 καὶ αὐγοῦ ἄσπρον] με τὸ ἄσπρον τοῦ αὐγοῦ M 9 το] τα B 10 ἱστήκει] ἱστίκη L εἱστήκει M στέκει A στίκει B 13 μετὰ] με MA 13 ὀξείδιν ὀλίγον] transp. A 13 το] om. B 13 πίῃ] πίνῃ MA καταπίνῃ B 15 πολλὴν ὥραν] transp. M 16 τὰ σπλάχνα γίνωνται σκληρά] γένωνται τὰ σπλάγχνα σκληρὰ B 16 γίνωνται] γένωνται A 16 βαλὼν] βάλε M 16 ἄρτου] ἄρτον L

εἰς τηγάνιν. καὶ ἂς βράσουν μετὰ νεροῦ. καὶ ἀπέκει βάλε μέλι καὶ μάλαξον αὐτό. καὶ ἅπλωνέ το εἰς τὸ πανὶν καὶ ἐπίθες. λινόσπορον κοπάνισον καλὰ καὶ κριθάλευρον. καὶ ἕψησον με τὸ νερόν. ὕστερον δὲ βάλε ἕψημα καὶ μάλαξον. καὶ ἅπλωνέ το εἰς τὸ πανὶν καὶ θέτε το. οὐ τάδε ποίει, καὶ ὅπου εὑρίσκει σκληρίαν ὄγκων.⌝

11 [10] ια′ πρὸς πόνον ἥπατος· ⌜σελίνου ῥίζας τρεῖς, καὶ σκόρδον κοπάνισον ἐν τῷ ἅμα. καὶ ἂς βράσουν με τὸ κρασὶν τὸ καλὸν καὶ ἅς το πίνῃ κατὰ πρωΐ. καὶ εἰς τοὺς δυσουριόντας αὐτὸ ποίει. εἰ δὲ πυρέσσει, βράζε τα με τὸ νερόν.⌝

12 [11] ιβ′ ὅταν γένηται ἡ σπλὴν σκληρά· ⌜ἔπαρον κραμβίου φύλλα καὶ καθάρισε τὰ ἀπὸ νεῦρα αὐτῶν καὶ κοπάνισε τὰ φύλλα μόνα καὶ ἀπέκει βάλε τριμμένον ἀρσενίκιν καὶ λίβανον τριμμένον. πρῶτον δὲ ἂς βράσουν τὰ φύλλα με τὸ νερὸν εἰς τὸ τηγάνιν. βάλε τὰ τρίμματα καὶ ἕνωσον καὶ μάλαξον αὐτὰ καὶ θὲς αὐτὰ ἕως ἡμέρας τρεῖς, καὶ τότε ἔκβαλε τὸ κατάπλασμα καὶ δὸς τὸν ὀξείδιν ὀλίγον καὶ ἂς ῥοφήσῃ. 2 ἄλλο. τὸ φλοῦδιν τῆς ἰτέας πλησίον τῆς ῥίζας ἔκδαρέ το. καὶ ἕψησέ το με τὸ κρασὶν τὸ καλόν καὶ ἂς βράσῃ μίαν. καὶ πάλιν ἐπίβαλε κρασὶν καὶ πάλιν ἂς βράσῃ καὶ ἐπιβάλε

12–14 πρῶτον δὲ... αὐτὰ] om. F

---

1 ἂς βράσουν] βράσον Σ 1 μετὰ νεροῦ] με τὸ νερὸν Σ 1 καὶ ἀπέκει] εἶτα A 2 μάλαξον αὐτό] μάλαξέ το Σ 2 ἐπίθες] θέσε το ἐπάνω οὕτως δὲ ποίει καὶ ὅταν εὕρεις σκηράν ὄγκων M οὕτως δὲ ποίει καὶ ὁταν εὕρης τόπον πρησμένον, σκληρόν A θέτε το B 2 λινόσπορον] λινόσπερμα A 3 κοπάνισον] κοπάνη.... L 3 καλὰ] αλὰ L 3 καὶ] βάλλε καὶ add. M 3 με τὸ νερόν] μετὰ νεροῦ M 4 μάλαξον] μάλαξέ το A 4 το] τα M 4 τὸ] om. AB 4–5 θέτε το] θές τα ἐπάνω M 5 οὐ τάδε... ὄγκων] om. M οὕτως δὲ ποίει, καὶ ὅταν εὕρης σκληρίαν ὄγκων AB 6 πρὸς πόνον ἥπατος] om. MB add. M²B² 6 σκόρδον] σκόροδον AB 7 ἂς βράσουν] βράσαι M 7 με τὸ] με M 7 τὸ] om. M 8 καὶ εἰς... ποίει] οὕτως δὲ ποίει καὶ εἰς τοὺς δυσουρίοντας M 8 δυσουριόντας αὐτὸ] δισουρίοντας τὸ B 9 πυρέσσει] πυρεύσσει B 9 βράζε] βράσε MA 9 τα] το A 9 με τὸ νερόν] μετὰ νεροῦ M μετὰ με τὸ νερὸν B 10 γένηται] om. B 10 ἡ] ὁ A om. B 10 σπλὴν σκληρά] σπλὴν σκληρὸς MA σπλήνα καὶ σκληρὰ B 10 κραμβίου] κραμπίου B 11 ἀπὸ] τὰ add. Σ 11 αὐτῶν] των A τοὺς B 11 μόνα] μόνον A 11–12 καὶ ἀπέκει] καὶ οὕτως M εἶτα A 13 εἰς τὸ τηγάνιν] εἶτα Σ 14 αὐτὰ] τὰ Σ 15 καὶ] om. A 15 τότε] εἶτα A 15 τὸν] τοῦ B 15 ὀξείδιν ὀλίγον] transp. Σ 16 ἰτέας] ἐταίας L 16 ῥίζας] ῥίζης AB 17 ἔκδαρέ το] ἔκβαλε M ἔκβαλέ το B 18 πάλιν] om. M

το ἕως τρεῖς φορᾶς καὶ δίδε τὸ καὶ ἂς το πίνῃ ἐπὶ ἡμέρας ἑπτά. 3 ἄλλο. ἀμμωνιακόν κοπανίσας ἐξάγιον α' κοπανίσας δὲ⌝ πάλιν ψιλὸν καὶ βάλε κρασὶν εἰς τὸ ἀμμωνιακὸν καὶ ἂς βραχῇ. καὶ πότιζέ το μετὰ κρασίν. 4 ἄλλο. ἀμμωνιακὸν κοπανίσας καὶ ἑνώσας μετ' ἐλαίου καὶ φλοῦν καππάρεως τρίψας καὶ μίξας τὸ ἀμμωνιακὸν ποίησον κοκκία ἴσα ἐρεβίνθου. καὶ ἅς τα καταπίνῃ ▷om. Mτὴν νύκτα ἤτοι τὸν δεῖπνον. καὶ μηδὲν δειπνᾷ. κουκία δὲ πέντε τὸν δείπνον καὶ ἀφ' ὅτου τὰ καταπίῃom. M◁ ἂς κεῖται εἰς τὸ μέρος τῆς σπλήνας τοῦ. 5 ἄλλο. κριθάλευρον ἕψησε με τὸ ὀξείδιν. ▷om. Aκαὶ ποίησέ το ὡς κόλλαν καὶ ἄλειφέ την. 6 ἄλλο. ⌜ρίζαν καππάρεως κοπανίσας καὶ ἀσβέστιν κρύον μάλαξέ τα με τὸ ὀξείδιν.om. A◁ καὶ κατάπασσε τὴν σπλήναν ἕως ὥρας ἑπτά.⌝ 7 ἄλλο. ▷om. Aβάλε ὀξείδιν δριμὺ καὶ φύλλα ποταμογείτονος ἤτοι καλαμίνθην καὶ νερὸν καὶ ἂς βράσουν. καὶ ἀπέκει βάλε κριθάλευρον εἶτα κατάπλασσεom. A◁ ▷om. MAκαὶ τὸ ζωμὶν τούτων τῶν βοτανῶν καὶ ἂς βράσῃ τὸ κριθάλευρον εἶτα κατάπλασσεom. MA◁ 8 ἄλλο. ἀπόβρεξε ἀμμωνιακὸν καὶ λίβανον με τὸ ὀξείδιν. εἶτα κοπανίσας καὶ ποιήσας ὡς κόλλαν. περίχρισον τὴν σπλήνα καὶ ἐπάνω θὲς στυππεῖα.

13 [12] ιγ' πυρία ὅταν ὀδυνῶνται τὰ νέφρα· ⌜ζεμάτισε πίτυρα καὶ βλησκούνιν καὶ σακελλίσας αὐτά, βρέξε σπογγάριν ▷om. Lκαὶ πυρίαζε τον τόπον πλὴν ζεστόν. πρῶτον δὲ ἅς τον ἰατρεύσουν κλυστήριν.⌝om. L◁

1 το] om. MB 1 τρεῖς φορᾶς] transp. M 1 το] om. MB 1 ἐπὶ] ἕως M 2 πάλιν] πάνυ M om. A καλῶς B 3 καὶ] om. M 3 εἰς τὸ ἀμμωνιακὸν] om. M 4 το] τὸν M 4 μετὰ κρασίν] om. M με τὸ κρασὶν AB 4 ἑνώσας] ἑ.ώσας L ἐσμίξας M μίξας A 5 φλοῦν] φλουδίου M φλοῦδαν AB 5 τρίψας καὶ μίξας τὸ ἀμμωνιακὸν] om. M 5 τὸ] με τὸ A 6 ἐρεβίνθου] μερεβίνθην M μέρεβίνθια A ἐρεβίνθα B 6 τα] om. M 7 τὸν δεῖπνον] ἀπὸ δεῖπνα A 7 καὶ μηδὲν δειπνᾷ. κουκία δὲ πέντε τὸν δείπνον] om. A 7 πέντε] τὴν νύκταν ἤγουν add. B 8 τὸ] om. M 9 ἕψησε] ἕψησον MB το add. A 10 την] om. M τὸ B 11 τα] om. M 11 με τὸ] μετ' B 11–12 κατάπασσε] κατάπλασσε MB 12 ὥρας] φορᾶς A, ἡμέρας B 12 ἑπτά] εὐδόμης B 13 καλαμίνθην] καλαμίνθης M καλαμίνθου B 14 εἶτα] καὶ οὕτως M om. B 14 κατάπλασσε] om. B 15 τούτων] om. B 16 ἀπόβρεξε] ἀπόβρεξον A 17 εἶτα] καὶ A 17 κοπανίσας] κοπάνισον M 17 ποιήσας] ποίησον M 17 ὡς] καὶ B 19 πίτυρα] πίτερα M 20 βλησκούνιν] φλησκούνιν A 20 βρέξε] βρέξον MB 20 σπογγάριν] σπογγα.. L

14 [13] om. ⌞ιδʹ εἰς τὸ ἀναβῆναι σάρκα· ⌜ῥητίνην ξηρὰν κοπάνισον καὶ ποίησον ὥσπερ ἄλευρον. καὶ πάσον τὴν πληγήν.⌝

15 [14] om. ⌞ιεʹ εἰς ἔμφραξιν ὠτίων· ⌜βάλε ἔψημα εἰς φλοῦδιν αὐγοῦ καὶ θές το φλοῦδιν εἰς τὰ καρβούνια καὶ ἂς χλιάνῃ. εἶτα βάλε εἰς τὸ ὠτίον. 2 ἄλλο. κοπάνισον κινάμωμον καὶ ποίησον ὅμοιον ἄλευρον. εἶτα ἔπαρον στυππεῖον καὶ ἅπλωσέ το. εἶτα χύσε ἐπάνω τοῦ στυππείου κρόκον αὐγοῦ καὶ ἐπάνω τοῦ αὐγοῦ τὸν κινάμωμον τὸν τριμμένον καὶ οὕτως ἐπίθες εἰς τὸ οὖς.⌝

16 [15] om. ⌞ιϛʹ εἰς πιτυρίδαν κεφαλῆς· ⌜ἐνδέχεται πρῶτον να κενώσῃ τὸ πᾶν σῶμα ἢ διὰ φλεβοτομίας ἢ να πίῃ βοήθημα, εἶτα κοπάνισον σεύτλου φύλλα καὶ τὸν ζωμόν του ἄλειφε τὴν κεφαλήν. ἢ λίβανον κοπάνισον καὶ μίξε με τὸν οἶνον καὶ με τὸ ἔλαιον καὶ ἄλειφε τὴν κεφαλήν. 2 ἄλλο. κοπάνισον λιθάργυρον καὶ πεπέρι καὶ λίβανον μερέαν τὸ καθ' ἕν. εἶτα βαλὼν ἔλαιον ἕνωσέ τα καὶ ἄλειφε τὴν κεφαλήν. 3 ἄλλο. βάλε ὕδωρ εἰς τζυκάλιν καὶ ἀγριαγγουρέαν κομμάτια μικρὰ καὶ λυπηνάρια κοπανίσας καὶ μίξας ἂς βράσουν. καὶ τὸν ζωμὸν ἐκείνων λοῦε τὴν κεφαλήν.⌝

17 om. ⌞ιζʹ πρὸς ὀδύνην κεφαλῆς· ἐὰν τὸ μέτωπον καίεται κοπάνι-

---

1 σάρκα] σάρξ F 1 ῥητίνην] ητίνην F 2 πληγήν] ἄλλο. ἀρετζίνην καὶ κερὶν καὶ ἐλά[δ] ἄλειφε πανὴν θέτε εἰς τὴν πληγήν add. F 18 cap. 17] om. F

---

1 τὸ] να add. A 1 ἀναβῆναι] ἀναβάσῃ A ἀναβήσασαι B 1 σάρκα] σάρξ B 1 ῥητίνην] ἐρητίνην M 2 ἄλευρον] ἀλεύριν AB 3 εἰς ἔμφραξιν] εἰς ἔφραξιν M ζινοτίων B 3 ὠτίων] ἤτοι ὅταν φραγῶσιν τὰ ὠτία AB 3 φλοῦδιν] φλοῦδον M 4 φλοῦδιν] φλοῦδον A 4 τὰ] τὸ B 4 καρβούνια] καρβούνα A καρβόνιν B 4 χλιάνῃ] χλιανθῇ A 4 βάλε] το add. B 5 ποίησον] om. B 6 ἄλευρον] ἀλεύρου A ἀλεύρου ποίησον τοῦτο B 6 εἶτα] καὶ ἀπίκυ B 6 ἅπλωσέ το] ἅπλωνέ τα ἶσον καὶ μίαν παλάμην A ἁπλωσέ τα ἶσα παλάμιν B 7 τοῦ στυππείου] scripsi τοῦστου στυππείου M τῶν στυππείων τὸν AB 7 κρόκον] τοῦ add. AB 7 αὐγοῦ] καὶ ἐπάνω τοῦ ῥιθέντα add. B 8 τὸν] om. A 8 οὕτως] om. A 8 οὖς] ὠτίον B 9 πιτυρίδαν] πητηρʹ. B 10 φλεβοτομίας] φλετομίας M 11 σεύτλου] σούτλου M 11 του] τῶν A τοὺς B 12 κοπάνισον] κοπανίσας B 12 καὶ] om. B 12 μίξε] μίξον A το add. B 13 κεφαλήν] αὐτοῦ add. B 13 πεπέρι] τεάφιν B 14 ἕν] ἕνα B 15 βάλε] βαλὼν B 15 εἰς] τὸ add. A 15 τζυκάλιν] τζουκάλιν B 16 κομμάτια] καίτια B 16 κοπανίσας] κοπάνισον B 16 μίξας] μίξον εἶτα B 17 ἐκείνων] τους B 17 λοῦε] ἀλούε M 18 κεφαλῆς] ὀφθαλμῶν A 18 μέτωπον] του add. A

---

18 πρὸς] The following two chapters are reversed in AB.

σον γλιστρίδας καὶ μίξας με τὸ κριθάλευρον καὶ με τὸ μέλι ἄλειφε τὸ μέτωπον καὶ ἐπάνω θὲς στυππεῖα καὶ σεύτλου φύλλα με τὸ κριθάλευρον καὶ μέλι καὶ αὐγοῦ κρόκον ἢ τὰ φύλλα τῆς βάτου ἢ τῆς ῥάμνου καὶ μίξε τα καὶ αὐτὰ με τὸ κριθάλευρον καὶ τὸ μέλι καὶ ἀμάραντα καὶ θέτε τα εἰς τὸ μέτωπον πλὴν κοπάνισον πρῶτον τὰ φύλλα καὶ τὰ ἀμάραντα εἶτα μίξον τὸ κριθάλευρον καὶ τὸ μέλι καὶ τὸν κρόκον τοῦ αὐγοῦ.

18 [16] om. ᴸιη′ εἰς ὀδύνην ὀφθαλμῶν· ⌜ἐνδέχεται τούτοις φλεβοτομία καὶ μετὰ ταῦτα βάλε εἰς τὸν ὀφθαλμόν του ἄσπρον αὐγοῦ ἢ γυναικεῖον γάλα. βάλε δὲ καὶ εἰς τζυκάλιν χαμαίμηλα καὶ τήλην καὶ ῥόδα καὶ νερὸν καὶ ἂς βράσουν καὶ τὸ νερὸν ἐκεῖνο χλιαρόν. πυρίαζε τοὺς ὀφθαλμοὺς με τὸ πανίν.⌝

19 om. ᴸιθ′ πρὸς ῥεῦμα ὀφθαλμῶν· ἔπαρε σίτου ἄλευρον καὶ βάλε τριμμένον λιβάνιν καὶ κομίδιν καὶ ἄσπρον αὐγοῦ καὶ μάλαξέ τα εἶτα ἄπλωσέ τα εἰς πανὶν καὶ θές τα εἰς τὸ μέτωπον. 2 ἄλλο. ἔπαρον σεμιδάλιν καὶ βάλε λίβανον τριμμένον καὶ σμύρναν καὶ ἄσπρον αὐγοῦ καὶ ἕνωσέ τα καὶ θές τα ἴσον τοῦ μέτωπου με τὸ πανίν. 3 ἄλλο. κοχλίου θαλασσίου ἢ γῆς ἔκβαλε τὸ κρέας. καὶ μίξον τριμμένον λίβανον καὶ κοπάνισον καὶ οὕτως ἄπλωνέ τα εἰς πανὶν καὶ θέτε τα

8 cap. 18] om. F 18 κοχλίου] χοχλύον F 18 ἢ γῆς] ἢ γήινων αὐτοῖ εἰς τὰ πέτρας εὑρισκοῦ (τ supra) 18 ἔκβαλε τὸ κρέας. καὶ] τούτων τὸ κρέας F

1 γλιστρίδας] κλειστρίδας B 1 μίξας] μίξε τας A μίξε αὐτὰς B 1 με τὸ] om. B 1 μέλι] ὀλίγον καὶ add. B 2–3 με τὸ κριθάλευρον καὶ μέλι] om. A 3 καὶ] τὸ add. B 5 ἀμάραντα] ἀμάραντον A 6 φύλλα] τῆς βάτου add. A 7 αὐγοῦ] καὶ θέτε τα add. A 8–9 φλεβοτομία] φλεβοτομεῖν B 9 καὶ μετὰ ταῦτα] εἶτα A 10 δὲ] om. B 10 καὶ] om. AB 10 τζυκάλιν] τζουκάλιν B 11 ῥόδα] ῥόδου B 13 ὀφθαλμῶν] ὅπου κρούῃ δὲ τὸ ῥεύμα τῶν ὀφθαλμῶν τοῦτο add. B 13 ἄλευρον] ἀλεύριν A 14 μάλαξέ τα] καλοὺς add. B 15 ἄπλωσέ τα] ἄπλωνέ τον B 15 εἰς] τὸ add. B 15 θές] θέτε A 15 ἔπαρον] ἔπαρε AB 16 σεμιδάλιν] σεμιγδάλιν B 17 καὶ ἕνωσέ τα] om. AB 17 τα ἴσον τοῦ μέτωπου με τὸ πανίν] τὸ εἰς τὸ πανὶν ἶσον τοῦ μετώπου AB 18 κοχλίου θαλασσίου] κόχλιον θέλεις θαλάσσιον AB 18 ἢ γῆς] θέλεις ὅπου γίνεται εἰς τὴν γῆν καὶ εὑρίσκεται εἰς τὴν πέτραν κολλημένος AB 18 ἔκβαλε τὸ κρέας. καὶ] τούτου τὸ κρέας A καὶ αὐτοὺς ἤγουν τὸ κρεάς .. B 18 μίξον] με τὸ add. AB 19 λίβανον] λιβάνιν B 19 καὶ] μάλαξον ἀκριβῶς καὶ add. AB 19 κοπάνισον] ἐν τῷ ἅμα AB 19 οὕτως] om. AB 19 τα] τὰ AB 19 εἰς] τὸ AB 19 τα] το AB

εἰς τὸ μέτωπον. 4 ἄλλο. λίβανον τριμμένον καὶ κομίδιν μάλαξον με τὸ ὀξείδιν. καὶ ἄλειφε τὸ μέτωπον.

20 [om. LM]κ′ εἰς ὀδύνην ὀφθαλμῶν· ζεμάτισε ψύλλεον τὸ λεγόμενον σέρφαν καὶ εἰς τὸν ζωμὸν ἐκείνου βάλε καταστατὸν καὶ ῥοδόσταγμα καὶ ἄλειφε τὸ μέτωπον. 2 ἄλλο. ἔπαρον στρύχνον τὸ λέγουσί τινες βρωμοβότανον, ὅπου ἔνεν τὰ φύλλα τοῦ μικρὰ μαυρούτζικα. κοπάνισον τούτου θέλεις τὰ κοκκία του θέλεις τὰ φύλλα των καὶ εἰς τὸν ζωμὸν τούτου βάλε ἅλας τριμμένον καὶ ἄλειφε τὸ μέτωπον. 3 ἄλλο. τετράγκανθον βάλε ἀφ' ἑσπέρας εἰς τὸ νερὸν ὅσον εἰς ὅτι να μοσχεύσῃ. καὶ τὸ πρωῒ σακέλλισον, εἶτα βάλε σέρφαν καὶ ῥοδόσταγμα καὶ ἄλειφε τὸ μέτωπον.

21 [17] [om. L]κα′ πρὸς ὑπόσφαγμα ὀφθαλμοῦ· ⌜ὑπόσφαγμα δὲ ἔνι ὅταν εἰς τὴν κόρην ἢ εἰς τὸ ἄσπρον τοῦ ὀφθαλμοῦ ἔνι ἶσα φακὴ κόκκινον. ἐνδέχεται δὲ τούτοις ἵνα στάξῃς εἰς τὸν ὀφθαλμὸν περιστερὰς αἷμα⌝ ἢ φάσσας ⌜ἢ ἅλας καππαδοκικὸν τριμμένον⌝ ἢ γάλα γυναικεῖον ἡνίκα τὸ εὐγάλη θερμόν. καὶ ὕσσωπον βράσε καὶ τὸν ζωμὸν πυρίαζε τοὺς ὀφθαλμοὺς ἐὰν καὶ οἱ δύο ὀφθαλμοὶ ἔχωσι τοιοῦτον κόκκινον. εἰ δὲ ὁ εἷς τὸ ἔχει πυρίαζε τὸν ἕνα ὀφθαλμόν.

22 [18] κβ′ [▷om. L]πρὸς ὠτία ῥέοντα ἕλκος· ⌜βάλε οἶνον χλίον καὶ μέλι καὶ ἕνωσον ταῦτα. εἶτα ἔπαρον κλυστήριν μὴ τοιοῦτον οἷον ἰατρεύουν διὰ τῆς ἕδρας ἀλλὰ τοιοῦτον ὅπερ ἔχουσι οἱ ἰατροὶ ὅπερ καλεῖται ὠτικόν. καὶ μὴ βάλῃς τὸν αὐλὸν ἀπέσω ἀλλὰ κράτει τον ἔξω ἐναντίον τῆς τρύπας καὶ σφίγγε το ἀσκόπουλον να εὐγένῃ ὁ

3 cap. 20] om. F 12 cap. 21] om. F 19 cap. 22] om. F

1 λίβανον] λιβάνιν A 1 μάλαξον] μάλαξέ τα AB 2 με τὸ] εἰς τὸ B 3–4 τὸ λεγόμενον σέρφαν] om. B 6 ἔνεν τὰ φύλλα τοῦ μικρὰ] ποιεῖ ὡσὰν σταφύλια μικρὰ B 7 τούτου] τοῦτον B 7 των] τοῦ B 9 ὅσον] ἧσον B 9–10 εἰς ὅτι] om. B 10 σέρφαν] ψίχεον B 12 πρὸς ὑπόσφαγμα ὀφθαλμοῦ] περὶ ὑποσφαγμάτων AB 12 ὑπόσφαγμα] ἀπόσφαγμα M 13 ἢ] εἰ B 13 ἶσα] ὅσον A ἧσα B 13 φακὴ] μεφακῆ M 14 ὀφθαλμὸν] του add. B 15 φάσσας] scripsi φάσας Σ 15 καππαδοκικὸν] καππαδόκιον A 16 εὐγάλη] εὐγάλλη M εὐκάλει B 16 τὸν] τὸ A 18 ἕνα] om. B 19 ῥέοντα] ὅταν ῥέουσι A 20 ταῦτα] αὐτὰ B 21 ἰατρεύουν] βάλλουν A 22 ὠτικόν] ὀτικῶν καὶ βάλε τὸ κρασὶν καὶ τὸ μέλι χλίον B 22 καὶ] ἀλλὰ AB 23 σφίγγε] σφάγγε M 23 εὐγένῃ] εὐφίνῃ B

ζωμός, πλὴν μὴ φοβηθεὶς. πάλιν γὰρ ἔξω εὐγένει τὸ οἰνόμελι μόνον να ἀπὸπλύνεις τὸ ὠτίον ἀπὸ τοῦ ἕλκους.⌝ 2 ἄλλο. βάλε σκουρίαν σιδέρου καὶ ὄξος καὶ ἂς ποιήσῃ ἐβδομάδαν εἶτα ἀπ' ἐκεῖνο στάζε εἰς τὸ ὠτίον ἀπέσω. 3 ἄλλο.om. L◁ ἀλόην τρίψε καὶ μίξον την με τὸ κρασὶν καὶ στάξον εἰς τὸ ὠτίον. 4 ἄλλο. ▷om. L⌜ἐλαίας φύλλα κοπάνισον καὶ τὸν ζωμὸν μίξε τον με τὸ μέλι. καὶ στάξον ὀλίγον εἰς τὸ ὠτίον.⌝ 5 ἄλλο. κεδρέαν μάλαξον με τὸ μέλι καὶ στάξον εἰς τὸ ὠτίον.om. L◁ 6 ἄλλο. ⌜ἀλόην τρίψε μόνην καὶ χύσε την εἰς τὸ ὠτίον. πλὴν ὀλίγην χωρὶς ὑγροῦ.⌝ 7 ἄλλο. ▷om. Bστιβῆς ἀφέψημα ἤτοι τὸν ζωμὸν αὐτῆς ἀφ' ὅτου βράσει με τὸ μέλι καὶ με τὸ κρασὶν χωρὶς νερόν. ἂς γένηται χλίον καὶ στάξε εἰς τὸ ὠτίον.om. B◁ 8 ἄλλο. ▷om. Σοἶνον καὶ μέλι μίξας τὰ δύο χλιαρὰ στάξαι εἰς τὸ ὠτίον.om. Σ ◁ 9 ἄλλο. ▷om. ABφακὴν βράσε με τὸ νερὸν καὶ τὸν ζωμὸν στάξον εἰς τὸ ὠτίον. 10 ἄλλο. ⌜ῥόδα ἑψήσας με τὸ νερὸν καὶ τὸν ζωμὸν στάξον εἰς τὸ ὠτίον.⌝om. AB◁

23 [19] om. Bκγ' πρὸς ὠτία, ὅταν ῥέουσιν αἷμα· πράσου ζωμὸν ἔνσταξε εἰς τὸ ὠτίον. 2 ἄλλο. τρυγέαν οἶνον τρίψε καὶ σμίξον με τὸ κρασὶν τὸ χλίον καὶ στάξον εἰς τὸ ὠτίον.

24 [20] om. Bκδ' πρὸς σκωλήκια τὰ ὄντα εἰς τὰ ὠτία· ⌜τὰ φύλλα τῆς καππάρεως κοπανίσας τὸν ζωμὸν στάξον εἰς τὸ ὠτίον.⌝ 2 ἄλλο.

---

16 ζωμὸν] με ἄσπρον αὐγοῦ add. F

---

1 φοβηθεὶς] φοβηθῆς MB 1 πάλιν] πάλαι B 1 γὰρ ἔξω εὐγένει] καὶ εὐγένει ἔξω A, δὲ ἔξω εὐγένει B 2 να] om. B 2 ἀπὸπλύνεις] ἀποπλούνεις M 2 σκουρίαν] σκωρίαν A, σκορέαν B 3 καὶ ὄξος] om. B 3 ἐβδομάδαν] ἐβδομάδα A, εὐδομάδαν B 3 ἐκεῖνο] ἐκείνου A 4 ἀπέσω] ὀλίγον add. AB 4 τρίψε καὶ] τρίψας Σ 4 μίξον] μίξε Σ 5 ἐλαίας] πτελαίας A ἐλέας B 6 ζωμὸν] τους add. B 6 ὀλίγον] om. B 7 ὠτίον] ὀλίγον ἀπέσω add. B 7 με τὸ μέλι] μετὰ μέλιτος B 8 ἀλόην τρίψε] transp. B 8 τρίψε] τρίψον AB 9 ὀλίγην] om. A 10 αὐτῆς] τῆς M τῆς στιβῆς A 11 νερόν] νεροῦ καὶ M νεροῦ A 11 καὶ] οὕτως add. M 11 στάξε] στάξον M 13 βράσε] βράσας M 14 ἑψήσας] ἕψησον M 17 τρίψε] τρίψον M 17–18 με τὸ] με M 20 καππάρεως] μαπάρεως L 20 ζωμὸν] τούτων add. MA 20 τὸ ὠτίον] τὰ ὠτία M τὸ οὖς A

---

4 ἄλλο] τούτων σμίξον με τὸ μέλι καὶ στάξαι εἰς τὸ ὠτίον ὀλίγον ἀπέσω text in L after a missing page. This fragment cannot be identified with certainty. 4 ἀλόην] The following three paragraphs are reversed in B. 5 ἐλαίας] The following two paragraphs are reversed in A.

▷Aἀνθρώπου οὖρον στάξον εἰς τὸ ὠτίον ἅμα τὸ κατουρήσει χλίον.A◁ 3 ἄλλο. καρπὶν ἄσπρον τρίψας καὶ κοπανίσας μίξον μετὰ κρασὶν καὶ στάξον εἰς τὸ ὠτίον. 4 ἄλλο. κεδρέαν μίξε μετὰ ὀξείδιν καὶ στάξε εἰς τὸ ὠτίον. 5 ▷Lπράσου ζωμὸν στάξον εἰς τὸ ὠτίον.L◁ 6 ἄλλο. ποταμογείτονα ἤτοι ⌜καλαμίνθην κοπανίσας χλωρόν, τούτου τὸν ζωμὸν στάξον εἰς τὸ ὠτίον. 7 ἄλλο. τῆς ῥοδακινέας τὰ φύλλα κοπανίσας στάξον τὸν ζωμὸν εἰς τὸ ὠτίον.

25 [21] om. Bκε΄ ὅταν ἐμπῇ τίποτε εἰς τὸ ὠτίον ἢ ψύλλος· ⌜ψύλλος μὲν ἐὰν πέσῃ, βάλε ἔλαιον εἰς τὸ ὠτίον πλὴν χλιαρόν. 2 ἐὰν δὲ ἄλλο τί ποτε πέσῃ, τὰ μάλια τῆς κεφαλῆς ἐκείνου τὰς ἄκρας τῶν τριχῶν, βάλε εἰς τὸ οὖς,⌝ πλὴν τυλιγμένας τὰς τρίχας, ⌜ὅσον να χωρήσῃ εἰς τὸ οὖς.⌝

26 [22,23] om. Bκϛ΄ ὅταν τὸ ὠτίον ἔχῃ ἦχον ἤτοι ἀγχόνην· κεδρέα σὺν ὄξει ἕψησέ την καὶ χλίον στάξε εἰς τὸ ὠτίον. 2 ἄλλο. βάτου βλαστάρια τρυφερὰ κοπανίσας τὸν ζωμὸν τούτων μίξε μέλι καὶ τάραξέ το καὶ στάξον εἰς τὸ ὠτίον πλὴν χλίον. 3 ἄλλο. ⌜βούτυρον μίξε με τὸ μέλι, καὶ χλιάνας στάξον εἰς τὸ ὠτίον. 4 ἄλλο. χήνειον ἀξούγγιν λυτὸν ἄπαστον στάξον εἰς τὸ ὠτίον.⌝ 5 ἄλλο. ἀψίνθιον βράσας καὶ τὸν ζωμὸν χλίον στάξον εἰς τὸ ὠτίον. 6 ἄλλο.

---

8 cap. 25] om. F 13 ἀγχόνην] βοὴν F

---

2 κοπανίσας] κοπάνισον M 2 μετὰ] με τὸ A 3 τὸ ὠτίον] τὰ ὠτία M 3 κεδρέαν] κεεραιαν L 3 μίξε] μίξας M 3 μετὰ] με τὸ A 3 ὀξείδιν] ὄξος A 3 καὶ] om. M 3 στάξε] στάξον MA 4 τὸ ὠτίον] τὰ ὠτία M 5 ἤτοι] del. A 5 καλαμίνθην] καλαμίνθαν M del. A 5 τούτου] καὶ M 6 τὸ ὠτίον] τὰ ὠτία M 6 φύλλα] χλωρὰ add. MA 7 στάξον τὸν ζωμὸν] τὸν ζωμὸν αὐτῶν στάξον MA 7 τὸ ὠτίον] τὰ ὠτία M 8 ἐμπῇ] πέσῃ M 8 ἢ ψύλλος] scripsi ἢ ψίλος L om. M ἢ A 8 ψύλλος] scripsi ψίλος LΣ 9 χλιαρόν] χρηστέλαιον add. MA 11 τυλιγμένας] τειλισμένας L τυλιμένας A 12 χωρήσῃ] χωρέσῃ L χωρέσει A 13 ὅταν τὸ ὠτίον ἔχῃ] πρὸς ὠτία ἔχοντα M 13 ἤτοι] ἤγουν M, om. A 13 ἀγχόνην] ἀγχον.. L om. A 13 κεδρέα] .υτὸς αὐτοῦ s.l. L³, κεδρέαν MA 14 την] om. M 14 χλίον] χλίαν M 14 στάξε] στάξον MA 14 τὸ ὠτίον] τὰ ὠτία M 15 κοπανίσας] καὶ add. M 15 τούτων] om. M τῶν A 15 μίξε] μίξον με M τὸ με τὸ add. A 16 το] τα M 16 τὸ ὠτίον] τὰ ὠτία M 16 πλὴν] ἂς ἔνε add. A 17 μίξε] μίξον M 17 τὸ ὠτίον] τὰ ὠτία M 17–18 χήνειον ἀξούγγιν λυτὸν] χηναρίου ἀξουγγίου λυτοῦ M χυνίου ἀξούγγιν A 18 ἄπαστον] om. MA 18 τὸ ὠτίον] τὰ ὠτία πλὴν ἄπαστον M πλὴν ἄπαστον add. A 19 βράσας] βράσον M 19 τὸν ζωμὸν] του add. MA 19 τὸ ὠτίον] τὰ ὠτία M

▷MA⌜κύμινον⌝ καὶ πήγανον ⌜καὶ ὄξος καὶ ῥοδέλαιον βάλε καὶ ἂς ἑψηθῶσι καὶ οὕτως στάξε χλίον.⌝ 7 ἄλλο. ὅταν τις οὔτε πολλὰ δύναται ἀκούσαι οὔτε ἔνι κωφὸς τελείως· ⌜ἐνδέχεται τοῦτον ἵνα πίῃ βοήθημα, ὕστερον δὲ καὶ τὴν ἱεράν. 8 ἄλλο. αἴγας στάξον οὖρον ἢ χολήν.⌝ 9 ἄλλο. τοῦ πηγάνου τὰ φύλλα κοπανίσας καὶ τὸν ζωμὸν μίξον με μέλι καὶ στάξον εἰς τὰ ὠτία.MA◁ 10 ἄλλο. ▷L ὀξείδιν καὶ ῥοδόμελι χλίον στάξον εἰς τὸ ὠτίον. 11 ἄλλο. ὀξείδιν καὶ μέλι χλιάνας στάξον εἰς τὸ ὠτίον.L◁ 12 ἄλλο. ⌜καστόριον τριμμένον μίξε με τὸ ἀνηθέλαιον καὶ χλίον ἐνσταζόμενον.⌝ 13 ἄλλο. ▷Lζόχια ζ' ἢ πέντε κοπάνισον καὶ τὸν ζωμὸν βράσον με τὸ ἐλάδιν. καὶ στάξε χλίον εἰς τὸ ὠτίον, οἷον δέχεται.L◁ 14 ἄλλο. ▷MAῥαφανίου ζωμὸν μίξον με τὸ ῥοδέλαιον καὶ χλίον στάξε εἰς τὰ ὠτία.MA◁ 15 ἄλλο. καστόριον τριμμένον καρπὶν ἀσπρὸν τριμμένον καὶ ὄξος καὶ ῥοδέλαιον ἑνώσας καὶ μίξας τὰ πάντα χλιανόμενα στάξον εἰς τὸ ὠτίον. 16 ἄλλο. ⌜χηναρίου χολὴν καὶ χρηστέλαιον ἴσα καὶ τὰ δύο χλιάνας στάξον εἰς τὸ ὠτίον.⌝ 17 ἄλλο. ▷Mπράσου τὰς κορυφὰς κοπανίσας καὶ τὸν ζωμὸν χλίον στάξον εἰς τὸ ὠτίον.M◁ 18 ἄλλο. κρομμύδιν καὶ ὕσσωπον ἂς βράσουν με τὸ οὖρος τοῦ ἀνθρώπου. καὶ χλίον στάξον εἰς τὸ ὠτίον. 19 ἄλλο. ▷Mλαγωοῦ χολὴν καὶ γάλα γυναικεῖον στάξον εἰς τὰ ὠτία πλὴν χλίον.M◁ 20 ἄλλο. ταύρου χολὴν με τοῦ πράσου τὸν ζωμὸν μίξας στάξε εἰς τὸ οὖς. 21 ἄλλο.

---

2 ὅταν] πρὸς ἄνθρωπον μεσόκουφον F

---

1 κύμινον] βαλὼν κύμινον A 2 οὕτως] τότε A 2 χλίον] εἰς τὸ ὠτίον πλὴν χλίον A 2–3 δύναται] ἵνα add. A 3 ἀκούσαι] ἀκούσῃ LMA 3 τοῦτον] τούτῳ A 4 ὕστερον δὲ] καὶ ὕστερον A 4 αἴγας στάξον οὖρον] ἢ οὖρον αἴγας στάξον A 5 ζωμὸν] τούτου add. A 6 με] με τὸ A 6 τὰ ὠτία] τὸ ὠτίον A 9 μίξε] μίξον M 9 χλίον ἐνσταζόμενον] στάξον εἰς τὰ ὠτία (τὸ ὠτίον A) πλὴν χλίον MA 12 χλίον] χλιάναι καὶ A 12 εἰς τὰ ὠτία] πλὴν τὸ ῥεφάνιον κοπάνισέ το A 13 τριμμένον] καὶ add. MA 14 καὶ μίξας τὰ πάντα] om. M 14 τὰ] om. A 14 χλιανόμενα] χλιαινόμενα MA 16 τὸ ὠτίον] τὰ ὠτία M 18 ὕσσωπον] βάλε καὶ add. M 18 με τὸ οὖρος] με τὸν οὖρον M 19 χλίον] χλιάνας M 19 τὸ ὠτίον] τὰ ὠτία πλὴν χλίον M 19 λαγωοῦ] λαγοῦ A 20 τὰ ὠτία] τὸ ὠτίον A 21 τὸν] τὴν ῥίζαν ἤτοι τὸν A 21 ζωμὸν] τοῦ ὅλου πράσου add. A 21 μίξας] μίξον καὶ A 21 στάξε] στάξον MA 21 τὸ οὖς] τὰ ὠτία M τὸ ὠτίον A

---

2 ὅταν] Here begins an independent chapter in MA.

⌜σπόρον δάφνης ἤτοι τὰ κοκκία αὐτῆς κοπάνισον καὶ μίξας μετὰ οἴνου παλαιοῦ χλιάνας στάξον εἰς τὸ ὠτίον.⌝ 22 ἄλλο. A▷αἰγείαν χολὴν καὶ χυλὸν κρομμυδίων χλιανόμενα στάξον εἰς τὸ ὠτίον.A◁ 23 ἄλλο. στιβὴν καὶ κρασὶν καὶ μέλι ἂς βράσουν καὶ χλία γενόμενα ἂς στάξουν εἰς τὸ ὠτίον.

27 [24] [om. B]κζ′ πρὸς παρωτίδας ἤτοι τὰ ἐκβάλλουσιν εἰς ὠτία ὑποκάτω· ἃ καὶ ὁμοιάζουσιν ὡς ἂν χελωνικά· ⌜κριθάλευρον καὶ λινόσπορον καὶ τήλης ἄλευρον ἕψησε μετὰ μέλιτος καὶ κρασίου καὶ θές τα ἐπάνω. 2 ἄλλο. τὰ φύλλα τοῦ λαπάτου κοπανίσας σὺν ἀξουγγίου ἀναλάτῳ. καὶ ἅλας καὶ πράσα κοπανισμένα ἑνώσας τὰ πάντα ἐπίθες αὐτὰ ἐπάνω τῷ τόπῳ.⌝ 3 ἄλλο. ὕδωρ θαλάσσιον βράσας πυρίαζε με τὸν οἶνον. καὶ ἅλας χλιάνας πυρίαζε. 4 ἄλλο. ⌜λουπηναρίων ἀλεύριν καὶ ἕψημα καὶ βολβὸν ἑψημένον κοπανιστὸν κατάπλαττε.⌝

28 [25] [om. B]κη′ περὶ αἵματος πολλοῦ ῥυέντος ἐκ τῆς ῥινός· ⌜καλαμίνθιν ξηράνας τὰ φύλλα καὶ κοπανίσας φύσα διὰ καλαμίου εἰς τὴν ῥίναν.⌝ 2 ἄλλο. σπόγγον καινούργιον βρέξε εἰς κεδρίαν. καὶ ἀπ' ἐκεῖ καῦσε τὸ σπογγάριν καλὰ να φρύγῃ. εἶτα ποίησον αὐτὸ

---

11 ὕδωρ θαλάσσιον] θάλασσαν F 15 cap. 28] om. F

---

1 ἤτοι τὰ κοκκία αὐτῆς] om. M 1 αὐτῆς] τῆς δάφνης A 1–2 μετὰ οἴνου παλαιοῦ] οἶνον παλαιὸν καὶ MA 2 τὸ ὠτίον] ὠτία M 5 ἂς στάξουν] στάξον MA 5 τὸ ὠτίον] τὰ ὠτία M 6–7 ἤτοι τὰ ἐκβάλλουσιν εἰς ὠτία ὑποκάτω· ἃ καὶ ὁμοιάζουσιν ὡς ἂν] τὰ λεγόμενα M 6 τὰ ἐκβάλλουσιν εἰς] scripsi τὰ ἐκβάλωσιν εἰς L ὅταν εὐγένουσιν εἰς τὰ A 7 ἃ] om. A 7 ὁμοιάζουσιν] ὁμοιάζωσιν L 8 ἕψησε] ἕψησον MA 8 μετὰ μέλιτος] με τὸ μέλι MA 8 κρασίου] τὸ κρασὶν MA 9 θές] θέτε A 9 λαπάτου] λαπάθου MA 10 ἀξουγγίου] ἀξουγγίῳ MA 10 πράσα] σα L 10–11 τὰ πάντα] ταῦτα M, om. A 11 αὐτὰ] om. MA 11 τῷ τόπῳ] εἰς τὸν ὄγγον MA 11–12 βράσας] βράσον M 12 με τὸν] σπόγγον add. in marg. L[1] σπόγγον ἢ MA 12 ἅλας χλιάνας πυρίαζε] ἅλας ἂς βράσουν καὶ πυρίαζε τὸν τόπον M, καὶ ἅλας ἂς βράσουν καὶ πυρίαζε τὸν τόπον A 13 λουπηναρίων] scripsi λουπιναρίων L λυπηναρίων MA 13 κοπανιστὸν] κοπανίσας A 14 κατάπλαττε] κατάπλασσε MA 15–16 καλαμίνθιν] καλαμίνθης A 16 ξηράνας] ξηρὰς MA 16 καὶ] om. MA 16 φύσα] ἐνφύσησον A 17 βρέξε] βρέξον MA 17 κεδρίαν] κεδρέαν MA 17–18 καὶ ἀπ' ἐκεῖ] εἶτα A 18 να] ἵνα A 18 εἶτα] καὶ A 18 ποίησον] ποιήσας MA

ὅμοιον ἀλεύρου. ἐμφύσα δὲ καλάμου εἰς τὴν ῥίνα. 3 ἄλλο. ⌜ἐὰν τὸ δεξιὸν ῥουθούνιν ῥέῃ αἷμα πολὺ ἄμετρον ἡμέραν καὶ νύκταν δένε τὰς χεῖρας καὶ τοὺς πόδας τὰς δεξίας καὶ δεξιὸν μέρος ἤτοι τὸ ἧπαρ. θέτε σικύαν ἤτοι σικύασέ το χωρὶς να κόψῃς τὸν τόπον. ▹Lἐὰν τὸ ἀριστερὸν μέρος ῥέῃ τὸ αἷμα, ποίησον πάλιν τὰ ἀριστερὰ μέρη δεσμοὺς καὶ σικύας.⌝L◃.

29 [26] om. Bκθ′ πρὸς τὸ να γένωνται οἱ ὀδόντες ἀσπροί· ⌜κέρατον ἐλάφου καύσας ἐστ' ἂν γένηται ἄσπρον. εἶτα τρίψον αὐτὸ καὶ ποίησον οἷον τὸ ἀλεύριν καὶ ἄλειφε τοὺς ὀδόντας.⌝ 2 ἄλλο. ἅλας καὶ μαστίχην τριμμένα τρίβε τοὺς ὀδόντας. 3 ἄλλο. ⌜κριθάριν φρύξας καὶ ἅλας κοπανίσας τρίβε τοὺς ὀδόντας.⌝ 4 ἄλλο. ▹MAμαλάθρου σπόρον καὶ ἄνθη τριακονταφύλλων ἑνώσας καὶ τρίψας ἄλειφε τοὺς ὀδόνταςMA◃. 5 ἄλλο. ἅλας καὶ πέπερι καὶ κύπρον κοπανίσας καὶ τρίψας ἄλειφε τοὺς ὀδόντας. 6 ἄλλο. ▹om. Aκίσσηριν μετὰ τριανταφύλλων ξηρὸν ἑνώσας καὶ τρίψας, ἄλειφε τοὺς ὀδόντας.om. A◃

30 [27] om. Bλ′ πρὸς αἷμα ὅταν ῥέῃ ἀπὸ τὰ οὖλη· ⌜τὰ φύλλα τῆς ἐλαίας βράσας με τὸ νερὸν καὶ τὸν ζωμὸν ἐκεῖνον κράτει εἰς τὸ ▹Lστόμα πολλὴν ὥραν. 2 ἄλλο. τῆς μυρσίνης τὰς κορυφὰς καὶ τριανταφύλλων ἄνθος βαλὼν καὶ ὄξος καὶ κρασὶν ἃς βράσουν καὶ τὸ ζωμὸν διακράτει εἰς τὸ στόμα.⌝ 3 ἄλλο. λειχήνηςκορυφὰς καὶ κυδωνέας ῥίζαν καὶ ὄξος καὶ κρασὶν ἃς βράσουν καὶ τὸν ζωμὸν

18 τῆς μυρσίνης] μερσυνέας F

1 ἀλεύρου] ἀλεύρῳ M 1 δὲ] διὰ MA 1 ῥίνα] ῥίναν. ἐὰν δὲ τὸ ἀριστερὸν μέρος ῥέῃ τὸ αἷμα ποίησον τὰ ἀριστερὰ μέρη δεσμοὺς καὶ συκίασον MA 1 τὸ] δὲ τὸ A 2 ῥουθούνιν] ῥωθώνιον M μέρος A 2 ἡμέραν] τε add. M ἡμέρα A 2 νύκταν] νύκτα A 3 τὰς δεξίας] τὰ δεξιὰ A 3 τὸ] εἰς τὸ A 4 σικύαν] σικύας M 4 ἤτοι] καὶ A 4 σικύασέ] σικύαζε A 4 το] τον M 7 γένωνται] γένονται L γένουν A 8 ἂν] να add. MA 9 καὶ] οὕτως add. M 9 ὀδόντας] καὶ τρίβε πλὴν νηστικὸς add. MA 10 μαστίχην] μαστίχιν M 10 τριμμένα] τρίψον καὶ M τριμμένον A 13 καὶ] om. M 13 κύπρον] κύπεριν MA 14 μετὰ] om. M 14–15 τριανταφύλλων] τριαντάφυλλα M 15 ξηρὸν] ξηρῶν L ξηρὰ M 16 αἷμα ὅταν ῥέῃ] ὅταν ῥέῃ αἷμα M 16 οὖλη] οὖλα A 17 καὶ] om. MA 18 κορυφὰς] κόρφας L 20 λειχήνης] scripsi ληχ^{v} L 20 κορυφὰς] scripsi κορφὰς L

11 μαλάθρου] The following two paragraphs are reversed in A. 16 πρὸς] The following two chapters are reversed in M.

διακράτει εἰς τὸ[L◁] μέρος τῶν οὐλῶν ἔνθα πονεῖ πλὴν πολλὴν ὥραν. 4 ἄλλο. [▷MA]σχίνου κορυφὰς καὶ κυδωνέας ῥίζας καὶ ὄξος καὶ κρασὶν ἅς βράσουν καὶ τὸν ζωμὸν διακράτει εἰς τὸ στόμα.[MA◁]

31 [28] [om. B]λα΄ πρὸς σαπημένα οὔλη· ⌜κηκήδιν ποίησον οἷον τὸ ἀλεύριν καὶ σμύρνην ξηρὰν οἷον τὸ ἀλεύριν καὶ πάσσε τὰ οὔλη. 2 ἄλλο. κρόκον ξηρὸν κοπανίσας καὶ ἅλας ποιήσας ὅμοιον ἀλεύρου με τὸ μέλι μίξας ἄλειφε.⌝ 3 ἄλλο. [▷M]κρόκον ξηρὸν κοπανίσας καὶ ἅλας καὶ ποιήσας ὅμοιον ἀλεύρου μίξον με τὸ μέλι καὶ ἄλειφε τὰ οὔλη.[M◁]

32 [29] [om. B]λβ΄ πρὸς κατασχισμένα χείλη· ⌜τοῦ φοινικίου τὸ ὀστοῦν ἔχει ὡς ἂν τζύπαν. ταύτην ἔπαρον καὶ κόλλησον ἐκεῖ εἰς τὸν τόπον, ἔνθα ἔνι τὰ χείλη κατασχισμένα.⌝ 2 ἄλλο. ἀξούγγιν αἴγιον ἄλειφε τὰ χείλη. 3 ἄλλο. ⌜χήνειον ἀξούγγιν ἄλειφε τὰ χείλη.⌝ 4 ἄλλο. [▷om. A]μυελὸν βόϊον ἄλειφε τὰ χείλη.[om. A ◁] 5 ἄλλο. ἐλάφειον μυελὸν ἄλειφε τὰ χείλη. τὸ φλοῦδιν τοῦ αὐγοῦ ἔχει τζύπαν ἄσπρην. ταύτην ἔπαρον καὶ κόλλησον εἰς τὰ χείλη. 6 ἄλλο. [▷om. A]μόσχον ἀξούγγιν ἄλειφε τὰ χείλη.[om. A◁] 7 ἄλλο. τὰ φύλλα τῶν τριανταφύλλων χλωρὰ βρέξε τα εἰς τὸ κρασὶν καὶ κόλλησέ τα εἰς τὰ χείλη ἕνα ἕνα φύλλον. 8 ἄλλο. κηκίδιν καὶ μαστίχην [▷om. A]ποίησον οἷον τὸ ἀλεύριν. πάσσε τὰ χείλη. 9 ἄλλο. ⌜μαστίχην[om. A◁] τριμμένον μίξας με τὸ κρασὶν ἄλειφε τὰ χείλη.⌝ 10 ἄλλο. τοῦ ὀστρυδίου τὸ φλοῦδιν

2 σχίνου] σκήνου F

1 πολλὴν] om. A 1 ὥραν] ὥραν κράτει εἰς τὸ στόμα M ἅς το κρατῇ πολλὴν ὥραν A 4 κηκήδιν] κηκίδιον M 5 σμύρνην] μυρτίαν A σμύρναν M 5 πάσσε] πάσον M 6 ξηρὸν] om. MA 6 κοπανίσας] καὶ τριακονταφύλλων ἄνθος add. M 6 ἅλας] om. MA 6 ὅμοιον ἀλεύρου] οἷον τὸ ἀλεύριν MA 7 με τὸ μέλι μίξας ἄλειφε] πάσον τὰ οὔλη M με τὸ μέλι μίξας ἄλειφε τὰ οὔλη A 10–11 ὀστοῦν] ὀστέον MA 11 κόλλησον] κόλουσον L κόλλυσον M 12 ἔνι] εἶναι M 12 αἴγιον] αἴειον M 14 βόϊον] βόειον M 14–15 ἐλάφειον μυελὸν] transp. M μυελὸν ἐλάφου A 15 ἄλειφε τὰ χείλη] transp. M 15 φλοῦδιν] φλούδιον A 15 τζύπαν] τζύππαν M 15 ἄσπρην] om. M 16 κόλλησον] κόλυσον LM 16 μόσχον] μόσχου M 17 ἄλειφε τὰ χείλη] τὰ χείλη ἄλειφε M 17 τριανταφύλλων] τριακονταφύλλων M 18 βρέξε] βρέξον A 18 τα] om. A 18 κόλλησέ τα] κόλησές τα L κόλυσές τα M 18 ἕνα] καὶ add. M, ἕναν καὶ A 19 μαστίχην] μαστίχιν M 19 ποίησον] ποιήσας M 20 ἀλεύριν] ἄλευρον M 20 πάσσε] πάσον M 20 μαστίχην] μαστίχιν M 21 ὀστρυδίου] scripsi ὀστριδίου L ὀστρουδίου M 21 φλοῦδιν] φλούδιον M

καῦσε τὸ καὶ ποίησέ το ὅμοιον ἀλεύρου καὶ πάσσε το εἰς τὰ χείλη.

33 [30] om. Bλγ′ εἰς γλῶσσαν ὅταν πρησθῇ· ⌜γλιστρίδας ἂς μασσᾶται.⌝ 2 ἄλλο. τὰ φύλλα τῆς παλιουρέας ἂς μασσᾶται. 3 ἄλλο. ▷om. A⌜φύλλα τρυφερὰ ἐλαίας ▷om. M ἂς μασσᾶται.om. A◁ 4 ἄλλο. ⌜τὰ φύλλα τὰ τρυφερὰ τῆς ἐλαίαςom. M◁ κοπάνισον καὶ μίξον αὐτὰ με τὸ ἔλαιον καὶ ἄλειφε τὴν γλῶσσαν. 5 ἄλλο. ἔψημα μίξας με τὸ ῥοδόσταγμα κράτει εἰς τὴν γλῶσσαν πολλὴν ὥραν.⌝ 6 ἄλλο. ψύλλεον με τὸ ῥοδόσταγμα μίξας κράτει εἰς τὸ στόμα τὸν ζωμόν. 7 ἄλλο. ▷L⌜δροσάτον κράτει εἰς τὸ στόμα.⌝L◁

34 [31] om. ABλδ′ ὅταν βρωμῇ τὸ στόμα· ⌜ῥόδα ξηρὰ καῦσας καὶ ποιήσας οἷον τὸ ἀλεύριν τρίβε τοὺς ὀδόντας. 2 ἄλλο. στάχος βράσον με τὸ κρασὶν καὶ κράτει εἰς τὸ στόμα.⌝ 3 ἄλλο. στάχος μασοῦ τὸ πρωΐ. 4 ἄλλο. σχίνου ἄνθος τρίβε τὰ οὖλη. 5 ἄλλο. κίσσηριν ποιήσας οἷον τὸ ἀλεύριν, μίξας με τὸ μέλι ἄλειφε τοὺς ὀδόντας καὶ τὰ οὖλη.

35 [32] om. Bλε′ περὶ συνάγχης· ⌜αὐτίκα φλεβοτόμησον αὐτόν. μὴ ἐκβάλλης δὲ εἰς μίαν πολὺ αἷμα καὶ λειποθυμήσῃ καὶ ἀποθάνῃ. εἰ μὴ ὀλίγον καὶ πάλιν ...† πόσας ἡμέρας φλεβοτόμησον τοῦτον κἂν πυρέσῃ, κἂν οὐ πυρέσῃ. καὶ οἶνον μὴ πίῃ καὶ ἄντλησε καὶ τοὺς πόδας αὐτοῦ. καὶ ἂς ἔνι καὶ ζεστὰ καὶ ἄνεμος μὴ τὸν κρούῃ. ἐὰν δὲ οὐδὲν ὠφεληθῇ φλεβοτόμησον τοῦτον ὑποκάτω τῆς γλώσσης ἢ τὴν

3 παλιουρέας] παλιρέας F   10 στόμα] πράσον με τὸ κρασὶν κράτει εἰς τὸ στόμα add. F   13 σχίνου ἄνθος] χοίνου ἄνθη F

1 καῦσε] καῦσον M   1 τὸ] om. M   1 ποίησέ το] ποίησον M   1 ὅμοιον] οἷον A   1 ἀλεύρου] ἀλεύριν A   1 πάσσε το εἰς] πάσον M   2 εἰς] τὴν add. A   2–3 μασσᾶται] μασίεται A   3 παλιουρέας] παλουρέας A   3 μασσᾶται] μασίεται A   5 κοπάνισον] καλὰ add. M κοπανίσας καλὰ A   5 καὶ] om. A   6 γλῶσσαν] γλῶτταν M   8 τὸ στόμα τὸν ζωμόν] τὴν γλῶσσαν A   10 ξηρὰ καῦσας] ξηροκαῦσας M   13 τὸ] om. M   13 σχίνου ἄνθος] ἀγχίνου ἄν$^{\theta}$ L^{4} χοίνου ἄνθεις M   13 τρίβε] πρωΐ add. M   13 κίσσηριν] κισσήριον M   16 περὶ] πρὸς M   16 συνάγχης] συνάγχην M, συνάχης A   16 μὴ] καὶ μὴ A   17 ἐκβάλλης] ἐκβάλη M   17 δὲ εἰς μίαν] om. A   17 λειποθυμήσῃ] λυποθυμήσῃ L ὀλιγοθυμήσει A   17–18 εἰ μὴ] ἀμὴ M ἀλλὰ ὀλίγον A   18 ...†] κἂν M μεκαν A   18 φλεβοτόμησον] βλεβοτόμησον M   19 πυρέσῃ] τε πυρέσσει A   19 πίῃ] πίνῃ MA   19 ἄντλησε] ἄντλησον MA   19 καὶ] om. A   20 καὶ] om. MA   21 τοῦτον ὑποκάτω τῆς γλώσσης] τὴν γλῶσσαν του ὑποκάτω M αὐτὸν εἰς τὴν γλῶσσαν ἀποκάτω A   21 ἢ] εἰς add. A

γλῶσσαν ἐπάνω. κατάκοψον με τὸ ξυράφιν. καὶ τὴν κοιλίαν ἰάτρευε καθ' ἑκάστην κλυστήριν καὶ εἰς τὸν τράχηλον θέτε μαλία βρέμενα με τὸ ἔλαιον τὸ θερμόν.¬ 2 ἄλλο. κριθάλευρον καὶ λινόσπορον κοπανιστὸν καὶ φοινίκια κοπανισμένα καὶ ψιχίας ψωμίου ποίησον κατάπλασμα. καὶ θέτε τὸ εἰς τὸν τράχηλον. 3 ἄλλο. φοινίκια ἂς βράσουν μετὰ νερὸν καὶ φακήν. καὶ τὸν ζωμὸν κράτει εἰς τὸ στόμα καὶ κλουκούνιζέ το. 4 ἄλλο. ▷M⌐γάλα γλυκὺ κράτει εἰς τὸ στόμα πλὴν χλίον.¬M◁ 5 ἄλλο. κόστον καὶ ἄνθος τριανταφύλλων, μέλι καὶ κρασὶν ἂς βράσουν. 6 ἄλλο. ▷om. A πιτύρων ζωμὸν κράτει εἰς τὸ στόμα.om. A◁ 7 ἄλλο. τοῦ σκύλλου τοῦ ἄσπρου τὸ κόπρον ξήρανέ το καὶ τρίψε το καὶ μίξε το μετὰ μέλι καὶ ὄξος ὀλίγον. καὶ ἄλειφε τὸν τράχηλον. 8 ἄλλο. τῶν ἀγρίων χελιδονίων τὰ κόπρα καῦσε τα. καὶ τὴν στάκτην τούτων μίξε την με τὸ μέλι καὶ ἄλειφε τῆς γλώσσης τὴν ῥίζαν. 9 ἄλλο. ⌐πηγάνου φύλλα χλωρὰ κοπανίσας καὶ τὸν ζωμὸν τούτων μίξας με τὸ γάλα ἄλειφε τὴν ῥίζαν τῆς γλώσσης.¬ 10 ἄλλο. ▷Lμελάνθιν ἔψησε με τὸ μέλι καὶ τὸ κρασὶν καὶ ἄλειφε τῆς γλώσσης τὴν ῥίζανL◁.

36 [33] om. B λϛ' ὅπου πτύῃ αἷμα· ⌐φλεβοτόμει αὐτὸν καὶ τοὺς πόδας καὶ τὰς χεῖρας καὶ ἄλειφε ἔλαιον. ἡ δὲ τροφὴ αὐτοῦ ἂς ἔνι αὐστηρή. 2 ἄλλο. ▷Lμαλία ἄπλυτα βρέξε τα εἰς τὸ ὀξείδιν τὸ χλίον

---

18 ὅπου πτύῃ αἷμα] πρὸς αἵματος πτύση ἐὰν ἔνε ἀπὸ τοῦ σικῶ^τ F

---

1 ἐπάνω] ἀπάνω A 1 με τὸ] αὐ αὐτὴν με τὸ A 1 ξυράφιν] ξουράφιν ὀλίγον ὀλίγον κατασχίσε την M ξυράφιν ὀλίγον ὀλίγον κατάσχιέ την A 1–2 ἰάτρευε] την add. A 2 κλυστήριν] κλυστήριον A 2–3 μαλία βρέμενα] transp. MA 3 καὶ] om. A 4 ψωμίου] ἄρτου M 4 ποίησον] ποιήσας MA 5 καὶ θέτε] θές MA 5 τράχηλον] του add. MA 6 μετὰ] με τὸ MA 6 κράτει] ἂς κρατῇ A 7 καὶ κλουκούνιζέ το] om. MA 7 γλυκὺ] μυαικὺ M 8 κόστον] κώνστον L 8 τριανταφύλλων] τριάκονταφύλλων A 9 βράσουν] καὶ τὸν ζωμὸν τῶν κράτει εἰς τὸ στόμα add. A 10 τοῦ] om. M 10 τοῦ] om. M 10 τὸ] τὸν M 10 κόπρον] τοῦτο ἄσπρον add. A 10–11 ξήρανέ το] ξήρανον M 11 τρίψε το] τρίψον M 11 μίξε το] μίξον M 11 μετὰ] με τὸ MA 12 χελιδονίων] χελιδόνων MA 12 τὰ κόπρα] τὴν κόπρον MA 12 καῦσε τα] καῦσον M καῦσε την A 13 τούτων] om. M, του A 13 μίξε την] μίξον M 13–14 τῆς γλώσσης τὴν ῥίζαν] τὴν ῥίζαν τῆς γλώσσης A 15 τὴν ῥίζαν τῆς γλώσσης] τῆς γλώσσης τὴν ῥίζαν M 18 ὅπου πτύῃ αἷμα] πρὸς πτύοντα αἷμα M 18 φλεβοτόμει] φλεβοτόμησον MA 19 καὶ τὰς χεῖρας καὶ] om. MA 19 ἔλαιον] παλαιὸν add. MA 20 αὐστηρή] αὐστηρὰ οἷον MA

καὶ τὸ ῥοδέλαιον καὶ πυρίαζε τὸ στῆθος.[L◁] 3 ἄλλο. ῥοΐδια ὄξυνα, μῆλα, ἀκροάγουρα καὶ ἀπίδια ὄξυνα. 4 ἄλλο. βάτου φύλλα κοπανίσας τὸν ζωμὸν μίξε τὸν με τὸν βῶλον τὸν τριμμένον καὶ πότισον αὐτόν.⌝ 5 ἄλλο. δὸς αὐτὸν καὶ τὴν θηριακήν. νεόκαυστος δὲ ἂς ἔνι πλὴν με τὸ κρασὶν τὸ ὀλίγον. 6 ἄλλο. καὶ εἰς τὸ στῆθος πυρία με τὸν σπόγγον να βράσῃς ὀξείδιν καὶ κρασίν. 7 ἄλλο. [▷A]⌜μαλία ἄπλυτα βρέξον εἰς τὸ ὀξείδιν τὸ χλίον καὶ τὸ ῥοδέλαιον καὶ πυρίαζε τὸ στῆθος.⌝[A◁] ἄλλο. 8 [▷om. A]θὲς δὲ καὶ τὸ ἔμπλαστρον τὸ λεγόμενον δι' ἰτεῶν εἰς τὸ στῆθος.[om. A◁]

37 [34] [om. B]λζ′ περὶ φλεγμονῆς βυζίων· ⌜βαλὼν ὀξείδιν καὶ κρασὶν ἂς βράσουν. εἶτα ἔπαρε σπόγγον καινούργιον καὶ βρέξε τὸν ἐκεῖ καλά. εἶτα ἀπότλιψέ τον καλὰ καὶ θές τον εἰς τὰ βυζία ἐπάνω. καὶ δῆσέ τον καὶ τὴν αὔριον πάλιν οὕτως.⌝ 2 ἄλλο. ψιχίας ἄρτου βράσας με τὸ ὀξείδιν καὶ τὸ κρασίν. εἶτα βάλε φοινίκια κοπανισμένα εἰς τὰς ψιχίας καὶ μάλα[▷om. L]ξέτας καλά. εἶτα ἄπλωνέ τα εἰς πανὶν καὶ θές τα εἰς τὰ βυζία ἐπάνω. 3 ἄλλο. [▷om. A]ψιχέας με τὸ ψύλλεον κατάπλασσε.[om. A◁] 4 ἄλλο. ⌜ψιχίας ἂς βράσουν με τὸ ὀξείδιν καὶ τὸ κρασὶν εἶτα βάλε γλιστρίδας κοπανισμένας καὶ μάλαξον ἀκριβῶς καὶ θὲς ἐπάνω. 5 ἄλλο. κριθάλευρον ἢ τήλης ἀλεύριν ἕψησον μετὰ νερόν. καὶ ἔπειτα βάλε μέλι καὶ μάλαξον καὶ θὲς ἐπάνω.⌝[om. L◁]

38 [35] [om. LB]λη′ ὅπου ξερᾷ εἴτι ἐὰν φάγῃ· ⌜ἀλόην μαστίχην στύρακα λάδανον λίβανον ἀψινθίας σπόρον κοπανιστὸν ἕψημα ὀλίγον πολλὰ ὀλίγον ἀλεύριν καθαρὸν σιταρίου καὶ οἰνάνθην ὀλίγον καὶ ποίησον ἔμπλαστρον καὶ θές το εἰς τὸ στῆθος του. 2 ἄλλο. τοῦ

---

21 ὅπου ξερᾷ εἴτι ἐὰν φάγῃ] πρὸς ἐμέτον F

---

2 ἀκροάγουρα] ἀκροάγκουρα M ἄγουρα A 2 καὶ] om. M 2 ὄξυνα] ἰδὲ ἔνε ἀπὸ τοῦ φλέγμωνος ἂς τρώγῃ γλυκεία καὶ κω[λξὰ] add. F 3 μίξε τὸν] μίξον MA 4 αὐτὸν] τούδε MA 5 καὶ εἰς τὸ στῆθος] om. A 5 πυρία] πυρίαζε M πυρίαζε τὸ στῆθος A 6 σπόγγον] ἤτοι add. M 8 τὸ] om. M 9 δι' ἰτεῶν] scripsi διιτεὼν L δι' ἰτθῶν M 10 περὶ φλεγμονῆς] πρὸς φλεγμονὴν M 11 ἔπαρε] ἔπαρον MA 12 καλά] καλά add. M 12 ἐπάνω] om. A 13 τὴν αὔριον πάλιν] πάλιν τὴν αὔριον M 13 οὕτως] ποίει add. M 13 βράσας] βράσον M 16 θές τα] θέτε το A 16 θές τα] ςσριπσι ψίλεον M 17 ἂς] ἄρτου A 19 θὲς] τα add. A 19 μετὰ] με τὸ A 20 ἔπειτα] ὕστερον A 20 καὶ] τοῦτο καὶ A 21 ἐὰν] om. A 22 ἀψινθίας] ἀψιθίας A 23 πολλὰ ὀλίγον] om. A 23 καὶ] om. A

ἡδυόσμου τὸν ζωμὸν καὶ τοῦ ῥοδίου τὸν ζωμὸν μίξον καὶ μέλι καὶ ἂς βράσουν καὶ ἂς ἑψηθῇ ὥς ποτέ να γένῃ παχὺ ὀλίγον. καὶ ἂς τρώγῃ τὸ βραδὺ ὅταν θέλῃ κοιμηθῇ καὶ μηδὲν δειπνᾷ. ἰάτρευέ τον καὶ κλυστήριον.⌝

39 [36] ^om. B^λθ′ ▷om. L^πρὸς δυσεντερίας· ⌜λημνίαν σφραγίδα ἄς την πίνῃ τριμμένην με τὸ νερὸν τὸ χλίον. 2 ἄλλο. νερὸν χλίον τὸ θέλει ἰατρευθὴν. βάλε ἀπέσω τριμμένην τὴν λημνίαν σφαγίδα καὶ κρούε τον κλυστήριον πλὴν χωρὶς ἅλατος καὶ ἐλαίου καὶ μέλιτος. 3 ἄλλο. βάτος καὶ ἵππουρις τὸ λεγόμενον πολυκόμπιν ἂς βράσουν με τὸ κρασὶν καὶ ἄς το πίνῃ. 4 ἄλλο. συκαμινέας καρπὸς ἄγουρος ξηρανθεὶς ποίησον αὐτὸν οἷον τὸ ἀλεύριν καὶ ἄς το πίνῃ με τὸ νερὸν τὸ χλίον. 5 ἄλλο. ἰάτρευε δὲ καὶ κλυστήριον ἤτοι βάλε ὀρύζην καὶ ἂς βράσῃ με τὸ νερὸν εἶτα ἔχε λυτὸν ἀξούγγιν αἴγειον ζεστὸν καὶ βάλε το εἰς τὸν ζωμὸν τῆς ὀρύζης καὶ ἰάτρευέ τον⌝ πλὴν χωρὶς ἅλατος καὶ μέλιτος καὶ ἐλαίου. 6 ἄλλο. βάτου καρπὸς ἄωρος ξηρανθεὶς ποίησον αὐτὸν οἷον τὸ ἀλεύριν καὶ πίε το με τὸ νερὸν τὸ χλίον. 7 ἄλλο. βάλε εἰς τὸ νε^om. L◁^ρὸν τὸ χλίον ψιμμίθιν τριμμένον. καὶ καταστατὸν καὶ ταράξε τὸ καὶ ἰάτρευέ τον κλυστήριν χωρὶς ἐλαίου μέλιτος καὶ ἅλατος. 8 ἄλλο. ▷om. A⌜ἂς πίνῃ δὲ καὶ τὴν θηριακὴν καὶ ἄς τον ἰατρεύουν καὶ κλυστήριν με τὴν θηριακὴν ἤγουν χλίον νερὸν με τὴν θηριακήν. ἂς ταράξῃ με τὸ νερὸν καὶ ἂς ἰατρευθῇ χωρὶς μέλιτος ἅλατος καὶ ἐλαίου.^om. A◁ 9 ἄλλο. βαλὼν μυρσίνας

---

5 λημνίαν] λιμνέαν F

---

23 ὀλίγον] καὶ οἶνον ὀλίγον A 1 ζωμὸν] κοπανισμένον add. A 1 μίξον] μίξε A 2 ποτέ] τε A 2 γένῃ] γένηται A 3 θέλῃ] να add. A 3 καὶ] ἀλλὰ A 4 κλυστήριον] κλυστήριν A 5 πρὸς] περὶ A 5 δυσεντερίας] δυσυσεντερίας M 5 λημνίαν] λημνίας M λιμνίων A 5 σφραγίδα] σφαγίδα M 5–6 την πίνη τριμμένην] del. M² 6 θέλει] θέλη A 7 τριμμένην τὴν] om. A 7 λημνίαν] λιμνίαν A 7 σφαγίδα] σφραγίδα τριμμένην A 9–12 βάτος... χλίον] del. M² et add. in marg. ἐὰν φλεβοτομη^ς οχρῆ εχει κλυ^στρ ἔχε τοῦτο καὶ εἰς τὰ ἔμπροσθεν 9 βάτος] βάτον A 9 ἵππουρις] ἵππουριν A 11 ἄς το πίνη] πίε το A 12 δὲ] τοῦτον add. A 12 ἤτοι] οἷον A 14 ὀρύζης] ῥίζης M 15 μέλιτος καὶ ἐλαίου] transp. A 15 βάτου καρπὸς] del. M² 16 ποίησον αὐτὸν] del. M² 16 τὸ] om. A 16 πίε το] ἄς το πίῃ A 16 με τὸ νερὸν τὸ χλίον] del. M² 17 ψιμμίθιν] ψιμμίθιον MA 18 τὸ] τὰ M 18 κλυστήριν] κλυστήριον M 18 ἐλαίου] καὶ add. A 20 καὶ ἄς τον ἰατρεύουν καὶ κλυστήριν με τὴν θηριακὴν] om. M 21 με τὸ νερὸν] om. M

ῥοϊδίου φλούδια βάτου φύλλα ἀψίνθιον καὶ κρασίν καὶ ἂς βράσουν. εἶτα βρέχε σπόγγον καινούργιον εἰς τὸ κρασὶν ἐκεῖνο χλίον. καὶ πυρίαζε τὴν ἕδραν.⌝ 10 ἄλλο. ποίησον ἔμπλαστρον καὶ βάλε ἀλόην ἐξάγια β' λίβανον ἐξάγια δύο καὶ ἔψημα.

40 [36] [om. B]μ΄ πρὸς βιασμούς· ⌜ἰάτρευε τούτους κλυστήριν ἤτοι βάλε χλίον κρασὶν καὶ μέλι καὶ πηγανέλαιον καὶ βῶλον τριμμένον χωρὶς ἅλας καὶ ἑνώσας τὰ πάντα ἰάτρευε. 2 ἄλλο. βάτου φύλλα μυρσύνης φύλλα σχίνου φύλλα δάφνης φύλλα ῥοϊδίου φλούδια κηκίδιν τριμμένον. ἔψησε πάντα ταῦτα με τὸ νερὸν. εἶτα ἔπαρε σπόγγον καινούργιον καὶ βρέχε τον ἐκεῖ εἰς τὸν ζωμόν. καὶ πυρίαζε τὴν ἕδραν αὐτοῦ καὶ ἂς ἔνι ζεστά. κρύον μὴ πίνῃ, ἄνεμος μὴ τὸν κρούῃ. μηδὲ φλεβοτομηθήσει.

41 [37] [om. B]μα΄ περὶ κωλικῆς διαθέσεως. ⌜ἰάτρευε δὲ τοὺς τοιούτους κλυστήριν ἔχων πηγανέλαιον καὶ καστόριον τριμμένον χωρὶς ἅλας μέλιτος καὶ ἐλαίου.⌝ 2 ἄλλο. βαλὼν κύμινον τριμμένον πηγάνου φύλλα ἀξούγγιν χηναρίου ἢ ὄρνιθος εἰς χρηστέλαιον καὶ ἂς βράσουν καὶ τὸ ἐλάδιν ἐκεῖνο ἰάτρευε τὸ κλυστήριν. βάλε εἰς ἀγγεῖον πολλὰ μετὰ νερὸν μὴ πολλὰ χλίον καὶ χρηστέλαιον καὶ ἔμπασε τὸν ἄρρωστον. ἀπέστω εἰς τὸ θερμὸν καὶ ἂς ἀργήσῃ. 3 ἄλλο. ⌜βάλε ἀψίνθιον εἰς νερὸν καὶ ἂς βράσῃ καὶ ζωμόν του πότιζέ τον καθ' ἡμέραν τὸ πρωῒ χλίον.⌝ 4 ἄλλο. καστόριον τριμμένον μετὰ

15 ἐλαίου] να ἀλείψει τὸ ὑπογάστριον add. F

22 μέλιτος] καὶ add. M 1 βάτου] φαβάτων L 2 εἰς τὸ κρασὶν ἐκεῖνο] om. M 2 χλίον] τὸ χλίον A 4 λίβανον ἐξάγια δύο] om. MA 4 ἔψημα] καὶ λίβανον ἐξάγια δύο add. A 5 τούτους] τοῦτον M τον A 5 κλυστήριν] κλυστήριον MA 7 ἅλας] scripsi ..λας L ἅλατος MA 7 τὰ πάντα] ταῦτα M 7–8 μυρσύνης] σχίνου MA 8 σχίνου] μυρσίνης MA 9 τριμμένον] καὶ add. M 9 ἔψησε] ἔψησον A 9 πάντα ταῦτα] transp. MA 9 ἔπαρε] ἔπαρον A 10 καὶ] εἶτα MA 11 αὐτοῦ] τοῦ MA 11 ἄνεμος μὴ τὸν κρούῃ] om. MA 12 φλεβοτομηθήσει] φλεβοτομηθῇ A 13 περὶ κωλικῆς διαθέσεως] πρὸς κωλικὴν διάθεσιν M 13–14 δὲ τοὺς τοιούτους] τούτους MA 14 κλυστήριν] κλυστήρια M κλυστήριον A 15 ἅλας] ἅλατος M ἅλατος καὶ A 16 χηναρίου] χηνάριον M 16 ὄρνιθος] ὀρνίθης A 16 εἰς] καὶ A 17 τὸ] τον A 17 κλυστήριν] κλυστήριον M 18 πολλὰ] om. A 18 μετὰ] μέγα MA 18 νερὸν] χλίον καὶ add. A 19 ἔμπασε] ἔμβρασε M σέβασε A 19 ἀπέστω] ἀπέσω MA 19 θερμὸν] λουτρὸν εἶτι ἰς τὸ θερμὸν ἐκεῖνο MA 19 ἀργήσῃ] ἀρτήσῃ L 20 καὶ] τὸν add. MA 21 καθ' ἡμέραν] om. MA

νερὸν τὸ χλίον πότιζε κατὰ πρωΐ. βάλε πέπερι τριμμένον ὀλίγον καὶ ὄξος ὀλίγον καὶ μέλι ὀλίγον καὶ πότισον αὐτόν. 5 ἄλλο. ⌜πότισον δὲ αὐτὸν καὶ τὴν πικρὰν ὡσεὶ ἐξαγίου τὸ ἥμισυ με τὸ κρασὶν τὸ χλίον.⌝ 6 ἄλλο. χηναρίου ἀξούγγιν ἢ ὄρνιθος λυτὸν ἄπαστον πολὺ ἰάτρευε κλυστήριν χωρὶς ἅλατος μέλιτος καὶ ἐλαίου. 7 ἄλλο. τὸ ζωμὶν τοῦ γυμνουκρίθου τὸ λεγόμενον χυλάριν τὸ πίνουν οἱ πυρέσσοντες βάλε τὸ καὶ ἂς χλιανθῇ καὶ βάλε πηγανέλαιον καὶ ἰάτρευε κλυστήριν. 8 ἄλλο. ▷om. A χρηστέλαιον πολὺ μόνον χλίον ἰάτρευε κλυστήριν χωρὶς μέλι ἅλας καὶ ἔλαιον. om. A◁ 9 ἄλλο. ⌜βάλε τήλην, θέλεις τὰ κοκκία της, θέλεις τὰ φύλλα της, θέλεις οὕτως ὡς ἔνι. βάλε ταύτην εἰς χρηστέλαιον πολὺ καὶ ἂς βράσῃ καὶ ἀφ' ὅτου χλιανθῇ, ἰάτρευε κλυστήριν χωρὶς ἅλας μέλι ἔλαιον καὶ νερόν.⌝ 10 ἄλλο. βάλε τὸν χυλὸν τοῦ γυμνοκρίθου καὶ ἂς χλιανθῇ. καὶ ἀπέκει βάλε ῥοδέλαιον καὶ ἰάτρευε τὸν κλυστήριν χωρὶς ἅλας μέλι ἔλαιον καὶ νερόν.

42 [38] om. B μβ′ περὶ ἑλμίνθων ἤτοι σκωληκίων ἐν τῇ κοιλίᾳ· ⌜πότιζε τοῦτον τῶν χαμαιμήλων τὸ ζωμίν.⌝ 2 ἄλλο. ἐλάφου κέρατον ῥίνισέ το. καὶ τὸ ῥίνισμα ἐκεῖνο πότισέ το με τὸ χλίον τὸ νερόν. 3 ἄλλο. ⌜ῥοδέλαιον με τὸ ἔψημα πότισον.⌝ 4 ἄλλο. ▷MA ἀψίνθιον βραστὸν με τὸ νερὸν τὸ χλίον πότισον. MA◁ 5 ἄλλο. ⌜ἡδύοσμον ἔψ-

---

21 μετὰ] με τὸ MA 2 πότισον] πότιζε M πότιζον A 2 αὐτόν] ἄλλον add. M 2–3 πότισον δὲ] πότιζε M 3 τὸ] om. M 5 κλυστήριν] κλυστήριον A 5 ἅλατος] om. MA 5 ἐλαίου] καὶ ἅλατος add. MA 7 ἂς χλιανθῇ] ἀχλιανθῇ M 7 ἰάτρευε] τον add. A 8 κλυστήριν] κλυστήριον M χωρὶς ἅλατος καὶ μέλιτος καὶ ἐλαίου add. A 9 μέλι] om. M 9 ἅλας] ἅλατος μέλιτος M 9 ἔλαιον] ἐλαίου M 10 θέλεις] θέλης MA 10 θέλεις] θέλης A 10 θέλεις] θέλης A 11 βάλε ταύτην εἰς χρηστέλαιον] καὶ βάλε χρηστέλαιον A 11 ταύτην] την M 12 κλυστήριν] κλυστήριον M 12 ἅλας μέλι ἔλαιον καὶ νερόν] ἅλατος μέλιτος καὶ ἐλαίου καὶ νεροῦ M ἅλατος μέλιτος ἐλαίου καὶ νεροῦ A 13 τὸν] om. A 13 καὶ ἀπέκει] εἶτα A 14–15 ἅλας μέλι ἔλαιον καὶ νερόν] ἅλατος νεροῦ καὶ μέλιτος καὶ χρηστελαίου MA 16 περὶ ἑλμίνθων] πρὸς ἕλμινθας M 16 σκωληκίων] σκωλήκων MA 16–17 πότιζε τοῦτον] πότισον τούτους MA 17 τὸ ζωμίν] τὸν ζωμὸν A 18 ῥίνισέ το] ῥίνισον M ῥίνισέ το A 18 ῥίνισμα] ῥίνισμαν M 18 ἐκεῖνο] ἐκεῖνον M 18 πότισέ το] πότισον M πότισέ τον A 19 ῥοδέλαιον] ῥοδελαίου A 19 με τὸ] τὸ A

ησε μετὰ νεροῦ καὶ πότισε χλίον. 6 ἄλλο. ἀλόην τρίψον καὶ πότισον με τὸ νερὸν τὸ χλίον.⌝ 7 ἄλλο. κουκουνάρια κοπάνισον καὶ ποίησον οἷον τὸ ἀλεύριν. καὶ πότισε μετὰ ῥοδέλαιον. 8 ἄλλο. ἕψησον τὸ λεγόμενον σκυλλοκρόμμυδον με τὸ νερὸν καὶ τὸν ζωμὸν τούτου μίξον μὲ ὀξείδιν ὀλίγον καὶ πότισον. 9 ἄλλο. ποταμογείτονα ἤτοι καλαμίνθην πότισον τὸν ζωμόν του. 10 ἄλλο. ταυρίαν χολὴν με τὸ χλίον τὸ νερὸν πότισον. 11 ἄλλο. ⌜μελάνθιν κοπανιστὸν πότισον μετὰ χλίον τὸ νερόν. 12 ἄλλο. μυελὸν ἐλάφου ἄλειψον τὴν ῥάχιν καὶ τὴν κοιλίαν.⌝ 13 ἄλλο. σελίνου φύλλα βράσας τὸν ζωμόν του πότισον χλίον.

43 [39] om. Bμγ′ ὅταν ἔχουν οἱ νεφροὶ λίθους· ⌜ἀγρίαν καὶ ἀδίαντον βράσας με τὸ νερὸν πότισον χλίον νήστικον. 2 ἄλλο. ἐρεβίνθιν μαῦρον βράσον με τὸ νερὸν καὶ πότισον πρωῒ νῆστις. 3 ἄλλο. λοῦέ τον δὲ συχνὰ καὶ ἡνίκα εὔγῃ ἀπὸ τὸ λουτρόν, πότιζέ τον τῆς ἀγρίας τὸν ζωμὸν καὶ τοῦ ἀδιάντου ἢ τοῦ ἐρεβινθίου τοῦ μαύρου τὸ νερόν.⌝ 4 ἄλλο. βαλσάμου τὸν σπόρον κοπάνισον καὶ πότισον μετὰ οἴνου χλίου. 5 ἄλλο. ▹MA⌜ἀνδράχνης σπόρον ἤτοι γλιστρίδας κοπάνισον

15 ἢ] καὶ ψαλιδοβότανον add. F 15 νερόν] ἄλλο. κταπόδι βρέχε με νερὸν ἀφ' ἑσπέρας καὶ τὸ πρωῒ βράσαι⊤ μετ' ἐκύνου τοῦ νεροῦ. ἂς τρώγῃ ὀκτάποδι ἂς πίνῃ καὶ τὸν ζωμὸν add. F 16 ἄλλο] ἂς τρόγει καρδαμηλίδας καὶ σέλινα καὶ κρίταμα ὁμὰ καὶ ψυμένα ἂς ἀπέχει ἀπὸ τυρήναν μυζήθραν καὶ γάλα καὶ ὀψάρην γαράτον καὶ ἀπὸ τὰ ζυμαρικά. ἄλλο. κοπάνιζε περδικούλην καὶ κουδουμεν (το supra) καὶ σέλινον καὶ ζεστανέτο εἰς τιγάνην καὶ θές τα εἰς τὸ ὑπογάστριον add. F

196.20–197.1 ἕψησε] ἕψησον MA 1 μετὰ νεροῦ] με τὸ νερὸν M 1 καὶ πότισε χλίον] τὸ χλίον πότισον M 1 πότισε] πότισον A 1 τρίψον] τρίψας MA 1–2 καὶ πότισον με τὸ νερὸν τὸ χλίον] με τὸ νερὸν πότισον χλίον M μετὰ νερὸν τὸ χλίον πότισον A 2 κουκουνάρια] κωνάρια M κονάρια A 3 πότισε] πότισον A 4 τὸν ζωμὸν τούτου] τὸ νερὸν τοῦ A 5 μὲ] με τὸ A 5 ὀξείδιν] τὸ add. A 5 καὶ] μὲ μέλι ὀλίγον καὶ add. M μέλι ὀλίγον καὶ add. A 6 ταυρίαν] αὐρίαν M 7 μελάνθιν] μελάνθιον MA 8 μετὰ] με τὸ M με A 8 τὸ] om. MA 8 ῥάχιν] ῥάχην MA 9 βράσας] καὶ add. M βράσον καὶ A 9 του] αὐτῶν MA 10 πότισον] πλὴν add. M 11 ὅταν] ὁπόταν M 11 ἔχουν] ἔχουσιν MA 12 νήστικον] νῆστιν M νηστικὸς A 12 ἐρεβίνθιν] ἐβίνθια M ἐρεβίνθια A 13 μαῦρον] μαῦρα MA 13 βράσον] βράσας M 13 πότισον πρωῒ νῆστις] πρωῒ πότισον M πότισον κατὰ πρωῒ A 14 τὸ λουτρόν] τοῦ λουτροῦ M 14 τον] om. M 15 νερόν] ζωμίν A 16 βαλσάμου] βαρσάμου A 16 τὸν] om. MA 16–17 οἴνου χλίου] transp. M

καὶ πότισον μετὰ χλίου οἴνου.⌝[MA◁] 6 ἄλλο. βλησκούνιν βράσας μετὰ οἴνου καὶ νεροῦ πότισον αὐτὸν νῆστιν. 7 ἄλλο. μολοχίου ἀγρίου τὸν σπόρον κοπάνισον καὶ πότισον μετὰ οἴνου. 8 ἄλλο. ⌜φλεβοτόμησον τοῦτον καὶ ἃς ἔνι ζεστὰ καὶ κρύον μὴ πίνῃ καὶ ἄνεμος μὴ τὸν κρούῃ.⌝

44 [om. B]μδʹ πρὸς δυσουρίαν· ἔπαρε τὸ λεγόμενον τειχοδαίμονα. πουλίτζιν γὰρ ἔνι μικρούτζικον ἀκρόξανθον καὶ περιπατεῖ εἰς τὰ κλαδία καὶ εἰς τοὺς φράκτας καὶ εἰς τὰς τρύπας. ἔνι γὰρ ἶσον καρύδιν μικρόν. βάλε αὐτὸ εἰς τζουκάλιν. μαδάρισέ το δὲ πρῶτον καὶ ἐξεντέρισέ το καὶ βάλε καὶ καβούρους ἀπὸ τοῦ ποταμίου καὶ κρασὶν καὶ ἃς βράσῃ. καὶ τὸ πρωῒ πότιζε νῆστιν χλίον. 2 ἄλλο. ἀγρίας καὶ ἀδιάντου ζωμὸν πότιζε.

45 [40] [om. B]μεʹ πρὸς κατουρούντα ἕλκος· ⌜καῦσον νάρθηκα καὶ στάκτην ἐκείνου πότισον μετὰ ἔψημα χλίον. 2 ἄλλο. βράσον μέλι καὶ ὕστερον ἀφ' ὅτου γένηται μήτε πολλὰ ξηρὸν μήτε πολλὰ ὑγρόν, ἔχε ἕτοιμον λινόκουκον κοπανιστὸν οἷον τὸ ἀλεύριν καὶ ἀγριαγγουρέας σπόρον τριμμένον καὶ καταστατὸν καὶ κουκουναρία τριμμένα, στάχος, σελινόσπορον οἷον τὸ ἀλεύριν καὶ βάλε τὰ εἰς τὸ μέλι. καὶ ἕνωσε αὐτὰ καὶ ἄς το τρώγῃ τὸ βραδὺ ὀλίγον ἶσα καρύδιν.⌝ 3 ἄλλο. μαγγούνας καὶ τοῦ γλυκορίζου τὸν σπόρον, μίξας με τὸ ζωμὶν τοῦ σχίνου, ἄς το πίῃ. 4 ἄλλο. ⌜ἀμύγδαλα πικρὰ τρῶγε τὴν νύκτα καὶ μὴ δείπνας. 5 ἄλλο. βράσε μέλι καὶ βάλε

17 ἤτοι] ἢ τῆς A 1 χλίου οἴνου] transp. A 2 μολοχίου] μολογχίου M 3 τὸν] om. A 4 κρύον] .... A 4 πίνῃ] κρύον add. A 4 καὶ] μηδὲ A 5 μὴ] om. A 6 ἔπαρε] ἔπαρον MA 8 τὰς] om. MA 8–9 ἶσον καρύδιν] ἰσακρίδην L 9 μικρόν] καὶ add. A 9 βάλε] δὲ add. M 9 τζουκάλιν] τζυκάλιν MA 9 μαδάρισέ] μαδαρήσε A 9 δὲ] om. MA 10 ποταμίου] ποταμοῦ MA 11 καὶ ἃς βράσῃ] om. A 11 βράσῃ] βράσουν M 12 πότιζε] πότισον M πότισέ τον A 13 πρὸς κατουρούντα] ὅπου κατουροῦσιν A 14 μετὰ] με τὸ MA 15 ὕστερον] om. M 16 λινόκουκον] λινόκουκκον M 16 κοπανιστὸν] κοπανιστὸ A 17 καὶ κουκουναρία] κωνάρια MA 18 τὸ] om. A 18 ἀλεύριν] ψιλὸν add. A 19 μέλι] ἀπέσω add. MA 19 ἕνωσε] ἕνωσον MA 19 το] om. A 19 τὸ] κατὰ A 19 ἶσα] ἶσον MA 20 καρύδιν] καρίσιον M καρυδίου A 20 μαγγούνας] σπόρον add. MA 21 μίξας] om. MA 21 με τὸ] με τὸν A 21 ζωμὶν] ζωμὸν A 21 το πίῃ] πίνῃ MA 22 τρῶγε] ἃς τρώγῃ A 22 δείπνας] δειπνὰ A

ἀπέσω ἀμύγδαλα πικρὰ καὶ κρόκον κοπανιστὸν καὶ ἄς τα τρώγῃ τὸ βραδύ.⌝

46 [41] om. Bμϛ′ ὅπου κατουρῶσιν αἷμα· ⌜φλεβοτόμει τούτους ἀπ’ ἀγκῶνος καθόλου φλέβα. 2 ἄλλο. ῥεπανίου σπόρον πότισον μετὰ ὀξείδιν καὶ μέλι ὀλίγον.⌝ 3 L▷λαγωοῦ πυτίαν πότισον μετὰ ὀξείδιν καὶ μέλι ὀλίγον.L◁

47 [42] om. Bμζ′ πρὸς στραγγουρίαν· ⌜στραγγουρία δὲ ἔνι, ὅταν κατουρῇ καὶ στάζῃ ὀλιγούτζικον μετὰ πόνου καὶ ἀνάγκης καὶ βίας. φλεβοτόμει τὴν καθόλου φλέβαν.⌝ 2 ἄλλο. ἔπαρον ἐχίνου τὸ δέρμα ἤτοι τὸ λεγόμενον σκαντζόχοιρον. ὅπου ἔχει ὅλον τοῦ τὸ δέρμα, ὡς ἂν σουβλία μακρά, τούτου τὸ δέρμα ἔπαρον καὶ θές το εἰς τὰ καρβούνια τὰ ζώντανα καὶ κάπνιζέ τον εἰς τὰ αἰδοῖα, ὅπου οὐ δύναται κατουρίσειν. 3 ἄλλο. ▷MA⌜λοῦε δὲ τοῦτον συχνῶς.MA◁ ἡ δὲ τροφή του ἂς ἔνι εὔχυμος ἤτοι κουβίδια λυθρινάριαστακοὶ καλαμάρια καβούρους ποταμίους καὶ πέστρουβας.⌝ 4 ἄλλο. θρύμβον παλαιὸν ἔψησον με τὸ κρασὶν τὸ παλαιόν. ἐξ αὐτοῦ δὲ πότισον κοχλιάρια δ’. 5 ἄλλο. ῥεπάνια κοπάνισε χωρὶς τὰ φύλλα τους. εἶτα βάλε τὸν ζωμόν τους εἰς τζουγκάλιον καὶ κρασὶν καὶ ἂς βράσουν. εἶτα σακέλλισον αὐτὰ δυνατὰ καὶ πότισον αὐτὸν ἡμέρας τρεῖς. 6

---

22 βράσε] βράσον A 1 κοπανιστὸν] κοπανιστᾶ MA 1 τα] τας M 3 ὅπου κατουρῶσιν] πρὸς κατουροῦντα M 4 καθόλου] καθολικὴν MA 4 φλέβα] φλέβαν A 4 ῥεπανίου] ῥεφανίου A 4 μετὰ] μὲ M με τὸ A 5 ὀλίγον] om. M πλὴν βλέπε μὴ βάλῃς πολὺ μέλι add. A 7 δὲ] om. A 8 ὀλιγούτζικον] ὀλιγούτζικον add. M 9 καθόλου] καθολοκὴν A 9 φλέβαν] φλέβα M 9 ἐχίνου] scripsi τοῦ σχίνου LA ἐχοίνου M 10 τὸ λεγόμενον] τὸν λέγουν M 10 ὅλον τοῦ] τὸ ὅλον M 10 τὸ δέρμα] σῶμα A 11 σουβλία] σουγλία L 11 μακρά] παροξυντά add. MA 11 το] om. MA 12 καρβούνια] κάρβουνα MA 12 κάπνιζέ τον] κάπνιζε M κάπνισον A 12 εἰς] om. MA 12 ὅπου] οὔτινος MA 13 κατουρίσειν] κατουρεῖν A 13 δὲ τοῦτον] τὸν τοιοῦτον A 13 συχνῶς] συχνὰ A καὶ add. M 14 ἡ δὲ] ἡ MA 14 του] αὐτοῦ MA 14 κουβίδια] κωμίαδια A 14 λυθρινάρια] scripsi λιθρινάρια LMA 14 στακοὶ] ἀστακοὶ M 15 καβούρους ποταμίους] καβοῦροι ποτάμιοι MA 15 πέστρουβας] πέστρουβαι MA 16 με τὸ κρασὶν τὸ παλαιόν] μετὰ οἴνου παλαιοῦ MA 17 ῥεπάνια] ῥεφάνιν MA 17 κοπάνισε] κοπάνισον MA 17 τους] om. M αὐτοῦ A 17 εἶτα] καὶ A 18 τους] αὐτῶν M αὐτοῦ A 18 εἰς] τὸ add. A 18 τζουγκάλιον] τζουκάλιν M τζυκάλιν A 18 καὶ] βάλε καὶ add. A 18 κρασὶν] οἶνον MA 18 βράσουν] βράσο M 19 αὐτὰ δυνατὰ] ἰσχυρῶς MA

ἄλλο. κολοκύνθιν ξηρὸν κοπάνισον οἷον τὸ ἀλεύριν ἴσα ἕνα κοχλιάριον. καὶ πότιζε μετὰ κρυοῦ νεροῦ. 7 ἄλλο. ⸢δενδρολίβανον ἂς βράσῃ με τὸ κρασὶν καὶ πότιζε αὐτὸν καθ' ἑκάστην ἡμέραν χλίον εἰς ἡμέρας δέκα.⸣ 8 ἄλλο. ἀγρίας ζωμὸν πότιζε διὰ τάχους.

48 [om. B]μη′ ὅπου κατουρεῖ τὰ ῥοῦχα τὴν νύκτα παρὰ τὸ θέλημα· πετενοῦ γούργουρον καύσε τον. εἶτα τρίψε τον καὶ ποίησέ τον ψιλὸν οἷον τὸ ἀλεύριν καὶ πότισον αὐτὸν τὸ πρωῒ νῆστιν μετὰ κρύου ὕδατος. 2 ἄλλο. λαγωοῦ ὀρχίδια ξύσε τα με τὸ μαχαίριν καὶ τὸ ψιλότερον πότισον με τὸν οἶνον τὸ χλίον. 3 ἄλλο. χοίρου φούσκαν ὅπου ἔχει τὸ κατούρημα ταύτην καῦσε εἶτα τρίψε αὐτὴν ὡς ἀλεύριν καὶ πότισον με τὸ κρασὶν τὸ χλίον κατὰ πρωῒ νῆστιν.

49 [43] [om. B]μθ′ πρὸς φλεγμονὴν ἥπατος τὸ λεγόμενον συκώτιν· οὕτως δὲ θέλεις νοήσαι, ⸢ὅτι φλεγμαίνει τὸ ἧπαρ, ὅτι γίνεται πόνος εἰς τὸ δεξιὸν μέρος, ὁ δὲ πόνος κρατεῖ ἐς τὸν ὦμον καὶ εἰς τὸ τέλος τῶν πλευρῶν ἤτοι τὰ κοντὰ πλεύρα. γίνεται δὲ πυρετὸς ὀξὺς καὶ βήχαν ἔχει καὶ τίποτε οὐ πτύει καὶ δίψαν καὶ ἀνορεξίαν ξερᾷ καὶ χολήν. φλεβοτόμει τοῦτον εὐθὺς ἀπ' ἀγκῶνος δεξιοῦ φλέβαν

---

19 πότισον] πότιζε M 1 κοπάνισον] καὶ ποίησον add. MA 1–2 ἴσα ἕνα κοχλιάριον. καὶ πότιζε μετὰ κρυοῦ νεροῦ] καὶ πότιζε αὐτὸ με τὸ νερὸν τὸ κρύον ἴσα ἐν κοχλιάριον M 1 ἕνα] ἓν A 2 πότιζε] πότισον A 2 μετὰ κρυοῦ νεροῦ] με τὸ νερὸν τὸ κρύον A 3 ἂς] om. A 3 βράσῃ] βράσον A 3 καθ' ἑκάστην ἡμέραν χλίον] om. A 4 εἰς] ἕως MA 4 ἀγρίας] πλὴν χλίον ἀγρίας A 4 πότιζε] πότισον MA 4 διὰ τάχους] om. MA 5 ὅπου κατουρεῖ τὰ ῥοῦχα τὴν νύκτα παρὰ τὸ θέλημα] πρὸς κατουροῦντα τὴν νύκτα τὰ ῥοῦχα M 5 τὰ ῥοῦχα] om. A 5 θέλημα] του add. A 6 πετενοῦ] πετεινοῦ M 6 καύσε τον] καῦσον MA 6 ποίησέ τον] ποίησον M 7 ψιλὸν] om. MA 7 τὸ] om. M 7 ἀλεύριν] ψιλὸν add. A 7 νῆστιν] νηστικὸν M 7–8 μετὰ κρύου ὕδατος] με τὸ χλίον τὸ νερὸν M με τὸ νερὸν τὸ χλίον A 8 ξύσε τα] ξύσον M 9 τὸ] τὸν A 10 κατούρημα] scripsi κατούρημαν L του add. MA 10 καῦσε] καῦσον A 10 τρίψε] τρίψον MA 10 ὡς] οἷον M καὶ ποίησον οἷον τὸ A 10 ἀλεύριν] ψιλὸν add. A 11 πότισον] πότιζε M 11 τὸ] πλὴν δὲ A 11 κατὰ πρωῒ] om. A 11 νῆστιν] νηστικὸν M καὶ ἂς ἔνι νηστικὸς A 12 πρὸς] περὶ LA 12 φλεγμονὴν] φλεγμονῆς A 12 τὸ λεγόμενον] ἤτοι M 12 συκώτιν] συκωτίου M 13 δὲ θέλεις] ὀφείλει A 13 νοήσαι] νοήσει L 13 ὅτι] ὅταν A 14 ἐς] ἕως MA 14 τὸν ὦμον] τῶν ὠμῶν A 15 ἤτοι] εἰς add. M 15 κοντὰ πλεύρα] κοντοπλεύρια A 16 βήχαν ἔχει] ἔχει βίαν A 16 καὶ] ἔχει add. A 17 τοῦτον] τούτους MA 17 εὐθὺς] om. A

καθόλου. εἰς δὲ τὸ ἧπαρ ἐπάνω θὲς κατάπλασμα. ἤγουν βάλε κριθάλευρον καὶ τήλης ἀλεύριν καὶ νερὸν καὶ ἕψησέ το καὶ βάλε καὶ λινόσπορον κοπανιστὸν οἷον τὸ ἀλεύριν καὶ φοινίκια κοπανισμένα καὶ ἂς βράσουν. εἶτα ἅπλωνέ τα εἰς τὸ πανὶν καὶ θέτε τὸ εἰς τὸ ἧπαρ.⌝ 2 ἄλλο. ἀψίνθιον καὶ ῥόδα βράσον με τὸ νερὸν καὶ πότισον αὐτὸν πρωῒ νῆστιν. 3 ἄλλο. $^{\triangleright\mathrm{MA}}$σχίνου ἄνθος καὶ πετροσέλινον βράσον με τὸ νερὸν καὶ πότισον αὐτὸν τὸ πρωῒ νῆστιν. 4 ἄλλο. πολυπόδιον κοπάνισον καὶ βράσον με τὸ νερὸν καὶ πότισον αὐτὸν νῆστιν$^{\mathrm{MA}\triangleleft}$. 5 ἄλλο. κροῦε δὲ αὐτὸν καὶ ⌜κλυστήριν οἷον νερὸν θερμὸν ἔλαιον μέλι καὶ ἅλας.⌝

50 [44] $^{\text{om. B}}$ν′ ὅταν γένηται ἀπόστημα εἰς τὸ ἧπαρ· $^{\triangleright\mathrm{L}}$πολυπόδιον κοπάνισον καὶ βράσον με τὸ νερὸν καὶ πότισον αὐτὸν νῆστιν.$^{\mathrm{L}\triangleleft}$ ἄλλο. 2 ⌜κάπνεον βάλε καὶ ἂς βράσῃ με τὸ νερὸν καὶ πότισον αὐτὸν νῆστιν.⌝

51 [44] $^{\text{om. B}}$να′ πρὸς ἧπαρ, ὅταν γένηται σκληρὸν καὶ ὡς ἂν πέτρα δυνατή· ⌜ἔπαρον ἀμμωνιακὸν καὶ ἀξούγγιν ἐλάφου καὶ ταύρου καὶ ποίησον ἔμπλαστρον καὶ θές το εἰς τὸ ἧπαρ. 2 ἄλλο. ἀγρίας καὶ ἀδιάντου ζωμὸν πότισον.⌝

52 [45] $^{\text{om. B}}$νβ′ περὶ ὑδέρων ἤτοι ὅταν πρησθῇ τὸ ὅλον σῶμα καὶ ἡ κοιλία καὶ οἱ πόδες καὶ οἱ ὄρχεις· ⌜ἐπὶ τούτων ἐνδέχεται, ἵνα ποτίζῃς τζουκνίδας σπόρον κοπανιστὸν με τὸ νερὸν τὸ χλίον.⌝ 2 ἄλλο. κράμβης ζωμὸν πότισον. 3 ἄλλο. ⌜βαλὼν σέλινον ἵππουριν τὸ λεγόμενον πολυκόμπιν καὶ ἄσαρ καὶ ἂς βράσουν με τὸ νερὸν καὶ πότιζον καθ’ ἑκάστην ἡμέραν τὸ πρωῒ νῆστιν πλὴν χλιερόν. 4 ἄλλο. πότιζον δὲ κατὰ δύο ἡμέρας τὴν καλουμένην πικρὰν ὡσεὶ ἐξαγίου τὸ ἥμισυ, πλὴν με τὸ χλίον τὸ κρασὶν. 5 ἄλλο. πότισον καὶ τὴν θηριακὴν μετὰ οἴνου χλίου.⌝ 6 ἄλλο. περιστερῶν κόπρα κοπανισμένα

---

17 φλέβαν] φλέβα M 1 καθόλου] καθολικὴν A 2 ἕψησέ] ἕψησον M 4 τὸ] τὰ MA 6 αὐτὸν] om. A 6 πρωῒ νῆστιν] om. MA 6 σχίνου] σχοίνος M 7 αὐτὸν τὸ πρωῒ νῆστιν] om. A 8 πότισον] πότισέ A 9 νῆστιν] ὄντα add. A 10 ἔλαιον] καὶ add. M 15 καὶ] om. A 16 δυνατή] δυνατὸν A 19 πρησθῇ] πρησθῶσι A 19 τὸ ὅλον] ὅλα A 19 σῶμα] τοῦ σώματος M 21 με τὸ] μετὰ M 21 τὸ] om. M 22 βαλὼν] λαβὼν A 22–23 ἵππουριν τὸ λεγόμενον] καὶ τὸ λεγόμενον ἵππουριν ἤτοι A 24 πότιζον] πότιζε M πότισον A 24 τὸ πρωῒ] om. A 24 χλιερόν] χλιαρὸν M χλίον A 25 δὲ] καὶ add. M 25 καλουμένην] λεγομένην A 26 τὸ] om. M 26 πότισον] ποτίζων A

καὶ λεῖα γενόμενα οἷον τὸ ἀλεύριν καὶ τεάφην ψιλὴν καὶ ἀμμωνιακόν. μίξε τὰ πάντα καὶ βαλὼν δαφνέλαιον μίξε τα καὶ ἄλειφε τὴν κοιλίαν πλὴν χλιαρὸν καὶ τὴν σπλήναν του καὶ τὸ ἧπαρ καὶ τὸν στόμαχον ἀπὸ ἐπάνω εἰς κάτω. 7 ἄλλο. θὲς καὶ τὸ ἔμπλαστρον εἰς τὸ ἧπαρ τὸ λεγόμενον παρὰ τῶν ἰατρῶν ἱκέσιον ἔμπλαστρον καὶ ἄλλο δι' ἰτεῶν. ⌜νερὸν μὴ πίνῃ μηδόλως μηδὲ μαγειρίαν τρώγῃ. μηδὲ δειπνᾷ μηδὲ κάθηται ἀλλὰ ἂς κοπιάζῃ ὅσον δύναται μηδὲ πρωῒ τρώγῃ, ἀλλὰ τὸ γεῦμα πλὴν ὀλίγον ἂς τρώγῃ. καὶ ἂς πίνῃ καὶ κρασὶν ἢ ἕνα ἢ δύο με τὸ θερμόν. ἡ δὲ τροφή του τοιαύτη. φακὴν μὴ τρώγῃ μηδὲ φάβα μηδὲ λάχανον οἷον καὶ ἔνι μηδὲ παστὸν ὀψάριν μηδὲ ἐρεβίνθια μηδὲ φασούλια μηδὲ ὀπωρικὰ τοῦ κόσμου ὅλου μὴ δὲ μονόκυθρον μηδὲ λαρδία μηδὲ παστὰ κρέη μηδὲ ἐλαφινὸν μηδὲ τυρία μήτε παστὰ μήτε τρυφερᾶ. εἰ μὴ πετραίους ἰχθύας. αὐγοῦ τζικὸν ῥοφητὸν πλὴν μόνον τὸν κρόκον. πρόβειον κρέας ὀπτὸν ψαχνόν. τὰ πάντα ὀπτὰ καὶ πᾶν ὀψάριν χλωρὸν ἐὰν οὐκ εὑρίσκῃ πετραίους ἰχθύας ἂς τρώγῃ αὐτὸ καὶ μόνον. 8 ἄλλο. πότιζε αὐτὸν καθ' ἑκάστην πρωῒ ἀφινθίου ζωμόν. βράσε τὸ ἀψίνθιν με τὸ νερὸν καὶ πότιζε αὐτὸν τὸ πρωῒ πλὴν χλίον.⌝

53 [47] om. Bνγ' ἐὰν πρησθῇ ἡ σπλῆνα· ⌜φλεβοτόμησον αὐτὸν εἰς τὸ χέριν τὸ ἀριστερὸν ἢ εἰς τὴν ἄκραν τοῦ χερίου τοῦ ἀριστεροῦ ἢ εἰς τὸν μικρὸν τὸν δάκτυλον. 2 ἄλλο. κριθάλευρον βράσον με τὸ νερὸν καὶ ἅπλωνέ το εἰς τὸ πανὶν καὶ θέτε τὸ εἰς τὴν σπλῆναν

27 κόπρα] κόπρια MA 1 τεάφην] τεάφιν MA 1 ψιλὴν] ψιλὸν M 2 μίξε τὰ πάντα] μίξον M μίξε πάντα A 2 μίξε τα καὶ] om. M 3 κοιλίαν] τοῦ add. A 3 του] om. M 3–4 τὸν στόμαχον] του add. A 4 ἐπάνω εἰς] πάνου ἕως M ἄνω ἕως A 4 θὲς] δὲ add. A 4 τὸ] ἕτερον A 5 ἧπαρ] του add. M 6 ἄλλο δι' ἰτεῶν] scripsi ἄλλον ἠδιητέον L ἄλλα ἰδιιτθῶν M ἰδιητθῶν A 7 κάθηται] καθέζεται M 7 μηδὲ] τὸ add. A 9 ἕνα] ἕναν ποτήριν A 9 του] om. MA 10 φάβα] φάβατα MA 10 ἔνι] ἂν ἔνι MA 10 ὀψάριν] ὀψάρην A 11 μηδὲ] μὴ M 11 ὅλου] ὅλα A 12 λαρδία] λαρδὶν A 12 κρέη] κρέει L 12 ἐλαφινὸν] ἐλάφεια M ἐλάφια A 13 τυρία] τηρία M 14 τζικὸν] om. M 15 ψαχνόν] παχὺ A 15 ὀψάριν] ὀψάριον MA 15 οὐκ] οὖν MA 16 αὐτὸ καὶ μόνον] αὐτὰ μόνον M αὐτοὺς καὶ μόνους καὶ κρόκον αὐγῶν A 17 πρωῒ] πρωΐαν MA 17 βράσε] βράσον M 17 ἀψίνθιν] ἀψίνθιον M 18 πότιζε αὐτὸν τὸ πρωῒ πλὴν χλίον] τὸ πρωῒ πότιζε αὐτὸν χλίον MA 19 ἡ σπλῆνα] ὁ σπλήν M 19–20 εἰς τὸ] ἀπὸ MA 20 χέριν] τὴν χεῖραν A 20 τὸ ἀριστερὸν] τὴν ἀριστερὰν A 21 τὸν] om. MA 22 τὴν] τὸν A

πλὴν βάλε χοίρειον ἀξούγγιν ἢ ὄρνιθος καὶ ἀφ’ ὅτου τὰ ἑψήσῃς βάλε χηναρίου ἀξούγγιν, πλὴν δὲ ἄπαστα μὴ ἔχων ἅλας μηδόλως. 3 ἄλλο. τήλης ἀλεῦρον, καρδαμόσπορον κοπανιστὸν ψιλὸν οἷον τὸ ἀλεύριν καὶ ἰσχάδων ζωμὶν καὶ βράσων ταῦτα καὶ ἅπλωνέ τα εἰς τὸ πανὶν καὶ θέτε τα εἰς τὴν σπλῆναν τοῦ ἀρρώστου⌝ κατὰ πρωΐ. 4 ἄλλο. λινόσπορον ψιλὸν καὶ κόπρον αἰγίαν ποίησον κατάπλασμα καὶ θέτε το εἰς τὴν σπλῆναν, ἀντὶ δὲ νεροῦ βάλε ζωμὶν ἰσχάδων.

54 [46,47] om. B νδ′ ὅταν γένηται ἡ σπλῆνα σκληρά· ⌜βάλε ἀμμωνιακὸν κοπανιστὸν τὸ βραδὺ εἰς τὸ ὀξείδιν τὸ δριμὺ καὶ ἂς βραχῇ καλὰ ἔστ’ ἂν λυθῇ. τότε ἄλειφε τὴν σπλῆναν πολλὴν ὥραν καὶ καθ’ ἑκάστην ἡμέραν τὸ πρωΐ πλὴν χλίον.⌝ 2 ἄλλο. ἀμμωνιακὸν ἑξάγιον α’ κοπανίσας οἷον τὸ ἀλεύριν. πότιζον μετὰ χλίου οἴνου τὸ πρωΐ. 3 ἄλλο. τὰ φλούδια τῆς καππάρεως κοπάνισον καὶ βάλε τα εἰς τὸ ὀξείδιν. καὶ τάραξέ τα καὶ ἄλειφε τὴν σπλῆναν τὸ πρωΐ πλὴν χλίον. 4 ἄλλο. ⌜ἔπαρε βοτάνιν τὸ λεγόμενον παρὰ τῶν ἰατρῶν κενταύριον παρὰ δὲ τῶν χωρικῶν καλεῖται ἀθάλη. ταύτην βράσον με τὸ νερὸν καὶ πότιζον αὐτὸν κατὰ πρωΐ νῆστιν. 5 ἄλλο. βάλε ὀξείδιν ὀλίγον καὶ κρασὶν καὶ ἀγρίου λαπάτου τὴν ῥίζαν καὶ ἂς βράσῃ. καὶ πότισον αὐτὸ τὸ πρωΐ νῆστιν πλὴν χλίον. 6 ἄλλο. ἂς κοπιάζῃ δὲ ὅσον δύναται καὶ τὰ ψυχρὰ μὴ τρώγῃ. θὲς δ’ ἐπάνω εἰς τὴν σπλῆναν

22 σπλῆναν] σπλῆνα A 1 ἢ] καὶ ἀξούγγιν M καὶ καὶ ἀξούγγιν A 2 μηδόλως] οὐδ’ ὅλως M 3 ἀλεῦρον] ἀλεύριν MA 3–4 ψιλὸν οἷον τὸ ἀλεύριν] om. A 4 ζωμὶν] ζωμὸν MA 4 καὶ] om. MA 4 βράσων] βράσον MA 4 ἅπλωνέ] θέτε A 5 πανὶν] ἀπλωτὰ add. A 5 καὶ] εἶτα A 5 τα] αὐτὰ A 5 τὴν σπλῆναν] τὸν σπλῆνα MA 5 τοῦ ἀρρώστου] ὡσὰ ἔμπλαστρον πλὴν δὲ A 6 καὶ] om. M 6 αἰγίαν] αἴγιον καὶ M αἴγειον A 7 σπλῆναν] σπλῆνα MA 7 ζωμὶν] ζωμὸν M 8 ὅταν] ἐὰν M 8 ἡ σπλῆνα] ὁ σπλὴν MA 8 σκληρά] ξηρὸς MA 9 βραδὺ] βραχὺ M 9 ὀξείδιν] ὄξος MA 10 ἔστ’] ἕως οὗ A 10 ἂν] να A 10 λυθῇ] λύσῃ A 10 τότε] καὶ τότε MA 10 σπλῆναν] σπλῆνα MA 11 τὸ πρωΐ] om. A 12 πότιζον] πότιζε M 12 χλίου οἴνου] χλίον οἶνον M πλὴν add. A 13 τα] om. MA 14 ὀξείδιν] ὄξος MA 14 σπλῆναν] σπλῆνα M 14 πλὴν] om. A 15 ἔπαρε] ἔπαρον MA 16 καλεῖται] om. MA 16 ἀθάλη] ἀθάλιν M ἀθάλλη A 17 πότιζον] πότισον A 17 νῆστιν] om. A 17 ὀξείδιν] ὄξος MA 17 ὀλίγον] om. A 18 ἀγρίου λαπάτου τὴν] ἀγριολαπάθου M ἀγριολαπάθου A 18 βράσῃ] βράσουν MA 18–19 πότισον] πότιζον M 19 αὐτὸ τὸ] αὐτὸν καθ’ ἑκάστην MA 19 πρωΐ] πρωΐαν A 19 νῆστιν] νηστικὸν MA 19 πλὴν] ἂς ἔνι add. A 19 δὲ] om. A 20 τὰ] om. A

τὸ ἔμπλαστρον τὸ λεγόμενον παρὰ τῶν ▷om. Lἰατρῶν κιτρίνην καὶ ἡ οὐσία μέλαινα.⌝om. L◁

55 [48] om. LBνε´ πρὸς ἴκτερον τὸν λέγουσι οἱ χωρικοὶ χρυσιασμόν· ⌜ἴκτερος ἔνι ὅταν ὅλον τὸ σῶμα καὶ οἱ ὀφθαλμοὶ γένωνται κίτρινοι πολλά. ἀρχὴν τοὺς τοιούτους φλεβοτόμησον, καὶ πότισον καὶ βοήθημα. 2 ἄλλο. βάλε λαπάθου ῥίζαν καὶ νερὸν καὶ ἂς βράσουν καὶ τότε ἂς πίνῃ ἀπ' ἐκεῖνο τὸ νερόν. 3 ἄλλο. ἀψίνθιον βράσον με τὸ νερὸν καὶ πότιζε κατὰ πρωῒ νηστικόν.⌝ 4 ἄλλο. χρυσολαχάνου σπόρον κοπάνισον καὶ μίξον με τὸ νερὸν τοῦ γλυκέου γάλακτος. καὶ πότισον τὸ πρωῒ χλίον. 5 ἄλλο. ⌜πότισον τὸ πρωῒ τὴν πικρὰν με τὸ κρασίν.⌝ 6 ἄλλο. κινάμωμον ποίησον οἷον τὸ ἀλεύριν καὶ πότισον με τὸ νερὸν καὶ ὀλίγον ὀξείδιν καὶ μέλι ὀλίγον. ἐὰν δὲ οἱ ὀφθαλμοὶ κίτρινοι, βάλε γάλα γυναίκειον οὕτως ὡς τὸ ἐκβάλλῃ. 7 ἄλλο. βάλε γάλα γυναικεῖον εἰς τὴν μύτην τοῦ ἀπέσω. 8 ἄλλο. μελάνθιον κοπάνισον καὶ μίξον με τὸ ὄξος καὶ τάραξέ το. εἶτα σακέλισον αὐτὸ με τὸ πανὶν καὶ τὸν ζωμὸν στάξον εἰς τὴν μύτην του. κυκλάμινον κοπάνισον καὶ τὸν ζωμόν του βάλε εἰς τὴν μύτην του. 9 ἄλλο. ⌜κριθάλευρον καὶ τήλης ἄλευρον καὶ λινόσπορον ποίησον οἷον τὸ ἀλεύριν. καὶ μίξε τα καὶ βράσε τα με τὸ κρασὶν καὶ τὸ ὄξος καὶ ἀπλωνέ τα εἰς τὸ πανὶν καὶ θέτε το εἰς τὸ δεξιὸν μέρος ἤτοι εἰς τὸ ἧπαρ καὶ κρασὶν μὴ πίνῃ μηδὲ εἰς ἥλιον κάθηται.⌝

56 [49] om. LBνϛ´ ὅταν ὁ ὀμφαλὸς τοῦ ἀνθρώπου εὔγῃ καὶ γένηται

---

3 πρὸς ἴκτερον] This chapter is listed in the pinax of L. 22 ὅταν] This chapter is listed in the pinax of L; it is striking that MA and N have an incorrect reading, 'eye' instead of 'navel'; only L is correct. Maybe the archetype of both braches of the transmission had an abbreviated form of the word that was then independently misread by several scribes.

---

20 δ'] om. M, δὲ A 1–2 ἡ οὐσία μέλαινα] οὐσίαν μέλαιναν A 3 πρὸς ἴκτερον] περὶ ἰκτέρου A 3–4 τὸν λέγουσι οἱ χωρικοὶ χρυσιασμόν] om. A 3–4 χρυσιασμόν] scripsi χρισιασμόν M 7 τὸ νερόν] ἀντὶ νεροῦ A 8 πότιζε] πότιζον A 8 νηστικόν] scripsi νηκόν M νήστην A 9 γλυκέου γάλακτος] γάλακτος τοῦ γλυκέου A 10 τὸ πρωῒ] καὶ A 11 κρασίν] τὸ πρωῒ add. A 12 ὀξείδιν] ὄξος A 12 δὲ] ἔνεν add. A 13 γυναίκειον] χλίον add. A 13 ἐκβάλλῃ] εὐγάλῃ. βάλε το δὲ συχνῶς καὶ πολλάκις A 15 μίξον] μίξε το A 18 ἄλευρον] ἀλεύριν A 20 το] τα A

---

12 ἐὰν] Here starts a new chapter in A.

μέγας· ⌜τρυγέαν οἴνου τρίψον καὶ ῥόδα καὶ κηκίδιν λεῖον μίξε με τὸ κρασὶν καὶ θές τα ἐπάνω. 2 ἄλλο. ἀφ' ὅτου δὲ θέσης ἐκεῖνα, θὲς ἐπάνω σπόγγον καινούργιον μάλαξέ τον δὲ πρῶτον με τὸ ὀξείδιν καὶ τὸ κρασὶν καὶ δένε το⌝ δυνατά. ἂς ἔνι δεδεμένον ἕως ἡμέρας ε'.

57 [50] om. LBνζ' πρὸς φλεγμονὴν ὀρχιδίων ἤτοι ἐὰν πρησθῶσιν· ⌜φλεβοτόμει τοὺς τοιούτους κάτω εἰς τὰ στραγάλια τῶν δύο ποδῶν. 2 ἄλλο. καθάρισον σταφίδας ἀπὸ τῶν κοκκίων αὐτῶν καὶ κοπάνισον καὶ κύμινον καὶ ποίησον οἷον τὸ ἀλεύριν. εἶτα βάλε τα εἰς φαβάτων ἀλεύριν καὶ ἕψησον ταῦτα με τὸ νερὸν καὶ ἁπλωνέ τα εἰς τὸ πανὶν καὶ θέτε τα ἐπάνω.⌝ 3 ἄλλο. ἀσφοδελέας ῥίζαν ἕψησον με τὸ νερὸν εἶτα κοπάνισέ την καλὰ καὶ μίξε την με τὸ κριθάλευρον καὶ ἕψησον με τὸ νερὸν καὶ ἅπλωνέ τα εἰς τὸ πανὶν καὶ θέτε τα εἰς τὰ ὀρχίδια. 4 ἄλλο. ⌜λινόσπορον ποίησον ψιλὸν οἷον τὸ ἀλεύριν καὶ ἕψησον με τὸ ἕψημα ἢ τήλης ἀλεύριν ἕψησον με τὸ μέλι καὶ τὸ κρασὶν ἢ κρίνου ῥίζαν ἕψησον με τὸ κριθάλευρον καὶ τὸ ἕψημα θέτε το ἐπάνω.⌝ 5 ἄλλο. κολοκύνθης ὠμῆς ξύσματα μίξον με τὸ νερὸν καὶ τὸ κριθάλευρον καὶ ἕψησον καὶ θέτε τα ἐπάνω. 6 ἄλλο. ⌜ἔμπλαστρον τὸ λεγόμενον παρὰ τῶν ἰατρῶν ἀνετὴν ταύτην τίθει ἐπάνω καὶ ἄλλο ἔμπλαστρον τὸ λεγόμενον διὰ χυλοῦ.⌝

58 [51] om. LBνη' πρὸς γονόρροιας ἤτοι χωρὶς συνουσίαν ῥέῃ ὁ σπόρος του παρὰ τὸ θέλημά του καὶ χωρὶς ὀρέξεως γυναικός· ἐπὶ τούτων ἡ θεραπεία ἔνι τοιαύτη ⌜να κοπιᾷ πάντοτεν ἀνάπαυσιν να

---

6 πρὸς] This chapter is listed in the pinax to L. 21 πρὸς] This chapter is listed in the pinax to L.

---

22 ὀμφαλὸς] scripsi ὀφθαλμὸς MA cf. pinax to L 1 λεῖον] καὶ add. A 1 μίξε] τα add. A 2 ἀφ'] καὶ ἀφ' A 2 δὲ] om. A 2 ἐκεῖνα] ἐκεῖνο A 6 φλεγμονὴν ὀρχιδίων] ὀρχίδια A 6 ἤτοι ἐὰν] ὅταν A 6 πρησθῶσιν] προσθῶσιν ἤτοι φλεγμαίνουσιν A 8 τῶν κοκκίων αὐτῶν] τὰ κοκκία τῶν A 8–9 κοπάνισον] αὐτὰς add. A 9–10 εἶτα βάλε τα εἰς φαβάτων ἀλεύριν] om. A 11 ἐπάνω] εἰς τὰ ὀρχίδια A 11 ἀσφοδελέας] ἀσφοδελαίας A 13 τα] το A 14 τὰ ὀρχίδια] ὄρχεις A 14 ψιλὸν οἷον τὸ] ὡς A 17 μίξον] μίξε τα A 18 νερὸν καὶ τὸ] om. A 18 ἕψησον] ἕψησέ τα με τὸ νερὸν A 18 ἐπάνω] om. A 20 ἄλλο] ἕτερον A 21 πρὸς] περὶ A 21 ἤτοι] ὅταν add. A 21 συνουσίαν] ἤγουν χωρὶς να πέση με τὴν γυναῖκαν add. A 21–22 ὁ σπόρος του] om. A 22 του] ὁ σπόρος του πυκνὰ πυκνὰ add. A 22 καὶ] om. A

μὴ ἔχῃ μηδεμίαν ὥραν καὶ ἡ τροφὴ του ξηρὰ ἤγουν ἄρτον νερὸν κρέας ψαχνὸν να τὸ ἐψίνῃ δυνατὰ να μὴ ἔχῃ τίποτε λιπαρότητα τὰ λιπαρὰ να μὴ τρώγῃ μηδὲ πηκτὴν μηδὲ αὐγὸν ἄσπρον.⌝

59 [52] om. LB νθʹ ὅταν καυλώσῃ παρὰ τὸ θέλημα του· θεραπεία δὲ ἐπὶ τούτων ποτίζειν τοῦ πηγάνου τὸν ζωμὸν καὶ τρώγειν καὶ τοῦτο συνεχῶς. 2 ἄλλο. ὠφέλιμος δὲ ἔνι ἐπὶ τῶν τοιούτων καὶ ὁ ἔμετος τοῦ να ξερᾷ πρὸ τροφῆς χυμούς τινας να κοπιᾷ καὶ πολλὰ να τρίβῃ δὲ καὶ τοὺς πόδας καὶ τὰς χεῖρας.

60 [53] om. LB ξʹ πρὸς φλεγμονὴν δακτυλίων· ⌜ἰντυβίου χυλὸν βάλε καὶ ῥοδέλαιον καὶ ψιμμίθιν καὶ λίβανον τριμμένον καὶ ποίησον ὡσὰν ζυμάριν καὶ θές το με τὸ πανίν. 2 ἄλλο. ψιχίας ἄρτου καὶ βάλε καὶ κρόκον κοπανιστὸν καὶ ἔψημα καὶ μάλαξον αὐτὰ καὶ θές τα ἐπάνω.⌝

61 [54] om. LB ξαʹ πρὸς ἐξοχάδας· ⌜ἐὰν εἶναι πρησμέναι, ἄλειψον αὐτὰς ῥοδέλαιον καὶ ἀφ' ὅτου ἀλείψεις τὸ ῥοδέλαιον πάσσε ἐπάνω κοπανισμένον ψιμμίθιον ψιλὸν καὶ λιθάργυρον ἢ μαστίχιον. 2 ἄλλο. ἐὰν δὲ εἶναι πρησμέναι πάλιν φλεβοτόμησον αὐτόν.⌝ 3 ἄλλο. ἐὰν δὲ μὴ εἶναι πρησμέναι, ἔπαρον κυπαρίσσου φύλλα καὶ καύσε τα καὶ ποίησον οἷον τὸ ἀλεύριν. καὶ τότε πλύνε τας με τὸ κρασὶν τὸ χλίον εἶτα πάσον τὰ φύλλα τὰ κοπανισμένα ἐπάνω. 4 ἄλλο. τὰ ὀστέα τῶν φοινίκων καῦσον καὶ τὴν στάκτην ἐκείνων πάσσε πλὴν πρῶτον ἀπόνιψέ τα με τὸ κρασίν. 5 ἄλλο. αὔγου φλούδιον κοπάνισον καὶ ποίησον ψιλὸν καὶ κατάπλασσε ἐπάνω. 6 ἀλλο. ▷Aπρινοκούκιν κοπάνισον ψιλὸν καὶ κατάπλασσε ἐπάνω.A◁ 7 ἄλλο. ⌜μελάνθιον

---

4 ὅταν] This chapter is listed in the pinax to L. 9 πρὸς] This chapter is listed in the pinax to L. 13 πρὸς] This chapter is listed in the pinax to L.

---

22 ὀρέξεως γυναικός] χωρὶς να ἔνεν καυλωμένος A 1 μηδεμίαν] κἂν μίαν A 3 αὐγὸν ἄσπρον] ἄσπρον αὐγοῦ A 4 ὅταν] περὶ πριασμοῦ ἤτοι ὅταν A 4 καυλώσῃ] δυνατὰ add. A 4–5 θεραπεία δὲ ἐπὶ τούτων] om. A 5 ποτίζειν] πότιζέ τον A 5 τρώγειν] τρώγει A 5 τοῦτο] τοῦτον A 6 δὲ] om. A 6 τῶν τοιούτων] τούτοις A 7 τοῦ] om. A 7 πολλὰ] καὶ add. A 9 δακτυλίων] scripsi δακτύλων MA 9 χυλὸν] χυλὸς M 11 βάλε καὶ] om. A 12 ἐπάνω] εἰς τοὺς δακτύλους add. A 14 καὶ ἀφ' ὅτου ἀλείψεις τὸ ῥοδέλαιον] εἶτα A 15 μαστίχιον] μαστίχην A 16 εἶναι πρησμέναι πάλιν] πάλιν ἔνι πρησμέναι A 17 μὴ εἶναι πρησμέναι] ἐὰν δὲ ἔνι ἀφλέγμανται A 19 πάσον] πάσσε A 19 ἐπάνω] om. A 21 φλούδιον] φλούδια A 22 καὶ ποίησον ψιλὸν] ψιλὰ A

κοπάνισον καὶ ποίησον ψιλὸν οἷον τὸ ἀλεύριν καὶ πρῶτον πλύνε με τὸ κρασὶν εἶτα πάσσε τὸ μελάνθιον. 8 ἄλλο. τῆς ἐλαίας τὰ ὀστέα καῦσον καὶ κολοκύνθιν ξηρὸν καὶ τὴν στάκτην ἐκείνων πάσον πλὴν πρῶτον ἀπόπλυνέ τας με τὸ κρασὶν τὸ χλίον. 9 ἄλλο. ἐὰν δὲ εἶναι ἐξοχάδες καὶ κρέμανται ἔξω φουσκομέναι καὶ πρησμέναι ἀπὸ τοῦ αἵματος ἄλειψον αὐτὰς κυκλαμίνου χυλόν. 10 ἄλλο. ἐὰν δὲ πολὺ ῥέῃ αἷμα ἀπ' αὐτὰς βάλε σχίνου φύλλα μυρσίνης φύλλα βάτου φύλλα ῥόδιν κοπανιστὸν ἤτοι τὰ φλούδια κηκίδιν κοπανισμένον τριακοντάφυλλα ξηρὰ κοπανισμένα καὶ ὄξος ὀλίγον καὶ κρασὶν καὶ ἂς βράσουν. εἶτα βρέχε σπόγγον καινούργιον εἰς τὸν ζωμὸν ἐκείνων καὶ πυρίαζε αὐτὰς πυκνά.⌝

62 [54] [om. LB]ξβ′ πρὸς ἕδραν ὅταν εὐγένῃ ἔξω· ⌜πρῶτον μὲν ἀποσφούγγισέ την με τὸ κρασὶν εἶτα ἔχε τα κοπανισμένα ψιλὰ ταῦτα καὶ πάσον ἐπάνω ἤτοι ψιμμίθιον κηκίδιον βῶλον ποίησον πάντα ψιλὰ καὶ μίξε τα καὶ πάσσε τα ἐπάνω.⌝ τὸ κατούρημα τοῦ ἀρρώστου ὅπου ἔχει τὰς ἐξοχάδας ἔπαρον αὐτὸ ἡνίκα τὸ κατουρίσαι ζεστὸν καὶ ἄντλει τὰς ἐξοχάδας τούτου πυκνὰ καὶ καθημέραν. 2 ἄλλο. τοῦ κοχλίου τὸ φλοῦδιν καῦσον καὶ ποίησον οἷον τὸ ἀλεύριν μίξον με τὸ λίβανον τὸ τριμμένον καὶ πάσσε ἐπάνω ▷[A]πλὴν πρῶτον ἀπόπλυνέ τας μετὰ χλίου οἴνου. μίξον δὲ τὸ ἀλεύριν τοῦ κοχλίου μὲ λιβάνιν τριμμένον.[A]◁. 3 ἄλλο. ὀστρυδίων φλούδια καῦσον καὶ ποίησον

12 πρὸς] This chapter is listed in the pinax to L.

22 κατάπλασσε] πάσον A 1 καὶ ποίησον] om. A 1 πλύνε] πλύνας A 2 μελάνθιον] ἐπάνω add. A 3 καῦσον] καύσε τα A 3 κολοκύνθιν] κολοκύνθην A 3 πάσον] πάσσε A 4 ἐὰν] πρὸς ἐξοχάδας ἐὰν A 4–5 δὲ εἶναι] ἔνι A 5 ἐξοχάδες] ἐξοχάδαι A 5 κρέμανται] κρέμμωνται A 6 κυκλαμίνου] κυκλαμιναίας A 7 ῥέῃ αἷμα] transp. A 8 ῥόδιν κοπανιστὸν ἤτοι τὰ φλούδια] ῥοϊδίου φλοῦδιν καὶ A 8 κοπανισμένον] κοπανιστὰ A 10 ἐκείνων] ἐκεῖνον A 12 εὐγένῃ] εὔγῃ A, s.l. A[2] add. ὁ κάθισμα 12–13 ἀποσφούγγισέ] ἀποσφόγγισέ A 13 τα] om. A 13–14 κοπανισμένα... ἤτοι] om. A 14 βῶλον] καὶ κοπάνισον καὶ A 16–17 ἔπαρον αὐτὸ ἡνίκα τὸ κατουρίσαι ζεστὸν καὶ ἄντλει τὰς ἐξοχάδας] M in marg. 16 κατουρίσαι] κατουρεῖ A 17 πυκνὰ] πυκνὰ add. A 18 φλοῦδιν] λέπος A 18 καῦσον] καῦσας A 18 ποίησον] ποιήσας ψιλὸν A 18–19 μίξον... καὶ] om. A 19 πάσσε] τὸ add. A 21 ὀστρυδίων] scripsi ἰστριδίων M ὀστροδίων A

4 ἐὰν] Here starts a new chapter in A.

οἷον τὸ ἀλεύριν καὶ μίξε τα μετὰ ῥοδελαίου καὶ ἄλειφε τὴν ἕδραν. 4 ἄλλο. κηκίδια κοπάνισον ψιλὰ καὶ ῥοϊδίου φλοῦδιν καὶ λίβανον καὶ λιθάργυρον. τὰ πάντα κοπάνισον καὶ ποίησον οἷον τὸ ἀλεύριν καὶ μίξε τα καὶ πάσσέ τα εἰς τὴν ἕδραν. πλὴν ἀποπλύνεται μετὰ οἴνου χλίου.

63 [55] om. LBξγ′ ὅταν τῆς γυναικὸς τὰ συνήθεια κρατοῦνται τὰ λεγόμενα παρὰ τῶν ἰατρῶν ἔμμηνα καταμήνια, παρὰ δὲ τῶν γυναικῶν γυναικεῖα καλοῦνται. τί δὲ να ποιήσουν να κινήσουν πάλιν· ⌜φλεβοτόμησον ταύτην εἰς τοὺς ἀστραγάλους τῶν δύο ποδῶν⌝ καὶ συκίασέ την εἰς τὰ μερία ἀλλὰ ἐκεῖ ὅπου τὴν συκίαν θέλει να μὴ τὸν κόψης τὸν τόπον ἐκεῖνον ὡς τὸ ποιοῦν πολλοὶ καὶ φρόνιμοι καὶ χωρικοὶ εἶτα ⌜πότισον αὐτὴν τὴν λεγομένην ἱεράν. πότιζε καὶ καστόριον τριμμένον με τὸ κρασὶν τὸ χλίον.⌝ 2 ἄλλο. ἀψίνθιον βράσον με τὸ κρασὶν τὸ χλίον καὶ πότισον αὐτήν. 3 ἄλλο. πήγανον κοπάνισον καὶ τὸν ζωμὸν πότισον. ⌜ἂς ἔνι δὲ ζεστὰ καὶ λοῦέ την συνεχῶς πλὴν εἰς καιρόν.⌝

64 [56] om. LBξδ′ περὶ ἰσχιάδος ἤτοι τὸ λεγόμενον περινέφριον· ⌜τὴν ἀρχὴν ἡνίκα γένηται, φλεβοτόμησον αὐτὸν ἀπὸ τοῦ ἀγκῶνος τὴν φλέβαν τὴν λεγομένην καθολικήν.⌝ 2 ἄλλο. κοπάνισον λυπηνάρια καὶ τὸ ἀλεύριν ποίησον ψιλὸν καὶ βράσε με τὸ ὄξος. ὕστερον δὲ βάλε καὶ μέλι καὶ ἅπλωνέ τα εἰς τὸ πανίν. καὶ θέτε το ἐπάνω ἔνθα ἔνι ἡ ὀδύνη. 3 ἄλλο. ἀλεύριν ῥητίνης ἕψησον με τὸ ὄξος. ὕστερον δὲ βάλε καὶ μέλι καὶ θέτε το ἐπάνω μετὰ πανίου ἔνθα ἔνι ὁ πόνος. 4 ἄλλο. κοπάνισον τὸν λεγόμενον ποταμογείτονα ἤτοι καλαμίνθην πλὴν τὰ φύλλα καὶ θέτε τα ἐπάνω ἔνθα ἐστὶν ὁ πόνος. 5 ἄλλο.

---

6 ὅταν] This chapter is listed in the pinax to L. 17 περὶ] This chapter is listed in the pinax to L.

---

21 καῦσον] καύσε τα A 1 μετὰ] τοῦ add. A 2 κοπάνισον] κοπανισμένα A 2 καὶ] ῥοΐδην καὶ add. A 3 οἷον τὸ] ὡς A 4 ἀποπλύνεται] ἀπόπλυνέ την A 6 κρατοῦνται] ἤγουν add. A 7 καταμήνια] om. A 7 παρὰ δὲ] καὶ παρὰ A 7–8 γυναικῶν] χωρικῶν A 8 καλοῦνται] om. A 8 ποιήσουν] ποιήσεις A 8 πάλιν] om. A 9 ἀστραγάλους] ἀστραγγάλους A 10–12 ἀλλὰ... ἐκεῖ] om. A 14 τὸ χλίον] καὶ ἂς χλιαίνῃ A 15 δὲ] καὶ add. A 17 ἰσχιάδος] ἰσχιάδες M 17 ἤτοι τὸ λεγόμενον περινέφριον] om. A 18 τοῦ] om. A 20 ἀλεύριν] τῶν add. A 20 βράσε] τα add. A 21 το] τα A 22 ὄξος] ὀξείδιν A 24 ποταμογείτονα] ποταμογείτονον M

⌜πότισον δὲ αὐτὸν καὶ καστόριον τριμμένον μὲ ὀλίγον ὀξείδιν καὶ μέλι.⌝ 6 ἄλλο. βράσον βοτάνιν τὸ λεγόμενον παρὰ τῶν ἰατρῶν κενταύριον, παρὰ δὲ τῶν χωρικῶν ἀθάλις καλεῖται. τοῦτο βράσον με τὸ νερὸν καὶ ἔπαρε τὸν ζωμόν του καὶ βάλε μέλι καὶ ἔλαιον καὶ ἅλας καὶ ⌜ἰάτρευέ τον κλυστήριν ἢ τὸ ζωμὶν τῶν λυπηναρίων.⌝ ἀφ' ὅτου ἐψηθῶσιν καλὰ βάλε και ἰάτρευσον. 7 ἄλλο. πρὸ τροφῆς δὲ ἄλλης νηστικὸς ἂς τρώγῃ ῥεπάνιν καὶ ἀφ' ὅτου τὸ φάγῃ μετὰ ὥραν πολλὴν ἄς τα ξερᾷ ἵνα ἐκβάλῃ χυμούς. 8 ἄλλο. ⌜βάλε καστόριον τριμμένον καὶ πέπερι καὶ εὐφόλβιον τριμμένον καὶ πύρεθρον καὶ κηρὸν καὶ χαλβάνιν καὶ πηγανέλαιον καὶ μίξε τα καὶ ἄλειφε τὸν τόπον χλίον τοῦτο ἔνθα ἔνι ὁ πόνος. πλὴν καθ' ἑκάστην ἡμέραν δευτέραν.⌝ 9 ἄλλο. τὸ λεγόμενον ἔμπλαστρον παρὰ τῶν δαφνῶν ἅπλωνέ το εἰς τὸ πανὶν καὶ ἐπίθες εἰς τὸν τόπον ἔνθα ἔνι ἡ ὀδύνη. 10 ἄλλο. ⌜ἔμπλαστρον τὸ λεγόμενον ἰβεριάδα ταύτην μάλαξον μετὰ χοιρίου ἀξουγγίου καὶ θέτε την ἐπάνω εἰς τὸ πανὶν ▷om. Aκαὶ ἐπίθες εἰς τὸν τόπον ἔνθα ἔνι ἡ ὀδύνη⌝om. A◁ 11 ἄλλο. κοπάνισον σινάπιν ψιλὸν καὶ ἰσχάδια καὶ μίξον καὶ θές τα ἐπάνω εἰς τὸν τόπον. 12 ἄλλο. ἀφ' ὅτου δὲ εὔγῃ ἀπὸ τοῦ λουτροῦ ἄλειψον τὸν τόπον ναρδέλαιον ἢ πηγανέλαιον καὶ ὕστερον πάσσε ἐπάνω ἔνθα ἔνι ὁ πόνος. τεάφιν ψιλὸν καὶ ξηρὸν καὶ πίσσαν τριμμένην ▷Aπλὴν δὲ ἂς ἔνε πολλὰ ψιλὰ ὡσὰν σκόνηA◁.

65 [57] om. LBξε' ὅταν οἱ δάκτυλοι κατασχισθῶσιν· ὑγροπίσσιν μίξον μετὰ ῥοδελαίου καὶ ἄλειφε τὰ κατεσχισμένα μέρη. 2 ἄλλο. σκύλλαν ἤτοι σκυλλοκρόμμυδον βράσον με τὸ ἔλαιον τὸ χρηστὸν καὶ ἔκβαλέ την σκύλλαν καὶ ῥίψε την καὶ εἰς τὸ ἔλαιον ἐκεῖνο

22 ὅταν] This chapter is listed in the pinax to L.

25 ἐστὶν] ἔνι A 1 καὶ] om. A 2 παρὰ τῶν ἰατρῶν] om. A 3 ἀθάλις] ἀθάλην A 3 καλεῖται] om. A 4 ἔπαρε] ἔπαρον A 4 τὸν ζωμόν] τὸ ζουμὶν A 5 τὸ ζωμὶν τῶν λυπηναρίων] τῶν λυπηναρίων τὸ ζουμὶν A 6 δὲ] om. A 7 νηστικὸς ἂς τρώγῃ] transp. A 7 ῥεπάνιν] ῥεφάνιν A 7 μετὰ] με A 9 εὐφόλβιον] φόλβιον A 9 τριμμένον] τριμμένα A 11 τοῦτο] om. A 12 δευτέραν] δεύτερον A 13 ἅπλωνέ το] ἅπλωνέτω M 13 ἐπίθες εἰς τὸν τόπον] θέτε το ἐπάνω A 14 μετὰ] τοῦ add. A 15 εἰς τὸ πανὶν] πλὴν δὲ μετὰ πανίου A 17 καὶ] om. A 17 ἐπάνω] om. A 20 ψιλὸν] τριμμένην A 20 ξηρὸν] ξηρὰν A 20 καὶ] om. A 24 σκυλλοκρόμμυδον] σκυλλοκρέμμιδον M 24 βράσον] βάλε A 25 καὶ] βράσον καὶ add. A

βάλε κέρατον αἴγας καυμένον καὶ τριμμένον καὶ χήνειον ἀξούγγιν καὶ μίξε τα καὶ ἄλειφε τους τόπους ἔνθα εἶναι κατεσχισμένοι. 3 ἄλλο. ⌜ψιμμίθιον ψιλὸν λιθάργυρον ψιλὸν κολοφωνίαν ψιλὴν λίβανον ψιλὸν καὶ ἔλαιον χρηστὸν καὶ ἀξούγγιν χοίρειον μίξον ἐν τῷ ἅμα καὶ ἄλειφε τοὺς τόπους⌝ τοὺς κατεσχισμένους καθ' ἑκάστην ἡμέραν ταχὺ καὶ βραδὺ πλὴν ἐν ἡλίῳ καὶ ἑστίᾳ.

66 [58] [om. LB]ξϛʹ πρὸς ἄρρωστον, ὅταν ἀπὸ πολλοῦ χρόνου κείμενος εἰς κρεβάτιν καὶ πληγωθῶσι κόκαλά του σὺν τὰ πλευρά του· ⌜ἀνάλυσε κηρὶν καὶ βάλε ἔλαιον καὶ λιθάργυρον τριμμένον καὶ ἕνωσον καὶ ἄλειφε τοὺς τόπους.⌝ 2 ἄλλο. χαλβάνιν κοπάνισον καὶ ποίησον ψιλὸν πολλὰ και ἕνωσον με τὸ μέλι καὶ ἄλειφε τοὺς τόπους. 3 ἄλλο. μαλία καῦσε καὶ τὴν στάκτην ἐκείνων πάσσε εἰς τὰς πληγάς. 4 ἄλλο. ⌜τρυγέαν κρασίου ἔψησον [▷om. A]μετὰ τοῦ νεροῦ εἰς τὸ τηγάνιν καὶ ἄλειφε τοὺς τόπους.⌝ 5 ἄλλο. τρυγέαν ἔψησον[om. A◁] μετὰ τοῦ μέλιτος καὶ ἄλειφε τοὺς τόπους.

67 [59] [om. LB]ξζʹ πρὸς πληγὰς ὅπου γίνωνται εἰς τοὺς πόδας δι' ὅτι συμβαίνει καὶ φορεῖ τὰ ὑποδήματα του μήναν ἢ δύο καὶ οὐδὲν ἐξυπολύεται δι' ἀναγκαίας δουλείας καὶ ἀγωνίσεις καὶ τριβόμενα τὰ ὑποδήματα εἰς τοὺς πόδας ποιοῦσιν πληγάς· φλεμμόνιν αἴγας ἢ ἀρνίου θέτε εἰς τὴν πληγήν. 2 ἄλλο. [▷A]⌜πετζὶν καύσας καὶ τὴν στάκτην ἐκείνου πάσε εἰς τὴν πληγήν.[A◁] ἀφ' ὅτου δὲ ἀποπρησθῶσιν αἱ πληγαί, καῦσον κηκίδιν καὶ πάσον τὴν πληγήν.⌝ 3 ἄλλο. ἀκακίαν τὴν λεγομένην τρίψον καὶ μίξον μετὰ ὄξους καὶ ἄλειφε τὰς πληγάς.

68 [60] [om. L]ξηʹ πρὸς ἥλους ἤτοι τὰ γινόμενα εἰς τὰς χεῖρας καὶ

---

7 πρὸς] This chapter is listed in the pinax to L. 16 πρὸς] This chapter is listed in the pinax to L. 24 πρὸς] This chapter is listed in the pinax to L.

---

25 εἰς] om. A 2 εἶναι] ἔνι A 3 κολοφωνίαν] scripsi cf. ℵ 57 καμφωνίαν M καλλιφονέαν A 4 μίξον] μίξε τα A 6 ταχὺ καὶ βραδὺ] πρωῒ καὶ ἑσπέρας A 6 καὶ ἑστίᾳ] scripsi καὶ ἡστίᾳ M ἢ εἰς τὸ πῦρ A 7 χρόνου] καιροῦ A 8 εἰς] τὸ add. A 8 πληγωθῶσι] τα add. A 8 κόκαλά] κώκαλά M 8 σὺν] καὶ A 9–10 ἕνωσον] μίξε τα A 11 ψιλὸν] ψιλλὴν M 11 πολλὰ] om. A 11 ἕνωσον] μίξε το A 12 μαλία] θαλλία M 18 δουλείας] δουλίας A 19 φλεμμόνιν] scripsi φλεμμόνην MA 22 καῦσον] καῦσε A 22 πάσον] πάσσε A 22–23 ἀκακίαν τὴν λεγομένην] transp. A 23 τρίψον] τρίψε τὴν A 23 μίξον] μίξε την A 23 μετὰ] ὀλίγου add. A

τοὺς πόδας ἃ καλοῦσιν οἱ ἰδιῶται κότζια· ⌜κοπάνισον μελάνθιον⌝ καὶ ποίησον ψιλὸν καὶ μίξον με τὸν οὖρον τοῦ μικροῦ παιδίου ▷om. B καὶ θέτε ἐπάνω τοὺς ἥλους. om. B◁ 2 ἄλλο. κανθαρίδας καὶ καστόριον τριμμένον μίξε τα καὶ θέτε τα ἐπάνω εἰς τοὺς ἥλους. 3 ἄλλο κανθαρίδες δὲ λέγονται ὅταν τὸ καλοκαίριν εὑρίσκονται εἰς τὰς στράτας μουμούδια ὥσπερ σφῆκες καὶ κάθηνται εἰς τὰς κοπρίας καὶ τηλίσσουν τὴν κόπρον ἀλλήλων καὶ ποιοῦσιν αὐτὴν στρογγύλην ἶσα καρύδιον καὶ σύρουσιν ταῦτα εἰς τὰ ποδάριά των. ταύτην δὲ τὴν κανθαρίδαν ἔπαρον οὕτως ὡς ἔνι καὶ τζάκισέ την πέτραν καλὰ καλά. εἶτα μίξε την ἀλλήλων καὶ θέτε ἐπάνω εἰς τοὺς ἥλους.

69 [61] om. L ξθ′ πρὸς παρωνυχίας· ⌜ἡ παρονυχία ἀπόστημά ἐστιν παρὰ τὴν ῥίζαν τοῦ ὄνυχος συνιστάμενον. 2 ἄλλο. κηκίδιν κοπάνισον καὶ ποίησον οἷον τὸ ἀλεύριν καὶ μίξε με τὸ μέλι καὶ ἄλειφε τὸν τόπον.⌝ 3 ἄλλο. φακὴν ψιλὴν μάλαξον με τὸ νερὸν καὶ θὲς ἐπάνω. 4 ἄλλο. ⌜φοινίκια κοπάνισον καὶ μίξον με τὸ κρασὶν καὶ μάλαξον καὶ θές τα ἐπάνω.⌝ 5 ἄλλο. τριακοντάφυλλα χλωρὰ μάλαξον με τὸ κρασὶν καὶ θές τα ἐπάνω. 6 ἄλλο. ▷A ὅταν τὰ ὀνύχια ἔχωσιν αἷμα

11 πρὸς παρωνυχίας] This chapter is listed in the pinax to L.

24 καὶ] εἰς add. A 1 ἃ] om. B 1 καλοῦσιν οἱ ἰδιῶται] διώται δὲ καλοῦσιν B 1 μελάνθιον] μελάνθιν AB 2 μίξον] μίξε το A 2 με τὸν οὖρον] μετὰ τοῦ οὔρου A 2 τοῦ μικροῦ παιδίου] τοῦ παιδίου ἥως τοῦ μικροῦ B 3 τοὺς] εἰς τοὺς A 5 κανθαρίδες] κανθαρίδας B 5 ὅταν] ἐκεῖνες ὁπου A 5 τὸ καλοκαίριν εὑρίσκονται] transp. A 6 τὰς στράτας] ταῖς στράταις A 6 μουμούδια] βουβούδια MB μαμούδην A 6 ὥσπερ] ὡς περὶ M 6 κάθηνται] κάθονται AB 7 ἀλλήλων] om. A ἄλλήλως B 7 στρογγύλην] στρογγύλον B 8 ἶσα] ἶσον A 8 καρύδιον] καρυδίου A καρύδια B 8 σύρουσιν] συννοῦσιν B 8 των] τος B 8 ταύτην] ταῦτα A 9 τὴν κανθαρίδαν] om. A 9 οὕτως] om. A 9 την] τα με τὴν A 10 καλά] om. AB 10 την] τα A 10 ἀλλήλων] ἀλλήλως B 10 ἥλους] ἤγουν εἰς τὰ λεγόμενα κότζια add. A εἶτα τὰ λεγόμενα κότζια παρὰ τῶν χορητῶν B 11 πρὸς παρωνυχίας] scripsi περὶ παρονυχίας A πρὸς παρονυχίδας MB 11 ἡ] ἔπ’ M om. B 12 παρὰ] ὑπὸ A 12 συνιστάμενον] ἐνισταμένην M συνισταμένους B 13 ποίησον οἷον τὸ ἀλεύριν καὶ μίξε με τὸ μέλι καὶ] om. A 13 μίξε] τὸ add. B 14 τόπον] πλὴν δὲ μίξετο μέλι add. A 14 μάλαξον] μάλαξε B 14 θὲς] την add. B 15 φοινίκια] κοινίκια M 15 μίξον] μίξε τα A 15 καὶ μάλαξον] om. A καὶ μάλαξέ τα B 16 τα] om. A 16 ἐπάνω] πλὴν δὲ μάλαξέ τα καλὰ A

ἔπαρον ἄλευρον σίτου ψιλὸν καὶ μίξε με τὴν πίσσαν καὶ θέτε την ἐπάνω.[A⊲]

70 [62] [om. L]ο′ πρὸς πτερύγιον ὄνυχος ἢ χειρῶν ἢ ποδῶν· ⌜πτερύγιον ἔνι ὅταν εἰς τὸ πλάγιον τοῦ ὄνυχος αὐξηθῇ σάρξ. ἀρσενίκιν ψιλὸν πάσσε ἐπάνω. εἶτα δένε ἐπάνω σπόγγον καινούργιον βρασμένον με κρασίν. 2 ἄλλο. ῥοϊδίου φλοῖα ξήρανον καὶ κοπάνισον καὶ ποίησον οἷον τὸ ἀλεύριν καὶ μίξας μάλαξον καὶ θέτε τα ἐπάνω.⌝

71 [om. L]οα′ [⊳om. B]πρὸς ὀνύχια ὅταν συνθλασθῶσιν ἀπὸ λίθου[om. B⊲]. μυρσίνης φύλλα τρυφερὰ καὶ ῥοϊδέας φύλλα τρυφερὰ κοπάνισον καλὰ καὶ θές τα ἐπάνω. 2 ἄλλο. [⊳om. A]ἔπαρον σίτου ἄλευρον καὶ μίξον με τὴν πίσσαν καὶ θέτε ἐπάνω.[om. A⊲]

72 [63] [om. L]οβ′ [⊳om. B]πρὸς ἐὰν εἶναι οἱ ὄνυχες λεπροὶ ὥσπερ ψωριασμένοι· ⌜ἔπαρον τοῦ δρυὸς τὸν ἰξὸν καὶ θές τον ἐπάνω.⌝[om. B⊲] 2 ἄλλο. τήλεως ἀλεύριν καὶ κανθαρίδα κοπάνισον καὶ μίξον με τὸ ἀλεύριν τῆς τήλης καὶ ὄξος ὀλίγον καὶ μάλαξέ τα καὶ θές τα ἐπάνω.

73 [63] ογ′ [⊳om. L]πρὸς κελεφίασιν· ⌜φλεβοτόμει τοὺς τοιούτους καὶ μᾶλλον ὅταν ἔνι τὸ ἔαρ ἤγουν τὸ μέσον τοῦ μαρτίου μηνὸς καὶ τὸν ἀπλίλλιον καὶ τὸν μάϊον καὶ πότιζέ τους καὶ τὴν ἱερὰν ἀφ' ὅτου δὲ φάγῃ ῥεφάνια καὶ τὴν τροφὴν αὐτῶν ἅπασαν τότε κελεύει ἵνα ξερᾷ καὶ εὐγάνει παχεὶς χυμοὺς καὶ ὑγροὺς καὶ κολλητικούς.

---

3 πρὸς πτερύγιον] This chapter is listed in the pinax to L.

---

16 μάλαξον] μάλαξέ τα A μάλαξε B 3 πρὸς πτερύγιον] περὶ πτερυγίου A 3 ὄνυχος ἢ χειρῶν ἢ ποδῶν] om. A ὀνύχων ἢ ποδῶν B 4 εἰς] om. A 4 τὸ πλάγιον] τοῦ πλαγίου B 4 ὄνυχος] ὀνύχου B 4 σάρξ] μακρὰ add. A μικρὴ add. B 4 ἀρσενίκιν] ἀρσενίκιον B 5 πάσσε] πάσον A 5 εἶτα δένε ἐπάνω] om. M 5 δένε] δὲ ἔναι B 5–6 βρασμένον] βρεμένον A 6 με κρασίν] μετὰ οἴνου A ἐπάνω B 6 φλοῖα] φλυὰς M φλό[δ] B 7 μίξας] μίξε το με τὸ μέλι καὶ AB 7 μάλαξον] μάλαξέ τα A 7 θέτε] θὲς B 8 ἀπὸ λίθου] ἢ δώσει τὰ πέτρα ἢ ξύλον A 9 ῥοϊδέας] ῥοϊδίας AB 9 τρυφερὰ] δροσερὰ A 10 ἔπαρον] παρον M 12 πρὸς] om. A 12 εἶναι] ἔνεν A 12 οἱ ὄνυχες] om. A 14 ἀλεύριν] ἀλεύρου B 14 κανθαρίδα] κανθαρίδαν A 14 μίξον] μίξε τα B 15 τα] καλὰ add. A 15 θές] θέτε A 16 πρὸς κελεφίασιν] om. B 16 κελεφίασιν] κελεφίας A 17 ὅταν] ὅτε A 17 ἤγουν] ἀπὸ add. A 17 μηνὸς] om. A 19 αὐτῶν] αὐτοῦ A 20 παχεὶς χυμοὺς] transp. B

---

12 ἐὰν] This chapter is part of the preceding one in A.

τοὺς δὲ τοιούτους πότιζε om. L◁ ὀξείδιν μικτὸν με τὴν κεδρέαν.⌝ 2 ἄλλο. ▷om. M κοπάνισον τὰ φύλλα τῆς κράμβης καὶ ἔκβαλε τὸν ζωμόν της καὶ πότιζε ▷om. B πρωῒ καὶ βραδύ. om. M◁ 3 ἄλλο. ἐλαφίου κέρατον καῦσε καὶ ποιήσας ψιλόν πότιζε om. B^{a}◁ μετὰ οἴνου. 4 ἄλλο. τὸ κέρατον τοῦ ἐλέφαντος ῥίνησε καὶ τὰ ῥινίσματα πότισον με τὸ χλίον κρασίν. 5 ἄλλο. κύμινον κοπάνισον καὶ ποίησον πολλὰ ψιλόν μίξε τὸ με τὸ μέλι. καὶ κρασὶν ὀλίγον. καὶ πότιζε αὐτὸν τὸ πρωῒ πλὴν χλίον. 6 ἄλλο. ⌜πότιζε δὲ καὶ θηριακὴν με τὸ κρασίν. ὡσεὶ κοκκία ἕξ.⌝ 7 ἄλλο. καλαμίνθην κοπάνισον χλωρὸν καὶ τὸν ζωμὸν πότισον. 8 ἄλλο. ⌜ἀφ' ὅτου δὲ λουθῇ καὶ εὔγῃ ἔξω, ἄλειφέ τον ἢ μυρσινέλαιον ἢ οἰνάνθην ἢ σχίνου ἔλαιον ἀμμωνιακὸν ἃς βραχῇ με τὸ κρασὶν ἔστ' ἂν λυθῇ καλὰ καὶ ἄλειφε.⌝ 9 ἄλλο. ▷om. B^{a} ἐνόσῳ δὲ λούεται, ἄλειφε τὸ ὅλον σῶμα καὶ τὴν κεφαλὴν ἀμμωνιακὸν βρεμένον καλὰ με τὸ ὀξείδιν. om. B^{a}◁ 10 ἄλλο. κοπάνισον χλωρὴν τήλην καὶ τὸν ζωμὸν ἄλειφε τὴν κεφαλὴν καὶ τὸ ὅλον σῶμα. 11 ἄλλο. ▷A τεάφιν καὶ ἔλαιον καὶ ἅλας καὶ κύμινον καὶ ἃς βράσουν καὶ ἄλειφε. A◁ 12 ἄλλο. ▷om. L λαπάθου ῥίζαν ἔψησον με τὸ ὀξείδιν καὶ τὸν ζωμὸν ἄλειφε τὴν κεφαλὴν καὶ ὅλον τὸ σῶμα. om. L◁ 13 ἄλλο.

---

20 κολλητικοὺς] εἰσλυτικοὺς B 1 πότιζε] τους add. A 1 ὀξείδιν] ὄξος MB 1 μικτὸν] σμισητόν L 1 κεδρέαν] κεδρέ... L 2 κοπάνισον] κοπάνιζε B 2 τῆς] δάφνης κάλιον τῆς add. B 2 ἔκβαλε] εὔγαλε B 3 της] τοῦ B^{a} 3 ἐλαφίου] scripsi ἐ..αφίου L ἐλάφου MAB^{b} 4 καῦσε] καῦσον MAB^{b} 4 ποιήσας] ποίησον AB^{b} 4 ψιλόν] οἷον τὸ ἀλεύριν καὶ A 5 τὸ] om. B^{a} 6 χλίον] τὸ add. MA 6 ποίησον] ποιήσας MA 6 πολλὰ ψιλόν] transp. M πολὴν ψιλὸν B^{a} 7 καὶ] τὸ add. MAB^{b} με add. B 7 πότιζε] πότιζον MA 8 χλίον] ζεστὸν B^{a} 8 πότιζε] πότισον A 8 δὲ] om. AB 8 καὶ] τὴν add. M αὐτὸν A 8–9 με τὸ κρασίν. ὡσεὶ κοκκία ἕξ] transp. MAB 10 πότισον] πότιζε B 11 οἰνάνθην] ἀνιθέλαιον MAB 11 σχίνου ἔλαιον] σχινέλαιον MB^{b} σχίνεον A κανόλαιον B^{a} 11 ἀμμωνιακὸν] μανιακὸν B^{a} 12 ἔστ' ἂν] ὅταν B^{a} 12 λυθῇ] να λύσῃ A 13 τὸ ὅλον] transp. MA τὸ add. B^{b} 14 ὀξείδιν] ἀλειφόμενον add. M 14–15 χλωρὴν τήλην] ἤγουν τοῦ μουχου^{τρ} add. B^{a} 14 χλωρὴν] χλωρὸν B 15 ζωμὸν] της add. MA του B 15 τὸ ὅλον] transp. AB^{b} τὸ B^{a} 15 σῶμα] τὸ ὅλον add. B^{a}

---

3 ἐλαφίου] The following text is transmitted twice in manuscript B. The sigle B^{a} designates the first and B^{b} the second version of the text. If both aggree in a reading, the sigla B is used.

φύλλα ξηρὰ συκῆς κοπάνισον καὶ ποίησον οἷον τὸ ἀλεύριν. καὶ βάλε τεάφην ψιλὴν καὶ ὀξείδιν καὶ τάραξέ τα καὶ ἄλειφε. 14 ἄλλο. ⌜κοπάνισον κομίδιν ψιλὸν καὶ κηκίδιν ψιλὸν οἷον τὸ ἀλεύριν. καὶ τῆς σηπίας τὸ ὀστοῦν καὶ ποίησον οἷον τὸ ἀλεύριν καὶ μίξε τὰ καὶ μάλαξέ τα με τὸ ὀξείδιν καὶ ἄλειφε τὴν κεφαλὴν καὶ ὅλον τὸ σῶμα.⌝

74 [64] οδ΄ πρὸς ψώρας· ⌜τοὺς τοιούτους ἐνδέχεται φλεβοτομία καὶ βοήθημα ἔχων τὸ μέλαν πλέον τῶν ἄλλων εἰδῶν.⌝ 2 ἄλλο. ἐλαίας δάκρυον ποίησον ψιλὸν καὶ κομίδιν καὶ καππάρεως τῆς ῥίζης τὸ φλοῦδιν ξήρανέ το. καὶ μίξον τὰ πάντα με τὸ ὀξείδιν καὶ ἄλειφέ τον εἰς τὸν ἥλιον ἔνθα ἔνι ἡ ψώρα. 3 ἄλλο. ⌜βάλε λιθάργυρον καὶ ἀγριοσταπίδαν καὶ τεάφην καὶ ποίησον τὰ πάντα ψιλά. καὶ βαλὼν ἔλαιον καὶ ὀξείδιν καὶ ὑδράργυρον καὶ δαφνέλαιον μίξε τὰ ὅλα καὶ ἄλειφε εἰς ἥλιον, ἔνθα ἔνι ἡ ψώρα.⌝ 4 ἄλλο. βάλε τριαντάφυλλα καὶ φύλλα δάφνης καὶ ἕψησον εἶτα βάλε κηρὸν καὶ μίξε τα καὶ ἄλειφε τὴν ψώραν πλὴν τὰ φύλλα τῆς δάφνης καὶ τὰ ῥόδα ἔκβαλε. 5 ἄλλο. ⌜τρίψον λιθάργυρον καὶ ῥούδιν καὶ ψιμμίθιν καὶ συρικὸν καὶ ὑγροπίσσιν καὶ ὀξείδιν καὶ ἄλειφε τὴν ψώραν.⌝ 6 ἄλλο. κοπάνισον σελίνου τὰ φύλλα καὶ εἰς τὸν χυλὸν ἐκεῖνον βάλε τριμμένην

18 τὸν ζωμὸν ἄλειφε] transp. AB 1 ξηρὰ συκῆς] transp. MB[b] 1 συκῆς] συ.. L 1 καὶ] τῆς συπίας τὸ ὀστοῦν καὶ add. A 1 ποίησον] αὐτὰ add. A 1 οἷον] ὡς B[b] 1 τὸ] om. B[a] 1 ἀλεύριν] ἀλεύρον B 1–2 καὶ βάλε] κα L 2 τεάφην] .....φὴν L τεάφιν MAB[a] 2 ψιλὴν] ψιλὸν MB 2 τάραξέ τα] τὰ ......τα L τάραξέ το AB 3 κομίδιν] κικίδην A 3 ψιλὸν] om. M πολλὰ add. AB[b] πολλὴν add. B[a] 3 κηκίδιν] κομίδιν A 3 ψιλὸν οἷον τὸ ἀλεύριν] om. M ψιλὸν πολλὰ A 3 ψιλὸν] om. B 4 σηπίας] σιπέας L συπίας AB[b] σιπίας B[a] 4 ὀστοῦν] στέον L 4 καὶ] om. M 4 ποίησον] ποίησέ τα M αὐτὰ add. A 4 τὸ] om. A 4–5 καὶ μάλαξέ τα με τὸ ὀξείδιν] om. M 5 σῶμα] πλὴν με τὸ ὄξος τὰ μάλαξε add. M 6 πρὸς] περὶ Σ 6 ψώρας] ψωρῶν A 7 τὸ μέλαν] τὸν μέγαν M 9 τὸ φλοῦδιν] τὸν φλοῦδον M τὸ φλοῦδον καὶ A 9 ξήρανέ το] ξήρανον M ξήρανέ τον A 11 ἀγριοσταπίδαν] ἀγριοσταφύδας M ἀγριοσταφίδαν A ἀγριοσταφίδα B 11 τεάφην] τεάφιν Σ 11 βαλὼν] βάλε MA 12 ὀξείδιν καὶ ὑδράργυρον καὶ δαφνέλαιον] om. M 12 μίξε] καὶ μίξε AB 13 εἰς] τὸν add. B 13 ἔνθα ἔνι ἡ ψώρα] om. A 13 τριαντάφυλλα] τριάκοντάφυλλα Σ 14 δάφνης] scripsi δαφν... L καὶ ἔλαιον add. Σ 15 πλὴν] ἔκβαλε add. A 15 ῥόδα] φύλλα B 15 ἔκβαλε] om. A 16 ῥούδιν] scripsi ῥούδην L ῥοΐδιν MB ῥοΐδην A 16 συρικὸν] συριακὸν A 18 σελίνου τὰ φύλλα] transp. Σ 18 εἰς] om. A 18 ἐκεῖνον] ἐκείνου M

5 σῶμα] Here ends the text transmitted twice in B.

τεάφην καὶ ζυγέαν καὶ ὀξείδιν καὶ μίξας τὰ πάντα ἄλειφε εἰς ἥλιον. 7 ἄλλο. ⊳om. M κοπάνισον τοῦ κρίνου τὴν ῥίζαν καὶ ποίησον ψιλὴν καὶ καρδάμου σπόρον καὶ βάλε μέλι καὶ μίξας ἄλειφε εἰς ἥλιον om. M⊲. 8 ἄλλο. ⌜ἀρσενίκιν ποίησον ψιλὸν καὶ μίξον με τὸ κρασὶν ἄλειφε εἰς ἥλιον.⌝

75 [64] οε′ ⊳om. B πρὸς λέπραν· ⌜αἶρας ἀλεύριν καὶ καρδάμου τὸν σπόρον ψιλὸν βάλε καὶ ὀξείδιν καὶ ἄλειφε εἰς ἥλιον. om. B⊲⌝ 2 ἄλλο. τὴν ῥίζαν τοῦ ἀσφοδέλου κοπάνισον καὶ τὸν χυλὸν τῆς ῥίζας μίξον με τὸ ὀξείδιν καὶ ἄλειφε ἢ εἰς ἑστίαν ἢ εἰς ἥλιον. 3 ἄλλο. ⌜⊳om. B τεάφην τριμμένην καὶ ὀξείδιν ἄλειφε εἰς ἥλιον. om. B⊲⌝

76 [65] οϛ′ πρὸς πόνον νευρῶν· ⌜ἀσβέστιν μίξε με τὴν τρυγέαν καὶ ἄλειφε.⌝

77 [65] οζ′ εἰς συκάμνιον· ⌜πεντάνευρον κοπάνισον καὶ μίξον με τὸ μέλι καὶ θέτε τὸ ἐπάνω.⌝

78 [66] οη′ ⊳om. B πρὸς ὀφθαλμούς, ὅπου οὐ βλέπωσι τὴν νύκτα· om. B⊲ ⌜πράσον κοπάνισον καὶ τὸν ζωμὸν τοῦτον μίξον με τὸ μέλι καὶ στάξον εἰς τοὺς ὀφθαλμούς.

79 οθ′ ⊳om. B πρὸς πόνον ἡμικράνου· om. B⊲ κοπάνισον ψιχίας ἄρτου καὶ μίξε με τὴν χολὴν τοῦ βοϊδίου καὶ ὀξείδιν καὶ ἄλειφε ἔνθα πονεῖ.

80 [66] π′ πρὸς λειχήνας· ⌜λαπάτου ῥίζαν τὸ φλοῦδιν αὐτῆς πλύνον. εἶτα θὲς αὐτὸ ἐπάνω τοῦ κεραμιδίου, καὶ ἀποκάτωθεν τοῦ κεραμιδίου

---

18 τριμμένην] τριμμένον M 1 τεάφην] τεάφιν Σ 1 ζυγέαν] ζυγίαν MA 2 ψιλὴν] ψιλὸν A 3 εἰς] τὸν add. B 4 κρασὶν] καὶ add. MA 4–5 εἰς ἥλιον] om. A εἰς τὸν ἥλιον B 6 πρὸς λέπραν] περὶ λεπρῶν A 6 αἶρας] ἦρας L 6 τὸν] τὸ A 8 ἀσφοδέλου] ἀσφεδίλλου M ἀσφοδίλλου A 8 ῥίζας] ῥίζης MB 8–9 μίξον με τὸ ὀξείδιν] σμῖ ....ξύδην L μίξε με τὸ ὀξείδιν AB 9 ἢ] om. AB 9 εἰς] om. B 9 ἑστίαν] ὑστίαν L ἱστίαν A om. B 9 ἢ] om. B 9 εἰς] τὸν add. B 9–10 τεάφην] τεάφιν M 10 τριμμένην] τρίψον M 10 εἰς ἥλιον] ἐν ἡλίῳ MA 11 ἀσβέστιν] ἄσβεστον MA 11 μίξε] μίξον MA 11 με τὴν] με B 11 τρυγέαν] τριγίαν L 13 εἰς συκάμνιον] συκάμινα B 14 τὸ] om. M τὰ A 15 ὅπου] ὅταν M 15 οὐ] οὐδὲν A 16 πράσον] πράσα Σ 16 τοῦτον] om. MA τους B 16 μίξον] μίξε B 17 στάξον] στάξε Σ 19 μίξε] μίξον Σ 19 βοϊδίου] βοὸς Σ 19 καὶ] μίξε τα καὶ add. AB 19 ἄλειφε] τὴν κεφαλὴν add. MB 19 ἔνθα πονεῖ] om. M 20 λαπάτου] λαπάθου Σ 20 ῥίζαν] ῥίζας M ῥίζης A 20 αὐτῆς πλύνον] πλύνας M πλύνας ἄλειφε A τῆς σπλῆνας B s.l. add. ῥίζας 21 εἶτα θὲς αὐτὸ ἐπάνω τοῦ κεραμιδίου] θές το φλοῦδιν εἰς τὸ κεραμίδιν MB πλὴν θές το φλοῦδιν εἰς τὸ κεραμίδην A 21 καὶ] om. B

βάλε καρβούνια, ὅπως ξηρανθῇ. εἶτα τρίψον αὐτὸ καὶ βάλε ὀξείδιν καὶ ἄλειφε.⌝ 2 ἄλλο. μελάνθιν κοπάνισον καὶ μίξον με τὸ ὀξείδιν καὶ θέτε ἐπάνω. 3 ἄλλο. ▷Aμελάνθιν κοπάνισον καὶ μίξε το με τὸ ὀξείδιν καὶ ἄλειφε. θέτε το καὶ ἐπάνω.A◁ 4 ἄλλο. ⌜τρίψον καρπὶν ψιλὸν καὶ μίξε το με τὸ ὑγροπίσσιν καὶ ἄλειφε.⌝ 5 ἄλλο. δαμασκηνέας κομίδιν ἀνάλυσέ το με τὸ ὀξείδιν καὶ ἄλειφε. 6 ἄλλο. ⌜καππάρεως φύλλα βάλε καὶ τεάφης με τὸ λίβανον τριμμένα καὶ βάλε καὶ ὀξείδιν καὶ ἄλειφε.⌝ 7 ἄλλο. ▷MBτεάφιν μίξον με τὸ ἔλαιον καὶ τὸ ὀξείδιν ἄλειφε.MB◁ 8 ἄλλο. ▷MAἀμμωνιακὸν βάλε εἰς τὸ ὀξείδιν καὶ ἂς λυθῇ καὶ τότε ἄλειφε.MA◁.

81 [67] πα′ πρὸς ἀπορύφας· ⌜ἐνδέχεται τοὺς τοιούτους φλεβοτομία μία καὶ δύο. καὶ εἰς τὴν ἀρχὴν ὅταν ἔνι ἀγουρὸς ἢ ἀπορυφή, βρέξε σπόγγον καινούργιον εἰς ὀξείδιν καὶ θέτε τον ἐπάνω καὶ δένε τον. τοῦτο δὲ ποίησον καὶ μίαν καὶ δύο καὶ τρεῖς⌝ διὰ τὸ να ἀποκρούσῃ τὸ ῥεῦμα. 2 ἄλλο. ▷L⌜μάλια προβάτου βρέξε εἰς χλίον ῥοδέλαιον καὶ θέτε τὰ ἐπάνω.L◁ 3 ἄλλο. ἀφ' ὅτου δὲ ἴδης ὅτι ὡριμάσαι θέλει, μάλαξε ἀμμωνιακὸν με τὸ μέλι καὶ θές το ἐπάνω.⌝ 4 ἄλλο. ▷LBτοῦ χρυσολαχάνου τὰ φύλλα ἢ τὸν σπόρον κοπάνισον καὶ μίξον με τὸ μέλι καὶ θέτε το ἐπάνω. 5 ἄλλο. ⌜κοπάνισον ῥητίνην καὶ μίξε ὀξείδιν καὶ μάλαξον καὶ θέτε ἐπάνω.⌝LB◁ 6 ἄλλο. ▷om. Bπρόπολιν μάλαξον με τὸ ὀξείδιν καὶ θέτε ἐπάνω.om. B◁ 7 ἄλλο. ⌜προζύμιν

21 ἀποκάτωθεν τοῦ κεραμιδίου] ὑποκάτω εἰς τὸ κεράμιον M ὑποκάτω A om. B 1 εἶτα] τὸ φλούδιον καὶ ἀφ' ὅτου ξηρανθῇ MB τὸ φλοῦδιν καὶ ἀφ' ὅτου ξηρανθῇ A 1 τρίψον αὐτὸ] τρίψε το Σ 2 μελάνθιν] μελάνθιον MB 2 καὶ μίξον] καὶ μίξε το A 3 θέτε] θές το M ἄλειφε A 3 ἐπάνω] om. A 4 καρπὶν] πὴν L 5 μίξε] μίξας MA 5 το] om. A 5 με τὸ ὑγροπίσσιν] μετ........σιν L με ὑγροπίσσιν MA 5 ἄλειφε] τὴν λέπραν add. MB 5 δαμασκηνέας] δαμασκηναίας M 6 ἀνάλυσέ] λυσον A 6 το] om. A τὰ B 6 ἄλειφε] τὴν λέπραν add. MB τὴν λειχήναν add. A 7 βάλε] om. MA 7 τεάφης] τεάφιν MB τεάφην A 7 με τὸ] με τὸν A 7 καὶ] om. Σ 7 καὶ] om. A 8 ὀξείδιν] καὶ add. B 9 ἄλειφε] τὴν λέπραν add. B 9 τὸ] om. B 11 φλεβοτομία] καὶ add. Σ 12 ἀγουρὸς] ἄωρος Σ 13 εἰς] τὸ add. A 13 θέτε] θὲς M 14 τρεῖς] τρὶς A 14 τὸ] om. A 16 ἐπάνω] ἐπ... L 16 ὡριμάσαι] scripsi ὀριμασ.L ὡριμάσιν M ὡριμάσειν A ὀρημάσει B 17 μάλαξε] μάλαξον A 17 το] om. A 18 μίξον] μίξε τα B 19 το] τὰ B 19 μίξε] το με τὸ add. B 20 μάλαξον] μάλαξέ την B 20 θέτε] θές την B 21 θέτε] θές την M θὲς A

μάλαξον με τὸ μέλι καὶ θέτε. 8 ἄλλο. ▷om. B σῦκα κοπάνισον καὶ μάλαξον με τὸ μέλι καὶ θέτε.⌝om. B◁ 9 ἄλλο. πίσσαν μάλαξον με τὸ μέλι καὶ τὸ ἔλαιον καὶ θέτε.

82 [68] πβʹ ⌜ὅταν τῆς πληγῆς ἡ οὐλὴ γένηται μαυρή· τρίψε λιθάργυρον καὶ πέπερι ἄσπρον καὶ μίξε τα με τὸ ῥοδέλαιον καὶ ἄλειφε τὴν πληγήν.⌝ 2 ἄλλο. ἀειδάρας γάλα ἄλειφε τὸν πόνον. 3 ἄλλο. ⌜εὐζώμου σπόρον κοπάνισον ψιλὸν καὶ μίξε τον με τῆς αἴγας τὴν χολὴν ἢ τοῦ βοϊδίου. βάλε δὲ καὶ λιθάργυρον καὶ ἄλειφε τὴν πληγήν.⌝ 4 ἄλλο. ἀμμωνιακὸν κοπάνισον καὶ ποίησον ψιλὸν καὶ μίξε το με τὴν χολὴν τοῦ χοίρου καὶ ἄλειφε τὴν πληγήν. 5 ἄλλο. λιθάργυρον τριμμένον καὶ τοῦ ἡδυόσμου τὰ φύλλα. ξήρανέ τα καὶ τρίψον καὶ μίξον με τὸ μέλι καὶ ἄλειφε τὸν τόπον. 6 ἄλλο. ⌜τεάφην τριμμένην μίξον με τὸ ὀξείδιν τὸ ἄσπρον καὶ μάλαξον καὶ ἄλειφε τὴν πληγήν. πλὴν ἄλειφε πυκνά.⌝

83 [69] πγʹ ὅταν αἱ τρίχες τῆς κεφαλῆς πίπτουσιν· κραμβίου φύλλα ξηρὰ κοπάνισέ τα ψιλὰ καὶ μάλαξέ τα με τὸ νερὸν καὶ θές το εἰς τὸν τόπον ἐκεῖνον, ὅπου αἱ τρίχαι πίπτουσιν. 2 ἄλλο. ⌜ἀλόην

21 προζύμιν] θὲς προζύμιν καὶ B 1 μάλαξον] μάλαξέ το B 1 με τὸ μέλι] μετὰ μέλιτος M με τὸ ὀξείδιν A μέλι B 1 θέτε] ἐπάνω add. M θὲς ἐπάνω πλὴν πολλάκις A θές τὰ ἐπάνω B 1 κοπάνισον καὶ] om. MA 2 μέλι] μέλιτος A 2 θέτε] ἐπάνω add. MA 2 πίσσαν] πίσαν L 2 με τὸ] μετὰ τὸ M μετὰ τοῦ A 3 μέλι] μέλιτος A ἔλαιον B 3 τὸ ἔλαιον] ἐλαίου A τὸ μέλι B 3 θέτε] θές την ἐπάνω MB ἐπάνω add. A 4 ὅταν τῆς πληγῆς ἡ οὐλὴ] πρὸς οὐλὴν πληγῆς ἐὰν M 4 μαυρή] .... B 4 τρίψε] τρίψον M 5 πέπερι ἄσπρον] transp. A 5 μίξε τα] μίξον M μίξε A 6 ἀειδάρας] γαειδάρας L 6 πόνον] τόπον Σ 7 μίξε τον] μίξον M 7 με τῆς αἴγας] μετὰ αἰγὸς M με τῆς αἰγὸς A μετὰ τῆς αἰγὸς B 8 δὲ] om. Σ 8 καὶ] μίξε τα καὶ add. Σ 9 ἀμμωνιακὸν] ἀμμωνιακὴν L 10 μίξε το] μίξον M τὴν add. A 10 τὴν πληγὴν] om. A 11 ἡδυόσμου τὰ] ἥδιος .. L 11 ξήρανέ τα] ξήρανον Σ 12 τρίψον καὶ] scripsi τρίψον .... L κοπάνησέ τα ψιλὰ καὶ Σ 12 μίξον] μίξε τὰ A μίξε B 12 τόπον] .... L 12 τεάφην] τεάφιν MB 13 τριμμένην] τριμμένον MB 13 μίξον] μίξε A μίξε το B 13 με τὸ] με A 13 ὀξείδιν] ἄσπρον A 13 τὸ ἄσπρον] ὀξείδιν A 13 καὶ] τάραξέ το καὶ add. MA 13 μάλαξον] μάλαξέ το MB 14 τὴν] om. A 14 πληγήν] ἔνθα ἔνι ἡ οὐλὴ ἡ μελαινὴ καὶ μαύρη add. MB ἔνθα ἔνι ἡ οὐλὴ ἡ μελαινή A 14 ἄλειφε] τὸ add. A 14 πυκνά] καὶ οὕτως πᾶν ὀφέλιμον B 15 ὅταν] θεραπεία ὅταν AB πρὸς τὸ ὅταν M 15 αἱ τρίχες τῆς κεφαλῆς πίπτουσιν] transp. M 15 κραμβίου] κράμβης Σ 16 κοπάνισέ τα] κοπάνισον M καὶ ποίησέ τα add. B 16 το] τα Σ 17 ἐκεῖνον] om. Σ

τρίψον καὶ μίξον με τὸ κρασὶν τὸ μαῦρον καὶ αὐστηρὸν καὶ ἄλειφε τὸν τόπον. ▷Bλάδανον κοπάνισον καὶ μάλαξέ το με τὸ κρασὶν καὶ ἄλειφε τὸν τόπον.⌉B◁ 3 ἄλλο. ▷ Σσεύτλου φύλλα κοπάνισον καὶ τὸν ζωμὸν αὐτοῦ μίξον με τὸ χρηστέλαιον καὶ ἄλειφε τὴν κεφαλήν. 4 ἄλλο. λάδανον μάλαξον με τοῦ σχίνου τὸ ἔλαιον καὶ ἄλειφε.Σ ◁

84 πδ′ δίχα τὸ να μὴ πέσουν ποτὲ αἱ τρίχαι τῆς κεφαλῆς, ποίησον τοῦτο· πολύτριχον βράσον με τὸ κρασὶν καὶ με τὸν ζωμὸν ἐκεῖνου μάλαξον λάδανον καὶ ἄλειφε τὰς τρίχας. 2 ἄλλο. βοτάνιν τὸ λεγόμενον ποτυρίδα ἔπαρον αὐτὸ καὶ κοπάνισον καὶ τὸν ζωμόν του μίξε μετὰ ἐλαίου καὶ ἄλειφε τὴν κεφαλήν.

85 [70] πε′ ἐὰν θέλῃς ἵνα αὐξήσουν αἱ τρίχαι τῆς κεφαλῆς, ποίησον τοῦτο· ⌈λάδανον ἀνάλυσε με τὸ κρασὶν καὶ ἂς μοσχεύσῃ καλὰ τὸ λάδανον καὶ ἄλειφε τὰς τρίχας. 2 ἄλλο. βάλε εἰς τζυκάλιν βροχῆς νερὸν καὶ πολύτριχον καὶ λάδανον καὶ κολοκύνθην ξηρὰ κομμάτια. καὶ βάλε τὰ εἰς ἀγγεῖον ὑέλινον καὶ ἂς κρέμαται εἰς τὸν ἥλιον ἡμέρας εἴκοσιν. ταράσσε τὸ δὲ δεύτερον τὴν ἡμέραν. καθ' ἑκάστην δὲ ἡμέραν τάρασέ το καὶ βάλε τὸ κτένιν σου εἰς τὸν ζωμὸν ἐκεῖνον καὶ κτενίζου.⌉

---

17 ὅπου] ἔνθα A 1 μαῦρον καὶ] om. MA μαῦρον τὸ B 2 τὸν τόπον] om. M 4 αὐτοῦ] τοῦ A 4 χρηστέλαιον] ὀλίγον add. B 5 ἄλειφε] τὴν κεφαλὴν add. A 6 δίχα τὸ να] πρὸς τὸ M 6 δίχα] διὰ A εἰς B 6 ποτὲ] om. B 6 τρίχαι] τρίχες Σ 7 τοῦτο] οὕτως A 7 με τὸν] τὸν AB 8 ἐκεῖνου] ἐκεῖνον Σ 8 μάλαξον λάδανον] μάλαξε M μάλαξον AB 8–9 βοτάνιν τὸ λεγόμενον ποτυρίδα] transp. MB 9 ποτυρίδα] ποτηρίδα AB 9 ἔπαρον αὐτὸ] .... L om. A 9 καὶ] om. A 10 του μίξε] .....ξαι L του μίξον M 10 μετὰ ἐλαίου] με ἔλαιον M ἔλαιον AB 10 τὴν κεφαλήν] om. M 11 ἐὰν θέλῃς ἵνα αὐξήσουν αἱ τρίχαι τῆς κεφαλῆς] πρὸς αὔξησιν τριχῶν κεφαλῆς M 11 τρίχαι] τρίχες A 11 τῆς κεφαλῆς] om. B 11–12 ποίησον τοῦτο] om. Σ 12 λάδανον ἀνάλυσε] ἀνάλυσον λάδανον MA 12–13 τὸ λάδανον] ὁ λάδανος MB om. A 13 καὶ] οὕτως add. M 13 τζυκάλιν] τζουκάλιν B 14 κολοκύνθην] κολοκύνθαν M 14 ξηρὰ κομμάτια] ξηρὰν κομμάτια M ξηρὸν κομμάτια A ξηρὰν μάτια B 15 κρέμαται] scripsi κρεμετι.. L κρέμμαται M κρέμμωνται A 16 εἴκοσιν] κ' ἢ λ' A 16 ταράσσε τὸ δὲ δεύτερον τὴν ἡμέραν] καὶ Σ 17 δὲ] om. Σ 17 τάρασέ] τάρρασέ Σ 17 καὶ] ἀφ' ὅτου πληρωθῶσιν αἱ ἡμέραι add. MA εἶτα ἀφ' ὅτου πληροθῶσιν αἱ ἡμέραι B 17 βάλε] βάφε AB 17 κτένιν] scripsi κτένην LAB κτένιόν M 18 κτενίζου] ἐκεῖνον δὲ τὸν ζωμὸν τάραττέ το δεύτερον τὴν ἡμέραν add. Σ

86 [71] πϛʹ ἐὰν θέλῃς να γένωσιν μαῦραι αἱ τρίχαι, ποίησον τοῦτο· ἤγουν ⌜βάλε εἰς τζυκάλιν κρασὶν αὐστηρὸν καὶ κηκίδια τριμμένα καὶ τοῦ σιδήρου τὸ γίωμαν καὶ ῥινίσματα σιδήρου καὶ βάλε ὀξείδιν καὶ μίξε τα. καὶ βάλε τα εἰς ἀγγεῖον ὑέλινον καὶ ἂς κρέμαται εἰς τὸν ἥλιον ἡμέρας δέκα. εἶτα ἄλειφε τὰς τρίχας τῆς κεφαλῆς πρωῒ καὶ ἂς ἔνι οὕτως εἰς πάλιν τὸ πρωῒ καὶ τότε κτενίζου.⌝

87 πζʹ ἐὰν θέλῃς να ποιήσῃς ξανθὰς τρίχας· λυπηνάρια ὠμὰ βάλε εἰς νερὸν καὶ ἂς βρέχηται νυχθήμερον καὶ τότε ἄλειφε τὰς τρίχας καὶ κτενίζου. 2 ἄλλο. τριαντάφυλλα ἄσπρα καὶ κόκκινα καὶ κηκίδιν ὀλίγον τριμμένον καὶ χελιδονέαν καὶ πολύτριχον καὶ λουπηνάρια ὠμὰ καὶ ψιμμίθιν καὶ νερὸν βάλε εἰς ἀγγεῖον ὑέλινον καὶ ἂς κρέμαται εἰς τὸν ἥλιον ἡμέρας εἴκοσιν. καὶ τάρασσε τὸ δεύτερον τὴν ἡμέραν. ἄφ' ὅτου δὲ πληρωθῶσιν αἱ ἡμέραι, βρέχε σπόγγον καινούργιον εἰς τὸ νερὸν ἐκεῖνο καὶ τρίβε τὰς τρίχας.

88 [72] πηʹ ἂν φθεῖρες πολλαί εἰσιν ἐν τῇ κεφαλῇ· ⌜λυπηνάρια ὠμὰ βράσουν με τὸ νερὸν καὶ ἀντὶ ἄλλου θερμοῦ λοῦέ τον μετ' ἐκεῖνο θερμὸν τῶν λουπηναρίων. 2 ἄλλο. ἀγριοσταπίδαν ἤτοι

---

1 ἐὰν θέλῃς] πρὸς M 1 θέλῃς] θέλουν B 1 γένωσιν] γενοῦν A 1 μαῦραι αἱ τρίχαι] transp. M 1 τρίχαι] τρίχες A τῆς κεφαλῆς add. B 1–2 ποίησον τοῦτο· ἤγουν] om. MA 2 τζυκάλιν] τζουκάλιν B 3 τὸ γίωμαν] τὸν ἰὸν Σ 3 ὀξείδιν] ὄξος M 4 μίξε] μάλαξέ A 4 καὶ βάλε τα] κ......... L καὶ βάλε A 4 κρέμαται] scripsi κρέμμετ. L κρέμμεται MB κρέμμανται A 5 δέκα] ιε' A 5 κεφαλῆς] τὸ add. Σ 6 καὶ] οὕτως add. M 6 οὕτως εἰς πάλιν τὸ] ἕως βραδὺ M οὕτως ἕως A οὕτως ἕως αὔριον B 7 ἐὰν θέλῃς να ποιήσῃς ξανθὰς τρίχας] ὅταν φθεῖραι πολλαὶ εἰσὶν εἰς τὴν κεφαλὴν M πρὸς τὰ να γένωσι τρυχαιῖς ξανθαῖς in marg. M^2 7 ὠμὰ] ξηρὰ B 8 εἰς] τὸ add. AB 8 βρέχηται] βρέχουνται M βρέχονται AB 9 τριαντάφυλλα] τριακοντάφυλλα Σ 9 ἄσπρα] ἄπρα M 10 χελιδονέαν] χεληδονίαν A 10 πολύτριχον] om. B 10 καὶ] om. B 10 λουπηνάρια] λυπηνάρια MA 11 ψιμμίθιν] ψιμμίνθην A 11 βάλε] καὶ βάλε τα AB 11–12 κρέμαται] κρεμμάται M κρέμμανται A 12 ἡμέρας] om. M 12 εἴκοσιν] λ' A 12 τάρασσε] τάραττε A 13 ἡμέραν] καὶ add. A 14 τρίβε] βρέχε Σ 15 ἂν] ὅταν AB 15 φθεῖρες] φθεῖραι AB 15 πολλαί εἰσιν] transp. B 15 ἐν τῇ κεφαλῇ] εἰς τὴν κεφαλὴν AB 15 λυπηνάρια] λ......ια L 16 βράσουν] βράσον Σ 16 ἀντὶ ἄλλου] ......αλλου L 16 λοῦέ τον] λούετρον M λοῦέ το B 16–17 μετ' ἐκεῖνο] μετακεῖνον M μετ' ἐκεῖνον B 17 θερμὸν] τὸν νερὸν M τὸ νερὸν A θερμῶν M 17 τῶν] om. B 17 λουπηναρίων] λυπηναρίων Σ

κονιδοβότανον κοπάνισον ψιλὸν καὶ μίξε το με τὸ χρηστέλαιον καὶ ἄλειφε τὴν κεφαλήν.⌝ 3 ἄλλο. κοπάνισον φύλλα κίσσου καὶ ἔκβαλε τὸν ζωμόν τους καὶ μίξε τον με τὸ μέλι καὶ ἄλειφε τὴν κεφαλήν. 4 ἄλλο. ▷L ῥεπανίου ἐλάδιν ἄλειφε τὴν κεφαλήν.L◁ 5 ἄλλο. ▷om. A ἀγριοσταπίδαν κοπάνισον καὶ ποίησον ψιλὴν καὶ μίξε την με τὸ ὀξείδιν καὶ τὸ χρηστέλαιον καὶ ἄλειφε τὴν κεφαλήν.om. A◁ 6 ἄλλο. ὑδράργυρον καὶ πτύσμα μάλαξον καὶ ἄλειφε τὴν κεφαλήν. 7 ἄλλο. ▷om. B⌜ὑδράργυρον καὶ ἔλαιον καὶ ὀξείδιν μίξε τα καὶ ἄλειφε.⌝om. B◁ 8 ἄλλο. ▷B ὑδράργυρον μίξε με τὸ κρασὶν καὶ τάραξέ την καὶ ἄλειφε.B◁ 9 ἄλλο. ▷L⌜ὑδράργυρον μίξον με χοίρεον ἀξούγγιν λυτὸν καὶ ἄλειφε.⌝L◁ 10 ἄλλο. ▷om. B κύπεριν κοπάνισον καὶ ποίησον ψιλὴν καὶ μάλαξέ την με τὸ κρασὶν καὶ ἄλειφε.om. B◁ 11 ἄλλο. προβάτου χολὴν μίξε με τὸ κρασὶν καὶ ἄλειφε. ⌜τοῦτο γοῦν ὠφελεῖ καὶ εἰς τὰς κόνυδας.⌝

89 [73] πθʹ ⌜πρὸς τὰ λεγόμενα γλυκέα τὰ γίνονται εἰς τὴν κεφαλήν, ▷om. MA ἅτινα καλοῦσιν αἱ γυναῖκες γλυκέας·om. MA◁ πότισον πρότερον βοήθημα καὶ φλεβοτόμησον αὐτὸν κρανιακὴν φλέβα. 2 ἄλλο. ἐκεῖ δὲ ὅπου εἰς τὴν κεφαλὴν φαίνονται εἰς ἕνα τόπον καὶ πάλιν εἰς ἄλλον ὡς ἂν πληγαὶ καὶ ἡ πληγὴ ἐκείνη ἔχει τρύπας πολλάς, βάλε λυπηνάρια καὶ ἂς βράσουν με τὸ νερὸν καὶ πυρίαζε τοὺς

---

17 ἀγριοσταπίδαν] ἀγριοσταφίδαν Σ 1 κονιδοβότανον] κονηδοβότανον A κονιβότανον B 1 ψιλὸν] om. M 1 μίξε το] μίξον M 3 ἔκβαλε] εὔγαλε MA 3 τὸν] τὸ M 3 τους] τοῦ MA 3 μίξε τον] μίξον M μίξε το B 4 κεφαλήν] του add. M 5 ἀγριοσταπίδαν] ἀγριοσταφίδαν MB 5 ψιλὴν] ψιλὸν B 5 μίξε την] ἕνωσον M 6 ὀξείδιν] ὄξος MB 7 ὑδράργυρον] λιθάργυρον A 7 πτύσμα] πτῶμα B 8 ἔλαιον] ἐλα.... L 8 μίξε τα] ἕνωσον M μάλαξον ὁμάδην A 8 ἄλειφε] τὴν κεφαλὴν add. A 12 μάλαξέ την] μαλα..... L 12 με τὸ κρασὶν] om. M 12 ἄλειφε] τὴν κεφαλήν add. MA 13 μίξε] ἕνωσον M 13 καὶ] τάραξέ την καὶ add. MA 13 ἄλειφε] τὴν κεφαλὴν add. MA 13 τοῦτο γοῦν ὠφελεῖ] τὰ αὐτὰ δὲ ποίει M τὰ αὐτὰ δὲ ποίησον AB 13 καὶ] ὁμοίως A 13 εἰς] ὅταν εἶναι MB ὅτε ἔνι A 14 τὰς κόνυδας] κονίδες εἰς τὴν κεφαλὴν M κονίδες ἐν τῇ κεφαλῇ A 15–16 εἰς τὴν κεφαλήν] ἐν τῇ κεφαλῇ B 16 γλυκέας] γλυκέα B 16–17 πότισον πρότερον] ἅτινα ὀφείλει πρότερον ποτίσαι A 17 φλεβοτόμησον] φλεβοτομῆσαι A 17 φλέβα] φλέβαν AB 18 ὅπου] om. MA 18 φαίνονται] φαίνεται MA 19 τρύπας] τρίχας Σ

τόπους ἔνθα ἔνι ἡ πληγή, πλὴν πρῶτον ξύρησον τὰ μαλία.⌝ 3 ἄλλο. ▷Lβάτου φύλλα βράσε με τὸ κρασὶν καὶ πυρίαζε τὸν τόπον.L◁ 4 ἄλλο. ▷ Σ⌜μυρσίνην βράσον με τὸ κρασὶν καὶ πυρίαζε τοὺς τόπους.⌝Σ ◁ 5 ἄλλο. 6 ἄλλο. φακὴν ἕψησον με τὸ νερὸν καὶ κρασὶν ὀλίγον καὶ ὀξείδιν καὶ πυρίαζε τὸν τόπον. 7 ἄλλο. λιθάργυρον ποίησον ψιλὸν πολλὰ καὶ μίξε το με τὸ κρασὶν τὸ αὐστηρὸν καὶ ἄλειφε τὴν κεφαλήν, ἔνθα αἱ πληγαί εἰσίν. πλὴν πρῶτον πυρίαζε τοὺς τόπους ὡς προεῖπόν σοι. 8 ἄλλο. ⌜καῦσε χαρτὶν βαμβακερὸν καὶ τὴν στάκτην ἐκείνου μίξον με τὸ αὐστηρὸν κρασὶν καὶ μάλαξέ το ἀκριβῶς καὶ ἄλειφε τὴν κεφαλήν.⌝ 9 ἄλλο. κοπάνισον πηγάνου φύλλα καὶ μυρσίνης καὶ ἀγριοσταπίδαν καὶ καλακάνθην καὶ ὀξείδιν καὶ μίξε τὰ πάντα καὶ ἄλειφε τὰς πληγάς. 10 ἄλλο. τεάφην ψιλὴν ποιήσας μίξον με τὸ χρηστέλαιον καὶ ὀξείδιν καὶ ἄλειφε 11 ἄλλο. ▷MB⌜λιθάργυρον ποιήσας ψιλὸν βάλε καὶ ἅλας καὶ κρασὶν καὶ ἔλαιον καὶ ὀξείδιν καὶ ἄλειφε τὴν κεφαλήν.⌝MB◁

90 [74] ϙ′ ⌜περὶ ἀλωπεκίας καὶ ὀφιάσεως· ἀλωπεκία λέγεται, ὅταν αἱ τρίχες τῆς κεφαλῆς πέσωσι. ὀφίασις δὲ λέγεται, ὅταν γένηται ἡ

19–20 πολλάς] ὡσὰν κῶλον βελονίου add. M ὡσὰν κῶλον βελωνίου AB 1 ἡ πληγή] om. A 1 πλὴν πρῶτον ξύρησον τὰ μαλία] om. A 1 μαλία] ἔνθα ἔνι ἡ πληγή add. M ὅπου ἔνι ἡ πληγὴ B 3 βράσον] βράσε A 3 τοὺς τόπους] ἔνθα ἔνι ἡ πληγὴ add. A 4 φακὴν ἕψησον] .....ψησον L βάτου φύλλα ἕψησον B 4 με τὸ] μετὰ M 4 καὶ] τὸ add. A 4 ὀλίγον] om. A 5 τὸν τόπον] τοὺς τόπους ἔνθα εἶναι αἱ πληγαί MB τοὺς τόπους ἔνθα ἔνι αἱ πληγαὶ A 6 μίξε το] μίξον M 7 αἱ πληγαί εἰσίν] transp. MA 7 τοὺς τόπους] τὸν τόπον Σ 7–8 ὡς προεῖπόν σοι] καθῶς παρ' ὀπίσω σοι εἴ.. ποταπὴν πυρίας θέλεις πυριάζειν τοὺς τόπους M καθὼς ἑρμηνεύθης ἄνωθεν A καθῶς ἄνωθεν σοι εἶπω B 8 καῦσε] παῦσον M s.l. κ M² καῦσον A ποταπὴν πυρίαν θέλης ἐπιριάζειν τοὺς τόπους καῦσαι B 8 βαμβακερὸν] βαμπακερὸν A παμπακινὸν B 9 μίξον] μίξε τὴν B 9 με τὸ] μετὰ A 9 αὐστηρὸν κρασὶν] transp. B 9 το] om. M τα B 9 ἀκριβῶς] μετὰ ἀκριβείας A 11 ἀγριοσταπίδαν] ἀγριοσταφίδας Σ 11 καλακάνθην] καλακάθιν M καλακάνθην A καλαμίνθιν B 11 μίξε] μίξον Σ 12 πάντα] μετ' ἀλλήλων add. MA ἀλλήλοις B 12 πληγάς] τῆς κεφαλῆς add. Σ 12 τεάφην] τεάφιν Σ 12 ψιλὴν] ψιλὸν Σ 12 μίξον] μίξε το B 13 με τὸ] μετὰ B 13 χρηστέλαιον] ἔλαιον τὸ χρηστὸν Σ 13 ἄλειφε] τὴν κεφαλὴν add. Σ 14 ποιήσας] ποίησον B 14 καὶ] μίξε τὰ ὅλα καὶ add. B 15 τὴν κεφαλήν] τὰς πληγὰς B 16 περὶ] πρὸς M 16 ὀφιάσεως] ὀφ...σεως L 17 τρίχες] τ..... L 17 πέσωσι] πίπτουσιν Σ

κεφαλὴ λεῖα να μὴ ἔχῃ πούποτε τρίχαν. 2 ἄλλο. ἐὰν ἴδῃς τὸ δέρμα τῆς κεφαλῆς, ὅτι ἔνι ὡς ἂν μέλαν, πότισον βοήθημα ἔχον πλέον τῶν ἄλλων εἰδῶν τὸ μέλαν. ἐὰν δὲ ἴδῃς τὸ δέρμα τῆς κεφαλῆς λευκόν, πότιζε αὐτὸν τὴν ἱεράν. ὁμοίως καὶ ὅταν ἔνι τὸ δέρμα τῆς κεφαλῆς μέλαν. ἐὰν δὲ ἴδῃς τὸ δέρμα μήτε πολλὰ μέλαν μήτε πολλὰ λευκόν, ἀλλὰ ἔνι μέσον, πότιζε αὐτὸν ἀλόην τριμμένην με τὸ κρασὶν καὶ τὴν λεγομένην πικρὰν ▹om. Bὡσεὶ κοκκία ιβ' καθ' ἑκάστην ἡμέραν πλὴν με τὸ κρασίν.om. B◃ 3 ἄλλο. βάλε εἰς τζυκάλιν βλησκούνιν καὶ θύμον καὶ ὕσσωπον. καὶ σινάπιν καὶ μέλι ὀλίγον καὶ ὀρίγανον καὶ ὀξείδιν καὶ ἂς βράσῃ καὶ τὸ ὀξείδιν ἐκεῖ κατὰ πρωΐ. ἄς το βάνῃ εἰς τὸ στόμα κατὰ πρωΐ καὶ ἄς το κρατῇ πολλὴν ὥραν. τοῦτο δὲ ποίει καθ' ἑκάστην ἡμέραν. 4 ἄλλο. ἔπαρε χοντρὸν πανὶν παχὺ καὶ βρέξε τὸ εἰς ὀξείδιν δριμὺ καὶ τρίβε τὸ δέρμα τῆς κεφαλῆς αὐτοῦ. τρίβε τόδε ἔστ' ἂν γένοιτο ὁ τόπος κόκκινος.⌝ 5 ἄλλο. καλάμου ῥίζαν ξήρανον καὶ ψιλὴν ποιήσας μίξε τὸ με τὸ ἔλαιον τὸ ἀπομένει εἰς τὸ λυχνάριν,

17 δὲ λέγεται] δ..... L 1 πούποτε] πούπετε M πούπεται A πόποτε B 1 τρίχαν] κἂν τὸ λεγόμενον διὰ ἰατρίαν add. M, δι' ἰατρείαν add. A τριχίτζαν καὶ τὸ λευόμενον δι' ἰατρείαν B 1 ἐὰν] δὲ add. M 2 ἔνι] ἔνεν A 2 πότισον] αὐτὸν add. M 2 ἔχον] ἔχων MA 3 κεφαλῆς] ὅτι ἔνι add. Σ 4 αὐτὸν] om. M τὸν AB 4 τὸ δέρμα τῆς κεφαλῆς] τὸ αὐτὸ καὶ A 5 δὲ ἴδῃς τὸ δέρμα] τὸ δέρμα τῆς κεφαλῆς ἴδῃς M τὸ δέρμα τῆς κεφαλῆς ἴδῃς A δὲ τὸ δέρμα τῆς κεφαλῆς ἴδῃς B 6 αὐτὸν] τὸν MA τους B 6–7 με τὸ κρασὶν καὶ τὴν λεγομένην πικρὰν] om. M καὶ τὴν λεγομένην πικρὰν A πολλὰ με τὸ κρασὶν καὶ B et add. lac. 8 τζυκάλιν] τζουκάλιν B 8 βλησκούνιν] φλισκούνην A 10 βράσῃ] βράσουν Σ 10 τὸ ὀξείδιν ἐκεῖ κατὰ πρωΐ] τὸν ζωμὸν M 10 ἐκεῖ] ἐκεῖνο AB 10 το] om. M 10 βάνῃ] βάλῃ AB 11 στόμα] του add. Σ 11 κατὰ πρωΐ] ...... L om. AB 11 το] om. M 11 πολλὴν ὥραν] transp. B 11 τοῦτο δὲ] καὶ τοῦτο MB καὶ τοῦ A 12 ἡμέραν] om. A 12 ἔπαρε] ἔπαρον AB 12 βρέξε] βρέχε MA 12 εἰς] τὸ add. A 13 ὀξείδιν] τὸ add. A τοξίδιν B 13 αὐτοῦ] om. Σ 13 τρίβε] καὶ τρίβε τὸ δέρμα τῆς κεφαλῆς τρίβε τὸ M 13–14 τόδε ἔστ' ἂν] δὲ ἕως A ὅταν B 14 γένοιτο] γένῃ L να γένηται MA 14 ὁ τόπος κόκκινος] ἐρυθρὸς ὁ τόπος ἤτοι κόκκινος πολλὰ MB κόκκινος ὁ τόπος πολλὰ A 15 ψιλὴν ποιήσας] transp. MA ποίησον ψιλὸν B 15 μίξε τὸ] καὶ μίξε A καὶ μίξε την B 15 τὸ] ὁποῦ A 15 ἀπομένει] ἀπομείνῃ M ἀπομήνει A μένοντα B 15 εἰς τὸ] ἐν τῷ B

καὶ μὴ βάλλεις ἔστερον καθαρόν. 6 ἄλλο. ἀξούγγιν ἀρκουδίας ἀνάλυσέ το καὶ ▷om. B ἔχε τριμμένον εὐφόρβιον. καὶ μίξε καὶ om. B◁ ἄλειφε τὴν κεφαλὴν πλὴν πρῶτον τρίψε τὴν κεφαλὴν ὥς σε εἶπον ὀπίσω. 7 ἄλλο. τεάφην ψιλὴν ποιήσας μίξον με τὸ ὑγροπίσσιν καὶ ἄλειφε τὴν κεφαλήν. 8 ἄλλο. εὐζώμου σπόρον τριμμένον ψιλὰ μίξε το με τὸ ὑγροπίσσιν καὶ ἄλειφε τὴν κεφαλήν. 9 ἄλλο. ▷LB ἀξούγγιν ἄρκου καὶ πέπερι τριμμένον καὶ μελάνθιν ψιλὸν ποιήσας μίξον καὶ ἄλειφε τὴν κεφαλήν. LB◁ 10 ἄλλο. σκανζοχοίρου δέρμα καῦσον καὶ μίξον με τὴν κεδρέαν καὶ με τοῦ ἄρκου τὸ ἀξούγγιν καὶ ἄλειφε τὴν κεφαλήν. εἴ τι δὲ ποιεῖς εἰς τὴν ἀλωπεκίαν καὶ ὀφίασιν τὴν αὐτὴν θεραπίαν ποίει.

91 [75] ϙα´ ⌜περὶ κεφαλαλγίας· ἐὰν ὁ ἄρρωστος αἰσθάνηται ὅτι ἔχει βάρος ἡ κεφαλή του, σημεῖον ἔνι, ὅτι πλῆθος ἐστιν εἰς τὴν κεφαλήν του. καὶ πότισέ τον βοήθημα καὶ φλεβοτόμησον αὐτὸν

---

15 λυχνάριν] scripsi λιχνάριν L λυχνάριν ὅπου ἧπτε πρότερον ἡ φωτία ἐκεῖνο τὸ ἔλαιον M εἰς τὸ λυχνάρην ὁποῦ ἧπτεν ἡ φωτία ἐκεῖνο δὲ τὸ ἔλαιον A λυχνάρην ὅπου ἧπτε πρότερον τὴν φωτίαν ἐκεῖνο τὸ ἔλαιον B 1 βάλλεις ἔστερον] ἄλλο καλὸν MB ἄλλο A 1 καθαρόν] καὶ ἄλειφε τὴν κεφαλὴν A 1 ἀρκουδίας] ἀρκούδας A 2 ἀνάλυσέ το] ἀνάλυσον M ἀνάλυσε A 2 τριμμένον] καὶ add. M 2 εὐφόρβιον] ἐφόρβιον L ἐφόλβιον M βόλβιον A 2 μίξε καὶ] μίξας M μίξε το καὶ A 3 πλὴν] δὲ add. MA om. B 3 πρῶτον τρίψε τὴν κεφαλὴν] om. B 3–4 ὥς σε εἶπον ὀπίσω] καθὼν εἶπον σοι ὀλίγον παροπίσω M ὡσεὶ εἶπον καλὰ A ὡς εἶπον ὀλίγον παρεπίσω B 4 τεάφην] τεάφιν MB 4 ψιλὴν] ψιλὸν MB 4 μίξον] μίξε τὴν A μίξε το B 5 εὐζώμου] εὐσώμου B 5 σπόρον] σπέρμα MA 5 τριμμένον] τρίψον M 5 ψιλὰ] καλὰ M ψιλὸν καλὰ A καλὰ ψιλὰ B 5–6 μίξε το] καὶ ἕνωσον M 6 κεφαλήν] ὅλην add. B 7 ἄρκου] scripsi .κου L ἀρκούδας ἢ ἄρκου B 7 μελάνθιν] scripsi μέλα L μελάνθη B 7 μίξον] μίξε ὅλα B 8 ἄλειφε] ἀ........ L 8 σκανζοχοίρου] τὸν λεγόμενον σκατζόχοιρον τὸ M τὸν λεγόμενον σκαντζόχοιρον ὁποῦ ἔχει τὸ A τὸ λεγόμενον σκατζούχοιρον ὅπου ἔχει τὸ B 8 δέρμα] δτου add. M τοῦ ὡσὰν σούγλια τὸ δέρμαν ἐκείνου add. A του ὡσὰν σουγλεία παροξυντὰ ἐκείνου τὸ δέρμα add. B 8 καῦσον] καῦσε καλὰ M καύσε το πολλὰ A πολλὰ add. B 9 μίξον] τρίψον καὶ μίξον M μίξε το A μίξε την B 9 με τοῦ ἄρκου] με τοῦ ἄκρον L τῆς ἄρκου Σ 10 εἰς] om. M 10 τὴν] λεγομένην add. MB 10 καὶ ὀφίασιν] om. Σ 11 ποίει] καὶ ἐπὶ τῆς ὀφιάσεως. εἶπον σοι γὰρ τι λέγετε ἀλωπεκία καὶ τι λέγεται ὀφίασις add. MB καὶ ἐπὶ τῆς ὀφιάσεως καθὼς καὶ ἄνωθεν ἐδηλώθη σοι add. A 12 περὶ] πρὸς M 12 ἐὰν ὁ] πᾶς MA 13–14 πλῆθος ἐστιν εἰς τὴν κεφαλήν] ἔχει πλῆθος ἡ κεφαλὴ A 13 ἐστιν] ἔνι MB

καὶ πότισον αὐτὸν καὶ τὴν ἱερᾶν, ἣν ἐποίησεν ὁ ἰατρὸς ὁ λεγόμενος Λογάδιος.⌝

92 [75] ϙβʹ εἰ δὲ φαίνεταί τον ὅτι ὡς ἂν ἀποταυρίζεται ἡ κεφαλή του, σημεῖον ἔνι ὅτι ἀπὸ πνευμάτων ὀχλεῖται καὶ πότισον αὐτὸν τὴν λεγομένην πικρᾶν. 2 ἄλλο. πέπερι τριμμένον καὶ κυκλάμινον καὶ ἄς τα μυρίζηται.

93 [75] ϙγʹ ⌜εἰ δὲ φαίνεταί τον ὅτι κένταται τὸ δέρμα τῆς κεφαλῆς αὐτοῦ με τὸ βελόνιν καὶ ἔχει καύσωνα ἡ κεφαλή του πολύν· πότισον αὐτὸν βοήθημα ἔχον τὰ ξανθὰ πλείονα τῶν ἄλλων εἰδῶν τοῦ βοηθήματος. λοῦέ τον δὲ πυκνὰ εἰς γλυκέα νερά. εἰς δὲ τὰ λεγόμενα θέρμα μὴ λούεται.⌝ βλάπτει γὰρ τὴν κεφαλὴν τὰ τοιαῦτα θέρμα ὡς στυπτικά. 2 ἄλλο. ▷MB⌜ἄλειφε τὴν κεφαλὴν ῥοδέλαιον.MB◁ 3 ἄλλο. ἔλαιον ἀπὸ ἄγρας ἐλαίας ἔκβαλε καὶ μίξε με τὸ ὀξείδιν καὶ ἄλειφε τὴν κεφαλήν.⌝

94 [76] ϙδʹ ⌜ἐὰν δὲ φαίνεταί τον ὅτι τὸ δέρμα τῆς κεφαλῆς του ἔνι ψυχρὸν παγωμένον καὶ αἱμωδιασμένον· ἄλειφέ τον τοιοῦτον εἰς τὴν κεφαλὴν ἀνηθέλαιον χλίον, οἷον δὲ ἔλαιον εὕρης ἐξ αὐτῶν, βάλε τριμμένον καστόριον ἢ εὐφόλβιον ὀλίγον.⌝

---

14 αὐτὸν] τὸν A 1 πότισον αὐτὸν] πότισέ τον A 1–2 ὁ ἰατρὸς ὁ λεγόμενος Λογάδιος] ὁ Λογάδιος ὁ ἰατρος A 3 εἰ δὲ] πρὸς τὸ ἐὰν M 3 τον] ἄνθρωπον add. M 3–4 ὡς ἂν ἀποταυρίζεται ἡ κεφαλή του] τὸ δέρμα τῆς κεφαλῆς αὐτοῦ ἀποταυρίζεται M 4 ἀπὸ] ὑπὸ M 6 τα] .. L om. M το A 7 εἰ δὲ] πρὸς ὅτι M 7 κένταται] ....ται L 8 αὐτοῦ] om. A τοῦ B 8 με τὸ βελόνιν] με τὸ βε...ρην L μετὰ βελώνιν M με τὸ βελώνιν A 8 καύσωνα] καύσονα M καῦσον ἂν A 8–9 πότισον αὐτὸν βοήθημα] om. M δὸς αὐτὸν βοήθημα A 9 ἔχον τὰ] scripsi ἔχωντα L ἔχοντα AB 9 ξανθὰ] om. A 9–10 εἰδῶν τοῦ βοηθήματος] transp. A 10 δὲ] om. M καὶ add. A 11 γὰρ] οὖν A 13 ἔλαιον] τὴν κεφαλὴν ἄλειφε ἔλαιον B 13 ἀπὸ] τῶν add. B 13 ἄγρας] ἀώρων Σ 13 ἐλαίας] ἐλαιῶν Σ 13 μίξε] ἕνωσον M το add. A 13 ὀξείδιν] ὄξος M 15 τον] om. A 16 παγωμένον] πεπαγωμένον M πεπαγομένον AB 16 αἱμωδιασμένον] αἱμοδιασμένον L μοδιασμένον A 16 τοιοῦτον] om. M 16–17 εἰς τὴν κεφαλὴν] om. A 17 χλίον] ἢ πηγανέλαιον χλίον ἢ ναρδέλαιον χλίον add. MA ἢ πηγανέλαιον οὕτως χλίον ἢ ναρδέλαιον ὡς λέγω B 17 ἔλαιον εὕρης] εὕρεις MB εὕρης A 17 βάλε] καὶ add. MA 18 ἢ] καὶ A 18 εὐφόλβιον] ἐφ...βιον MB βόλβιον A

---

3 εἰ δὲ] This chapter is part of the preceding one in AB. 7 εἰ δὲ] This chapter is part of the preceding one in AB. 15 ἐὰν] This chapter is part of the preceding one in MA.

95 [77] ϙε´ ⌜ὅπου κεφαλαλγεῖ ἀπὸ στομάχου· αὐτὸν δὲ θέλεις νοήσειν ὅτι ἀπὸ στομάχου πάσχει, ὅτι ἡ ὀδύνη τῆς κεφαλῆς ὀξέα γίνεται, ὥσπερ ὥρας ὥρας. 2 ▷om. L κρούει τον σουγλέαν να εἴπης ὅτι κόπτεται ἡ καρδία τοῦ ἀπὸ τῶν ὀξέων πόνων τοῦ δέρματος τῆς κεφαλῆς καὶ αἱ φλέβες τοῦ μετώπου του πηδῶσι πολλὰ ἤτοι σφύζουσι καὶ οἱ μήνιγγοί του καὶ τὰ ὠτία του ἔχωσιν ἦχον ἤτοι τὸ ἀχὸν λεγόμενον. 3 ἄλλο. ἰάτρευε τοῦτον κλυστήριον καθ' ἑκάστην ἡμέραν ἤτοι βάλε θερμὸν μέλι ἅλας καὶ ἔλαιον τὴν δὲ κεφαλὴν ἄλειφε ἀνηθέλαιον μικτὸν μετὰ ὀλίγου ὄξους. 4 ἄλλο. οἰκονόμει δὲ τὸν τοιοῦτον ἄρρωστον πρῶτον καὶ ἂς ξερᾷ εἶτα πότιζέ τον τὴν καλουμένην πικράν. εἰς δὲ τὸν στόμαχόν του θέτε ἔμπλαστρον τὸ καλοῦμενον παρὰ τῶν ἰατρῶν διὰ σπερμάτων. 5 ἄλλο. ποίησον δὲ καὶ πυρίαν τοιαύτην ἤγουν βάλε εἰς τζυκάλιν κρασὶν παλαιὸν καλόν. εἰ δὲ οὐκ ἔχεις παλαιόν, βάλε στυφὸν ἀληθινὸν ἢ ἄσπρον. βάλε καὶ ἀψίνθιον ἢ τὸν σπόρον του ἢ τὰ ξύλα του καὶ τριψίδιν τριμμένον καὶ στάχος καὶ μαστίχιν καὶ ἂς βράσουν καλά. εἶτα πυρίαζε τὸν στομαχόν του ὡσεὶ πανία εἴκοσιν ἢ ιε'. εἶτα ἂς λούηται. 6 ἄλλο. ἡ τροφή του ἂς ἔνι τοιαύτη· τὸ παστὸν οἷον ἂς ἔνι μὴ τρώγῃ μήτε κρέας μήτε ὀψάριν μὴ δὲ ὄσπρια μηδὲ δειπνᾷ ἀλλὰ ἂς τρώγῃ μίαν τὴν ἡμέραν καὶ ἂς πίνῃ καὶ ἂς πίνῃ καὶ κρασὶν ὀλίγον.⌝om. L◁

---

1 στομάχου] στ.μαχοῦ L 2 νοήσειν] νοῆσαι MA νοήσεις B 2 πάσχει] ἔχει καὶ πάσχει M ἔχει A 2 ὅτι] ὅταν A 2 ἡ] om. B 2 ὀξέα] τοιαύτη Σ 3 γίνεται] γένηται B 3 ὥσπερ] ὀξέα ὥσπερ MB ὀξεία ὅτι A 3 ὥρας] om. MB 3 σουγλέαν] σουγλέας A 4 δέρματος] αἵματος B 5 κεφαλῆς] τοῦ add. B 5 του] om. AB 6 σφύζουσι] scripsi σφίζουσι MA σφίγγουσι B 6 μήνιγγοί του] scripsi μηνιγγίτου M μήνιγγοι A μιλιγγίτου B 6 ἔχωσιν] ἔχουσιν B 6 ἦχον] πολλὺν add. A 6–7 ἤτοι τὸ ἀχὸν λεγόμενον] om. A transp. B 7 ἰάτρευε] ἰάτρευσον A 7 κλυστήριον] κλυστήριν AB 8 ἡμέραν] om. A 8 ἤτοι] om. AB 8 βάλε] δὲ add. A 8 ἅλας] καὶ ἅλας A 9 ἄλειφε] την add. A 9 οἰκονόμει] εἰκονόμει M 10 πότιζέ] πότισον B 10 τον] om. B 12 ἄλλο] ἕτερον κάλιον B 13 εἰς] om. M 13 τζυκάλιν] τζουκάλιν B 14 παλαιόν] καλὸν add. B 14 στυφὸν] στηφὸν AB 15 ἢ] ἤγουν A 15 τριψίδιν] τριψίδην A τριδίψιν B 16 μαστίχιν] μαστίχην A 17 του] αὐτοῦ B 17 ὡσεὶ] om. B 17 πανία] παννία M 18 τροφή] δὲ add. B 18 του] αὐτοῦ B 18 οἷον ἂς] ἐὰν B 19 ὀψάριν] ὠψάρην A 19 μὴ δὲ] μήτε B 20 ἂς πίνῃ] καὶ κρασὶν ὁ πρὸς ἐὰν πίνῃ πολλὴ κρασὶν καὶ ἀπ' ἐκείνου κεφαλαλγῇ B 20 καὶ ἂς πίνῃ] om. AB 20 καὶ κρασὶν ὀλίγον] om. B

96 [78] ϘϚʹ ⊳om. L⌜πρὸς ἐὰν πίῃ κρασὶν πολὺ καὶ ἀπ' ἐκείνου κεφαλαλγεῖ· ἐὰν οὐδὲν ἐχώνευσεν ἀκμὴν τὸ κρασίν, ὅρισέ τον καὶ ἂς ξεράσῃ ἤγουν δὸς τον νερὸν χλίον καὶ ἂς πίῃ να τὸ ξεράσῃ. 2 ἄλλο. ἐὰν δὲ ἀφ' ὅτου ἐχώνευσεν πονῇ τὴν κεφαλήν, ἄλειφέ τον ῥοδέλαιον μόνον ἢ ὀξείδιν με τὸ ῥοδέλαιον.⌝ 3 ἄλλο. κοπάνισον τὰ φύλλα τοῦ κισσοῦ καὶ τὸν ζωμόν του ἄλειφε τὴν κεφαλήν. 4 ἄλλο. ⊳om. AB⌜κοπάνισον τοῦ κραμβίου τὰ φύλλα καὶ τὸν ζωμὸν ἄλειφε τὴν ⊲om. L κεφαλήν.⌝om. AB⊲ 5 ἄλλο. ἢ τὰ φύλλα κραμβίου κοπανισμένα μίξε τα με τὸ αὐγόν. καὶ θές τα εἰς τὸ μέτωπον αὐτοῦ. 6 ἄλλο. ⌜ἂς τρώγῃ δὲ τὴν κράμβην καλὰ ἑψημένην, ἔστ' ἂν ἡ κεφαλαλγία λυθῇ. 7 ἄλλο. ἂς τρώγῃ καὶ ῥοΐδια καὶ μῆλα καὶ ἀπίδια.⌝ 8 ἄλλο. ⌜ἐὰν δὲ θέλῃ να μὴ μεθύῃ ἂς τρώγῃ νῆστις, τὸ γουλὶν τοῦ κραμβίου⌝ ἢ ἀμύγδαλα πικρά.

97 [79,80] ϘΖʹ ἀποφλεγματισμοὶ καθαίροντες τὴν κεφαλήν· βάλε εἰς τζουκάλιον ὄξος καὶ ⌜ἔψημα καὶ ὕσσωπον καὶ βλησκούνιν καὶ ὀρίγανον καὶ σάμψυχον καὶ ἂς βράσουν καὶ τὸ ζωμὶν τούτων κράτει εἰς τὸ στόμα καὶ εὐγάνει πολλὰ ὑγρά.⌝ 2 ἄλλο. ⊳om. Σ⌜κοπάνισον

1–2 πρὸς ἐὰν πίῃ κρασὶν πολὺ καὶ ἀπ' ἐκείνου κεφαλαλγεῖ] om. B 1 ἐκείνου] ἐκεῖνο A 2 ἐχώνευσεν] ἐχόνευσε AB 2 ὅρισέ τον] ὥρισέ τους B 3 ἤγουν δὸς τον] δὸ τὸν δὲ A δὸς τὸ δὲ νερὸν B 4 ἐχώνευσεν] ἐχόνευσεν Σ 4 κεφαλήν] του add. A 5 με τὸ ῥοδέλαιον] ἀντάμου A 6 τοῦ] om. B 6 τὸν ζωμόν] τὸ ζουμὶν B 6 του] om. A τους καὶ B 8 κεφαλήν] .... L 8 ἢ] om. MA 8 κραμβίου] κράμβης M τῆς κράμβης AB 9 κοπανισμένα μίξε τα] κοπανισμ...ξέτα L κοπανισμένα μίξον A 9 με τὸ αὐγόν] μετὰ τοῦ αὐγοῦ τὸ ἄσπρον καὶ τὸν κρόκον M με τὸ ἄσπρον τοῦ αὐγοῦ καὶ τὸν κρόκον A μετὰ αὐγοῦ τὸ ἄσπρον καὶ τὸν κρόκον B 9 τα] τον A το B 9 μέτωπον] μ.... L 10 δὲ] καὶ MA 10 καλὰ ἑψημένην] transp. MA 10 ἔστ'] ἕως B 12 δὲ θέλῃ] θέλης MA δὲ θέλης B 12 μὴ] μηδὲν Σ 12 μεθύῃ] μεθύης AB 12 ἂς] om. A 12 τὸ] om. A 13 τοῦ κραμβίου] τῆς κράμβης Σ 13 πικρά] πέντε add. MA μικρὰ πέντε B 14 ἀποφλεγματισμοὶ] πρὸς ἀποφλεγματισμοὺς M εἰς ἀποφλεγματισμοὺς B 14 καθαίροντες] καθαίροντας M ἤτοι εἰς κάθαρσιν B 14 τὴν κεφαλήν] τῆς κεφαλῆς B 15 τζουκάλιον] τζυκάλιν Σ 15 καὶ] om. B 15 βλησκούνιν] φλησκούνιν A 16 ὀρίγανον] ὀριγάνιν M 16 ἂς βράσουν καὶ] om. A 16 τὸ ζωμὶν] τὸν ζωμὸν MA τὸ ζουμὶν B 16 τούτων] τοῦτο B 17 εἰς τὸ στόμα] ἐν τῷ στόματι Σ 17 καὶ] πλὴν δὲ ἂς βράσουν εἰς τὴν ἰστίαν καὶ A 17 εὐγάνει] scripsi εὐγάνῃ L να εὐγάλῃς MA να ἐκβάλῃς B

---

12 ἐὰν] Here starts a new chapter in A. In the pinax of L it is listed separately as well.

ἀγριοσταπίδα καὶ ποίησον πολλὰ ψιλὴν καὶ πέπερι τριμμένον. βάλε καὶ μαστίχην καὶ μίξε τὰ πάντα καὶ ἄς το μασσᾶται καὶ να εὐγάλη πολλὰ ὑγρά.⌝◁om. Σ 3 ἄλλο. ζίγγιβερ τριμμένον καὶ ἀγριοσταπίδα μίξε καὶ ἃς τὰ μασσᾶται. ▷om. A ζίγγιβερ δὲ λέγεται τὸ λεγόμενον ξυλοζιζίβερι.om. A◁ 4 ἄλλο. βάλε ἀγριοσταπίδαν καὶ ἕψημα καὶ ὄξος καὶ γλήχωνα καὶ θύμον καὶ ὀρίγανον καὶ ἃς βράσουν καὶ ἄς τα κρατῇ εἰς τὸ στόμα. καὶ να εὐγάνῃ πολλὰ ὑγρὰ καὶ σίελα.

98 [81] ϙη′ ἐὰν θέλῃς τὸ κεφάλιν σου ἵνα καθαιρίσῃς ἀπὸ τὰ ὑγρά· ⌜κοπάνισε τοῦ σεύτλου τὰ φύλλα καὶ τὸν ζωμὸν αὐτὸν χύσε τον εἰς τὴν μύτην τοῦ.⌝ 2 ἄλλο. ▷om. L μελάνθιν τρίψον ψιλὸν καὶ μίξον με τὸ γάλα τὸ γυναικεῖον καὶ βάλε εἰς τὴν μύτην του.◁om. L 3 ἄλλο. ⌜κοπάνισον κυκλάμινον καὶ τὸν ζωμόν του χύσε τον εἰς τὴν μύτην τοῦ. 4 ἄλλο. ▷om. L κοπάνισον μελάνθιν καὶ ψιλὸν ποιήσας μίξον μετὰ χρηστέλαιον καὶ ἃς ἀργήσει καὶ τότε χρίσε το εἰς τὴν ῥίναν του.⌝◁om. L

99 [82] om. B ϙθ′ ⌜ἐὰν θέλῃς να πτέρνῃς πολλά· καρπὶν ἄσπρον τρίψον καὶ καστόριον τριμμένον καὶ πέπερι ψιλὸν καὶ μίξε τα καὶ δένε τα εἰς τὸ πανὶν καὶ ἄς τα μυρίζηται.⌝ 2 ἄλλο. ▷om. MB κυκλά-

17 ὑγρά] νερὰ B 3 ζίγγιβερ] ζιζίβερ B 3 ἀγριοσταπίδα] ἀγριοσταφίδα AB 4 μίξε] μίξον M 4 καὶ ἃς] ..... L 5 ξυλοζιζίβερι] ξυλοτζιτζύβερη L ξυλοζιζίβερ B 5 ἀγριοσταπίδαν] ἀγριοσταφίδα Σ 5 καὶ ἕψημα] om. Σ 5 ὄξος] ὀξείδιν MB 6 γλήχωνα] ἤτοι βλησκούνιν add. M γλήχωνα A 6 ὀρίγανον] ὀριγάνιν ἣν λέγει διοσκορίδης κονιάνην M ὀρίγανιν A 6 καὶ ἄς τα] καὶ τὸν ζωμὸν ἃς M 7 στόμα] του add. Σ 7 εὐγάνῃ] εὐγάλῃ MA ἐκβάλει B 7 ὑγρὰ] πτύσματα Σ 7 σίελα] ἤτοι σάλια καὶ ὑγρά M ὑγρὰ A συέλλη ἤγουν σάλη ὑγρὰ B 8 τὸ κεφάλιν σου ἵνα καθαιρίσῃς] scripsi c.f. ind. L τὸ κεφάλιν σου L τὴν κεφαλὴν να τὴν καθάρις M τὴν κεφαλήν σου να τὴν καθάρῃς A να καθάρῃς τὴν κεφαλὴν B 8 τὰ ὑγρὰ] τῶν ὑγρῶν Σ 9 κοπάνισε] κοπάνισον Σ 9 καὶ] om. A 9 αὐτὸν] του Σ 9 χύσε τον] βάλε M θέτε τον A θές το B 10 εἰς τὴν μύτην] ἐπάνω B 10 τοῦ] σου A om. B 10 τρίψον ψιλὸν] ψιλὸν ποίησον AB 10 μίξον] μίξε το AB 11 καὶ] τὸ γάλα ἐκεῖνο add. AB 11 βάλε] το add. AB 12 του] αὐτοῦ A αὐτῆς B 12 χύσε τον] βάλε MA ἔκβαλε B 12–13 μύτην τοῦ] ῥίναν αὐτοῦ MB 13 μελάνθιν καὶ ψιλὸν ποιήσας] μελάνθιν ψιλὸν A 13 μίξον] καὶ μίξε το A μίξε τα B 14 μετὰ] με τὸ AB 14 ἃς ἀργήσει καὶ τότε] τὸ ἔλαιον ἐκεῖνο A ἃς ἀργήσει καὶ τὸ ἔλαιον B 14 ῥίναν] μύτην A 15 του] αὐτοῦ B 16 πτέρνῃς] πτάρις M πτάρῃς A πτάρῃ B 17 τρίψον] τρίψας MA 17 καὶ] om. A 17 μίξε τα] ἕνωσον M 18 δένε] δὲς M

μινον κοπάνισον καὶ τὸν ζωμόν του ἄς τον ἀνασύρῃ εἰς τὴν μύτην του.[om. MB]◁ 3 ἄλλο. ▷[om. A]⌜κηκίδιν τριμμένον δένε το εἰς τὸν πάνιν καὶ ἄς το μυρίζηται.⌝[om. A]◁ 4 ἄλλο. ▷[om. LA]κυκλάμινον κοπάνισον καὶ τὸν ζωμόν του ἄς το σύρῃ εἰς τὴν μύτην του.[om. LA]◁

100 [83] [om. B]ρ′ ⌜ἐὰν θέλῃς να παύσουν οἱ πταρμοὶ ποίησον τοῦτο· μυρίζου βασιλικόσπορον ἢ ψύλλεον μυρίζου συνεχῶς.⌝

101 [84] ρα′ ἐὰν ἀπὸ ψύξεως πονῇ τὸ κεφάλιν του· ⌜βάλε δάφνης κοκκία καὶ κοπάνισον καὶ ποίησον ψιλὸν καὶ πηγάνου σπόρον ὁμοίως βάλε καὶ δᾳδίου ῥίζαν καὶ ὀξείδιν, καὶ ἄς βράσουν. εἶτα βάλε κηρὸν ὀλίγον εἰς τὸ ζωμίν τους καὶ ἄλειφε τὴν κεφαλήν. 2 ἄλλο. ῥοδακίνου ὀστέον τζάκισον καὶ τὸ ἀπ' ἔσω τρίψε καὶ λίβανον τριμμένον καὶ ὀξείδιν καὶ μάλαξον αὐτὰ καὶ θές το εἰς τὸ μέτωπον με τὸ πανίν.⌝ 3 ἄλλο. καρδαμόσπορον τρίψον ψιλὸν καὶ ξηρὸν κόπρον περιστερᾶς καὶ ὄξος καὶ μάλαξον αὐτὰ καὶ θέτε το με τὸ πανὶν εἰς τὸ μέτωπον.

102 [84] ρβ′ ἐὰν δὲ πονεῖ τὸ κεφάλιν του ὅτι ἔδειρεν τὸν ὁ ἥλιος πολλά· ἔπαρε κοχλίους καὶ τζάκισέ τους. καὶ τὸ κρέας αὐτῶν μίξε το με τὸ λίβανον τὸ τριμμένον καὶ τὸ ὀξείδιν. καὶ θέτε τὰ εἰς τὸ

18 τὸ] om. M 2 κηκίδιν] κηκίδια M 2 τριμμένον δένε] τρίψον καὶ δέσε M 2 τὸν] om. M 5 ἐὰν θέλῃς να παύσουν οἱ πταρμοὶ] πρὸς ἐὰν θέλῃς να μὴ πτάρνῃς πολλὰ M 5 ποίησον τοῦτο] om. AB 6 μυρίζου] μυ...... L μιρίζου M 6 βασιλικόσπορον] σιλικόσπορον L βασιλικοῦ σπόρον Σ 6 μυρίζου συνεχῶς] om. M καὶ οὕτως ἔνι ὀφέλιμον B 7 ψύξεως] ψύχους ἤτοι ψύξεως A 7 τὸ κεφάλιν του] τὴν κεφαλὴν Σ ποίησον οὕτως add. B 8 καὶ ποίησον ψιλὸν] ἐν ὅλμῳ ἤγουν ἐνιγδίῳ M 8 ψιλὸν] ψιλὰ AB 9 βάλε] om. M 9 βάλε] καὶ add. B 10 τὸ ζωμίν τους] τὸν ζωμὸν M τὸν ζωμὸν τῶν A 11 ὀστέον] ὀστοῦν MB 11 τζάκισον] τζάκκισε M τζάκισε A 11 τρίψε] τρίψον M το add. AB 11–12 τριμμένον] om. M 12 καὶ] βάλε add. M 12 αὐτὰ] ταῦτα M 12 το] τα Σ 12–13 με τὸ πανίν] μετὰ πανίου Σ 13 τρίψον] τριμμένον AB 13 ψιλὸν καὶ ξηρὸν] καὶ M ψιλὸν καὶ A ξερὸν ψιλὸν καὶ ξηραν B 14 περιστερᾶς] ξηρὸν add. M περιστερῶν ψιλὸν A 14 καὶ] βάλε add. M 14 ὄξος καὶ] om. A 14 θέτε το] θές τα MB θέτε τα A 14–15 με τὸ πανὶν εἰς τὸ μέτωπον] transp. B 14 με τὸ πανὶν] μετὰ πανίου B 16–17 ἐὰν δὲ πονεῖ τὸ κεφάλιν του ὅτι ἔδειρεν τὸν ὁ ἥλιος πολλά] ἐὰν τον δείρει ὁ ἥλιος πολλὰ καὶ πονῇ τὴν κεφαλὴν B 16 ἐὰν δὲ] πρὸς ἐὰν M 16 τὸ κεφάλιν του] τὴν κεφαλὴν MA 16 ὅτι ἔδειρεν] ὅ..δειρεν L 17 ἔπαρε] ἔπαρον AB 17 κοχλίους καὶ] κο..... L 17–18 μίξε το] μίξον με τὸν M 18 με τὸ] με τὸν MA 18 τὸ] τὸν MA 18 τὸ ὀξείδιν] βάλε καὶ ὄξος M

μέτωπον με τὸ πανίν. 2 ἄλλο. τὸ κρέας δὲ τῶν κοχλίων κοπάνισέ το καλά, εἶτα μίξε το. 3 ἄλλο. τὰ ἔντερα τῆς γῆς ἤτοι τὰ σκουλήκια ὅπου πιάνουν τὰ ὀψάρια. κοπάνισέ τὰ σκουλήκια καὶ τριμμένον λίβανον καὶ καστόριον καὶ ὄξος καὶ μάλαξέ τα καὶ θέτε τὰ με τὸ πανὶν εἰς τὸ μέτωπον. 4 ἄλλο. ⌜εὐφόλβιον τριμμένον καὶ λίβανον καὶ ὄξος μίξε τα καὶ θέτε τα εἰς τὸ μέτωπον του με τὸ πανίν.⌝

103 [85] ργʹ ⌜περὶ κατάρρου· ὅταν ῥέῃ τὸ ὑγρὸν ἀπὸ τὸν οὐρανίσκον, ἐκεῖνο λέγεται κατάρρους. ὅταν δὲ ἀπὸ τὴν μύτην του, λέγεται κόρυζα καὶ ἐάν τον ἔδειρεν ἥλιος καὶ πολὺ καῦμα ἔνι καὶ ἐκεῖνο ὡς ἂν δριμὺ καὶ λεπτὸν καὶ φαίνεταί τον ὅτι ὡς ἂν καίει τον τοιοῦτον, ἄλειφε τὴν κεφαλὴν του ῥοδέλαιον ἢ ἔλαιον ἀπὸ ἐλαίας ἀγούρας καὶ λοῦε καὶ πυκνά. ἂς μυρίζωνται δὲ καὶ μελάνθιν καὶ κύμινον μικτὸν καὶ φρύξε τα ὀλίγον εἰς καινούργιον τζουκάλιν καὶ δένε τα εἰς τὸ πανίν. καὶ ἄς τον μυρίζηται.⌝

104 [85] ρδʹ ἐὰν δὲ ἀπὸ ψύξεως ἐγένητο, ἐκεῖνο τὸ ῥέει ἔνι παχὺ καὶ ψυχρόν. θεράπευε δὲ τοὺς τοιούτους οὕτως· ⌜ἂς τρώγωσιν

18 τὰ] τὸ AB 1 μέτωπον] om. M 1 τὸ] δὲ add. A 1 δὲ] om. MA 1 το] om. B 2 καλά] πρῶτον add. Σ 2 μίξε το] μίξον αὐτῷ με τὸ ἅλας M μίξον αὐτὸ με τὸ ἅλας A με τὸ ἅλας add. B 2 τὰ] δὲ add. A 2 σκουλήκια] scripsi σκουλίκια L σκωλήκια Σ 3 ὅπου πιάνουν] ὀποπιάνουν L 3 κοπάνισέ] κοπάνισον Σ 3 τὰ σκουλήκια] scripsi τὰ σκουλίκια L αὐτὰ MA 4 καὶ μάλαξέ τα] ἕνωσον M 4 θέτε] θὲς MB 4 τὰ] ἐπάνω add. B 4–5 με τὸ πανὶν εἰς τὸ μέτωπον] transp. B 5 τριμμένον] τρίψον M 6 ὄξος μίξε τα] μίξον με τὸ ὄξος M ὄξος ποίησέ τα οὕτως A 6 ὄξος] καὶ add. B 6 τὸ μέτωπον του με τὸ πανίν] om. A 6 του] om. M 7 κατάρρου] οὕτως ποίησον B 7–8 τὸν οὐρανίσκον] τοῦ οὐρανίσκου του M τὸν οὐρανίσκον του A 8 ἐκεῖνο] ἐκεῖνον AB 8 δὲ] ῥέῃ add. M 8 μύτην] ῥίναν MB 9 κόρυζα] κόριζα B 9 ἔδειρεν] ὁ add. MA 9 καὶ] om. MA 10 φαίνεταί τον] φαίνετον A 10 ὅτι] om. M 10 ὡς ἂν] om. A 10 καίει] κέῃ MB καίωει A 11 ἄλειφε] δὲ add. A 11–12 ἐλαίας ἀγούρας] ..... ἄγουρας L ἐλαιῶν ἀώρων Σ 12 λοῦε] τον add. A τους B 12 καὶ] om. A 12 πυκνά] πικρὰ M 12 μυρίζωνται δὲ] .... δὲ L μυρίζεται M 13 μικτὸν καὶ] scripsi σμικτ... καὶ L πλὴν πρῶτον M μικτὸν AB 13 ὀλίγον] om. M 13 εἰς] τὸ B 13 τζουκάλιν] τζυκάλιν MA 13 καὶ] τότε add. M · 14 τα] το A 14 τὸ] om. MA 15 δὲ] om. B 15 ἐγένητο, ἐκεῖνο τὸ] om. B 15–16 ἔνι παχὺ καὶ ψυχρόν] ὑγρὸν ἀπὸ τοῦ οὐρανίσκου B 16 οὕτως] om. B

15 ἐὰν] This chapter is a part of the preceding one in M.

ὀλίγον τὰ ψυχρὰ μή τα πίνουν. ἄνεμος μὴ τοὺς κρούῃ, μηδὲ καπνός. ἄλειφε δὲ τὴν κεφαλήν του ἀνηθέλαιον ἢ πηγανέλαιον ἢ ναρδέλαιον. τὴν δὲ μύτην του ἄλειφε τὴν ἀπέσω λίβανον τριμμένον ψιλὸν καὶ ἔλαιον χρηστὸν καὶ μίξε τα καὶ ἄλειφε ἀπέσω τὴν μύτην.⌝ 2 ἄλλο. μελάνθιν καὶ κούμινον φρύξε τα καὶ δένε τα εἰς τὸ πανὶν καὶ ἄς τα μυρίζηται. εἰς δὲ τὸ μέτωπον του θὲς ἔμπλαστρον τὸ λεγόμενον παρὰ τῶν ἰατρῶν ὀξηρὰ ἢ τὴν ἱκέσιον ἔμπλαστρον, ἀπὸ τὰ δύο οἷον. 3 ἄλλο. ⌜εἰς δὲ τὸ στῆθος αὐτοῦ ἄλειφε ἔλαιον τὸ λεγόμενον βαρσαμέλαιον ἢ πηγανέλαιον ἢ ἀνηθέλαιον ἢ ναρδέλαιον ἢ σάμψυχον καὶ βάλε καὶ χρηστέλαιον⌝ καὶ ἄς βράσουν. καὶ τὸ ἔλαιον ἐκεῖνο ἄλειφε τὸ στῆθος αὐτοῦ.

105 [86] ρε´ ὅταν αἱ τρίχες τῶν ὀφθαλμῶν πέσουν, ποίησε τοῦτο καὶ να αὐξήσουν· ⌜φοινίκου ὀστοῦν καῦσε καὶ τρίψε τὸ ψιλὸν καὶ μίξε το με τὸ ῥοδέλαιον καὶ ἄλειφε.⌝ 2 ἄλλο. φελλὸν ἀπὸ καλιγίου καὶ καύσε τον καὶ ποίησον αὐτὸν λεῖον καὶ μίξον το με τὸ δαφνέλαιον. καὶ ἄλειφε τὸν τόπον. 3 ἄλλο. φοινίκων ὀστέα βάλε εἰς κεδρέαν καὶ ἄς εἶναι δέκα πέντε ἡμέρας πλὴν τὰ ὀστέα κοπανισμένα. εἶτα βάλε ναρδέλαιον καὶ μίξε το με τὴν κεδρέαν καὶ ἄλειφε τὸν τόπον.

106 [87] ρϛ´ ἐὰν θέλῃς να ποιήσῃς τὰς τρίχας τῶν ὀφρύων μαῦρας·

---

16 τρώγωσιν] τρώγουν Σ 1 τὰ] δὲ add. MB 1 τα πίνουν] καταπίνουν A 1 τοὺς] τὸν A 1 καπνός] καὶ add. Σ 2 δὲ] om. Σ 3 μύτην] ῥίναν M 3 τριμμένον] τριμμένην A 3 ψιλὸν] ψιλὴν A 4 καὶ μίξε τα] ἐν τ' ἅμα M 4 καὶ ἄλειφε ἀπέσω τὴν μύτην] om. M 4 ἀπέσω] ἀπέσσε A 4 μύτην] του add. AB 5 κούμινον] κύμινον Σ 5 φρύξε] φρύξον M 5 τα καὶ δένε] καὶ δέσε M 5 τὸ] om. M 7 τὰ] τῶν MB 7 οἷον] εὕρεις add. MB εὕρῃς add. A 8 αὐτοῦ] om. A 8 ἔλαιον] .... L om. A 8–9 βαρσαμέλαιον] βαλσαμέλαιον Σ 9 ἢ] om. A 9 πηγανέλαιον] om. MA 9 ἢ ἀνηθέλαιον ἢ ναρδέλαιον ἢ] om. M 9 ναρδέλαιον] πηγανέλαιον A 10 καὶ] om. MB 10–11 τὸ ἔλαιον ἐκεῖνο ἄλειφε τὸ στῆθος αὐτοῦ] ἔπειτε ἄλειφε M ἄλειφε τὸ στῆθος AB 12 ὀφθαλμῶν] ὀφρύων AB 12 πέσουν] πέσουσιν M 12 ποίησε] ποίησον Σ 13 ὀστοῦν] στέον L 14 ἄλειφε] τὸν τόπον add. MA τὸ μέτωπον add. B 14 φελλὸν] φελὸν LM 14 καλιγίου] καλυγίου MB τὸ καλύγια A 15 καὶ καύσε τον] om. M καῦσε A καῦσε τον B 15 μίξον] σ.. L μίξε AB 15 το] τον A 16 τὸν τόπον] τὸ μέτωπον B 16 ὀστέα] στέατα L 17 εἶναι] ἕως add. M ἔνε A 17 δέκα πέντε ἡμέρας] transp. Σ 17 ὀστέα] στέατα LB 18 μίξε το με τὴν κεδρέαν] ἕνωσον M 18 το] τα B 18 κεδρέαν] κεδρ.... L 18–19 τὸν τόπον] τὸ μέτωπον B

⌜κέρατον τράγου καῦσε το καὶ ποίησον ψιλὸν καὶ μίξε το με τὸ ναρδέλαιον. ἄλειφε τὸν τόπον.⌝ 2 ἄλλο. ⌜λάδανον μάλαξον με τὸ κρασὶν. βάλε καὶ ἔλαιον ῥεπανίου χρίε. 3 ἄλλο. ▹om. A ἀξούγγιν ἀγριοαιγινὸν λῦσον καὶ χρίε.⌝ om. A◃ 4 ἄλλο. ἔνι χώρα ἡ λεγομένη Ποντικὸν πλησίον τοῦ λεγομένου Μωρέα καὶ γίνονται ἐκεῖσε καρύδια. καὶ βάλε ἕνα καρύδιν εἰς τζυκάλιν καινούργιον πλὴν δὲ ἀζάκιστον. καὶ θές το τζουκάλιν εἰς καρβούνια ζῶντα καὶ ἂς ἑψηθῇ τὸ καρύδιν ἀπέσω. εἶτα ἔκβαλέ το. καὶ τὰς σκελίδας τοῦ κοπάνισον οἷον τὸ ἀλεύριν καὶ μίξε τα μετὰ ἀξούγγιν λυτὸν ἀγριοαιγιδίου καὶ μάλαξον αὐτὰ καὶ μίξε τα καὶ ἄλειφε τὸν τόπον ὅπου αἱ τρίχες ξανθαὶ εἰς τὸ ὀφρύδιν καὶ εἰ θέλεις να γένουν μαῦραι.

107 [88] ρζ′ πρὸς τρίχας ὅταν εἶναι διπλαὶ εἰς τὰ ὀφρύδια καὶ θέλης να γένουν πρὸς μίαν ὥσπερ πάντων τῶν ἀνθρώπων· ἀνάσπασον τὰς διπλὰς τρίχας, εἶτα ἔπαρον τὸ αἷμα τοῦ λεγομένου τζυμουρίου καὶ ἄλειφε. τζιμούριον δὲ ἔνι, ὅπερ εὑρίσκεται εἰς τὰ τῶν ἀρνίων δέρματα μέσα εἰς τὰ μαλία ὥσπερ κόρυδαι. 2 ἄλλο. ⌜νυκτερίδα σφάξον

20 μαῦρας] ἔπαρον add. A 1 τράγου] καὶ add. A 1 καῦσε] καῦσον MB 1 το] om. B 1 ψιλὸν] πολλὰ add. A 1 μίξε το] μίξον M 2 ναρδέλαιον] καὶ add. MB 2 τὸν τόπον] τὸ μέτωπον B 2–3 μάλαξον με τὸ κρασὶν] scripsi μάλαξ... τὸ κρασίν L με τὸ κρασὶν μάλαξον Σ 3 βάλε καὶ] transp. Σ 3 ῥεπανίου] scripsi ῥεπαν... L ῥαφανίου καὶ MB ῥεφάνου καὶ A 4 ἀγριοαιγινὸν] ἀγριοαιγειδίου M ἀγριοαιγιδίου B 4 ἔνι χώρα] transp. A ἔνι ἡχώρα ἡχώρα B 5 Ποντικὸν] Ποντικὴ Σ 5 λεγομένου] om. A 5 Μωρέα] Μωρέως A 5 γίνονται] γίνεται MA 5 ἐκεῖσε] εἰς τὸν τόπον A 5–6 καρύδια] καρύδιν MA 6 ἕνα] ἕναν MA 6 τζυκάλιν] τζουκάλιν MB 6–7 ἀζάκιστον] ἀτζάκιστον MB τὰ καρύδια ἀτζάκιστον A 7 τζουκάλιν] τζυκάλιν MA 7 καρβούνια] κάρβουνα A 8 σκελίδας] κελίδας M χελίδας B 8 κοπάνισον] κοπάνισέ τας MB τζάκισε A 9 οἷον τὸ ἀλεύριν] ψιλὰς MB ψιλὰς ἤγουν κοπάνισέ τας A 9 μίξε τα] μίξε τας MA 9 μετὰ] με τὸ AB 9 ἀξούγγιν λυτὸν] transp. MA μέλι τὸ ἀξούγγιν τοῦ B 9 ἀγριοαιγιδίου] τοῦ ἀγριοαιγειδίου M τὸ ἄγριοαιγίδιον A 10 αὐτὰ] om. M 10 καὶ μίξε τα] om. A 10 ὅπου] εἶναι add. MB ἔνεν add. A 11 ξανθαὶ] ξάνθες B 11 τὸ ὀφρύδιν] scripsi τὸ ὀφρίδην L τὰ ὀφρύδια Σ 11 καὶ εἰ θέλεις να] ἵνα A 11 εἰ] om. M 11 γένουν] γένωνται Σ 13 να] ἵνα A 13 γένουν] γένονται M γένωνται A 13 τῶν] om. A 13 ἀνάσπασον] ἀνάσπασε B 14 διπλὰς] ..πλας L 14 τοῦ λεγομένου] ......μένου L 15 τζιμούριον] τζυμ..... L τζυμούριον M τζυμόυριν B 15 δὲ] οὐδὲ B 15 ὅπερ] ὅπου MA om. B 15 εὑρίσκεται] om. B 15–16 δέρματα] δερμάτια Σ 16 μέσα εἰς τὰ μαλία] μέσων τῶν μαλλίων MB μέσον τῶν μαλίων A

με τὸ χρυσάφιν καὶ τὸ αἷμα ἄλειφε τὸν τόπον.⌝ 3 ἄλλο. κουρούνας αἷμα ἄλειφε τὸν τόπον. 4 ἄλλο. βατράχου αἷμα ἄλειφε τὸν τόπον. 5 ἄλλο. λίβανον τριμμένον μίξε τὸ με τὸ αἷμα τοῦ βατράχου ἢ τῆς κουρούνας καὶ ἄλειφε τὸν τόπον. 6 ἄλλο. ⌜σκολοπένδριον θαλάσσιον κοπάνισον καὶ μίξε τα μετὰ χυλὸν κωνείου καὶ ἄλειφε τὸν τόπον.⌝ τὸ κώνειον δὲ ἔνι βοτάνιν.

108 [89] ρη′ ὅταν τὸ βλέφαρον τοῦ ἀνθρώπου ἔχῃ τὸ λεγόμενον κριθάριν ἢ τὸ λεγόμενον χαλάζιν, ποίησον θεραπείαν τοιαύτην· ⌜τεάφην τριμμένην ψιλὸν μάλαξον με τὸ ὀξείδιν καὶ ἄλειφε.⌝ 2 ἄλλο. μύρμηκος μεγάλου τὴν κεφαλὴν τζάκισον μόνην καὶ μάλαξέ την με τὸ δακτύλιόν σου καὶ ἄλειφε τὸν τόπον ▷[om. A] τὸ λεγόμενον σαγαπηνὸν, τρίψας καὶ ψιλὸν ποιήσας μίξε το με τὸ ὀξείδιν καὶ μάλαξέ το καὶ ἄλειφε τὸ τόπον, ἔνθα ἔνι τὸ πάθος.[om. A]◁

109 [90] ρθ′ ⌜ὅταν τὸ ἄσπρον τοῦ ὀφθαλμοῦ γένηται κόκκινον πολλὰ καὶ φαίνεταί τον ὅτι ἔνι μεγαλότερον τὸ ἄσπρον τοῦ ὀφθαλμοῦ παρὰ τὴν κόρην, ὅπερ αὐτὸ τὸ πάθος καλεῖται παρὰ τῶν ἰατρῶν χήμωσις. θεραπεύετε δὲ τοὺς τοιούτους πρῶτον μὲν φλεβοτομίαν κρανιακήν, ἀπέσω δὲ εἰς τὸν ὀφθαλμὸν βάλε ἄσπρον αὐγοῦ⌝ ἢ γάλα

---

16 κόρυδαι] κόρυδα MA κόριδα B 1 τόπον] πλὴν πρῶτον ἀνάσπασον τὰς διπλὰς τρίχας add. M πλὴν πρῶτον ἀνάσπασε τὰ διπλὰς τρίχας ἐκ τῶν ὀφρύων add. A πλὴν πρῶτον ἀνάσπασε τὰς διπλὰς τρίχας ὡς ἔφημεν B 1 κουρούνας] κωρώνης Σ 3 λίβανον] λιβαν.. L 3 βατράχου] scripsi βατραχ.. L προβάτου B 4 κουρούνας] κορώνης B 5 μίξε τα] μίξον Σ 5 μετὰ] με τὸν MA τὸν add. B 5 κωνείου] τοῦ κωνείου MA 6 κώνειον δὲ] transp. Σ 6 βοτάνιν] βότανον M βοτάνη AB 7 τὸ βλέφαρον τοῦ ἀνθρώπου] transp. M 8 ἢ] ἤγουν A 8 τὸ λεγόμενον] om. AB 8 χαλάζιν] χαλάζιον A λαλάζιν B 8 ποίησον θεραπείαν τοιαύτην] om. B ἤγουν add. MA 9 τεάφην] τεάφιν MB 9 τριμμένην] τριμμένον MB 9 ψιλὸν] ψιλὴν A 9 μάλαξον] om. Σ 9 ὀξείδιν] μάλαξον add. A μάλαξον αὐτὸ add. B 9 καὶ] om. M 9 ἄλειφε] πλὴν πρῶτον μάλαξον αὐτὰ add. M αὐτὴν add. A 10 μύρμηκος] μύρμιγγος A 10 τζάκισον μόνην] transp. MB καὶ μόνον τζάκισέ την A 11 δακτύλιόν] δάκτυλόν M δακτύλιν A 12 σαγαπηνὸν] σαγάπιον L 12 μίξε το] μίξον M 13 μάλαξέ το] μάλαξ.... L 13 τὸ τόπον, ἔνθα ἔνι τὸ πάθος] τὸν κρίθον M 14 ὀφθαλμοῦ] .... add. B 15 τον] του A om. B 15 ὅτι] om. B 15 ἔνι μεγαλότερον] μεγαλιότερον ἔνι M ἔνι A μεγαλότερον B 15–16 τοῦ ὀφθαλμοῦ] om. B μεγαλότερον add. A 16 παρὰ τῶν ἰατρῶν] om. B 17 χήμωσις] χύμωσις AB 17 θεραπεύετε] scripsi θεραπεύεται L θεραπεύε Σ

γυναικεῖον. 2 ἄλλο. ⌜κρασὶν δὲ μὴ πίνῃ. μὴ δὲ ἥλιος τὸν κρούῃ, μηδὲ ἄνεμος, μηδὲ κονιορτός, μηδὲ καπνός, μηδὲ λούεται,⌝ μηδὲ τίποτε ἀλείφει τὴν κεφαλήν του.

110 [91] ρι′ ⌜τὸ ἐμφύσημα ὄγκος ἔνι τοῦ βλεφάρου.⌝ ἡ δὲ θεραπεία ἔνι τοιαύτη· ⌜ὅταν τὸ βλέφαρον ἔχῃ κνησμάραν, χλίαινε ὄξος καὶ κρασὶν καὶ πυρίαζε μετὰ πανίου. 2 ἄλλο. βράσε φακὴν με τὸ νερὸν [▷om. B]καὶ ζωμὶν πυρίαζε τοὺς ὀφθαλμούς.⌝[om. B◁] 3 ἄλλο. [MB▷]ἐὰν ἔχεις χήνειον ἀξούγγιν ἢ ὄρνιθος θέτε τὸ βραδὺ εἰς τοὺς ὀφθαλμούς.[MB◁] 4 ἄλλο. [▷om. B]βάλε ῥόδα καὶ κρασὶν καὶ ἂς βράσουν καὶ πυρίαζε τοὺς ὀφθαλμούς.[om. B◁] 5 ἄλλο. ⌜βάλε κρασὶν γλυκὺ καὶ κρόκον καὶ ἂς βράσουν καὶ πυρίαζε τοὺς ὀφθαλμούς. τὸ αὐτὸ ἄλειφε καὶ τὸ βλέφαρον.⌝

111 [92] ρια′ ⌜περὶ σκληροφθαλμίας· σκληροφθαλμία ἔνι, ὅταν σαλεύῃ τὸν ὀφθαλμόν του μετὰ βίας καὶ μετὰ πόνου καὶ εἶναι καὶ κόκκινοι καὶ ξηροὶ ἀπέσω οἱ ὀφθαλμοὶ καὶ δάκρυον οὐδὲν ῥεῖ. πυρίαζε αὐτοὺς μὲ νερὸν ποτάμιον πλὴν χλίον. πυρίαζε δὲ μετὰ σπόγγου καινουργίου, εἶτα θέτε τὸν σπόγγον εἰς τὰ βλέφαρα.⌝ μάλαξον κρόκον αὐγοῦ με τὸ ῥοδέλαιον καὶ τὸ βραδὺ θέτε το με τὸ βαμβάκιν εἰς τοὺς ὀφθαλμούς. 2 ἄλλο. ⌜ἐὰν ἔχῃς χήνειον ἀξούγγιν ἢ ὀρνιθιτικὸν θέτε το τὸ βραδὺ εἰς τοὺς ὀφθαλμούς. τὴν δὲ κεφαλὴν ἄλειφε χρηστέλαιον.⌝

---

18 εἰς] om. B 1 δὲ] om. A 1 μὴ] om. B 1 μὴ δὲ ἥλιος τὸν κρούῃ] om. B 1 τὸν] om. A 1 κρούῃ] om. A 2 μηδὲ ἄνεμος] om. A 4 τὸ ἐμφύσημα ὄγκος ἔνι τοῦ βλεφάρου] om. M et add. in marg. πρὸς ὅταν τὸ βλέφαρον ἀγγωθῇ 4 βλεφάρου] φλεβάρου B 4–5 ἡ δὲ θεραπεία] transp. MA θεραπεία B 5 κνησμάραν] κνισμάραν LMA 5 χλίαινε] λίαινε L χλίενε AB 6 βράσε] βράσον Σ 7 ζωμὶν] scripsi .ωμὴν L τὸν ζωμὸν MA 8–9 ἢ ὄρνιθος θέτε τὸ βραδὺ εἰς τοὺς ὀφθαλμούς] ὁμοίως ποίει B 9 βάλε ῥόδα] βα .... L 9–10 ἂς βράσουν καὶ] om. MA 10 πυρίαζε] πυρι..... L 11 κρόκον] οἷον βάλουσιν εἰς τὸ γλυκόζυνον add. MB 13 περὶ] πρὸς M 13 σκληροφθαλμίας] κληροφθαλμίας καὶ ξηροφθαλμίας καὶ ἐκτροπίας M σκληρότητος ἡ δὲ B 14 σαλεύῃ] σαλέβῃ M 14 του] αὐτοῦ MA 14 μετὰ] om. A 16 αὐτοὺς] τους B 16 μὲ] μετὰ M με τὸ AB 16 ποτάμιον] τὸ ποτάμιον AB 16 πυρίαζε δὲ] om. Σ 17 εἶτα θέτε] πυρίαζε καὶ θὲς M πυρίαζε καὶ θέτε AB 17 βλέφαρα] του add. A 18 θέτε το] θές τα M 18–19 βαμβάκιν] παμπάκην A 19 χήνειον] χηνὸς MB χυνὸς A 20 ὀρνιθιτικὸν] scripsi ...νιθίτικον L ὄρνιθος Σ 20 θέτε το] θὲς M 20 τὸ] om. B 21 χρηστέλαιον] ..στέλαιον L

112 [93] ριβʹ ⌜περὶ ξηροφθαλμίας· ξηροφθαλμία ἔνι, ὅτι ἔχουσιν μὲν κνησμάραν τὰ βλέφαρα τῶν ὀφθαλμῶν καὶ ῥεῦμα ῥέουσι. ἡ θεραπία δὲ τούτων εἰσὶ τὰ λουτρὰ πυκνά.⌝

113 [94] ριγʹ ⌜ἐκτρόπιον ἔνι, ὅταν τὸ βλέφαρον τὸ κάτω ἐξηστράφῃ. θεράπευε δὲ αὐτὸν οὕτως· μολύβδιν καῦσε καὶ μίξον καὶ τεάφην καὶ ἄλειφε τὸ βλέφαρον.⌝

114 [95] ριδʹ ⌜αἰγίλωψ ἔνι, ὅταν πλησίον τῆς μύτης εἰς τὴν ἀπέσω μερέαν εἰς τὴν ἄκραν τοῦ ὀφθαλμοῦ γένηται ἀπόστημα. κρίνου ῥίζαν κοπάνισον καὶ μίξον με τὸ μέλι καὶ θέτε το ἐπάνω. 2 ἄλλο. λουπηνάρια κοπάνισον καὶ πᾶσσε εἰς τὴν πληγήν.⌝ 3 ἄλλο. πηγάνου φύλλα κοπάνισον καὶ μίξε τα με τὴν χολὴν τοῦ ταυρίου καὶ ποίησον ὡς πήττας καὶ θέτε τας ἐπάνω. 4 ἄλλο. κοχλίου τὸ κρέας κοπάνισον καὶ μίξον μετὰ ἀλόης τριμμένης καὶ μάλαξέ το καὶ θέτε το ἐπάνω.

115 [96] ριεʹ ⌜θεραπεία ὅταν αἱ τρίχαι τῶν βλεφάρων πέσουν· πρῶ-

---

1 περὶ ξηροφθαλμίας] om. MA 1 ξηροφθαλμία] δὲ add. Σ 1 ὅτι] ὅτε MA 2 κνησμάραν] κν.... L κνισμάραν M 2 ῥεῦμα] ῥευ... L οὐ add. A 3 δὲ] om. A 3 τούτων] τοιούτων M 3 εἰσὶ] ἐστὶ A 3 λουτρὰ] τὰ add. AB 4 τὸ] τοῦ B 4–5 ἐξηστράφῃ] ἐξιστράφῃ A 5 αὐτὸν] τοὺς τοιούτους M 5 μολύβδιν] μολούβιν M μολύβην A μολύβδον B 5 καῦσε] καῦσον MA καύσας B 5 μίξον] μίξε το B 5–6 καὶ τεάφην] με τὸ τεάφιν MB με τὴν τεάφην A 7 αἰγίλωψ ἔνι] om. MB 7 ὅταν] πρὸς ὅταν εἰς τὸν μέγαν κανθὸν τοῦ ὀφθαλμοῦ M 7–8 πλησίον τῆς μύτης εἰς τὴν ἀπέσω μερέαν] transp. M εἰς τὸν ἀπέσω μερέαν B 7 μύτης] καὶ add. A 8 εἰς τὴν ἄκραν τοῦ ὀφθαλμοῦ] om. MB 8 εἰς] om. A 8 ἀπόστημα] τὸ λέγουσιν οἱ ἰατροὶ αἰγίλωπα add. M αἰγίλοψ ἔνι ὅταν πλήσιον τῆς μίτης καὶ εἰς τὴν ἀπέσω μερέαν τὴν ἄκραν τοῦ ὀφθαλμοῦ γένηται ἀπόστημα add. B 8–9 κρίνου ῥίζαν κοπάνισον] transp. M 9 καὶ] πάσσε εἰς τὴν πληγὴν ἤγουν εἰς τὸ ἀπόστημα add. A 9 μίξον] πλὴν μίξε τὸ κοπανισθὲν ἐκεῖνο A 10 λουπηνάρια] λυπηνάρια MA 11 μίξε τα] μίξον M 11 ταυρίου] τάρου M ταύρου A 12 ὡς πήττας] ὡσπίτας M ὡ πίττας A ὡς πίτας B 12 τας] τα B 12 κοχλίου] τὸ χλίον A κοχλίων B 13 μίξον] μίξε A μίξε το B 13 μετὰ ἀλόης τριμμένης] με τὴν εἰρημένην χολὴν M με τὴν τριμμένην χολὴν A με τὴν τριμμένην ἀλόην B 13 μάλαξέ το] μάλαξον M 13 θέτε το] θὲς M 14 θεραπεία ὅταν] θεραπεία ... L πρὸς ὅταν M 14 τρίχαι] τρίχες AB 14 πέσουν] πέσωσιν MA

---

1 περὶ ξηροφθαλμίας] This chapter is part of the preceding chapter in M 4 ἐκτρόπιον] This chapter is part of the preceding chapter in M. 10 πηγάνου] The following paragraphs are reversed in M.

τον φλεβοτόμησέ τον εἶτα λοῦσέ τον, εἶτα ▷om. Bἂς πίνῃ ἀπόζεμα. εἶτα ἂς πίνῃ βοήθημα ἔχον τὰ ξανθὰ πλείονα παρὰ τὰ ἄλλα εἴδη τοῦ βοηθήματος. διότι ἀπὸ δριμέου χυμοῦ ἤτοι ἀπὸ ξανθῆς χολῆς γέγονε τοῦτο να πέσουν αἱ τρίχες τῶν βλεφάρων. 2 ἄλλο. τῶν φοινίκων τὰ ὀστέα κοπάνισον καὶ ποίησον ψιλὰ πολλὰ καὶ καστόριον τριμμένον καὶ κρόκον οἷον βάνουν εἰς τὸ γλυκόξυνον. καὶ ἄρκου ἀξούγγιν λυτὸν μίξε τα καὶ ἄλειφε τοὺς τόπους τῶν βλεφάρων, ὅπου αἱ τρίχες ἔπεσαν. πλὴν πρόσεχε να τὰ ἀλείφῃς ἐπιτηδείᾳ τὰ βλέφαρα. μὴ πῶς ἡ ἀλειφὴ ἔμπῃ εἰς τὸν ὀφθαλμὸν καὶ πλείονα βλάψῃ παρ' οὗ να ὠφελήσει. 3 ἄλλο. πράσα δὲ μὴ τρώγῃ μηδὲ κρόμμυα μηδὲ σκόρδα μηδὲ πολὺ κρασὶν μηδὲ πέπερι μηδὲ ὄσπρια. ἂς λούηται δὲ ἀρεά.⌝om. B◁

116 [97] om. Bρις′ ⌜πρὸς τὸ ὅταν ἀπὸ τὸν μέγαν κανθὸν ῥέῃ τὸ δάκρυον παρὰ τὸ θέλημα.⌝ ὅπερ τὸ πάθος τοῦτο καλεῖται παρὰ τῶν ἰατρῶν ῥυὰς ἤτοι ἀπὸ τοῦ ῥέειν· ▷om. Lθεράπευε δὲ τοὺς τοιούτους οὕτως· ἤτοι ⌜ἔπαρον σπόγγον καινούργιον καὶ βρέξον εἰς τὴν κεδρέαν καὶ τὴν στάκτην τοῦ σπόγγου πάσε ἐκεῖ ἔνθα ῥέῃ τὸ δάκρυον.

234.14–235.1 πρῶτον] .... L 1 φλεβοτόμησέ τον] φλεβοτόμησον Σ 1 λοῦσέ τον] λου ...... L 1 εἶτα] καὶ A εἰθ' οὕτως B 2 ἂς πίνῃ] scripsi ...ινει L πότισέ τον M πότισον αὐτὸν A 2 ἔχον τὰ] scripsi ἔχων τὰ L ἔχοντα M 2 παρὰ] om. A 2 τὰ ἄλλα εἴδη] τῶν ἄλλων εἰδῶεν A 3 δριμέου] δριμαίου M δριμοῦ A 4 τοῦτο] τὸ add. M ὅτι add. A 4 πέσουν] πέσωσιν A 6 βάνουν] βάνο M βάλλουσιν A 7 μίξε] καὶ μίξε MA 8 ἔπεσαν] ἔπεσον A 8 να τὰ] ἵνα A 8–9 ἐπιτηδείᾳ τὰ] scripsi ἐπὶ τῇ δείατα L ἐπὶ τήδια M ἐπὶ τήδεια τὰ A 9 ἡ ἀλειφὴ ἔμπῃ] εἰσέλθῃ ἡ ἀλειφὴ A 9 ἔμπῃ] εἰσέλθῃ M 10 ὠφελήσει] scripsi ὀφελέσει L ὀφελήσῃ MA 11 κρόμμυα] κρομίδια A 11 σκόρδα] σκόροδα A 11 πολὺ κρασὶν] πολὺ add. A 11 μηδὲ] ... L 13–15 πρὸς τὸ ὅταν ἀπὸ τὸν μέγαν κανθὸν ῥέῃ τὸ δάκρυον παρὰ τὸ θέλημα.⌝ ὅπερ τὸ πάθος τοῦτο καλεῖται παρὰ τῶν ἰατρῶν ῥυὰς ἤτοι ἀπὸ τοῦ ῥέειν] text of the index of L ῥυὰς λέγεται, ὅταν ἀπὸ τὴν ἄκραν τοῦ ὀφθ...... L ῥοιὰς λέγεται, ὅταν ἀπὸ τῆς ἄκρας τοῦ ὀφθαλμοῦ πλήσιον τῆς μύτης ὁ λεγόμενος κανθὸς καὶ ἡ ἄλλη ἄκρα τοῦ ὀφθαλμοῦ λέγεται μικρὸς κανθὸς ῥέῃ παρὰ τὸ θελημά του τὸ δάκρυον πάντοτε A 13 τὸ] om. M 14 παρὰ τὸ θέλημα] om. M 16 ἤτοι] om. A 16–17 κεδρέαν] καὶ καῦσε τον add. A

13–15 πρὸς τὸ ὅταν ἀπὸ τὸν μέγαν κανθὸν ῥέῃ τὸ δάκρυον παρὰ τὸ θέλημα.⌝ ὅπερ τὸ πάθος τοῦτο καλεῖται παρὰ τῶν ἰατρῶν ῥυὰς ἤτοι ἀπὸ τοῦ ῥέειν] Because this sentence is only half extant in L, which is damaged here, I copied the index of L to reconstruct the wording. M, which would step in, has a very similar text.

2 ἄλλο. ▷om. Aἔπαρον πάλιν σπόγγον καινούριον καὶ βρέξε τον εἰς τὴν κεδρέαν καὶ τὴν στάκτην τοῦ σπόγγου ἀφ' ὅτου τὸν καύσης μίξε τηνom. A◁ με τὸ αἷμα τοῦ δρακοντίου καὶ μάλαξον αὐτὰ καλῶς ▷om. Aκαὶ ἄλειφε τὸν τόπον ἔνθα ῥέη τὸ δάκρυον.om. A◁ κρασὶν μὴ πίνῃ μηδὲ τὰ θερμαίνοντα οἷον κρόμμυα, πράσα καὶ τὰ λοιπά, μηδὲ λούεται μηδὲ ἥλιος τὸν κρούῃ μηδὲ ἄνεμος.om. L◁⌝

117 [98] om. Lριζ′ πρὸς ὅταν ὁ ὀφθαλμὸς ἔχει ἄσπρον· ⌜σηπίας ὄστρακον καῦσον καὶ τὴν στάκτην του μίξον με τὸ μέλι τὸ καλὸν καὶ βάλε εἰς τὸν ὀφθαλμόν.⌝ 2 ἄλλο. αἰλούρου χολὴν μίξον με τὸ μέλι τὸ λεγόμενον ἀκάπνιν καὶ μάλαξέ τα καλὰ καὶ βάλε εἰς τὸν ὀφθαλμόν. 3 ἄλλο. ⌜ἡ δὲ τροφή του σπαστρικὴ ἤτοι τὰ πάντα χλωρά. παστὰ μὴ τρώγῃ μήτε χοίριον μήτε ὀψάρια μήτε ὄσπρια μήτε λάχανα μήτε τὸ λεγόμενον μονόκυθρον μήτε τὸ ἄσπρον τοῦ αὐγοῦ εἰ μὴ τὸν κρόκον μόνον.⌝

118 [99] om. Bριη′ ▷om. L⌜πρὸς πτερύγιον ἐν ὀφθαλμῷ· πτερύγιον ἔνι ὅταν ἀπὸ τοῦ μεγάλου κανθοῦ καὶ τοῦ ὀφθαλμοῦ αὐξάνῃ σὰρξ μικρὰ καὶ στενὴ καὶ ἁπλώνῃ ἕως τὴν κόρην. αὐτὸ δὲ καλεῖται πτερύγιον. καὶ ποίησον θεραπείαν τοιαύτην· ἤτοι ἔπαρον τὸν λεγόμενον μαγνῆτιν καὶ τρίψον καὶ ποίησον ψιλὸν πολλὰ καὶ κρόκον τριμμένον καὶ μέλι καλὸν καὶ ἕνωσον καὶ μάλαξον ταῦτα καλὰ καὶ ἄλειφε τὸν τόπον ἐκεῖνον ἔνθα ἔνι ἡ σάρξ.⌝ 2 ἄλλο. στρύχνον τὸ λεγόμενον βρωμοβότανον κοπάνισον καὶ μίξον με τὸ ἀλεύριν τῶν λυπηναρίων καὶ με μέom. L◁λι καλὸν ἀκάπνιν καὶ μίξε καὶ μάλαξέ τα καλὰ καὶ ἄλειφε τὴν σάρκαν πλὴν ἐκείνην καὶ μόνην.

---

7 ὅταν] This chapter is listed in the pinax to L.

---

17 δάκρυον] πλὴν δὲ μίξε τὴν add. A 3 δρακοντίου] δράκοντος A 5 θερμαίνοντα] τρώγει add. A 6 ἄνεμος] μηδὲ καπνὸς add. A 7 πρὸς] om. A 7 σηπίας] σιπέας A 8 καλὸν] ἀκάπνην καὶ μάλαξέ τα καλὰ A 9 βάλε] τὰ add. A 9 αἰλούρου] scripsi ἐλούρου M ἀλεποῦς A 10 μέλι τὸ λεγόμενον] om. A 10 ἀκάπνιν] μέλι add. A 10 βάλε] τα add. A 11 του] ἂς ἔνε A 11 τὰ] om. A 16 ἔνι ὅταν] ἤτοι A 17 στενὴ] στένει A 18 ἤτοι ἔπαρον] om. A 19 μαγνῆτιν] μαγνῖτην A 19 καὶ] om. A 20 ἕνωσον] μίξε A 20 μάλαξον] μάλαξε A 23 με] om. A 23 καλὸν] om. A 23 καὶ μίξε] scripsi καὶ μίξ........ L om. MA 24 ἄλειφε τὴν σάρκαν πλὴν] ἄλειφ.....κα πλὴν L ἄλειφε τὸν τόπον M ἄλειφε τὴν σάρκαν A 24 ἐκείνην] ἐκεῖνον M 24 καὶ] om. A 24 μόνην] μόνον M om. A

119 [100] ριθ′ περὶ ἀμαυρώσεως καὶ ἀμβλυοπίας· ⌜ἀμαύρωσις λέγεται, ὅταν χωρίς τινος πάθους οὐ βλέπει ὁ ἄνθρωπος οὐδὲ ὅλως.⌝ θεράπευε τοὺς τοιούτους οὕτως· ⌜πρῶτον μὲν φλεβοτομία καὶ λουτρά. ἂς πίῃ καὶ βοήθημα καὶ ἄνεμος μὴ τὸν κρούῃ, μηδὲ ἥλιος μηδὲ κονιορτὸς μηδὲ καπνὸς μηδὲ ὄσπρια τρώγῃ μηδὲ παστὰ μηδὲ λάχανα, εἰ μὴ χλωρὴν μαγειρίαν καὶ ὀπτὰ καὶ κρόκον μόνον αὐγοῦ καὶ κρασὶν ὀλίγον ζεστόν.⌝

120 [100] ρκ′ ⌜ἀμβλυοπία δὲ ἔνι, ὅταν πότε μὲν βλέπει ὀλίγα ποτὲ δὲ ὀλίγου τζικόν, ποτὲ δὲ οὐδ' ὅλως· φλεβοτόμησον αὐτὸν εἰς τὴν ἄκραν τοῦ ὀφθαλμοῦ πλήσιον τῆς μύτης ἣν λέγουσι οἱ ἰατροὶ κανθόν.⌝ βάλε εἰς τοὺς μήλιγγας βδέλλας ἐὰν καὶ οἱ δύο ὀφθαλμοὶ ἔχωσι τὸ πάθος. εἰ δὲ πάσχει ὁ εἷς, ποίησον εἰς τὸν ἕνα. 2 ἄλλο. ἰάτρευε τοῦτον καὶ κλυστήριν καθ' ἑκάστην ἡμέραν. 3 ἄλλο. ⌜ἄλειφε δὲ ἔξω τοὺς ὀφθαλμοὺς ▹om. Bἔλαιον παλαιὸν καὶ μέλι καλόν. 4 ἄλλο. πρῶτον δὲ ἀποσφούγγιζε τοὺς ὀφθαλμοὺςom. B◃ μετὰ νερὸν καὶ ἅλατος πλὴν ζεστὰ με τὸ σπογγάριν.⌝ ὕστερον δὲ κοπάνισον τοῦ μαλάθρου τὰ φύλλα. καὶ εἰς τὸν ζωμὸν ἐκείνων βάλε κρόκον κοπανισμένον καὶ σκρόφας χολὴν καὶ μέλι καὶ πέπερι τριμμένον. καὶ ἀμμωνιακὸν τριμμένον καὶ μίξε τα ὅλα καὶ ἄλειφε τοὺς ὀφθαλ-

---

1 περὶ ἀμαυρώσεως καὶ ἀμβλυοπίας] πρὸς πτερύγιον ἐν ὀφθαλμῷ M, corr. M[2] πρὸς ἀμλιωπίαν ἐν ὀφθαλμῶ καὶ ἀμαύρωσιν M om. B 1 ἀμαύρωσις] δὲ add. MB 2 τινος πάθους] πάθος A 3 θεράπευε] θερπαπεύομεν Σ δὲ add. AB 3 οὕτως] om. MA 3 μὲν] ἐν A ἐν add. B 3–4 λουτρά] λουτρῷ Σ 4 μηδὲ] μὴ ὁ B 5 μηδὲ] καὶ B 6 μόνον] om. MA 8 δὲ ἔνι] εἰσὶ B 8 βλέπει] βλε... L 8 ὀλίγα] ὀλίγον Σ 9 δὲ ὀλίγου] om. B 9 τζικόν] τζικον MA om. B 9 ποτὲ δὲ] ...πο... L om. B 9 φλεβοτόμησον] θεραπεύομεν δὲ τοὺς τοιούτους πρῶτον μὲν φλεβοτομία A 9 αὐτὸν] om. A 10 πλήσιον] scripsi πλη...L ὅπου πλήσιον M ὁποῦ ἔνι σιμὰ εἰς A ὅπου ἔνι σημὰ εἰς B 10 τῆς] τὴν AB 10 μύτης] ῥινὸς M μύτην AB 10–11 ἣν λέγουσι οἱ ἰατροὶ κανθόν] om. M τὴν λεγομένην παρὰ τῶν ἰατρῶν μέγα κανθὸν AB 10–11 κανθόν] scripsi ...νθόν L 11 μήλιγγας] μήνιγγας τὰς λεγομένας Σ 11 βδέλλας] βδέλα̣ L ἀβδέλας B 11 ἐὰν καὶ] ..... L ....ε εἰς τοὺς μήλιγγας βδέλας in marg. L[2] 12 ποίησον] βάλε A 12 ἕνα] μήνιγγας τὰς βδέλας add. M μήνιγγα τὰς βδέλλας add. AB 13 τοῦτον] τὸν Σ 13 ἑκάστην] om. MA 15 ἀποσφούγγιζε] ἀποσφọ́γγιζε M 15 νερὸν] νεροῦ Σ 16 σπογγάριν] σπόγγον Σ 17 τὸν ζωμὸν] τὸ ζουμὶν AB 17 ἐκείνων] ἐκεῖνον MB ἐκεῖνο A 18 σκρόφας] σκόρφας MA 19 καὶ ἀμμωνιακὸν τριμμένον] om. B 19 μίξε τα] ἕνωσον ταῦτα M συνένωσον αὐτὰ B 19 ὅλα] ὁμοῦ add. B

μοὺς ἔξω. 5 ἄλλο. ▷om. Bκοπάνισον φύλλα πηγάνου καὶ μαλάθρου καὶ τὸν ζωμὸν αὐτῶν μίξον με τὸ μέλι τὸ ἀκάπνιν. καὶ μάλαξον καὶ ἄλειφε τοὺς ὀφθαλμούς.om. B◁ 6 ἄλλο. ⌜ἂς τρώγῃ δὲ καὶ τὸ γουλὶν τῆς κράμβης ὠμόν. βλάπτει δὲ αὐτοὺς ἄνηθον φακὴ ἔλαιαι μαῦραι μαρούλια πράσα βασιλικὰ λάχανα σέλινα. 7 ἄλλο. μελάνθιν κοπάνισον καὶ μίξον αὐτὸ με τὸ νερὸν καὶ μάλαξον αὐτὸ καλὰ καὶ ἄλειφε τοὺς ὀφθαλμούς.⌝

121 ρκα´ ὅταν ὁ ὀφθαλμὸς ἔχῃ ἄσπρον· φλεβοτόμησον αὐτὸν κρανιακήν. πότισον δὲ καὶ βοήθημα. συκίασε δὲ τοῦτον εἰς τὴν κεφαλήν. ἂς πίνῃ δὲ κρασὶν τρικέραστον. 2 ἄλλο. τρίψον ἀγριοσταπίδαν καὶ πέπερι καὶ μαστίχην καὶ κηρὶν καὶ μίξε τὰ ὅλα καλὰ καὶ ἂς μεμασσᾶται. 3 ἄλλο. βάλε ὀξείδιν εἰς τζουκάλιν καὶ θύμον καὶ ὀρίγανον καὶ γλήχωνα καὶ ὕσσωπον καὶ ἂς βράσουν καλά. καὶ τὸ πρωῒ ἂς βάνῃ ἀπ᾽ ἐκεῖνο εἰς τὸ στόμα αὐτοῦ καὶ ἄς το κρατῇ πολλὴν ὥραν καὶ τότε ἄς το πτύῃ. καὶ να εὐγάνει πολλὰ πτύσματα καὶ σίελα ἤτοι σάλια. εἰς ἥλιον δὲ μὴ κάθηται μηδὲ ἄνεμος τὸν κρούῃ μηδὲ καπνός. 4 ἄλλο. πότιζε δὲ τοὺς τοιούτους τὴν λεγομένην πικρὰν με τὸ κρασὶν τὸ χλίον. 5 ἄλλο. μάλαθρον κοπάνισον καὶ τὸν ζωμὸν αὐτοῦ μίξον μὲ ὀλίγον χρηστέλαιον καὶ μέλι καὶ ἄλειφε ἔξω τοὺς

237.19–238.1 τοὺς ὀφθαλμοὺς ἔξω] transp. A 1 μαλάθρου] μαλα.. L 2 μίξον] μίξε A 2 με τὸ μέλι] μετ..λι L μέλι M 2 ἀκάπνιν] ἀκάπνην L ἀκάνιν M 2 μάλαξον] μάλαξέ τα A 3 δὲ καὶ] .. L 3–4 τὸ γουλὶν τῆς κράμβης] κραμπὴν τοῦ γουλίου L 4 βλάπτει δὲ αὐτοὺς] scripsi βλα. δὲ αὐτοὺς L βλάπτει δὲ MB βλάπτει τὸν δὲ A 4–5 ἔλαιαι μαῦραι] ἐλέαι.. L 5 μαρούλια] μαρούλιν B 5 πράσα] πράσον A καὶ add. B 5 λάχανα] καὶ add. B 5 μελάνθιν] μελάνθη M 6 αὐτὸ] om. M τὸ A 6 αὐτὸ] om. MA αὐτὰ B 7 τοὺς ὀφθαλμούς] om. M τοὺς A 8 αὐτὸν] om. M 9 κρανιακήν] φλέβα add. M 9 πότισον δὲ] πότισέ τον A 9 συκίασε] συνκίασε L 9 δὲ τοῦτον] τὸν A 10 δὲ] καὶ add. MA 10–11 ἀγριοσταπίδαν] ἀγριοσταφίδα MB ἀγριοσταφύδα A 11 πέπερι] πεπέρην A 11 μαστίχην] μαστίχιν MB 11 κηρὶν] κηρὸν Σ 11 μίξε τὰ ὅλα καλὰ] ἕνωσον ταῦτα M 12 μεμασσᾶται] τὰ μασσᾶται Σ 12 τζουκάλιν] τζυκάλιν M τζυκάλην A 13 ὀρίγανον] ὀρίγανιν A 13 γλήχωνα] γλήχων M 14 βάνῃ] βάλει A βάλουν B 14 ἐκεῖνο] ἐκεῖνα MB 14 αὐτοῦ] του MA 14 το] τα M 15 το] τα M ομ. B 15 εὐγάνει] εὐγάλῃ Σ 16 ἤτοι σάλια] om. A 16 δὲ] om. Σ 16 τὸν] ... L τοῦ B 16–17 μηδὲ καπνός] om. M 17 πότιζε] πότισον A 17 δὲ] om. B 17 τοὺς τοιούτους] τούτους A 18 τὸν ζωμὸν] τὸ ζουμὶν B 19 αὐτοῦ] τοῦ Σ 19 μίξον] μίξε τα B

ὀφθαλμούς. 6 ἄλλο. ▷Lκαρπὶν ἄσπρον τρίψον καὶ ποίησον ψιλὸν πολλά. καὶ βάλε μέλι καὶ μίξε καὶ ἄλειφε ἔξω τοὺς ὀφθαλμούς.L◁ 7 ἄλλο. βοτάνιν τὴν λεγομένην καυκαλίδα κοπάνισον καὶ τὸ ζωμὶν μίξον με τὸ μέλι. ἄλειφε ἔξω τοὺς ὀφθαλμούς.

122 [101] ρκβʹ ⌜ὅταν ὑποκάτω τοῦ κάτω βλαφάρου γένηται μέλαν. τοῦτο δὲ γίνεται τὸ πάθος καὶ μᾶλλον εἰς τὰς γυναῖκας καὶ εἰς τὰ παιδία, ὅπερ τὸ πάθος τοῦτο καλεῖται παρὰ τῶν ἰατρῶν ὑπώπιον.⌝ ποίησον δὲ θεραπείαν τοιαύτην ἤγουν· ⌜ῥεπάνιν κοπάνισον. καὶ θέτε ἐπάνω.⌝ ὅταν δὲ τρώγῃ τὸν τόπον τῆς πληγῆς τὸ ῥεπάνιν σήκοσέ το. 2 ἄλλο. ⌜τύρον τρυφερὸν κοπάνισον καὶ θές το ἐπάνω εἰς τὴν πληγήν.⌝ 3 ἄλλο. κοπάνισον ἀψίνθιον χλωρὸν καὶ ἔκβαλε ▷om. Bτὸ ζουμίν του καὶ βάλε κηρίον καὶ λύσε το. εἶτα βάλεom. B◁ χαμαιμηλέλαιον καὶ ῥοδέλαιον καὶ χήνειον ἀξούγγιν καὶ ὄρνιθος, εἶτα βάλε τὸ χυλὸν τοῦ ἀψινθίου καὶ τάρασσέ τα ἐστ' ἂν κρυώσῃ. εἶτα θέτε το εἰς τὴν πληγήν. 4 ἄλλο. ▷om. Bἀψινθίου χλωροῦ τὰ φύλλα κοπάνισον καὶ θέτε τα ἐπάνω. 5 ἄλλο. ⌜λιθάργυρον τρίψον καὶ μίξον με τὸ μέλι καὶ θέτε το εἰς τὴν πληγήν.⌝om. B◁ 6 ἄλλο. ▷om. Σἀνάλυσε κηρὶν καὶ βάλε χαμεμηλέλαιον καὶ σινάπιν τριμμένον καὶ μίξον καὶ θέτε τὰ εἰς τὴν πληγήν.om. Σ ◁ 7 ἄλλο. ▷om. MA⌜κύμινον τριμμένον δύσε τὸ εἰς τὴν πανὶν καὶ βρέχε τὸ εἰς τὸ ὀξείδιν ζεστὸν καὶ πυρίαζε τὸν τόπον.om. MA◁ 8 ἄλλο. ἀνάλυσε κηρὶν καὶ βάλε ῥοδέλαιον καὶ ἀψίνθιν τριμμένον καὶ λίβανον. καὶ ἡδύοσμον καὶ

---

19 μὲ] μετὰ Σ 3 βοτάνιν] βοτάνην A 3 καυκαλίδα] καυχαλίδα M καυκαλίδαν AB 3 τὸ ζωμὶν] τὸν ζωμὸν τῆς MA τῆς add. B 4 μίξον] μίξε A μίξε το B 4 μέλι] καὶ τάραξέ τα καὶ add. Σ 5 ὑποκάτω τοῦ κάτω βλαφάρου] τὸ κάτω βλέφαρον M τὸ ὑποκάτω τοῦ βλεφάρου A ὑποκάτω τῶν βλεφάρων B 6 δὲ] om. B 6 γίνεται τὸ πάθος] transp. B 6 εἰς] om. B 8 δὲ] om. B 8 τοιαύτην ἤγουν] om. A 8 ῥεπάνιν] ῥεφάνιν M om. A 9 θέτε] θές το Σ 9 ἐπάνω] εἰς τὴν πληγὴν B 9 τῆς πληγῆς] om. A 9 ῥεπάνιν] ῥεφάνιν MA 10 σήκοσέ το] συκίασέ τον A 10 τύρον] τυρ.. L μύρον M .υρὸν B 10 το] τον A 10–11 ἐπάνω εἰς] ..... L 11 ἀψίνθιον] ἀψίνθιν AB 12 τὸ ζουμίν] τὸν ζωμόν MA 12 κηρίον] κηρὸν MA 12 λύσε] λύε MA 13 χαμαιμηλέλαιον] τὸ χαμαίμηλον A 13 χήνειον] χηνὸς MB χυνὸς A 14 τὸ] τὸν Σ 14 καὶ] ἕνωσε καὶ add. Σ 14 ἐστ' ἂν] ἕως ὅτου A 15 θέτε το] θές τα M θέτε τα add. A 16 θέτε τα] θές τα M 17 μίξον] τρίψον A 17 καὶ] τάραξέ τα καὶ add. M 17 το] τα MA 18 σινάπιν] ἀψίνθην A

λιθάργυρον καὶ μίξε τὰ πάντα καὶ ἄλειφε τὴν πληγήν.⌝

123 [101] ρκγ´ ⌜ὅταν δὲ τὸ κάτω βλέφαρον γένηται κόκκινον ὥσπερ αἷμα ἢ ὥσπερ ξανθὸν τὸ χρῶμα τοῦ τόπου· κοπάνισον φύλλα ὑσσώπου καὶ σταπίδας ἐκ τῶν κοκκίων αὐτῶν καὶ θέτε τα εἰς τὴν πληγήν. 2 ἄλλο. ▷om. Bκοπάνισον ἀψίνθιον καὶ θέτε το ἐπάνω.om. B◁⌝ 3 ἄλλο. ▷om. MAὀρίγανον κοπάνισον καὶ μίξον μετὰ μέλιτος καὶ θέτε το ἐπάνω. 4 ἄλλο. βολβὸν κοπάνισον καὶ μίξον μετὰ μέλι. καὶ θέτε ἐπάνω.om. MA◁ 5 ἄλλο. ⌜κυκλάμινον κοπάνισον καὶ θέτε το εἰς τὴν πληγήν, ἔνθα ἔνι ὁ τόπος κόκκινος ἢ ξανθός. κρασὶν δὲ μὴ πίνῃ. μηδὲ ἥλιος τὸν κρούῃ μηδὲ ἄνεμος. φλεβοτόμησον δὲ τὸν τοιοῦτον κρανιακήν.⌝

124 [102] ρκδ´ ⌜ὅταν γένηται πόνος ἀπὸ κρυότητος· τρίψον καστόριον ψιλὸν πολλὰ καὶ μίξον με τὸ πηγανέλαιον στάξε εἰς τὸ ὠτίον.⌝ 2 ἄλλο. καστόριον ψιλὸν πολλὰ μίξε ▷om. MAτο με τὸ νάρδελαιον καὶ στάξον εἰς τὸ ὠτίον. 3 ἄλλο. καστόριον ψιλὸν μίξονom. MA◁ ▷om. Bμετὰ δαφνέλαιον καὶ στάξον εἰς τὸ ὠτίον. 4 ἄλλο. καστόριον ψιλὸν μίξονom. B◁ με τὸ ἀμαράκινον ἔλαιον καὶ στάξον εἰς τὸ ὠτίον. 5 ἄλλο. ⌜πέπερι ψιλὸν πολλὰ μίξον με τὸ ναρδέλαιον καὶ στάξον εἰς τὸ ὠτίον.⌝ 6 ἄλλο. ▷om. LAκαστόριον ψιλὸν μίξον με τὸ ἔλαιον τὸ λεγόμενον φουλιατὸν καὶ στάξον εἰς τὸ οὖςom. LA◁. 7 ἄλλο. ▷om. Aχρηστέλαιον μίξον με τὸ εὐφόλβιον τὸ τριμμένον καὶ

---

21 κηρὶν] κηρὸν Σ 1 λιθάργυρον] λιθά... L 1 καὶ μίξε] ...... L ἕνωσον M 1 τὰ πάντα] scripsi ṭὰ πάντα L ταῦτα M 1 ἄλειφε] θές τα εἰς M θέτε τα εἰς AB 2 τὸ] om. B 4 σταπίδας] σταφίδας Σ 4 ἐκ] ἐκτὸς Σ 4 κοκκίων] κουκκίων B 4 αὐτῶν] om. Σ 4 θέτε] θὲς M 5 θέτε] θὲς M 6 μετὰ μέλιτος] με τὸ μέλι B 6 θέτε] θὲς B 7 μετὰ] με τὸ B 7 θέτε] θές το B 8 κοπάνισον] καλὰ add. MB 8 θέτε] θὲς M 10 μηδὲ] μὴ B 10 τὸν κρούῃ μηδὲ ἄνεμος] ἢ ἄνεμος τὸν κρούῃ A 10 μηδὲ ἄνεμος] ὡς ἔφημεν καὶ εἰς ἕτερα B 10 τὸν τοιοῦτον] τοῦτον MA 11 κρανιακήν] φλέβα add. M 12 πόνος] εἰς τὸ ὠτίον add. MA 12 ἀπὸ κρυότητος] om. B 13 πολλὰ] om. A 13 μίξον] μίξε το B 13 στάξε] καὶ στάξον Σ 14 πολλὰ] ...λα L om. A 14 μίξε] μίξον M 15 ψιλὸν μίξον] ὑψ......ξον L ὁμοίως B 16 μετὰ] με τὸ MA 17 με τὸ] σὺν B 17 ἀμαράκινον ἔλαιον] ἔλαιον τὸ λεγόμενον ἀμαράκινον M ἔλαιον τὸ λεγόμενον φουλατὸν A ἐλαίῳ τὸ λεγομένῳ ἀμαράκινον B 18 πολλὰ] πολὺ B 18 μίξον] μίξε AB 19 μίξον] μίξε το B 20 φουλιατὸν] φαλιατὸν M 20 οὖς] ὠτίον B

---

18 πέπερι] This paragraph has a different location in MB.

στάξον εἰς τὸ ὠτίον.om. A◁ 8 ἄλλο. ▷om. LAἀποβάλσαμον· μίξον μετὰ ἔλαιον τὰ ἄνω εἰρημένα καὶ στάξον εἰς τὸ ὠτίον. om. LA◁ 9 ἄλλο. ⌜σκόρδον κοπανισμένον καλὰ βράσον με τὸ χρηστέλαιον. εἶτα ἂς χλιανθῇ καὶ στάξον εἰς ὠτίον.⌝ 10 ἄλλο. ▷om. Mκρομμύδιν κοπάνισε καλὰ καὶ βράσε τα εἰς τοῦ ὀμιδίου τὸ φλοῦδιν ἢ εἰς ὀστρυδίου με τὸ χρηστέλαιον καὶ στάξον εἰς τὸ ὠτίον.om. M◁ ⌜ἄνεμος δὲ μὴ τὸν κρούῃ, μηδὲ κρύα πίνῃ μηδὲ ὄσπρια ἢ παστᾶ τρώγῃ.⌝

125 [103] ρκε′ ⌜ὅταν γένηται πόνος εἰς τὸ ὠτίον διὰ θερμὴν δυσκρασίαν,⌝ οἷον ἀπὸ ἡλίου καὶ καύματος καὶ κόπου πολλοῦ· ⌜μίξον ἄσπρον αὐγοῦ με τὸ ῥοδέλαιον τὸ † ...ὺ καὶ στάξε εἰς τὸ ὠτίον.⌝ 2 ἄλλο. ἄσπρον αὐγοῦ μίξον με τὸ ῥοδέλαιον καὶ ὀξείδιν ὀλίγον καὶ χλίανε καὶ στάξε εἰς τὸ ὠτίον.

126 [103] ρκϛ′ ▷Lὅταν διὰ θερμὴν δυσκρασίαν γένηται εἰς τὸ ὠτίον ἀπέσω πόνοςL◁· ⌜ἄσπρον αὐγοῦ μίξον με τὸν χυλὸν τοῦ στρύχνου, ὃ λέγεται παρὰ τῶν χωρικῶν βρωμοβότανον, καὶ χλίανε τὸν ζωμὸν τούτου μετὰ τοῦ αὐγοῦ καὶ στάξε εἰς τὸ ὠτίον.⌝ 2 ἄλλο. ▷om. Bἀσφοδέλου ῥίζαν κοπάνισον καὶ τὸν ζωμὸν του μίξον με ὀλί-

---

21 μίξον] μίξε το B 1 μίξον] μίξε B 2 ἔλαιον] ἐλαίου A ἐλαίῳ B 2 τὰ] τοῦ A 2 ἄνω εἰρημένα] ἄνωθεν λεγόμενα B λεγομενου ἀμαράκη A 3 σκόρδον] σκόρωδον A σκόροδα B 3 κοπανισμένον] κοπάνισον Σ 3 καλὰ] καὶ M καὶ add. AB 3 βράσον] βράσε B 3–4 εἶτα ἂς χλιανθῇ καὶ] καὶ ἀφ' ὅτου χλιάνῃ Σ 4 εἰς] τὸ B 4 κοπάνισε] κοπάνισον B 5 ὀμιδίου] ὀμυδίου LA 5 φλοῦδιν] ὄστρακον B 5 ὀστρυδίου] ὀστριδίου B 6 δὲ] om. MA 7 μηδὲ ὄσπρια ἢ παστᾶ τρώγῃ] om. A 8 γένηται πόνος] transp. MB 8 διὰ] ἀπὸ A 8–9 θερμὴν δυσκρασίαν] θερμῆς δυσκρασίας A θερμοῦ δυσκρασίας B 9 οἷον] ἤγουν MA 9 πολλοῦ] καὶ διὰ τοῦτο ἐγένετο ὁ πόνος εἰς τὸ ὠτίον αὐτοῦ add. M καὶ διὰ τοῦτο ἐγένετο ὁ πόνος εἰς τὸ ὠτίον add. AB 10 με τὸ] καὶ Σ 10 τὸ † ...ὺ] χλίον Σ 10 στάξε] στάξον Σ 11 αὐγοῦ μίξον] ....ξον L 11 ὀξείδιν ὀλίγον] transp. MA 12 χλίανε] τα add. MA 12 στάξε] στάξον Σ 14 με τὸν] με τὸ Σ 14 τοῦ] λεγομένου παρὰ τῶν ἰατρῶν add. MB 15 ὃ λέγεται παρὰ] παρὰ δὲ MB om. A 15 τῶν χωρικῶν βρωμοβότανον] om. A 15 καὶ χλίανε] αὐτὸ γοῦν κοπάνισον καὶ MB καὶ κοπάνισον καὶ A 16 τούτου μετὰ τοῦ αὐγοῦ] τοῦ μίξον με τὸ ἄσπρον τοῦ αὐγοῦ MB 16 καὶ] καὶ χλίανέ τα καὶ MaB 16 εἰς τὸ ὠτίον] om. B 17 ἀσφοδέλου] ἀσφοδελέας MA 17 του] om. M 17 με] μετὰ M

---

13 ὅταν] This chapter is a part of the preceding one in the Σ group.

γον ῥοδέλαιον καὶ χλιάνας στάξε εἰς τὸ ὠτίον.[om. B◁] 3 ἄλλο. ἀμυγδαλέλαιον στάξε εἰς τὸ ὠτίον. 4 ἄλλο. [▷om. AB]γάλα γυναικεῖον μίξον[om. AB◁▷om. B] με τὸ ῥοδέλαιον καὶ ὀλίγον ὄξος καὶ χλιάνας στάξον εἰς τὸ ὠτίον.[om. B◁] 5 ἄλλο. [▷om. AB]ἔψημα μίξε με τὸ ῥοδέλαιον καὶ χλιάνας στάξε εἰς τὸ ὠτίον.[om. AB◁] 6 ἄλλο. [▷B]ἔψημα χλίανε το καὶ ὀλίγον ὄξος καὶ χλίανέ τα καὶ στάξον εἰς τὸ οὖς.[B◁] 7 ἄλλο. γάλα γυναικεῖον μίξον με τὸν χυλὸν τοῦ βρωμοβοτάνου καὶ χλιάνας στάξον εἰς τὸ ὠτίον. 8 ἄλλο. ἀσφοδέλου τὸ ζουμὶν μίξον με τὸ ῥοδέλαιον καὶ γάλα γυναικεῖον στάξον εἰς τὸ ὠτίον ⌜εἰς ἥλιον δὲ μὴ κάθηται μηδὲ ἄνεμος τὸν κρούῃ.⌝

127 [104] ρκζ′ ⌜ὅταν γένηται πόνος εἰς τὸ ὠτίον διὰ παχεῖς χυμούς. θέλεις δὲ νοήσειν ὅτι διὰ παχεῖς χυμοὺς ἔνι ὁ πόνος, ἀπὸ τοῦ εἶναι βάρος εἰς τὴν κεφαλὴν καὶ τὰ ὠτία ἔχουσιν ἀχόν. ἐὰν δὲ βάρος ἡ κεφαλὴ οὐκ ἔχει, καὶ φαίνεται ὡς ἂν ἀποταυρίζεται τὸ δέρμα τῆς κεφαλῆς καὶ μόνον σημεῖον, ἔνι τοῦ ὠτίου καὶ μόνον τὸ πάθος καὶ οὐχὶ τῆς κεφαλῆς. ποίησον δὲ θεραπείαν τοιαύτην· ἤγουν βάλε ὀξείδιν καὶ μέλι καὶ χλιάνας στάξε εἰς τὸ ὠτίον. 2 ἄλλο. προβάτου χολὴν μίξον με τὸ ἀμυγδαλέλαιον καὶ χλιάνας στάξε εἰς τὸ ὠτίον. 3 ἄλλο. πράσα κοπάνισον καὶ τὸ ζουμὶν τούτων μίξον με τὸ μέλι τὸ καλὸν καὶ χλιάνας στάξον εἰς τὸ ὠτίον.⌝ 4 ἄλλο. κρομμύδιν

241.17–242.1 ὀλίγον ῥοδέλαιον] ὀλίγου ῥοδελαίου MA 1 καὶ χλιάνας] om. M 1 στάξε] στάξον M 2 στάξε] στάξον Σ 3 ὄξος] ὀξείδιν A 3–4 χλιάνας στάξον] χλίανέ τα καὶ στάξον MA 4 ὠτίον] οὖς M 4 μίξε με τὸ] χλίανον καὶ μίξον με τὸ μέλι καὶ τὸ M 5 στάξε] στάξον M 5 ὠτίον] οὖς M 7 με τὸν χυλὸν τοῦ βρωμοβοτάνου] με τοῦ βρωμοβοτάνου τὸν ζωμὸν ἤγουν με τοῦ στρύχνου A 7 με τὸν χυλὸν] ζωμὸν M με τὸ χλίον B 8 καὶ χλιάνας] χλίον καὶ A 8 χλιάνας] om. MB 8 ὠτίον] οὖς MB 8 τὸ ζουμὶν] ζωμὸν MB τὸν ζωμὸν A 9 με τὸ] με M 9 ῥοδέλαιον] ὀλίγον ῥοδέλαιον Σ 9 στάξον] καὶ στάξον Σ 9 ὠτίον] οὖς MA 10 εἰς ἥλιον δὲ μὴ κάθηται μηδὲ ἄνεμος τὸν κρούῃ] ἥλιος μὴ τὸν κρούῃ A 10 εἰς] τὸν add. B 10 δὲ μὴ κάθηται] ....θηται L 10 ἄνεμος] παντελὸς B 12 νοήσειν] νοῆσαι MA 13 εἰς τὴν κεφαλὴν] ἐν τῇ κεφαλῇ Σ 13 ἀχόν] ἀγχών L ἦχον A 13–14 ἡ κεφαλὴ οὐκ ἔχει] transp. B 14 καὶ] om. A 14 φαίνεται] ὅτι add. MB τὸν add. A 14–15 τὸ δέρμα τῆς κεφαλῆς] transp. M 15 ἔνι] ὅτι add. B 15 καὶ μόνον] ἔνι A add. B 15 μόνον] σμεῖον ἔνι τοῦ ὠτίου add. M μόνου B 15 καὶ] μόνον add. M 16 δὲ] om. AB 16–17 ὀξείδιν] ὄξος M 17 στάξε] στάξον Σ 18 μίξον] μίξε A 18 στάξε] στάξον MB 19 τὸ ζουμὶν] τὸν ζουμὸν M 19 τούτων] αὐτῶν B

κοπάνισον καὶ τὸ ζωμὶν αὐτοῦ μίξον με τὸ μέλι καὶ στάξον εἰς τὸ ὠτίον καὶ χλιάνας. 5 ἄλλο. δρακοντίας βοτάνης τὸ ζουμὶν μίξον με τὸ μέλι καὶ στάξον εἰς τὸ ὠτίον.

128 [105] ρκη′ ⌜ὅταν τὸ ὠτίον φλεγμαίνῃ,⌝ ποίησον θεραπείαν τοιαύτην· ἤγουν ⌜φλεβοτόμησον αὐτὸν καὶ πυρίαζε τὸν με τὸ χρηστέλαιον καὶ τὸ ἕψημα καὶ χλιάνας βρέξε μαλία πρόβεια ἀπὸ τὸν τόπον τῶν ὀρχιδίων, οὕτως ὡς ἔνι ἄπλυτα. καὶ πυρίαζε τὸν τόπον. 2 ἄλλο. στάξον δὲ εἰς τὸ ὠτίον ἀπέσω χήνειον ἀξούγγιν λυτὸν χλίον.⌝ 3 ἄλλο. ▷om. A ἀλωποὺς ἀξούγγιν ἄπλυτον χλίον στάξον εἰς τὸ ὠτίον.om. A◁ 4 ἄλλο. καστόριον τριμμένον πολλὰ μίξον με τὸ γάλα γυναικεῖον καὶ χλιάνας στάξον εἰς τὸ ὠτίον. 5 ἄλλο. ⌜▷om. A ἕψημα χλίανε καὶ στάξον εἰς τὸ ὠτίον.om. A◁ 6 ἄλλο. ▷MA κρασὶν δὲ πολὺ μὴ πίνῃ, σχηδὸν δὲ εἰπεῖν μὴ δὲ ὅλως. ἃς ἔνι δὲ καὶ ζεστὰ ἥλιος μὴ τὸν κρούῃ μηδὲ ἄνεμος μηδὲ καπνός.MA◁⌝

129 [106] ρκθ′ ⌜ἐὰν θέλῃς να αὐξήσουν τὰ ὀδόντια τοῦ παιδίου,⌝ ποίησον θεραπείαν τοιαύτην· ἤγουν ἄλειψον τὰ οὖλη τῶν ὀδόντων σκύλλας γάλα καὶ ἐξέρχονται οἱ ὀδόντες. 2 ἄλλο. ⌜λαγωοῦ μυελὸν ἀπὸ τὸ κεφάλιν του ἄλειψον τὰ οὖλη καὶ ἐξέρχονται οἱ ὀδόντες.⌝ 2 ἄλλο. κατούρημα παιδίου μικροῦ ἄλειψον τὰ οὖλη πλὴν μικτὸν με

19 μίξον] μίξε το B 1 τὸ ζωμὶν] τὸν ζωμὸν MA 1 αὐτοῦ] τούτων M τούτου A του B 1 μίξον] μίξε το B 1 μέλι] τὸ καλὸν add. Σ 1 καὶ] χλιάνας add. Σ 2 καὶ χλιάνας] om. Σ 2 δρακοντίας] δ...... L 2 τὸ] τὸν A 2 ζουμὶν] ζωμὶν M ζωμὸν A 2 μίξον] στάξον A στάξε μικτὸν B 3 καὶ] ... L om. AB 3 στάξον] om. AB 3 εἰς τὸ] om. B 3 ὠτίον] οὺς M om. B 4 ὅταν τὸ ὠτίον φλεγμαίνῃ] πρὸς φλεγομὴν ὠτίου M 4 ὠτίον] ὀπίον B 4 φλεγμαίνῃ] ἐγβαίνῃ B 6 καὶ] om. A 6 βρέξε] βρέξον MA 6 πρόβεια] προβάτου Σ 6–7 τὸν τόπον] τοῦ τόπου Σ 7 ἔνι] εἶναι MA ἐν B 7 τόπον] καλῶς add. A 8 στάξον] στάξε AB 8 χήνειον ἀξούγγιν λυτὸν] λυτὸν ἀξούγγιν χηναρίου Σ 9 ἀλωποὺς] ἀλώπεκος MB 9 ἄπλυτον] λυτὸν B 9 χλίον] om. MB 10 ὠτίον] χλίον add. MB 10 πολλὰ] καλὰ M 11 γάλα] τὸ add. Σ 12 χλίανε] χλιάνας MB 12 καὶ] om. MB 13 δὲ] om. A 13 μὴ] μηδὲ B 13 σχηδὸν δὲ εἰπεῖν μὴ δὲ ὅλως] om. A 13 δὲ] om. B 13 καὶ] om. A 14 ζεστὰ] ὁ add. B 14 μὴ] δὲ οὐδόλως B 14 ἄνεμος μηδὲ καπνός] καπνὸς μήδε ἄνεμος A 15 τὰ] οἱ AB 15 ὀδόντια] δόντια M ὀδόντες AB 15 τοῦ παιδίου] om. B 16 ἤγουν] om. A 16 ἄλειψον] ἄλειο L ο corr. φε L^2 17 σκύλλας] add. M 17 ὀδόντες] τοῦ παιδίου add. M 18 τὸ κεφάλιν του] τὸν ἐγκέφαλον Σ 18 οὖλη] τοῦ παιδίου add. Σ 18 καὶ ἐξέρχονται] ......χόνται L 18 οἱ ὀδόντες] om. A

τὸ γάλα τῆς σκύλλας.

130 [107] ρλα´ ⌜ὅταν γένηται ὄγκος ὑπὸ τὴν γλῶτταν. ὁ δὲ ὄγκος ἐκεῖνος ὑποκάτω τῆς γλώσσης γίνεται πλέος εἰς τὰς φλέβας τῆς γλώσσης. ποίησον θεραπείαν τοιαύτην· ἤγουν κηκίδιν τριμμένον ψιλὸν μίξον με τὸ ἕψημα καὶ ἄλειφε τὸν ὄγκον.⌝ 2 ἄλλο. ῥόβιν ψιλὸν πολλὰ καὶ κοσκινησμένον μίξε το με τὸ μέλι καὶ ἄλειφε τὸν ὄγκον ἔνθα ἔνι εἰς τὰς φλέβας τῆς γλώσσης. 3 ἄλλο. τριακονταφύλλων σπόρον κοπάνισον καὶ ψιλὸν ποιήσας μίξον με τὸ μέλι καὶ ἄλειφε τὸν ὄγκον. 4 ἄλλο. ⌜ἐλαίας φύλλα κοπάνισον καὶ τὸν ζωμὸν ἄλειφε τὴν γλῶσσαν ▷om. Aἔνθα ἔνι τὸ πάθος.⌝ 5 ἄλλο. τὸ ζωμὶν τῶν φύλλων τῆς ἐλαίας, κράτει πολλὴν ὥραν εἰς γλῶσσαν.om. A◁

131 [108] ρλα´ ⌜ἄφθα ἔνι, ὅταν εἰς τὰ χείλη παιδίων ὅπου βυζάνουν φαγόνουνται ἢ δι’ ὅτι ἔνι τὸ γάλα τῆς γυναικὸς δριμὺ καὶ διὰ τοῦτο φαγόνουνται. τοῦτο δὲ γίνεται καὶ εἰς τοὺς τελείους.⌝ ποίησον δὲ θεραπείαν τοιαύτην· ἔπαρον φακὴν καὶ κοπάνισον καὶ μίξον με τοῦ

---

19 παιδίου μικροῦ] transp. Σ 1 σκύλλας] καὶ ἐξέρχονται οἱ ὀδόντες add. M καὶ ἐξέρχονται οἱ ὀδόντες εὐκόλως add. A καὶ ἄλειψον τὰ οὔλη καὶ ἐξέρχονται οἱ ὀδόντες σύντομον add. B 2 ὑπὸ] εἰς MA ἀπὸ B 2 τὴν] τῆς B 2 γλῶτταν] γλῶσσαν MA γλώττης B 2–3 ὁ δὲ ὄγκος ἐκεῖνος] om. M ὁ ὄγγος δὲ ὁ γινόμενος εἰς τὴν γλῶσσαν A γινόμενος εἰς τὴν γλῶσσαν B 3 τῆς γλώσσης] δὲ MA om. B 3 πλέος] πλεῖον M om. A πλέον B 4 γλώσσης] πλέον add. A χρὴ ποιεῖν οὕτως add. B 4 ποίησον] δὲ add. MA 4 κηκίδιν] κ...δην L 5 μίξον] μίξε A καὶ μίξε το B 5 ὄγκον] ὄγγον MA 5–6 ῥόβιν ψιλὸν] scripsi ῥοβιν...λὸν L πρόβιν ψιλὸν M ὀρόβιν ψιλὸν A τὸ ῥόβιν ψιλὸν B 6 πολλὰ] om. B 6 κοσκινησμένον] πολλὰ add. B 6 μίξε το] μί...το L μίξον M μίξε A 6 τὸν] τὸ A 7 ὄγκον] πρήσμαν A 7 ἔνθα] ὅπου B 7 εἰς τὰς φλέβας τῆς γλώσσης] ὁ πόνος A 8 κοπάνισον] κοπανίσας MA 8 μίξον] μίξε A μίξε το B 9 ἄλειφε] ἀ....ε L 9 ὄγκον] τόπον MA 9 ζωμὸν] αὐτῶν add. M τῶν add. A τῶν φύλλων add. B 10 τὴν γλῶσσαν] τὸν τόπον Σ 10 τὸ ζωμὶν] τὸν ζωμὸν MB 11 πολλὴν ὥραν] om. M 11 εἰς γλῶσσαν] scripsi ... γλῶσσαν L εἰς τὴν γλῶσσαν σου MB 12 ἄφθα ἔνι] πρὸς ἄβθας M περὶ ἄβθας· ἄβθα ἔνι A ἡ ἄφθα ἔνι B 12 εἰς τὰ] scripsi ọτ̣α L τὰ χείλη Σ 12 χείλη] scripsi ..... L τῶν μικρῶν M τῶν AB 12 βυζάνουν] βιζάνουν L βαζάνουν M 13 φαγόνουνται] φαγẹνονται M φαγώνουνται A 13–14 ἢ δι’ ὅτι ἔνι τὸ γάλα τῆς γυναικὸς δριμὺ καὶ διὰ τοῦτο φαγόνουνται] om. MA διὰ πολλὴν δριμύτητα τοῦ γάλακτος τῆς μητρὸς αὐτῶν διὰ γὰρ τούτων φαγόνουνται ἀπὸ τὸ δριμύ B 14 δὲ] om. B 14 τοὺς τελείους] ἀνθρώπους add. Σ 14 δὲ] om. MA 15 τοιαύτην] ἤγουν add. MB ἤγουν A

ἄρτου τὰς ψίχας καὶ μάλαξον αὐτὴν με τὸ κρασὶν καὶ ἄλειφε τὰ χείλη. 2 ἄλλο. ▷om. M⌜μυελὸν ἐλάφου ἄλειφε χείλη.⌝om. M◁ 3 ἄλλο. ▷MBμόσχου μυελὸν ἄλειφε τὰ χείλη.MB◁ 4 ἄλλο. ⌜ἀπιδίων ζωμὸν ἄλειφε τὰ χείλη.⌝ 5 ἄλλο. ▷om. Bἀντύβια κοπάνισον καὶ τὸν ζωμὸν ἄλειφε τὰ χείλη.om. B◁ 6 ἄλλο. κέρατον ἐλάφου καῦσας, τρίψον καὶ ἄλειφε τὰ χείλη. 7 ἄλλο. ⌜τὴν ῥίζαν τοῦ μαλάθρου καῦσε καὶ τὴν στάτην ἐκείνην πάσσε τὰ χείλη. 8 ἄλλο. κρασὶν δὲ πολὺ μὴ πίῃ. φλεβοτόμησον δὲ τούτους εἴπερ ἔνι μέγας.⌝

132 [109] ρλβʹ ⌜ὅταν φλεγμαίνουσι τὰ λεγόμενα σιαγόνες, ⌝ποίησον θεραπείαν τοιαύτην· ἤγουν φλεβοτόμησον τοῦτον κρανιακὴν καὶ ἀφ' ὅτου διάβουν δέκα ἡμέρες, φλεβοτόμησον αὐτὸν φλέβαν τὴν μέσην. τινὲς δὲ λέγουσι ταύτην καθόλου φλέβα βασιλικὴν ▷om. Aὡς ὑπερέχουσα αὐτὴ ἡ φλὲψ καὶ αὐθεντεύουσα τῶν ἄλλων φλεβῶν.om. A◁ οἱ δὲ τοιούτοι πολλὰ μὴ τρώγουν. μὴ δὲ πολλὰ πίνουν. 2 ἄλλο. βράσον φακὴν καὶ ῥόδα καὶ φοινίκια με τὸ κρασὶν καὶ ἀφ' ὅτου βράσουν καλὰ κράτει τὸ κρασὶν ἐκεῖνο εἰς τὸ στόμα πολλὴν ὥραν. 3 ἄλλο. ἀφ' ὅτου δὲ φλεβοτομήσῃς τοῦτον, μετὰ μίαν

15 μίξον] αὐτὴν add. M μίξε την AB 1 ψίχας] ψιχίας MA 1–2 τὰ χείλη] χλίον B 2 ἄλειφε] τὰ add. AB 4 ἀντύβια] ἰντύβια A 4 ζωμὸν] αὐτῶν add. MA 5–6 καῦσας, τρίψον καὶ ἄλειφε τὰ χείλη] τρίψον καὶ πάσον τὰ χείλη πλὴν δὲ τὸ κέρατον πρῶτον καύσε το Σ 6 καῦσε] καῦσον MB τὴν add. A 7 στάτην] στάκτην Σ 7 ἐκείνην πάσσε] πάσον MA 7 ἄλλο] ἑρμηνεία A 7 δὲ] om. Σ 7 πίῃ] πίνῃ Σ 8 δὲ] καὶ add. B 8 τούτους] τοῦτον Σ 8 εἴπερ ἔνι μέγας] ἐὰν ἔνι χρονῶν ιβ' ἢ ιε' ἢ κ' Σ 9 τὰ λεγόμενα] αἱ λεγόμεναι A αἱ φλεγομέναι B 9 σιαγόνες] ὅπου εἶναι τὰ γ̣ένια τῶν μαγούλων add. M ὁποῦ εἶναι τὰ γένεια τῶν μαγούλων add. A ὅπου εἶναι τὰ γένει τῶν μαγούλων B 10 τοιαύτην] om. A 10 τοῦτον] αὐτὸν M 10 κρανιακὴν] φλέβα add. M φλέβαν add. A 11 ἀφ' ὅτου διάβουν] ἀφ...βοῦν L 11 ἡμέρες] ἡμέραι MA 11 φλεβοτόμησον αὐτὸν] φλεβοτόμησ..... L φλεβοτόμησον τοῦτον A 11–12 φλέβαν τὴν] ἄλλην φλέβαν τὴν λεγομένην M 12 καθόλου φλέβα βασιλικὴν] καθικὴν τινὲς δὲ βασιλικὴν M καθολ.ην A καθολικὴν B 13 αὐτὴ] αὐτὴν M 13 αὐθεντεύουσα] scripsi αὐτητεύουσα L ἀφεντέβουσα M αὐθεντεῦσα B 13 ἄλλων] ὅλων add. B 14 οἱ] .. L 14 τρώγουν] φάγουν B 14 μὴ] .. L 14–15 πολλὰ πίνουν] transp. B 15 ῥόδα καὶ] om. B 16 ἀφ' ὅτου] ..ọ τ̣ọ̀ṿ L 16 τὸ κρασὶν ἐκεῖνο] transp. Σ 16–17 εἰς τὸ στόμα πολλὴν ὥραν] transp. A 16 εἰς τὸ στόμα] om. B 16–17 πολλὴν ὥραν] om. M 17 ἀφ' ὅτου] ...του L

4 ἀντύβια] The following two entries are reversed in A.

ἡμέραν, βαλὼν γλυκόριζαν καὶ ἕψημα καὶ σῦκα ἂς βράσουν καὶ ὕστερον βάλε ῥοδόσταγμα καὶ κράτει εἰς τὸ στόμα. 4 ἄλλο. βάλε ὕσσωπον καὶ ζωμὶν ῥοϊδίου καὶ μίξε τὰ καὶ ἂς βράσουν ὀλίγον. ὕστερον δὲ βάλε δροσάτον καὶ κράτει εἰς τὸ στόμα. 5 ἄλλο. ⌜γάλα γλυκὺ πρόβειον κράτει εἰς τὸ στόμα. 6 ἄλλο. βράσον νερὸν βροχῆς καὶ βρέχε σπόγγον καινούργιον εἰς τὸ νερὸν ἐκεῖνο καὶ πυρίαζε τὸν τόπον πολλὴν ὥραν.⌝ 7 ἄλλο. ▷om. A νερὸν βροχῆς ἂς βράσῃ καὶ βρέχε κυκλία ὠμὰ καὶ πυρίαζε τὸν τόπον. om. A◁ 8 ἄλλο. ⌜κριθάλευρον καὶ λινόσπορον ψιλὸν κοσκινισμένον βράσον με τὸ νερὸν καὶ ὀλίγον κρασίν. θέτε με τὸ πάνιν ἐπάνω εἰς τὸν τόπον. τοῦτο δὲ ποίησον ἡμέρας τρεῖς πρωῒ βραδύ.⌝ 9 ἄλλο. φοινίκια κοπάνισον καλά. μίξε το με τὸ ἀλεύριν τῆς τήλης καὶ βάλε νερὸν καὶ ὀλίγον κρασὶν καὶ ἕψησον καὶ θέτε εἰς τὸν τόπον. 10 ἄλλο. ⌜βράσε γλυκόριζον καὶ κρόκον καὶ κρασὶν καὶ ἂς βράσουν καὶ κράτει εἰς τὸ στόμα.⌝

133 [110] ρλγ΄ ⌜πρὸς βῆχας· ἂς καταπίνῃ κατὰ βραδὺ τὸν λεγόμενον στύρακα ἴσα ἐρεβίνθην καὶ μὴ δειπνᾷ. αὐτὸς δὲ ὁ λεγόμενος

17 φλεβοτομήσης] φλεβοτομήσεις MB 1 γλυκόριζαν] scripsi γλυκόρυζαν L γλυκόριζον Σ 1 σῦκα] καὶ add. A 2 κράτει] κρασὶν A 3 ζωμὶν] τὸν ζωμὸν τοῦ Σ 3 καὶ μίξε τὰ] om. M 4 δὲ βάλε δροσάτον] om. B 4 κράτει] scripsi ..άτει L μίξον τα καὶ κράτει M μίξε τα καὶ κράτει AB 5 πρόβειον] πρὸ .....ον L χλίον add. AB 5 κράτει] χλίον κράτει AB 5 στόμα] ἔνθα ἔνι ὁ τόπος add. M ἔνθα ἔνι ὁ πόνος B 5 βράσον νερὸν] βρά ..... L 6 καὶ] τότε add. Σ 6 βρέχε] τὸν add. MA τὸ B 6 σπόγγον] σπογγάριν B 6 καινούργιον] ...νούργιον L τὸν καινούργιον MA 6 εἰς τὸ νερὸν ἐκεῖνο] ἔστω B 6 πυρίαζε] πυρι... L 7 ὥραν] om. AB 7 βροχῆς] βρο.... L 7 καὶ] τότε βάλε add. M τότε add. B 8 βρέχε] om. M 8 καὶ] .. L 8 τόπον] ἔνθα ἔνι ἡ ὀδύνη add. M ἔνθα ὑπάρχει ἡ ὀδύνη B 9 κοσκινισμένον] scripsi κοσκινισμεμ.. L κοσκινησμένον καλὰ MB κοσκυνησμένα A 9 βράσον] καλὰ add. A καὶ ἂς βράσουν B 9 ὀλίγον] με τὸ M 10 θέτε] καὶ θὲς σὲ M καὶ θέτε A καὶ θέτε το B 10 τόπον] ἔνθεν ἔνεν ἡ ὀδύνη add. A 10 ποίησον] ποίη M 11 πρωῒ] καὶ add. MB om. A 11 βραδύ] om. A 11–12 μίξε το] καὶ μίξον M καὶ μίξε τα AB 13 ἕψησον] ἕψ... L 13 θέτε] θὲς M το add. A θές τα B 13 βράσε] βράσον Σ 13 γλυκόριζον] γλυ..ριζον L 14 κρόκον] οἷον βάνουν εἰς τὸ γλυκόξυνον add. MB οἷον βάννουν εἰς τὸ γλυκόξυνον add. A 14 καὶ] μέλι καὶ add. M μέλι καὶ A 14 κρασὶν καὶ] om. B 14 βράσουν] βρασ... L 15 βήχας· ἂς] βήχα L 16 στύρακα] ῥάκα L 16 ἴσα] ἴσον MA 16 ἐρεβίνθην] ἐρεβίνθια MB τερέβινθον A 16 μὴ] μηδὲ B 16 δειπνᾷ. αὐτὸς] δειπν.... L

στύρακας ὀφείλει εἰς τὸν κατάρρουν. 2 ἄλλο. τὸ πρωῒ πρὶν φάγει ▷om. L ἂς ἐπάρῃ ἰσχάδας ξηρὰς λιπαρὰς καὶ ἄς τα κατακόψῃ μικρὰ κομμάτια καὶ ἄς τα βάλῃ εἰς κρασὶν ἄκρατον καλὸν καὶ ἄς τα τρώγῃ. ἆφ' ὅτου δὲ τὰ πληρώσει ἐπάνω ἂς πίνῃ γλυκὺ κρασὶν ὀλίγον.⌝ 3 ἄλλο. τὰ δὲ σῦκα πέντε πέντε κατὰ πρωῒ εἰς δὲ τὸ γλυκὺ κρασὶν καὶ ἂς βάνῃ καὶ νερὸν ὀλίγον καὶ ἄς το πίνῃ. 4 ἄλλο. ⌜ποιοῦν δὲ καὶ τὴν λεγομένην γροῦταν,⌝ ἤγουν ἂς βάλῃ εἰς τὸ κεχρὶν ἐκεῖνο τὸ βράζῃ ὕσσωπον βλησκούνιν καλαμίνθην καὶ ὀρίγανον καὶ ἂς βράσουν καλὰ καὶ κένωσέ τα καὶ βάλε καὶ μέλι καλὸν καὶ τάραξέ τα καὶ ἄς τα φάγῃ. τὰ δὲ βοτάνια ἄς τα ῥίψῃ. αὐτὸν δὲ ἂς πίνῃ ἐστ' ἂν λυθῇ ὁ βῆξ, καὶ κρύα μὴ πίνῃ μηδὲ ἄνεμος τὸν κρούῃ μηδὲ ἥλιος. ἂς ἔνι δὲ ζεστὰ ὅσον δύναται καὶ ἂς πίνῃ θερμὸν κρασὶν καὶ μὴ δειπνᾷ καὶ ⌜ὅταν θέλει κοιμηθῇ ἂς τρώγῃ ἀμύγδαλα πικρὰ ε' κατὰ βραδύ. εἰ δὲ οὐκ ἔχει πικρὰ ἂς τρώγῃ γλυκέα δέκα καὶ ἂς πίνῃ θερμὸν κρασὶν ἐπάνω. εἰ δὲ οὐκ ἔχει οὐδὲ γλυκέα ἂς τρώγῃ φοινίκα τρία. 5 ἄλλο. βράσον μέλι καλὸν καὶ ἕψησον καὶ παστέλλωσον αὐτὸ να μὴ ἔνι πολλὰ ὑγρόν ἢ πολλὰ ξηρὸν ἀλλὰ μέσον. καὶ τότε βάλε ἀμύγδαλα γλυκέα κοπανισμένα ρ' καὶ κωνάρια κοπανισμένα κ' καταστατὸν κοπανισμένον καὶ πικρὰ ἀμύγδαλα κοπανισμένα εἴκοσι. καὶ ἐὰν οὐ πυρέσσῃ βάλε καὶ πεπέρι κοπανισμένον πλὴν χονδρόν. καὶ γλυκόριζον κοπανισμένον ψιλούτζικον πολλὰ καὶ λινόσπορον ψιλὸν ἥμισον κοχλιάριν καὶ φοινίκια κοπανισμένα λιπαρὰ καὶ φύλλα ξηρὰ κοπανισμένα

16 δὲ] om. B 1 ὀφείλει] ὠφελεῖ δὲ καὶ MB ὠφελεῖ καὶ A 1 πρὶν φάγει] π.. L νῆστιν A πρὶν φάγῃ B 3 εἰς] τὸ add. B 5 πέντε] om. A 6 κρασὶν] καὶ add. M 6 βάνῃ] βάννῃ AB 6 νερὸν ὀλίγον] transp. A 6 το] τον A 6 πίνῃ] om. AB 7 ποιοῦν] ποιοῦσι A 7 γροῦταν] scripsi γροῦτταν Σ 8 τὸ κεχρὶν ἐκεῖνο] transp. A 8 βλησκούνιν] γλησκούνιν AB 10 βοτάνια] βότανα B 11 ῥίψῃ] τρίψῃ B 11 αὐτὸν] αὐτὸ AB 11 ἂς] το add. AB 11 πίνῃ] ποιῇ B 11 ἐστ' ἂν] ὥστε να A 11 βῆξ] ὁ βῆχα A ὁ βῆχας B 11 καὶ] om. B 13 θερμὸν] ζεστὸν B 13 θέλει] να add. A 14 ἀμύγδαλα] om. B 14 ε'] om. B 14 κατὰ βραδύ] om. A 14 εἰ δὲ οὐκ ἔχει] om. B 14–15 πικρὰ ἂς τρώγῃ] transp. B 15 γλυκέα] πικρία B 16 τρώγῃ] τὸ βραδὺ add. AB 16 βράσον] βράσε A 19 κοπανισμένα κ'] καὶ αὐτὰ κοπανισμένα καὶ A κοπανισμένα καὶ B 20 εἴκοσι] om. A 21 χονδρόν] χοντρὸν AB 22 ἥμισον] ἥμισυ AB 22–23 κοχλιάριν] κοχλιαρέαν AB 23 φύλλα] τὰ φύλλα τὰ B

ὑσσώπου καὶ μαστίχιν καὶ ζίγγιβερ τὸ λεγόμενον ξυλοζιζίβερι καὶ αὐτὸ ψιλὸν πολλά. καὶ στάχος ψιλὸν πολλὰ καὶ τριψίδιν ψιλὸν τὸ λεγόμενον κινάμωμον καὶ κύμινον ψιλὸν ὀλίγον καὶ ἄσπρον πεπέρι τὸ τρωγώμενον καὶ τὸ λεγόμενον μακροπεπέρι πλὴν ψιλὸν πολλά. καὶ τζουκνίδας σπόρον κοπανισμένον καὶ ψιλὸν ποιήσας ἕνωσον τὰ πάντα καὶ βάλε τὰ ὅλα ὅταν ἐκβάλης ἔξω τὸ τζουκκαλόπουλον ὅπου ἔχει τὸ μέλι. καὶ ἀφ' ὅτου βάλης τὰ εἴδη ὅλα ἄλλο μηδὲν βράσης τὸ δὲ βραδὺ ἂς τρώγῃ ἀπ' ἐκεῖνα ἶσον καρύδιον. ⊳om. B καὶ ἀπέκει ἂς πίνῃ κρασὶν πολλὰ καλόν. om. B⊲ 6 ἄλλο. ἄλειφε δὲ τὸ στῆθος αὐτοῦ ὀλίγον βούτυρον προβάτου καὶ μή τον τρίβης μὴ δὲ ὅλως καὶ πλέον βλάψης παρ' ὠφελήσεις. 7 ἄλλο. ἄλειφε χήνειον ἀξούγγιν ἢ ὄρνιθος τὸ στῆθος του.⌝ 8 ἄλλο. κοπάνισον χαλβάνιν καὶ ποίησέ το ψιλὸν πολλά. καὶ καστόριον ψιλούτζικον καὶ μίξον καὶ βάλε καὶ ὀλίγον κρασὶν ὅσον δύναται να ποιήσης σκληρὸν τζυμάριν. καὶ τότε ποίησέ το κοκκία μικρὰ ἶσον ἐρεβίνθια καὶ μικρότερα καὶ τὸ βραδὺ ὅταν θέλῃ κοιμηθὴν ἂς τα καταπίνῃ καὶ μὴ δειπνᾷ ὅλως. ἐπ' ἄνω ἐκεῖνα ἂς πίνῃ ζεστὸν κρασὶν καλὸν αὐτὸ δὲ ἂς πίῃ κατὰ νύκταν οὕτως. 9 ἄλλο. ⌜εἰς δὲ τὸ στῆθος τοῦ θὲς ἔμπλαστρον διὰ βουτύρου καὶ ἄλλο ἔμπλαστρον ἡ λεγομένη ἀνηθίνη.⌝ καὶ θὲς ἀπὸ τῶν δύο

23 κοπανισμένα] om. B 1 ὑσσώπου] τοῦ ὑσσώπου ποίησον ψιλὰ AB 1 μαστίχιν] μαστίχων A 1 ξυλοζιζίβερι] ξυλοτζιτζίβερι A 1–2 καὶ αὐτὸ ψιλὸν πολλά] om. A 2 στάχος] ὁμοίως B 2–3 πολλὰ καὶ τριψίδιν ψιλὸν τὸ λεγόμενον] καὶ A 2 πολλὰ] om. B 3 ψιλὸν ὀλίγον] transp. B 4 τὸ τρωγώμενον] ὀλίγον ὅπου εὑρίσκεται εἰς αὐτὸ τὸ πεπέρι ὁποῦ (τὸ B) τρώγωμεν AB 4 καὶ] om. A 5 τζουκνίδας σπόρον] τζουκνιδόσπορον A 5 καὶ] om. A 5 ψιλὸν ποιήσας] πλὴν δὲ ὅλα ψιλὰ καὶ A 5 ποιήσας] ποίησον B 6 ἐκβάλης] εὐγάλης A 6 ἔξω] om. A 6 τζουκκαλόπουλον] τζυκαλόπουλλον A 7 βάλης] βάλει εἰς B 7 βράσης] βράσῃ B 8 ἐκεῖνα] ἐκεῖνο B 8 ἶσον] ὅσον B 8 καρύδιον] τρανὸν add. AB 10 τρίβης] τρίψης B 11 ὠφελήσεις] οὗ να ὠφελήσης A 11 ἄλλο. ἄλειφε] ἢ AB 12 τὸ στῆθος του] om. AB 13 ψιλούτζικον] καὶ πεπέρι ὀλίγον λιγούτζικον B 13 καὶ μίξον] om. A καὶ μίξε τὰ τρία B 14 κρασὶν] καὶ μίξε τὰ τρία add. A 14 ὅσον δύναται να ποιήσης σκληρὸν τζυμάριν] καὶ ποίησον ζυμάριν σκληρὸν A 14 καὶ τότε] εἶτα A 15 μικρὰ] ψιλὰ A 15 ἶσον] ὅσον A 16 κοιμηθὴν] να κοιμηθῇ A κοιμηθῆναι B 16 τα καταπίνῃ] κατὰ νύκταν ἐξ αὐτῶν add. A καταπίνῃ κοκκία τέσαρα B 16 ἐπ' ἄνω] καὶ ἐπάνω εἰς A 17 πίῃ] τὸ ποιεῖ A 18 θὲς] θέτε AB 18 ἔμπλαστρον] τὸ add. A 19 θὲς] θέτε AB

ἐμπλάστρων οἷον εὕρης εἰς τὸ στῆθος του με τὸ πανὶν ἢ με τὸ τὸ μάριν.[om. L◁]

134 [om. L][111] ρλδ´ ⌜πρὸς ὅσοι ἀπὸ φωνῆς μέγα βραγχιάζῃ ποίησον θεραπείαν τοιαύτην· ἤγουν λοῦέ τον συνεχῶς καὶ ἡ τροφὴ του μὴ ἔνι δριμέα οἷον πράσα καὶ τὰ τούτοις ὅμοια. μηδὲ παστὸν κρέας ἢ ὀψάριν εἰ μὴ πάντα χλωρά. 2 ἄλλο. ἃς τρώγῃ δὲ καὶ καταστατὸν καὶ τὰ ῥοφητὰ αὐγά. καὶ τὸ γλυκὺ γάλα τὸ πρόβειον χλίον.⌝

135 [112] [om. L]ρλε´ πρὸς ὅσοι εὑρεθῶσιν καὶ ἔχουσιν δύσπνοιαν καὶ οὐδὲν δύνανται να ἀνασάνουν καὶ μάλιστα ὅταν γένηται κρυάδα ἔχουσι κακῶς. ποίησον θεραπείαν τοιαύτην· ἤγουν σκυλλοκρέμμυδον τὸ λεγόμενον παρὰ τῶν ἰατρῶν σκύλλαν παρὰ δὲ τῶν λαγγιβάρδων σκα[λ]. ταύτην δὲ ἕψησον με τὸ ὀξείδιν. ἃς βάλῃ καὶ νερὸν καὶ ἄς το πίνῃ ἀντὶ κρασίου. 2 ἄλλο. ⌜τὴν σκύλλαν ἕψησον εἰς τὰ καρβούνια καὶ ἔπαρον τὰ φλούδια τῆς ἕνα ἕνα. καὶ βάφε τα εἰς τὸ μέλι καὶ ὄξος ὀλίγον καὶ ἄς τα τρώγῃ. 3 ἄλλο. πότισον αὐτὸν καὶ τὴν λεγομένην ἱεράν. 4 ἄλλο. ἃς τρώγῃ δὲ ῥαφάνια καὶ τότε ἄς τα ξερᾷ, ἵνα ἐκβάλῃ χυμούς. 5 ἄλλο. ποταμογείτονος σπόρον κοπάνισον καὶ ψιλὸν ποιήσας πότισον τὸ πρωΐ μετὰ οἴνου χλίου. 6 ἄλλο. ἐὰν ἔχῃ πολὺν καιρὸν ὅπου οὐδὲν ἐφλεβοτομήθῃ ἃς φλεβοτομηθῇ τότε. 7 ἄλλο. εἰς δὲ τὸ στῆθος τοῦ θὲς αἶρας ἄλευρον καὶ νερὸν καὶ μάλαξέ το καὶ βάλε καὶ σῦκα κοπανισμένα καὶ μάλαξον αὐτὰ

---

3 πρὸς] This chapter is listed in the pinax to L. 8 πρὸς] This chapter is listed in the pinax to L.

---

19 τῶν] τὰ A 1 του] om. A 2 μάριν] μα.. A 3 πρὸς] om. A 3 ὅσοι] ὅτου B 3 μέγα] μεγάλης AB 3 βραγχιάζῃ] scripsi βραχιάζει L βραχιάζουσι A βραχνιάζουσιν B 4 ἡ] om. B 4 τροφὴ] φωνὴ A 4 του] αὐτοῦ A 5 τούτοις ὅμοια] ἄλλα δριμέα AB 6 καὶ] τὸ add. A 6 καταστατὸν] ἀκαταστατὸν B 7 ῥοφητὰ] ῥοφιτὰ A 8 πρὸς] om. AB 8 εὑρεθῶσιν καὶ] om. A 9 γένηται] γίνεται A 10–11 σκυλλοκρέμμυδον] σκυλλοκρόμμυδον AB 11 παρὰ τῶν ἰατρῶν] om. A 11 σκύλλαν] σκῦλα. A 11–12 παρὰ δὲ τῶν λαγγιβάρδων σκα[λ]] om. A 12 σκα[λ]] λέγεται ἄσκαλα B 12 δὲ] om. B 12 ὀξείδιν] καὶ add. A 12 βάλῃ] βάνῃ B 13 κρασίου] κρασὶν A 13 σκύλλαν] σκύλην A 14 ἕνα] καὶ add. AB 15 τα] το B 16 δὲ] om. A 16 ῥαφάνια] ῥεφάνια AB 19 ὅπου] ὅτι A 19 ἐφλεβοτομήθῃ] ἐφλεβοτόμησεν AB 19 φλεβοτομηθῇ] φλεβοτομη[θ] A φλεβοτομήσει B 20 τότε] om. AB 20 δὲ] om. A 20 τοῦ] αὐτοῦ B 20 αἶρας] ἔρας A 20 ἄλευρον] ἀλεύριν AB

καὶ θὲς εἰς τὸ στῆθος του ▷om. B με τὸ πανίν.⌝ 8 ἄλλο. ὑσσώπου καρπὸν ποίησον ψιλὸν καὶ πότισον μετὰ οἴνου χλίου καὶ ⌜ἄλειφε τὸ στῆθος του om. B◁ ποτὲ μὲν ἀνηθέλαιον ποτὲ δὲ πηγανέλαιον.⌝ 9 ἄλλο. βράσον ἀψίνθιν με τὸ κρασὶν καὶ μίξον με τὸ καστόριον τὸ τριμμένον καὶ ποίησον ὡσεὶ κοκκία πέντε. καὶ πότιζον αὐτὸν κατὰ πρωΐ. 10 ἄλλο. ⌜ἀμμωνιακὸν τρίψον καὶ πότισον μετὰ οἴνου ζεστοῦ. 11 ἄλλο. ἂς ἔνι δὲ ζεστὰ ὅσον δύναται καὶ ψυχρὰ μὴ πίνῃ μηδὲ ἄνεμος τὸν κρούῃ.⌝

136 [123] om. L ρλϛʹ ⌜πρὸς ὅταν τις πτύῃ αἷμα πολὺ καὶ εἰς μίαν. σημεῖον ἔνι ὅτι ἐκόπη φλέβα ὥσπερ ἐάν τις πέσῃ ἀπὸ ὑψηλοῦ τόπου καὶ ξεράσῃ αἷμα ἢ πτύσῃ· πότισον αὐτὸν βῶλον μετὰ νερόν. εἰ δὲ ὅσον ἔνι καταρισμένος καὶ πτύσῃ αἷμα φλεβοτόμησον αὐτόν. 2 ἄλλο. ἐὰν δὲ πτύσῃ αἷμα σημεῖον ἔνι ὅτι ἀπὸ τῶν ἀναπνευστικῶν του ἔνι τὸ αἷμα.⌝ καὶ φλεβοτόμησον αὐτὸν. βρέχε καὶ σπόγγον καινούργιον εἰς τὸ ὀξείδιν τὸ χλίον καὶ πυρίαζε τὸ στῆθος. 3 ἄλλο. μαλία πρόβεια ἀπὸ τῶν ὀρχιδίων οὕτως ὡς εἶναι ἄπλυτα βρέξον εἰς ῥοδέλαιον χλίον καὶ πυρίαζε τὸν τόπον. 4 ἄλλο. ⌜μυρσίνης φύλλα βράσον με τὸ ὀξείδιν καὶ πυρίαζε τὸ στῆθος. 5 ἄλλο. οἶνον αὐστηρὸν χλίον βρέχε τὸν σπόγγον καὶ πυρίαζε. 6 ἄλλο. εἰς δὲ τὸ στῆθος θὲς ἔμπλαστρον τὸ λεγόμενον παρὰ τῶν ἰατρῶν δι' ἰτέων.⌝ 6 ἄλλο. ▷om. A πότιζε δὲ αὐτὸν βῶλον τριμμένον με τὸ νερὸν τὸ χλίον. om. A◁ 7 ἄλλο. ⌜λημνίαν σφραγίδα τρίψον καὶ πότισον ▷om. B με τὸ κρασὶν νῆστιν.⌝ 8 ἄλλο. ῥοϊδίου ζωμὸν πότιζε om. B◁ τὸν ἄρρωστον. 9 ἄλλο.

---

9 πρὸς] This chapter is listed in the pinax to L.

---

21 μάλαξέ το] μάλαξον A 1 θὲς] θέτε B 1 του] om. A 1 πανίν] πλὴν δὲ θές το πολλάκις φοραῖς add. A 3 του] om. A 4 μίξον] μίξε το AB 5 καὶ ποίησον] om. AB 7 καὶ] om. A 9 πρὸς] om. A 9 τις πτύῃ] τὸ αἷμα πτύῃ τις A 9 αἷμα πολὺ] transp. B 9 αἷμα] om. A 9 καὶ εἰς μίαν] ποίησον οὕτως ἐὰν B 10 σημεῖον ἔνι] transp. B 10 ἐάν] ὥσπερ B 10 τις πέσῃ] transp. AB 10 ὑψηλοῦ] ψυλοῦ B 11 μετὰ] με τὸ AB 12 καὶ πτύσῃ] τὸ πτύσῃ A πτύσῃ B 12 αὐτόν] καὶ ἔστι ὀφέλιμον πάνυ καὶ καλὸν B 13 τῶν ἀναπνευστικῶν] τὸ ἀναπνευστικὸν A 14 του] om. B 16 ὡς] ἂς B 16 εἶναι] ἔνι A 17 τόπον] πλὴν τὸ ῥοδέλαιον χλίον add. A 18 βράσον] βράσε A 20 τὸ] τὸν A 20 δι' ἰτέων] scripsi διιτθῶν M διιτεὸν B 21 αὐτὸν] αὐτοὺς B 21 τριμμένον] om. B 22 λημνίαν] λιμνίαν A ἡμνείαν B 22 τρίψον] τριμμένην AB 22 καὶ] om. AB 22 πότισον] πότιζε A 23 νῆστιν] κατὰ πρωΐ A 23 ζωμὸν] τὸν ζωμὸν A

βράσον φοινίκια με τὸ νερὸν καὶ τὸν ζωμὸν ἔχε τρίβον τριμμένον κρομμύδιν καὶ πότιζον αὐτόν.

137 [112] ρλζ′ ⊳om. L πρὸς πλευρίτιδα· ⌜ἐὰν ἔνι ὁ πόνος καὶ κρατεῖ τὸν εἰς τὸν ὦμόν, φλεβοτόμησον αὐτόν. εἰς δὲ τὴν κοιλίαν ἢ εἰς τὰ σπλάγχνα ἔνι ὁ πόνος πότισέ τον βοήθημα. 2 ἄλλο. βράσον τὸ λεγόμενον κενταύριον παρὰ δὲ τῶν ἰδιωτῶν λέγεται ἀθάλις. βράσον σὺν αὐτῷ καὶ πίτυρα καὶ τὸν ζωμὸν ἐκείνων ἰάτρευσον καὶ κλυστήριν μετὰ μέλιτος καὶ ἐλαίου καὶ ἅλατος. 3 ἄλλο. φρύξε κέγχριν εἰς τὸ τζουκάλιν καὶ πυρίαζε τὸν τόπον.⌝ 4 ἄλλο. μαλία ἄπλυτα βρέξον εἰς τὸ κρασὶν τὸ χλίον καὶ χρηστέλαιον καὶ πυρίαζε τὸν τόπον. 4 ἄλλο. ⌜βάλε τὰ λεγόμενα χαμαίμηλα καὶ ἀψίνθιον καὶ ἰσχάδας καὶ νερὸν καὶ ἃς βράσουν. καὶ βάλε τὸ νερὸν ἐκεῖνο εἰς τὸ κριθάλευρον καὶ εἰς τοῦ λινοσπόρου τὸ ἀλεύριν καὶ βράσε τα καὶ ἀπόπλυνέ τα εἰς τὸ πανὶν καὶ θές το ἐκεῖ ὅπου πονεῖ. ἀφ' ὅτου δὲ διάβῃ ἡ εὐδόμη ἡμέρα, ποίησον ἐλεκτὸν μετὰ ἀμυγδάλων γλυκέων καὶ τῆς τζουκνίδας τὸν σπόρον τριμμένον φρύξον κριθάριν καὶ πυρίαζε τὸν τόπον. 5 ἄλλο. ἔμπλαστρον τὸ λεγόμενον σάμψυχον ἐπίθες εἰς τὸν τόπον ἔνθα γίνεται ἡ ὀδύνη εἰς τὸ πλευρόν.⌝ 6 ἄλλο. τὸ ἔμπλαστρον τὸ λεγόμενον παρὰ τῶν ἰατρῶν πηγανερὰ ἐπίθες καὶ αὐτὸ εἰς τὸ πλευρὸν ἔνθα ἔνι ἡ ὀδύνη. 6 ἄλλο. ⌜ἐὰν δὲ ποιήσεις αὐτὰ ὅλα καὶ πάλιν ἔνι ἡ ὀδύνη εἰς τὸν τόπον ἐκεῖνον ἔνθα ἦν πρότερον συκίασέ τον ἐκεῖ ἐπάνω ὅπου πονεῖ πλὴν δὲ ἐπιτηδείᾳ. 7 ἄλλο. ἀνάλυσε χοίρειον om. L◁ἀξούγγιν καὶ κηρὶν λυτὸν καὶ βάλε τεάφιν

---

23 τὸν ἄρρωστον] om. A 1 βράσον] βράσε AB 1 τὸν ζωμὸν ἔχε τρίβον] καὶ ἔπαρον τὸν ζωμὸν καὶ ἔχε ἕτοιμον A τὸν ζουμὶν τους ἔχε ἕτοιμον B 3 πρὸς πλευρίτιδα] θεραπεία πλευρίτιδος A 4 ἢ] καὶ A 5 ἔνι ὁ πόνος] om. A 5 βράσον] βράσε AB 6 ἰδιωτῶν] ἰατρῶν B 6 ἀθάλις] αἰθάλη AB 6–7 βράσον] αὐτὸ βράσον B 7 σὺν αὐτῷ καὶ] μετὰ A καὶ B 7 τὸν ζωμὸν] τὸ ζουμὶν B 7 ἐκείνων] ἐκεῖνον A 7 ἰάτρευσον] ἰάτρευε B 7 καὶ] om. AB 8 κλυστήριν] κλυστήριον A καὶ add. B 9 κέγχριν] κεγχρὸν B 9 τζουκάλιν] τζυκάλιν A 10 πυρίαζε] ἄλειφε A 12 ἐκεῖνο] ὅλον add. B 13 εἰς] om. B 13 βράσε τα] ἃς βράσουν A 14 το] τα B 14 ὅπου πονεῖ] εἰς τὸν πόνον A 16 φρύξον] ῥύξον M 17 ἔμπλαστρον] δὲ add. B 17 λεγόμενον] παρὰ τῶν ἰατρῶν B 18 τὸ] καὶ τὸ B 19 παρὰ τῶν ἰατρῶν] om. A 21 ἔνι ἡ ὀδύνη] ἡ ὀδύνη ἐκείνη ἔνι B 21 ἦν πρότερον] ἔνι καὶ πρήν A 22 πλὴν δὲ] om. B 22 ἐπιτηδείᾳ] ἐπιτήδια M δὲ add. B 23 ἀξούγγιν] ..κὴν L 23 κηρὶν] scripsi ..ρὴν L κηρὸν Σ

καὶ κυπρινέλαιον. οὕτως γὰρ ὀνομάζεται καὶ βούτυρον. καὶ ἐρεβινθίνην τριμμένην καὶ ἀνάμιξέ τα καὶ μάλαξον αὐτὰ καὶ ποίησον ὡς κηρωτὴν καὶ θέτε το εἰς τὸν τόπον ἐκεῖνον ἔνθα ἔνι ἡ ὀδύνη τοῦ πλεύρου. 8 ἄλλο. βάλε εἰς μελίκρατον πήγανον καὶ ὕσσωπον καὶ καλαμίνθην καὶ τὸ λεγόμενον ἶριν. καὶ τὴν λεγομένην στοιχάδα καὶ τὸ λεγόμενον χαμαιπίτυν καὶ ῥέον τριμμένον ὀλίγον πλὴν ἀφότου βράσουν τὰ βότανα ὕστερον βάλε τὸ ζωμὶν τὸ ῥέον ὡσεὶ κοκκία ἓξ καὶ πότιζε καθ' ἑκάστην ἡμέραν ὀλίγον τὸ ῥέον †πλέο τερί..τζην δὲ τὸ ζωμὶν ἐκεῖνο.⌝

138 [113] ρλη´ πρὸς πνευμονίας. ⌜πάθος δὲ ἔνι ἡ πνευμονία ὅταν βήχῃ καὶ πτύῃ ἕλκος παχὺ κίτρινον. βρωμεῖ δὲ ποτὲ μὲν ὀλίγον ποτὲ δὲ πολὺ ποτὲ δὲ ὀλιγούτζικον· φλεβοτόμησον αὐτόν. ἀλλὰ πρόσεχε μὴ λειποθυμήσῃ. πότιζε δὲ αὐτὸν χυλὸν γυμνοκρίθου. ἂς τρώγῃ δὲ καὶ σεμιδάλιν. ποίησον δὲ αὐτοὺς θίασον μετὰ ἀμύγδαλα τὰ γλυκέα καὶ δροσάτον καὶ τὸ λεγόμενον ἀπόμελι.⌝ 2 ἄλλο. βράσον ὕσσωπον μετὰ ἰσχάδια καὶ †κάλαμον ...ζαν καὶ νερόν. καὶ ἂς βράσουν καὶ

---

23 τεάφιν] τε...... L 1] τριμμένη .... L τριμμένον Σ 1 γὰρ ὀνομάζεται] ....νομάται L 1–2 ἐρεβινθίνην] scripsi ερεβιν̣....ίνην L τερεβινθίνην Σ 2 μάλαξον αὐτὰ] ....ζον .υτὰ L 3 ὡς] .. L 3 θέτε το] θὲς M 3 ἐκεῖνον] om. A 3 ἔνθα] om. M 3 ἔνι ἡ ὀδύνη] καὶ ὁ πόνος B 4 τοῦ πλεύρου] om. A 4 μελίκρατον] μέλια ...υτὸν L τὸ μελίκρατον AB 5 καλαμίνθην] κα..ίνθ.. L 5 τὴν] τὸ A 5 λεγομένην] ...... L λεγόμενον A 5–6 στοιχάδα] scripsi στιχάδα LMB στιχάδαν A 6 τὸ] τὸν A 6 χαμαιπίτυν] scripsi χαμαιπίτην MB ...αιπι.... L χαμεπίτην A 7 ἀφότου] ..του L 7 βράσουν] καὶ ἔνι ἐπιτήδεια add. B 7 ὕστερον] δὲ B 7 τὸ ζωμὶν] εἰς τὸν ζωμὸν τῶν βοτάνων Σ 7 ῥέον] τριμμένον add. Σ 7 ὡσεὶ] καὶ ἂς εἶναι ὡς B 8 ὀλίγον] ὀρίον M 8 ῥέον] ῥε... L 8–9 †πλέο τερί..τζην] πλεωτερίτζιν M πλεοτερίτζην A πλεοτέριν B 9 δὲ] om. A 9 τὸ ζωμὶν] τὸν ζωμὸν AB 9 ἐκεῖνο] ἐκεῖνον A 10 πρὸς πνευμονίας] scripsi ...πνευμονίας L πρὸς περιπνευμονίας M εἰς πάθος τὸ λεγόμενον πνευμωνίαν A πρὸς πάθος τὸ λεγόμενον παρὰ τῶν ἰατρῶν περιπνευμονίαν ποίησον οὕτως B 10 πάθος] πάθους B 10 ἡ πνευμονία] scripsi ἡ πνευ...ία L τοῦ καλουμένου πνεύμονος M τοῦ καλούμενον πνεύμωνος AB 11 βήχῃ] .. L 11 κίτρινον] ...τρίνον L 11 ποτὲ] ...τε L 12 πολὺ] πολλὰ B 12 φλεβοτόμησον] ...... τὸ μὴ .. L 12 πρόσεχε] βλέπε MA 13 λειποθυμήσῃ] ...θ̣υ̣μήση L 13 αὐτὸν] scripsi αὐ..ν L τοὺς τοιούτους Σ 13 χυλὸν γυμνοκρίθου] χυλ....ν̣οκρίθου L γυλὸν γυμνοκρίθου M 14 σεμιδάλιν] ἐμηδ̣.....ν L 14 θίασον] θᾶσον A θέασον B 14 μετὰ ἀμύγδαλα] scripsi μετ.. ἀμιγ..αλα L μετὰ ἀμυγδάλων Σ 14 τὰ γλυκέα] γλυκέων Σ

ἄς τα πίνῃ. 3 ἄλλο. ⌜θέτε δὲ εἰς τὴν θώρακα τοῦ ἀρρώστου ἤτοι εἰς τὸ στῆθος κερωτὴν τὴν λεγομένην πηγανερὰν ἤτοι βάλε κηρὸν καὶ ἂς λυθῇ. ὕστερον δὲ βάλε χήνειον ἀξούγγιν ἢ ὄρνιθος ἢ ἐλάφιον μυελόν.⌝ ἀντὶ χρηστελαίου βάλε πηγανέλαιον μάλαξον καὶ τάραξον αὐτὰ ἔστ' ἂν λυθῶσιν ὅλα. ἐὰν δὲ οὐκ ἔχῃς πηγανέλαιον, βάλε τὸ λεγόμενον κυπρινέλαιον ἢ χρηστέλαιον.

139 [114,115] ρλθ' ⌜ἐμπύημα λέγεται, ὅταν ποιήσει ἡ πλευρίτις πολὺν καιρὸν καὶ βήχῃ καὶ πτύῃ ταχέως ὥσπερ ἕλκος καὶ βρωμεῖ τὸ πτύσμα· βάλε ὕσσωπον καὶ θύμον καὶ νερὸν καὶ ἂς πίῃ τοῦτο ὁ ἄρρωστος. 2 ἄλλο. βάλε μέλι καὶ ἂς βράσῃ καὶ βάλε τριμμένον γλυκορίζην καὶ σησάμιν τριμμένον εἰς τὸ μέλι ἀπέσω καὶ τάραξον αὐτὰ καὶ ἂς τρώγῃ κατὰ βραδὺ ὀλίγον ὀλίγον καὶ ἐπάνω ἂς πίνῃ κρασὶν ζεστόν.⌝ 3 ἄλλο. δίδε δὲ καὶ τὴν ἀντίδοτον τὴν λεγομένην διὰ πρασίου καὶ τὴν ἀντίδοτον λεγομένην διὰ ὀρόβων. βάλε μέλι, ἂς βράσῃ καὶ ὕστερον βάλε ἀμύγδαλα γλυκέα καὶ κουκουνάρια εἰς τὸ μέλι καὶ ἄς τα τρώγῃ κατὰ βραδὺ ὀλίγον ὀλίγον καὶ ἐπάνω ἂς

---

16 †κάλαμον ...ζαν] καλαμοκρίθου ῥίζαν M καλαμοκρίνου ῥίζαν AB 1 τα πίνῃ] scripsi τα ..νει L πίνει MB τὸ πίνει A 1 θέτε δὲ] θὲς M θέτε AB 1 τὴν] τὸν MB τὸ A 1 θώρακα] στῆθος A 1 ἤτοι] om. A 1–2 εἰς τὸ στῆθος] om. A 2 κερωτὴν] κέρωτ.. L 2 πηγανερὰν] πηγακρὰν M 2 κηρὸν] κη.... L 3 δὲ] om. A 3 βάλε] om. B 3 ἐλάφιον] ἐλαφίου M ἐλάφου AB 4 μυελόν] καὶ add. Σ 4 χρηστελαίου] ἐλαίου A 4 πηγανέλαιον] καὶ add. B 4 μάλαξον καὶ τάραξον] transp. M καὶ τάραξον καὶ μάλαξον A 5 αὐτὰ ἔστ' ἂν] scripsi αὐτ...ἔστ' ἂν L αὐτὰ να M αὐτὰ ἵνα A ἔστ' ̓̀αν B 5 ὅλα. ἐὰν δὲ οὐκ ἔχῃς] ἀντὶ δὲ M ὅλα ἀντὶ δὲ AB 5 πηγανέλαιον] scripsi πηγανέλαι.. L πηγανέλαιον ἐὰν οὐκ ἔχεις M πηγανελαίου ἂν οὐκ ἔχεις AB 5–6 τὸ λεγόμενον] om. A 6 ἢ] .. L 7 ἐμπύημα] ἐμπτύημα L πρὸς ἐμπτύημα ἐμπτύημα M περὶ ἐμπυήματος ἐμπύημα A πρὸς ἐμπτυήματα· ἐμπτύημα B 7 ὅταν] ὁτ... L 7 ἡ] ὁ A 8 καὶ] κ... L 8 βήχῃ] om. A 8 καὶ] om. A 8 ταχέως] παχέα Σ 8 ἕλκος] ἕλκ... L 9 πτύσμα] του add. MB πτύσμαν του A 9 ὕσσωπον καὶ] ὕσ.... L 9 θύμον] θήμον A 10 ὁ ἄρρωστος] ..ῤρωστος L 11 γλυκορίζην] γλυκόριζον Σ 11 ἀπέσω καὶ] om. A 12 τρώγῃ] αὐτὰ add. M ἀπ' αὐτὸ add. A 12 ὀλίγον] om. A 13 κρασὶν ζεστόν] scripsi κρασὶν .ετόν L ζεστὸν κρασὶν Σ 13 δὲ] om. B 13 τὴν λεγομένην] ..γομένην L 14 διὰ πρασίου καὶ τὴν ἀντίδοτον λεγομένην] om. Σ 14 διὰ] δι' AB 14 ὀρόβων] scripsi, cf. ῥόβων L ῥοίβαν M ὀρώβων A ὀρόβου B 15 ἀμύγδαλα γλυκέα] ἀμίδ̣αλα .λυκέα L 15 κουκουνάρια] κωνὰ M κωνάρια AB 16 τα] om. A

πίνῃ κρασὶν θερμὸν ἀλλὰ μηδὲν δειπνᾷ μηδ’ ὅλως.

140 [115] ρμ′ ⌜ὅπου δὲ ἔλθῃ ἀπὸ τὸ νόσημα τὸ λεγόμενον ἐμπύημα εἰς νόσον τοιαύτην τὴν λέγουσιν φθίσιν οἱ χωρικοὶ ἣν καλοῦσιν οἱ ἰατροὶ ἑκτικὸν νόσημα, ποίησον θεραπείαν τοιαύτην· ἤγουν τὴν λεγομένην ἀντίδοτον τοῦ ῎Εσδρα. θὲς δὲ ἐπάνω εἰς τὸ στῆθος αὐτοῦ κηρωτὰς τὰς λεγομένας διὰ βουτύρου καὶ κυπρίνου καὶ ἔμπλαστρον τὸ λεγόμενον Μνάσεον. 2 ἄλλο. πότιζε δὲ αὐτὸν τὸν ἄρρωστον τὴν ἀντίδοτον τὴν λεγομένην Μιθριδάτην καὶ τὴν θηριακήν.⌝

141 [115] ρμα′ ὅταν δὲ ἔλθῃ καὶ πονῇ τὸ κεφάλιν του καὶ ⌜πυρέσσῃ· ποίησον αὐτὸν χυλὸν με τὸ γυμνόκριθον. καὶ βάλε καὶ σταφίδας χωρὶς τὰ κοκκία καὶ ῝υρικὰ ζίντζιφα καὶ δαμάσκηνα καὶ φοινίκια δύο ἢ τρία. ὕστερον δὲ ἀφ’ ὅτου βράσῃ, βάλε μαστίχην καὶ σάχαρ. εἰ δ’ οὐκ ἔχει σάχαρ, βάλε μέλι ὀλίγον.⌝

16 ὀλίγον] ὀ...γον L om. A 1 κρασὶν] ζεστὸν A 1 θερμὸν] ζεστὸν M κρασὶν A 1 ἀλλὰ] ..αλά L 1 μηδ’] οὐδ’ B 1 ὅλως] μήτε πολὺν μήτε ὀλίγον add. A 2–4 ὅπου δὲ ἔλθῃ ἀπὸ τὸ νόσημα τὸ λεγόμενον ἐμπύημα εἰς νόσον τοιαύτην τὴν λέγουσιν φθίσιν οἱ χωρικοὶ ἣν καλοῦσιν οἱ ἰατροὶ ἑκτικὸν νόσημα] πρὸς ὅταν μεταπέσῃ ἀπὸ τοῦ ἐμπτυήματος καὶ ἔλθῃ εἰς τὸ πάθος τὸ λεγόμενον φθίσιν ὑπὸ τῶν χωρικῶν παρὰ τῶν ἰατρῶν λέγεται ἑκτικὸν νόσημα M 2 ὅπου δὲ] ὅταν B 2 ἔλθῃ] ..θῃ L 2 τὸ νόσημα] τοῦ νοσήματος AB 2 τὸ λεγόμενον] τοῦ λεγομένου A om. B 2 ἐμπύημα] scripsi ἐμπτύ..μα L ἐμπτυήματος AB 3 νόσον τοιαύτην τὴν λέγουσιν] τὴν νόσον τὴν λέγουν AB 3 φθίσιν οἱ χωρικοὶ] transp. A 3 φθίσιν] φθῆ... L 3 ἣν καλοῦσιν] om. AB 4 οἱ] δὲ add. AB 4 ἑκτικὸν νόσημα] scripsi ἑκ..κὸν νόσημα L λέγουσι τοῦτο ἕκτικὸν νόσημα A λέγουν ταύτην ἑκτικὸν νόσημα B 4 ποίησον] δὲ add. A 4 ἤγουν] .... L 5 τοῦ ῎Εσδρα] τοῦ ἔσ.ρα L ῎Εσδραν M 5 δὲ] om. MA 5–6 αὐτοῦ] του MB om. A ἤτοι τὸν θώρακα add. B 6 τὰς λεγομένας] om. A 6 διὰ βουτύρου] διαβ...τύρου L 7 πότιζε] πότιζον A 7–8 τὸν ἄρρωστον] om. A 8 ἀντίδοτον τὴν λεγομένην] transp. B 8 Μιθριδάτην] scripsi ἡμιεράτην L ἡμιθριδάτειον M ἡμιθριδάτινον A ἡμιεράτην B 9 τὸ κεφάλιν] τὴν κεφαλὴν MA ἡ κεφαλὴ B 9 πυρέσσῃ] πυρέξει M 10 χυλὸν] χ..λὸν L 10 με τὸ] με τὸν A 10 σταφίδας] ..φίδας L 11 τὰ] τῶν B 11 κοκκία] τῶν add. Σ κοκκίων B 11 ῝υρικὰ] ἤτοι add. Σ 11 ζίντζιφα] ζίζιφα M τζίτζιφα A τζίντζινφα B 11 δαμάσκηνα] δαμάσκινα A 12 βράσῃ] βράσουν M 12 μαστίχην] μαστίχιν MB 13 εἰ δ’ οὐκ ἔχει σάχαρ, βάλε] ἢ A βάλε B 13 ἔχει] ἔχεις M 13 μέλι ὀλίγον] transp. B

9 ὅταν] This chapter is in Σ part of the preceding chapter.

142 [115] ρμβ′ ⌜φθίσις δὲ λέγεται, ὅταν τὸ πλεμόνιν τοῦ ἀνθρώπου ποιήσῃ ἕλκος καὶ πτύῃ ἕλκος καὶ βρωμῇ καὶ μαραίνεται⌝ καὶ †πτ-...μωνεῖ τὸ σῶμα καὶ πτύῃ ἕλκος με τὸ ▷om. Lαἷμα, ὡσὰν ἀφρῶδες. ὅταν τοιοῦτον πτύῃ σημεῖον ἔνι ὅτι ὁ πνεύμων του σήπεται. ποίησον δὲ θεραπείαν τοιαύτην· ἤγουν ἀειδάρας γάλα καὶ γυναικεῖον ἂς πίνῃ καθὼς ὁ Γαληνὸς λέγει καὶ γράφει. ⌜θὲς δὲ ἐπάνω εἰς τὸν θώρακα κηρωτὰς διὰ βουτύρου.⌝ καὶ κηρωτὴν ὅπου γίνεται δαφνέλαιον ἀντὶ χρηστελαίου βάλε καὶ ποίησον καὶ βάλε κηρὶν καὶ ἂς λυθῇ. καὶ ὕστερον βάλε ἀντὶ χρηστελαίου δαφνέλαιον. 2 ἄλλο. ⌜ἄλειφε δὲ τὸν τοιοῦτον εἰς τὸ στῆθος πηγανέλαιον, ἀνηθέλαιον. ὕστερον δὲ θὲς καὶ τὸ ἔμπλαστρον τὸ λεγόμενον παρὰ τῶν ἰατρῶν Μνάσεον. 3 ἄλλο. ▷om. Bπότιζε δὲ αὐτοὺς τὸ λεγόμενον μελίκρατον με τὸ καταστατόν. ἢ ἂς βράζῃ γλυκόριζον με τὸ νερὸν καὶ ἄς το πίνῃ.⌝om. B◁ 4 ἄλλο. δὸς δὲ αὐτὸ καὶ βόϊον γάλα ἢ αἴγας. πλὴν ἄφες τὸ γάλα τῆς ἀειδάρας καὶ τῶν γυναικῶν. 5 ἄλλο. βάλε μέλι καὶ ἂς βράσῃ καὶ ἀμύγδαλα γλυκέα τριμμένα καὶ κωνάρια καὶ πιστάκια καὶ κρόκον τριμμένον καὶ ἀμύγδαλα πικρὰ καὶ σησάμιν καὶ φοινίκων σάρκας κοπανισμένας. 6 ἄλλο. ⌜πότιζε δὲ τοὺς τοιούτους τὴν ἀντίδοτον τὴν λεγομένην ἀθανασίαν καὶ τὴν λεγομένην ἰσόθεον καὶ τὴν λεγομένην ἀντίδοτον τοῦ Ἔσδρας καὶ τὴν θηριακὴν ὡσεὶ κοκκία τρία ἢ τέσσαρα με τὸ κρασὶν τὸ ὀλίγον τὸ πρωΐ.⌝om. L◁

---

1 φθίσις] πρὸς φθίσιν. φθίσις MB 1 δὲ] om. MA 1 πλεμόνιν] scripsi πλεμον.. L πλεύμονι MB φλεμόνιν A 2 ἕλκος] om. A πονῇ add. B 2–3 †πτ...μωνεῖ] πτενέβη M πτενεύει A πτενώνῃ B 3 σῶμα] του add. Σ 3 με τὸ] μ.. L με τὸν A μετὰ B 3 αἷμα] αἵματος καὶ B 4 ὁ] om. B 5 δὲ] οὖν B 6 δὲ] om. AB 6–7 τὸν θώρακα] τὸ στῆθος A 7 κηρωτὰς] κερωτὰς B 7–8 καὶ κηρωτὴν ὅπου γίνεται δαφνέλαιον ἀντὶ χρηστελαίου βάλε καὶ ποίησον] καὶ βάλε δαφνέλαιον ἀντὶ χρηστελαίου A 8 βάλε] om. A 8 κηρὶν] κηρὸν AB 9 ὕστερον βάλε ἀντὶ χρηστελαίου δαφνέλαιον] ἄλειφε A 10 πηγανέλαιον] ἢ add. B 10 ἀνηθέλαιον] om. A 11 ἔμπλαστρον] om. AB 11 ἰατρῶν] ἔμπλαστρον add. AB 14 αὐτὸ] αὐτὸν A 14 βόϊον] βόειον A 14 αἴγας] αἰγὸς A αἴγιον B 14–15 πλὴν ἄφες τὸ γάλα τῆς ἀειδάρας καὶ τῶν γυναικῶν] om. A 15 τῆς ἀειδάρας] τοῦ ἀειδάρου B 16 κωνάρια καὶ] om. A 18 τοιούτους] αὐτους A 19 λεγομένην] δι' ὀρόβον πάλιν δὲ πότιζε τούτους τὴν ἀντίδοτον τὴν λεγομένην add. B 20 καὶ] τὴν λεγομένην θεοδώρητον καὶ add. AB 20 Ἔσδρας] Ἔσδρα AB 20 ὡσεὶ] om. A 21 τὸ] om. A

143 [116] om. L ρμγʹ ⌜ὅταν ἀπὸ πυρετοῦ πρῶτον να θυμωθῇ καὶ οὕτως να πυρέξῃ ὁ ἄνθρωπος· ἀπὸ τῆς πρώτης ἡμέρας λοῦε καὶ ἄλειφε τὸ σῶμα αὐτοῦ ὅλον χρηστέλαιον καὶ ἀφ' ὅτου τὸν λούσῃς τρέφε τον⌝ καὶ ἐπὶ τὴν αὔριον ἅς ποιῇ οἷαν δουλίαν ἔχει. 2 ἄλλο. ⌜τοιαύτην δὲ θεραπίαν ποίησον ὅταν τις λυπηθῇ καὶ πυρέξῃ ἀπὸ φροντίδος ἢ ἀπὸ κόπου ἢ ἀπὸ ἡλίου πύρωμα. ποίησον τοιαύτην θεραπείαν οἷαν ἐποίησας ὅταν ἐθυμώθη.⌝

144 [117] om. L ρμδʹ πρὸς τριταῖον· ⌜ὁ τριταῖος πυρετός ἐστιν ὅταν μίαν ἡμέραν ἔχῃ ἄνεσιν τὴν δὲ ἄλλην πυρρέσῃ.⌝ καὶ δεῖ ἐπὶ τῶν τοιούτων ποιεῖν θεραπείαν τοιαύτην· ἤγουν ⌜μαρούλια ἰντύβια με τὸ ὀξείδιν⌝ καὶ ὕδωρ καὶ ὀξύγαλα. 2 ἄλλο. ἰατρεύε τον καὶ κλυστήριν. ⌜ἤτοι ἔλαιον μέλι ἅλας μολόχια⌝ καὶ πίτυρα βρασμένα τὸν ζωμὸν ἐκεῖνον βάλε ἀντὶ θερμοῦ ὕδατος. 3 ἄλλο. ⌜κρασὶν μὴ πίνῃ μήτε παστὰ τρώγῃ μήτε κρέας μήτε ὀψάριν μήτε ἄλλο τὶ παστόν. 4 ἄλλο. πότιζε δὲ τοὺς τοιούτους σελίνου ῥίζαν καὶ ἀνήθου ῥίζαν καὶ ἀγρίαν ῥίζαν πλυμένην καὶ κιχώρια καὶ εὐπατόριον καὶ ἵππουριν τὸ λεγόμενον πολυκόμπιν. καὶ δενδρολίβανον καὶ ἅς βράσουν καὶ ἅς πίνῃ αὐτὸ⌝ ἀντὶ νεροῦ. ἀφ' ὅτου δὲ βράσῃ ὕστερον βάλε μαστίχιν καὶ ⌜ἀντὶ μέλιτος σάχαρ. πότιζε δὲ τοὺς τοιούτους κατὰ πρωΐ

1 ὅταν] This chapter is listed in the pinax to L. 8 πρὸς τριταῖον] This chapter is listed in the pinax to L.

1–2 πυρετοῦ πρῶτον να θυμωθῇ καὶ οὕτως να] θυμοῦ AB 2 πυρέξῃ] πυρρέξει A 2 λοῦε] λούεται A λοῦσε τον B 3 αὐτοῦ] τοῦ B 3 τὸν λούσῃς] λουθῇ A 4 τρέφε] τρίφε A 4 ἐπὶ] om. B 4 ἅς ποιῇ οἷαν δουλίαν ἔχει] ἔστιν ὑγιῆς A 4 οἷαν] ἤτοι B 5 τοιαύτην δὲ] τὴν αὐτὴν A 5 θεραπίαν] θεραπείαν AB 6–7 τοιαύτην θεραπείαν οἷαν ἐποίησας ὅταν ἐθυμώθη] τὴν ἄνωθεν θεραπείαν καὶ ὑγιαίνει A καὶ ἰάτρευσέ τον οὕτως ἰάτρευσον καὶ τούτους τοὺς πυρέξαντας ἀπὸ ὑποθέσεως τοιαύτης B 8 πρὸς τριταῖον] om. A πρὸς τριταίον πυρετόν B 8 ὁ] om. B 8 ἐστιν] ἔνι AB 9–10 τῶν τοιούτων] τὸν τοιοῦτον M 10 ἤγουν] ἵνα τρώγει A 10 μαρούλια] καὶ B 11 ἰατρεύε τον] ἰάτρευσον A ἰάτρευσέ τον B 11–12 κλυστήριν] κλυστήριον A 12 μέλι ἅλας] transp. B 12 μολόχια] μωλόγχια B 13 ἐκεῖνον] ἐκείνων AB 13 κρασὶν] δὲ add. B 14 τὶ] τίποτας B 15 τοὺς τοιούτους] αὐτοὺς A 15 σελίνου ῥίζαν] σελινόριζα B 15 καὶ ἀνήθου ῥίζαν] om. AB 16 καὶ ἀγρίαν ῥίζαν] om. B 16 ἀγρίαν] ἀγρίας A 16 πλυμένην] πλημμένην A 16 εὐπατόριον] ἀπακτώριον A 16 ἵππουριν] ἵππουρον B 17 καὶ] νερὸν καὶ add. AB 18 ἅς] το add. AB 18 αὐτὸ] om. A

ἀψίνθιον βρασμένον με τὸ νερόν. 5 ἄλλο. ὅτε δέ ἐστιν καιρὸς τοῦ λουσθῆναι λοῦέ τον εἰς θερμὸν γλυκέων ὑδάτων⌝ μὴ τοιαύτα ἅτινα ὀνομάζουσιν θερμά. παρὰ δὲ τοῦ Γαληνοῦ καλοῦνται ἄλβουλα. 6 ἄλλο. ⌜ἐὰν δὲ ὁ τοιοῦτος ἄρρωστος εὑρεθῇ καὶ ἀγαπᾷ τὸ λουτρὸν καὶ δεύτερον τὴν ἡμέραν ἐὰν τοῦ λούης ἢ ἡμέραν παρ' ἡμέραν τίποτε οὐ μή τον βλάψης. 7 ἄλλο. πότιζε δὲ τοὺς τοιούτους καὶ ὑδροσάτον καὶ τὸ λεγόμενον ζουλάπιν μετὰ ψυχροῦ ὕδατος.⌝

145 [118] [om. L]ρμε' ⌜πρὸς τεταρταίον· ὁ δὲ τεταρταῖος πυρετός ἐστὶν ὅτε δύο ἡμέρας πυρέσσῃ μίαν δὲ ἔχει ἄνεσιν. γίνεται δὲ ἐπὶ μελαγχολικῷ χυμῷ,⌝ ὅστις ὑπάρχει ξηρὸς καὶ ψυχρός. γίνεται δὲ ὁ χυμὸς οὗτος ἔξω τῶν φλεβῶν. ποίησον δὲ θεραπείαν τοιαύτην· ἤγουν ⌜τὴν ἀρχὴν μὴ φλεβοτομεῖν μήτε καθαίρειν. εἰ δὲ ἔχει πολλὰς ἡμέρας ὅτι οὐκ ἐφλεβοτόμησεν ἂς ἐκβάλῃ ὀλίγον αἷμα καὶ ἐὰν ἴδης τὸ αἷμα ὅτι ἔνι μαῦρον ἂς ῥέῃ πλὴν δὲ βλέπε μὴ λυποθυμήσῃ.⌝ εἰ δὲ ἴδης ὅτι ἔνι ξανθὸν κράτησε τοῦτον εὐθὺς καὶ ⌜δός τον τροφὴν χρήσιμον οἷον ὀψάριν, κρέας πλὴν χλωρὸν⌝ καὶ τὰ ἑξῆς χλωρά. 2 ἄλλο. βράσον βλησκούνιν καὶ τὸ λεγόμενον ἐπίθυμον ἤτοι τὸ ἐπάνω τοῦ θυμαρίου. βράσον με τὸ νερὸν καὶ τὸν ζωμὸν τούτων βάλε καὶ μέλι καὶ ἔλαιον καὶ ἅλας καὶ κρούε τον κλυστήριν. καὶ

8 πρὸς τεταρταίον] This chapter is listed in the pinax to L.

19 δὲ] om. B 1 ἐστιν] om. AB 2 λουσθῆναι] λουθῆναι AB 2 λοῦέ τον] λοῦσε αὐτὸν B 2 ὑδάτων] καὶ add. A 3 ὀνομάζουσιν] λέγουσι A 3 ἄλβουλα] ἄλβουα M 4 ἄρρωστος] ἄνθρωπος A 4 καὶ ἀγαπᾷ] ἀπὸ B 5 τοῦ] τὸν AB 5 ἢ ἡμέραν παρ' ἡμέραν τίποτε] om. B 5 ἡμέραν] ἡμέρα A 5 ἡμέραν] ἡμέρα A 6 καὶ] om. B 7 ζουλάπιν] δρολάπιν A 7 ψυχροῦ ὕδατος] transp. B 8 πρὸς τεταρταίον] om. A πυρετὸν add. B 8–9 ἐστὶν] ἔνι B 9 ὅτε] ἐὰν add. B 9 ἡμέρας πυρέσσῃ] ἡμερέσει M 9 μίαν δὲ] καὶ μίαν A 9 ἐπὶ] ἀπὸ B 10 μελαγχολικῷ] μελαγχολικοῦ B 10 χυμῷ] θυμῷ M χυμοῦ B 10 ξηρὸς καὶ ψυχρός] transp. B 11 δὲ] om. A 12 μὴ] μήτε B 13 οὐκ] οὐ B 13 ἐφλεβοτόμησεν] φλεβοτόμησεν B 13 ἴδης] ὅτι add. B 14 ὅτι] om. B 14 βλέπε] πρόσεχε B 14 λυποθυμήσῃ] λυποθυμήσης B 15 ἔνι] τοῦτο add. B 15 κράτησε] κράτησον A 15 τοῦτον] τούτου M το B 15 εὐθὺς] om. A 15 τον τροφὴν] τον κρασὶν A αὐτοῦ τροφὴν B 16 ὀψάριν] ὀψάρια A 16 χλωρὸν] χλωρὰ A 16 ἑξῆς] τὰ πάντα add. B 17 βλησκούνιν] φλησκούνην A βλισκούνιον B 17 ἐπίθυμον] με τὸ νερὸν add. A ἐπιθύμιον με τὸ νερὸν B 17–18 ἤτοι τὸ ἐπάνω τοῦ θυμαρίου] om. A 18 βράσον με τὸ νερὸν] om. AB 18 καὶ] εἰς add. AB

ἀφότου τὸ ποιήσεις τὸ κλυστήριν αὐτὸ καὶ διαβάση τρεῖς ἡμέρας πότισέ τον τὸ λεγόμενον παρὰ τῶν ἰατρῶν διὰ τριῶν πεπερέων ἢ τὸ λεγόμενον διοσπολιτικόν. 3 ἄλλο. ⌜πέπερι τριμμένον πότιζε τὸν ἄρρωστον αὐτὸν με τὸ χλίον τὸ νερόν.⌝ 4 ἄλλο. πότιζε δὲ τὸν τοιοῦτον ἄρρωστον τοῦτο ἤγουν βάλε ἄσαρ καὶ ἀδίαντον καὶ σέλινον καὶ ἵππουριν τὸ λεγόμενον πολυκόμπιν καὶ ἰχώρια καὶ δενδρολίβανον καὶ εὐπατόριον καὶ νερὸν καὶ ἂς βράσουν. καὶ ἀφ' ὅτου βράσουν βάλε ὕστερον μαστίχιν καὶ ἀντὶ μέλιτος σάχαρ καὶ ἂς το πίνῃ. 5 ἄλλο. ⌜τὸ δὲ σπλάγχνον ἤγουν τὸν σπλῆνα βάλε ἔμπλαστρον τὸ λεγόμενον παρὰ τῶν ἰατρῶν κοπτὸν ἢ τὴν λεγομένην χρυσῆν καὶ τὰ λεγόμενα Πολυάρχια.⌝ 6 ἄλλο. δὸς καὶ ἀντίδοτον τὴν λεγομένην παρὰ τῶν ἰατρῶν διὰ θείου. καὶ ἄλλην ἀντίδοτον τὴν λεγομένην μοῦσαν. 7 ἄλλο. ⌜πρῶτον δὲ κένωσον μελαγχολικὸν χυμὸν διὰ γαστρὸς εἶτα πότισε τον τὴν ἀντίδοτον τὴν λεγομένην θηριακὴν καὶ τὸ φάρμακον τοῦ ὀποῦ τὸ διὰ Κυρηναϊκοῦ σκευαζόμενον. 8 ἄλλο. ἂς τρώγῃ δὲ καὶ τὰ παστὰ οἷον χοίρεια κρέη καὶ ὀψάρια καὶ σινάπιν καὶ ἄλλα θερμαίνοντα οἷον πέπερι σκόρδα⌝ καὶ τὰ ἄλλα τὰ τούτοις ὅμοια. τὸ μὲν νόσημα τοῦτο πολὺν χρόνον κρατεῖ εἰς τὸ σῶμα τοῦ ἀνθρώπου οὐ μὴν θάνατον προσφέρει.

146 [119] om. L ρμϛʹ πρὸς ἀφημερινοῦ πυρετοῦ· ⌜ἐπὶ μὲν τοῦ ἀφημερινοῦ πυρετοῦ κατ' ἀρχὴν μὲν πρόσφερε καὶ πότισον αὐτὸν τὸ λεγόμενον ὀξύμελι καὶ τοῖς διουρητικοῖς φαρμάκοις⌝ ταῦτα οἷον

---

20 πρὸς ἀφημερινοῦ πυρετοῦ] This chapter is listed in the pinax to L.

---

18 τούτων] τοῦτον B 1 τὸ ποιήσεις τὸ κλυστήριν αὐτὸ καὶ] om. A 1 τὸ] om. B 1 τρεῖς] scripsi τὸ εἰς M μετὰ τὸ κλυστήριν τρεῖς A 2 πότισέ] πότιζέ B 2 ἰατρῶν] τὸ add. B 4 αὐτὸν] om. A 4 τὸ] om. AB 4–5 τὸν τοιοῦτον ἄρρωστον] καὶ A 6 τὸ λεγόμενον] ἤγουν A 6 πολυκόμπιν] πολύκομπον A 6 ἰχώρια] κιχώριν B 7 εὐπατόριον] ἀπακτώριον A 8 μαστίχιν] μαστίχην A 8 σάχαρ] σάραχ M 8 το] om. A 9 τὸ δὲ] εἰς δὲ τὸ A 9 βάλε] θέτε B 12 διὰ θείου] δια θεῖα M 13 κένωσον] τὸν add. B 15 τὸ] om. A 15 Κυρηναϊκοῦ] Κυριναϊκοῦ MAB 16 καὶ] om. A 17 σκόρδα] σκόροδα A 17 τὰ] om. B 18–19 τὸ σῶμα τοῦ ἀνθρώπου] τὸν ἄνθρωπον A 19 μὴν] δὲ add. A μὴ B 20 πρὸς ἀφημερινοῦ πυρετοῦ] om. A 20 ἀφημερινοῦ πυρετοῦ] ἀφημερινὸν πυρετὸν B 20 μὲν] δὲ B 21 αὐτὸν] om. B

βάλε σελίνου ῥίζαν καὶ εὐπατόριον καὶ κάπνιον καὶ νερὸν καὶ κιχωρίων τὰς ῥίζας, καὶ ἀφ' ὅτου βράσουν ὕστερον βάλε ἀντὶ μέλιτος σάχαρ καὶ ἄς το πίνῃ. 2 ἄλλο. ▷om. A⌈πρῶτον δὲ ἂς πίνῃ τὸ λεγόμενον ὀξύμελι, εἶτα δίδε αὐτῷ om. A◁ τρώγειν ῥεφάνια καὶ ἂς ξερᾷ τὴν τροφὴν καὶ τὰ ῥεπάνια⌉ ὅπως ἐκβάλῃ χυμοὺς τινὰς γλίσχρους καὶ παχέους οἵτινες ποιοῦσιν τὸν ἀφημερινόν. ἂς τρώγῃ δὲ καὶ κύμινον καὶ πέπερι καὶ ἄλλα θερμαίνοντα. ψυχρὸν δὲ μὴ πίνῃ μηδὲ ψυχρὰ τρώγῃ ⌈καὶ ἂς ἔνι ζεστὰ πολλά.⌉

147 [120] ρμζ' ▷om. L⌈πρὸς νόσημα λεγόμενον πυρεκτική ὅταν γένηται νόσημα τοιοῦτον λοιμικὸν ὃ καλεῖται παρὰ τῶν χωρικῶν πυρεκτική, ποίησον θεραπείαν τοιαύτην· πρῶτον ἰατρεύε τον κλυστήριν. καὶ ἐὰν ἔχῃ πολλὰς ἡμέρας ὅτι οὐκ ἐφλεβοτόμησεν ἂς ἀποχύσῃ ὀλίγον αἷμα. βάλε δὲ σέλινα καὶ ἵππουριν τὸ λεγόμενον πολυκόμπιν καὶ ἰχώρια καὶ ἂς βράσουν με τὸ νερὸν καὶ ἂς πίνῃ.⌉ 2 ἄλλο. πότιζε αὐτὸν καὶ τῶν χαμαιμήλων τὸν ζωμὸν με τὸ νερὸν ἐκεῖνο ὅπου βράζῃ. 3 ἄλλο. ⌈ἐὰν δὲ καίηται ἐπέσω εἰς τὰ κούφη του πολλὰ πολλά, πότισέ τον κρύον νερὸν ὅσον θέλει ▷om. B ἵνα σβέσῃ τὴν φλόγα τὴν ἔχει.⌉om. B◁ 4 ἄλλο. ▷om. AB βράσε καὶ νερὸν om. AB◁ ▷om. A καὶ ἀφ' ὅτου βράσῃ ἂς γένηται κρύον πολλὰ καὶ ἂς πίνῃ. om. A◁ 5 ἄλλο. ▷B πότιζε τὸ λεγόμενον ἀπόμελι καὶ τὸ χυλάριν με τὸ γυμνόκριθον καὶ τὰς σταφίδας ἄνευ τῶν κοκκίων τους καὶ βάλε καὶ τζουκνίδας σπόρον

22 διουρητικοῖς] διουρικοῖς M διοριτικοῖς B 1 βάλε] βάλετε B 1 καὶ] ἀδιάντον ἄσαρ καὶ ἵππουριν καὶ A ἀδίαντα ἄσαρ τὸ λεγόμενον ἵππουριν τὸ λεγόμενον πολυκάμπη καὶ B 1 εὐπατόριον] ἀπακτώριον A 1 κάπνιον] κάπνεον B 3 σάχαρ] ζάχαρ A 4 αὐτῷ] αὐτὸν τοῦ M 4 τρώγειν] ἂς τρώγῃ A om. B 4 ῥεφάνια] ῥεπάνια AB 5 τὴν] καὶ τὴν AB 6 παχέους] παχεὶς B 6 ἀφημερινὸν] ἀμφημερινόν M 6 δὲ] om. AB 7 ἄλλα] ἄλας καὶ ὅσα B 7 ψυχρὸν δὲ μὴ πίνῃ] εἶτι δὲ ψυχρὸν μὴ πίνῃ A 8 καὶ] om. B 8 ἔνι] βλεπετε ἵνα ἔνι A δὲ καὶ add. B 9 πρὸς νόσημα λεγόμενον πυρεκτική] om. A 9 νόσημα] τὸ add. B 9 πυρεκτική] πυρεκτικῷ B 9–10 ὅταν γένηται νόσημα] om. A 10 τοιοῦτον] τὸ B om. A 10 λοιμικὸν] εἰς λοιμικὸν νόσημα A 10 παρὰ τῶν χωρικῶν] om. B 12 καὶ] om. A 12 ἔχῃ] καὶ add. B 13 τὸ λεγόμενον πολυκόμπιν] om. A 13 πολυκόμπιν] scripsi πολλοκόμπιν MB 14 ἰχώρια] κιχώρια B 14 ἂς] τα add. A 14 πότιζε] δὲ add. AB 16 ἐπέσω] πολλὰ add. A πολλὰ πολλὰ add. B 16 εἰς τὰ κούφη του πολλὰ πολλὰ] ὡσὰν ἵστα ἀπέσω εἰς τὰ κούφη τοῦ A ὡς ἔνι ἡ ἱστία ἀπέσω εἰς τὰ κούφη του B 17 κρύον νερὸν] transp. B 17 ὅσον θέλει] καὶ ἂς ἔνι πολλὸν κρύον A 17 ὅσον] νερὸν add. B

καὶ νερὸν καὶ ἂς βράσουν καὶ ὕστερον βάλε σάχαρ καὶ μαστίχιν καὶ ἂς πίνῃ.[B◁] 6 ἄλλο. ⌜πότιζε δὲ αὐτὸν καὶ τὸν λεγόμενον ἀρμένιον βῶλον τριμμένον με τὸ νερὸν τὸ χλίον⌝ πλὴν νηστικόν. 7 ἄλλο. [▷om. A]κοπάνισον ἀλόην καὶ [om. L◁]ἀμμωνιακὸν καὶ σμύρνην καὶ βάλε ὀλίγον καὶ σμίγε το με τὸ κρασὶν τὸ ὀλίγον καὶ πότιζε αὐτὸν καθ' ἑκάστην ἡμέραν νηστικὸν τὸ πρωΐ.[om. A◁] 8 ἄλλο. ⌜πότιζέ τον καὶ τὴν ἀντίδοτον τὴν λεγομένην θηριακὴν καὶ τὴν λεγομένην ἀθανασίαν με τὸ κρασὶν τὸ ὀλίγον τὸ πρωΐ νηστικόν.⌝

148 [121] ρμη' [▷om. B]πρὸς τὰς ἐν τοῖς τιθοῖς καρκινώδεις διαθέσεις, ἤτοι[om. B◁] ὅταν τὰ βυζία τῶν γυναικῶν ἢ τῶν ἀνδρῶν ἔχωσι τὸ πάθος τὸ λεγόμενον κάρκινον παρὰ τῶν ἰατρῶν. ποίησον δὲ θεραπείαν τοιαύτην εἰς τὸ πάθος αὐτό· ⌜βάλε λιθάργυρον τριμμένον καὶ χρηστέλαιον καὶ ὀλίγον ὄξος καὶ μίξε τὰ ἐν τῷ ἅμα καὶ ἄλειφε τὸ πάθος που ἔνι εἰς τὰ βυζία. 2 ἄλλο. βάλε τὸ λεγόμενον ἀγριοκάρδαμον τριμμένον ψιλὸν πολλὰ καὶ ψιμμίθιον τριμμένον καὶ μυρσινέλαιον καὶ μίξε τὰ καὶ ἄλειφε τὸ πάθος τῶν καρκίνων.⌝

149 [121] ρμθ' ⌜ἀφ' ὅτου δὲ τὸ πάθος αὐτὸ ὁ καρκίνος αὐτὸς ὑγιαίνῃ καὶ διάβῃ, ἐὰν θέλῃς τὰ βυζία της να εἶναι καλὰ να μηδὲν πάθωσι τί ποτε ἄλλο κακὸν ὅμοιον αὐτοῦ τοῦ καρκίνου καὶ χειρότερον,

19 ἂς] τὸ add. B 2 πότιζε] πότισον A 3 τριμμένον] om. B 3 νηστικόν] νηστικὸς A 4 σμύρνην] σμύρναν M σμύρνα B 4–5 βάλε ὀλίγον] βα.... L βάλε κρασὶν M 5 το] τα M 5 με τὸ κρασὶν τὸ ὀλίγον] om. M 5 καὶ] .. L 6 νηστικὸν τὸ πρωΐ] transp. MB 6 τον] δὲ M 8 τὸ πρωΐ νηστικόν] πλὴν νηστικὸς A 9 τοῖς] om. A 9 τιθοῖς] scripsi .ηθοῖς L τιθὰς M τιτθὰς A 10 ἢ] καὶ M 10 τὸ] om. B 11 κάρκινον] καρκίνων M 11 παρὰ τῶν ἰατρῶν] om. M 11 δὲ] om. Σ 12 εἰς τὸ πάθος αὐτὸ] ἤγουν MA οἷον add. B 13 ὀλίγον ὄξος καὶ] om. B 13 μίξε τὰ ἐν τῷ ἅμα καὶ] και ἕνωσον ταῦτα καὶ M om. B 13–14 ἄλειφε τὸ] om. B 14 που] ὅπερ Σ 15 ψιμμίθιον] ψιμμίνθιν B 15 τριμμένον] ψιλὸν πολλὰ add. MB ψιλὸν add. A 16 μίξε τὰ] μίξον αὐτὰ M 16 ἄλειφε τὸ πάθος] ἄλει......θος L 17–18 ἀφ' ὅτου δὲ τὸ πάθος αὐτὸ ὁ καρκίνος αὐτὸς ὑγιαίνῃ καὶ διάβῃ] om. M ἄφ' ὅτι ὑγιαίει ὁ καρκίνος B 17 αὐτὸς] om. A 18 ἐὰν] πρὸς ἐὰν M ἂν B 18 θέλῃς] θέλῃ B 18 τὰ βυζία] ὅτι τὰ βυζία M ὅτι A να εἶναι καλὸν τὰ βιζία B 18 της να] scripsi τῆς μ... L ταύτης να M om. A τῆς B 18 εἶναι καλὰ] om. AB 19 τί ποτε] τὰ βυζία A τίποτας B 19 ἄλλο] τίποτε add. A 19 αὐτοῦ] om. A

17 ἀφ'] This chapter is part of the preceding one in A.

ποίησον θεραπείαν τοιαύτην· κύμινον τριμμένον ψιλὸν πολλὰ μίξε το με τὸ νερὸν καὶ μάλαξέ το καὶ κατάπλασσε τὰ βυζία ἤτοι θέτε τὸ εἰς βυζία τῆς γυναικός, ἔνθα ἔνι τὸ πάθος. ἀφ' ὅτου δὲ θέσεις τὸ κύμινον ἐπάνω εἰς τοῦ κυμίνου, θέτε σπόγγον καινούργιον βρεμένον με τὸ ὀξείδιν καὶ τὸ κρασὶν καὶ δένε το με τὸ πανίν. ὅτε δὲ διαβάσῃ τρεῖς ἡμέρας σήκοσε τὸν κύμινον καὶ θὲς κοπανισμένον βολβὸν μαλαγμένον με τὸ μέλι καὶ θέτε τὸ ἐπάνω εἰς τὸ πάθος. δένε δὲ τοῦτον τὸν βόλβον με τὸ μέλι ἕως τρεῖς ἡμέρας πίνε δὲ τοῦτο τρεῖς φορὰς τὸν μήναν.⌝

150 [122] ρν′ ὅταν αἱ μασχάλαι των βρωμοῦσι καὶ ὀξύνουσι, ποίησον θεραπείαν τοιαύτην διὰ τὸ να μὴ βρωμῶσι· ἤγουν ⌜στυπτηρίαν ὑγρὰν τρίψε καὶ σμύρναν τριμμένην καὶ μίξε το με τὸ κρασὶν καὶ ἄλειφε τὰς μασχάλας καὶ αὐτὸ ποίησον καὶ ἐπὶ τῶν ἀνδρῶν τὰς μασχάλας ὅταν βρωμοῦσι καὶ ὑδρώνουσι.⌝ 2 ἄλλο. ▷AB τὸ αὐτὸ ποί-

19 καὶ χειρότερον] κ... χ.ιρότερον L ἢ χειρότερον M ἢ χεῖρον ἀλλὰ να ἔνε πάντοτε καλὰ A 1 τοιαύτην] ἤγουν add. Σ 1–2 μίξε το] μίξον M 2 με τὸ νερὸν] μετὰ νερὸν M 2 μάλαξέ το] μάλαξον M 2–3 ἤτοι θέτε τὸ εἰς βυζία τῆς γυναικός, ἔνθα ἔνι τὸ πάθος] om. M 3 εἰς] τὰ add. AB 3 τῆς γυναικός] om. A 3–4 θέσεις] θέσης Σ 4 ἐπάνω] ...νω L 4 εἰς] om. Σ 4 τοῦ κυμίνου] om. A 4 καινούργιον] om. A 5 βρεμένον] βεβρεγμένον M βρεμένον καινούργιον A 5 τὸ κρασὶν] scripsi τὸ κρασ... L με τὸ κρασὶν M om. A 5 καὶ] om. A 5 ὅτε] ὅταν Σ 6 διαβάσῃ] διαβ...σει Σ 6 σήκοσε] ἔπαρον MA ἐπάνω B 6 θὲς] τὸ add. M 7 βολβὸν] κολβὸν M 7 μαλαγμένον] om. A 7 με τὸ] δὲ τὸ M om. A δὲ με τὸ B 7 μέλι] ἕως ἡμέρας τρεῖς add. M ἕως τρεῖς ἡμέρας add. A 7–8 καὶ θέτε τὸ ἐπάνω εἰς τὸ πάθος. δένε δὲ τοῦτον] om. MA 8 τὸν βόλβον με τὸ μέλι] πλὴν δ` μαλαγμένον με τὸ μέλι αὐτὸν δὲ τὸν βόλβον πίε τον μαλαγμένον με τὸ μέλι A 8 ἕως τρεῖς ἡμέρας] om. MA transp. B 8 πίνε δὲ τοῦτο] scripsi ποίη... δὲ τοῦτο L πίε τον M om. A αὐτὸν δὲ τὸν βολβὸν πίνε τον με τὸ μέλι B 9 φορὰς] φοραῖς A 10 ὅταν αἱ] ὅ...... L πρὸς M 10 μασχάλαι] ἀμασχάλας γυνιακῶν M αἱ μασχάλαι τῶν γυναικῶν AB 10 των] ὅταν ἱδρώνωσιν καὶ M om. AB 10 βρωμοῦσι καὶ ὀξύνουσι] βρομωμῶσιν M βρωμοῦσι καὶ ἱδρώνουσι AB 10 ὀξύνουσι] scripsi ὀξυ...σι L 11 τοιαύτην] om. A 11 διὰ τὸ να μὴ βρωμῶσι] om. MA 11 μὴ] scripsi ... L 11 ἤγουν] ποίησον add. MB 12 τρίψε] ῥίψε L τρίψον MA 12 τριμμένην καὶ μίξε το] καὶ ἕνωσον M τριμμένην καὶ μίξε τα A καὶ συνένωσον αὐτὰ καὶ B 13 μασχάλας] ἀμασχάλας M 13–14 καὶ αὐτὸ ποίησον καὶ ἐπὶ τῶν ἀνδρῶν τὰς μασχάλας ὅταν βρωμοῦσι καὶ ὑδρώνουσι] ἐκεῖ ὅπου ἔνι ὁ βρῶμος Σ

ησον καὶ ἐπὶ τῶν ἀνδρῶν ὅταν αἱ μασχάλαι τῶν ἀνδρῶν βρωμῶσι καὶ ὑδρώνουσι.AB◁ 3 ἄλλο. ⌜λιθάργυρον τριμμένον καὶ μυρσίνην τριμμένην καὶ τὸ λεγόμενον παρὰ τῶν ἰατρῶν ἄμωμον. τρίψε τὰ ὅλα καὶ μίξε με τὸ κρασὶν το καλὸν καὶ ἄλειφε τὰς μασχάλας.⌝ 4 ἄλλο. ▷om. MAστυπτηρίαν ὑγρὰν τριμμένην καὶ σμύρνην καὶ ἄμωμον καὶ τὸ λεγόμενον παρὰ τῶν ἰατρῶν ναρδοστάχυον. τὰ πάντα δὲ τριμμένα καὶ μίξε τα με τὸ καλὸν τὸ κρασὶν τὸ μυρωδάτον καὶ ἄλειφε τὸν τόπον τῶν μασχαλῶν ἔνθα ἔνι ὁ βρῶμοςom. MA◁.

151 [124] ρναʹ ⌜ὅταν ὁ ἄνθρωπος ἐρεύγεται ὡσὰν αὐγὰ τηγάνου σημεῖον ἔνι ὅτι πολλὴ δυσκρασία ἔνε εἰς τὴν κοιλίαν του καὶ ποίησον θεραπείαν τοιαύτην· ἤγουν ἂς βρέχῃ τὸ ψωμὶν εἰς τὸ ὀξείδιν καὶ εἰς τὸ νερὸν τὸ μικτὸν καὶ ἄς το τρώγῃ. ἂς τρώγῃ δὲ καὶ ὀξύγαλα παχὺ πολλὰ ὄξυνον καὶ ἀντύβια μετὰ ὄξους δριμέου καὶ ἂς πίνῃ δὲ ἀντὶ οἶνον ὀξείδιν δριμὺ με τὸ νερόν.⌝

152 [125] ρνβʹ ὅσοι δὲ διὰ ψύξιν ἀνορεκτοῦσιν, ἤτοι ἀπὸ κρυότητος ἤγουν ἐκρύωσαν, ποίησον θεραπείανη τοιαύτην· ⌜ἤγουν ἂς πίνῃ κρασὶν παλαιὸν ἀκράτον⌝ ἢ με τὸ θερμὸν καὶ τὸν λεγόμενον κονδῖτον οἶνον ⌜καὶ ὕδωρ καὶ γάρος μικτὸν ἂς πίνῃ. 2 ἄλλο. βάλε δὲ

261.14–262.1 τὸ αὐτὸ ποίησον] τὴν αὐτὴν θεραπείαν ποίει M 1 καὶ] τὴν αὐτὴν θεραπείαν καὶ add. B 1 ὅταν αἱ μασχάλαι τῶν ἀνδρῶν] om. M 1–2 βρωμῶσι καὶ ὑδρώνουσι] om. M transp. B 2–3 μυρσίνην τριμμένην] σμύρναν M σμύρναν τριμμένην AB 3 καὶ] om. B 3 τρίψε] καὶ τρίψε MB 4 μίξε] ἕνωσον M μίξε τα AB 4 μασχάλας] καὶ ὑγιαίνει add. A 5 σμύρνην] σμύρναν B 6 ναρδοστάχυον] ναρδοστάχιον B 7 τὸ] om. B 7 μυρωδάτον] μυροδάτον B 9 ἐρεύγεται] ῥεύεται L ῥεύγεται B 9 ὡσὰν αὐγὰ] ......σιᾶ L 9 τηγάνου] τυγάνου L ποίησον εἰς αὐτὸν θεραπείαν τοιαύτην add. B 10 δυσκρασία] δυσκρασίαν AB 10 ἔνε] ἔνι θερμὴ MB ἔχει ἔσσω A 10 κοιλίαν] καρδίαν A 10 του] om. A 10 καὶ] om. M θερμὴν καὶ A 11 τοιαύτην] om. A 11 ὀξείδιν] ὄξος MA 12 τὸ] om. Σ 12–13 ὀξύγαλα] ὀξύγαλον M 13 παχὺ πολλὰ ὄξυνον] om. M 13 πολλὰ] καὶ add. A 13 ἀντύβια] ἰντύβια MA ἰντίβια B 13 μετὰ ὄξους] με τὸ ὀξείδιν Σ καὶ μαρούλιν καὶ πυκτῆς τρέμουσαν καὶ εἶτα ἄλλον θέλει φαγεῖν με τὸ ὀξείδιν add. B 13 δριμέου] τὸ δριμὺ Σ 13 καὶ] om. Σ 14 οἶνον] οἴνου A κρασίου B 14 δριμὺ] om. AB 15 ὅσοι] πρὸς ὅταν B 15 δὲ] om. Σ 15 ἀνορεκτοῦσιν] ἀνορεκτεῖ B 15–16 ἤτοι ἀπὸ κρυότητος] om. B 16 ἤγουν ἐκρύωσαν] ὅτι ἐκρύωσε MA om. B 16 ποίησον θεραπείανη τοιαύτην] καὶ θεράπευσον αὐτοὺς οὕτως A 17 κρασὶν] om. B 17–18 κονδῖτον] ἄκρατον A κονδίτην B 18 γάρος] ...ρὸς L 18 πίνῃ] ἀργὰ καὶ πρωῒ add. A νερὸν εἰς τὸ τζουκάλιν καὶ ἄνηθον καὶ πήγανον καὶ μακεδονίου ῥίζας καὶ ἂς βράσουν καὶ ἀφ' ὅτου βράσουν, βάνε εἰς ἀγγεῖον πότε πεπέρι τριμμένον ὀλίγον καὶ ἄς το πίνῃ κατὰ πρωῒ νηστικὸς ὅσον ποτήριον ▷om. A γεμάτον. τὸ δὲ ποτήριον ἂς ἔνι μεσαῖον. ἂς τρώγῃ δὲ καὶ τοῦτο τὸ βραδυ, ἤγουν om. A◁ βάνε μέλι καὶ ἂς βράσῃ καὶ ἂς γένηται ὡς ποιεῖς τὸ κυδωνάτον ἢ τὸ διὰ κιτρίων. καὶ τότε βάλε πεπέρι τριμμένον ▷L καὶ ἄσπρον πεπέρι τριμμένον οἷον τρώγομεν. καὶ μακρὸν πεπέρι τριμμένον καὶ L◁ ἄσπρον ▷om. A πέπερι τριμμένον εὑρίσκεται τοιοῦτον ἄσπρον πεπέρι εἰς τὸ πέπερι ὃν τρώγομεν. om. A◁ καὶ τάραξον αὐτὰ ἀκριβῶς καλά, εἶτα τὸ βραδὺ μὴ δειπνᾷ καὶ ἂς τρώγῃ ἀπ' ἐκεῖνο ἴσα καρύδιν καὶ μηδὲν τὸ μασσᾶται πολλὰ εἰ μὴ ἐὰν δύναται καὶ ἔνι ὁδός, ἄς το πίνῃ οὕτως. εἶτα ἂς πίνῃ ἐπάνω εἰς αὐτὸ θερμὸν κρασὶν παλαιὸν καὶ ἂς κοιμᾶται. αὐτὸ δὲ ἂς ποιῇ κατὰ βραδὺ πλὴν μὴ δειπνᾷ μηδόλως μηδέ τι ποτὲ ἄλλο τρώγῃ. ἢ πρὸτερον ἢ ὕστερον. ἵνα μὴ ἀντὶ τοῦ καλοῦ γένηται μεγάλη ζημία ▷LB ἂς προσέχῃ δὲ να μὴ δειπνᾷ. LB◁ 3 ἄλλο. ἂς τρώγῃ καὶ τοῦτο ἀφ' ὅτου πληρώσῃ ἐκεῖνο τὸ λεγόμενον παρὰ τῶν ἰατρῶν διὰ καλαμίνθην. ἤτοι βάλε μέλι καὶ ἂς βράσῃ ὥς σε παρήγγειλα ἄνωθεν ὀλίγον. εἶτα βάλε καλαμινθίου ἤτοι ποταμογείτονος τοῦ λεγομέ-

18 δὲ] om. Σ 1 νερὸν εἰς τὸ τζουκάλιν] transp. B 1 τὸ τζουκάλιν] τζυκάλιν Σ 2 ῥίζας] ῥίζαν A 2 καὶ ἀφ' ὅτου βράσουν] om. M 2 βάνε] βάλε Σ 2 πότε] om. MA τίποτε B 3 το] τα A 3 νηστικὸς] νῆστις M 3–4 ὅσον ποτήριον] ọσ̣......ιον L ὅσον ἥμισον ποτήριον A 4 ποτήριον] ποτήριν B 4 ἔνι μεσαῖον] ....εσέον L να γεμάτον B 5 καὶ] om. B 5 τὸ] om. B 5 βάνε] βάλε Σ 5–6 ἂς γένηται] ἐγένη L ἂς γένῃ B 6 κιτρίων] κιτρίνου M κυτρίου A κητυρίου B 6 καὶ τότε] om. A 8 πέπερι τριμμένον] om. M 9 εὑρίσκεται] γὰρ add. M δὲ B 9 ὃν] ὅπου M τὸ B 10 ἀκριβῶς] om. A 10 καὶ] ἀμὴ A 11 ἐκεῖνο] ἐκείνου A 11 ἴσα καρύδιν] ὅσον καρυδίου MA 11 μασσᾶται] μασάται M μασίεται A μασᾷ B 11 μὴ] μὶ M 12 καὶ ἔνι ὁδός] om. A 12 πίνῃ] καταπίνῃ AB 12 εἶτα] εἰς τὰ L 13 εἰς] om. A 13 αὐτὸ] αὐτὸν M om. A 13 ἂς] το add. Σ 14 πλὴν μὴ δειπνᾷ μηδόλως μηδέ τι ποτὲ ἄλλο τρώγῃ] om. M transp. B 15 ἢ πρὸτερον ἢ] ἢ ....η L πρῶτον ἢ A 15 τοῦ] om. Σ 15 γένηται] .....τ̣α̣ι̣ L 15–16 μεγάλη ζημία] transp. MA ζημίαν ἄλλὰ B 16 μὴ] μηδὲν B 17 τῶν ἰατρῶν] τ... L 18 καλαμίνθην] καλαμίνθης M 18 ἤτοι] ἤγουν B 18 βράσῃ] ...σει L 18 σε παρήγγειλα ἄνωθεν] ἄνωθεν εἴρηκα M ἄνωθεν εἶπον B 19 καλαμινθίου] καλαμίνθιν B 19 ἤτοι] εἶτα B

νου τὰ φύλλα ξηρὰ τετριμμένα ὁμοίως καὶ φύλλα γληχουνίου καὶ πετροσελίνου ξηροῦ φύλλα τριμμένα ἤτοι μακεδονησίου καὶ τοῦ λεγομένου θύμου τὰ φύλλα ξηρὰ τετριμμένα καὶ πέπερι τριμμένον ὀλίγον καὶ τάραξον αὐτὰ με τὸ μέλι καὶ ἂς τρώγῃ αὐτὸ ὡς τὸ ἄλλο ὅπερ σε εἶπα.⌝ ἂς τρώγῃ δὲ καὶ τοῦτο τὸ λεγόμενον παρὰ τῶν ἰατρῶν διοσπολίτην ἤγουν βάλε μέλι καὶ ἂς βράσῃ ὡς σε εἶπον ὀλίγον ὀπίσθεν. εἶτα βάλε πέπερι τριμμένον οἷον τὸ τρώγομεν καὶ φύλλα τριμμένα πηγάνου καὶ κύμινον τριμμένον καὶ ξυλοτζιτζίβεριν καὶ τὸ λεγόμενον παρὰ τῶν ἰατρῶν νίτρον καὶ μίξε τα με τὸ μέλι. καὶ τάραξον αὐτὰ καλὰ καὶ ἂς ποιῇ καὶ αὐτὸ ὡς σε εἶπον καὶ τὸ ἄλλο ὀλίγον ὄπισθεν πλὴν πρόσεχε να μὴ δειπνᾷ. 4 ἄλλο. ⌜ἂς πίνῃ δὲ καὶ τὴν ἀντίδοτον τὴν λεγομένην διὰ μόσχου ὡσεὶ κοκκία δ' με τὸ κρασὶν με τὴν ἀντίδοτον τὴν λεγομένην θηριακὴν καὶ ἂς τρώγῃ καὶ σκόρδα ὠμὰ καὶ ἑψημένα.⌝

153 [126] ρνγ´ ὅσοι δὲ διὰ πολλὰ ὑγρὰ εἰς τὸν στόμαχον εἶναι καὶ δι' ἐκεῖνα τὰ ὑγρὰ ἀνορεκτεῖ. ποίησον θεραπείαν τοιαύτην· ⌜ἤγουν πότιζε αὐτὸν καλὰ πρωῒ πικρὰν ὡσεὶ κοκκία η' με τὸ κρασίν. 2

263.19–264.1 ποταμογείτονος τοῦ λεγομένου τὰ φύλλα] transp. B 1 ὁμοίως] βάλε add. B 1 γληχουνίου] γλήχωνος ἤτοι βλησκουνίου M γλησχωνίου ὅπερ λέγουν οἱ ἰατροὶ γλίσχωνα B 2 μακεδονησίου] μ... κεδονισίου L μακαιδονισίου B 2–3 τοῦ λεγομένου] om. M 3 τετριμμένα] τριμμένα MB 5 ὅπερ] .. L 5 σε] σοι MB 5 εἶπα] εἶπον MB 6 βράσῃ] βράσουν B 6 σε εἶπον] εἶπον σοι MB 7 ὀπίσθεν] ἔμπροσθεν M 7 εἶτα] om. A 7 τριμμένον] om. MA 7 οἷον] ὅπερ B 7 τὸ] om. Σ 7–8 φύλλα τριμμένα] φ.....μμένα L om. A 8 τριμμένα] om. B 8 πηγάνου] φύλλα τριμμένα add. A 8 ξυλοτζιτζίβεριν] ξυλοζιγγίβερι A 8 καὶ] om. A 9 μίξε τα] ἔνωσον ταῦτα M 10 αὐτὰ καλὰ] αὐτὸν καλῶς B 10 ποιῇ] το τρώγῃ A 10 σε εἶπον] εἶπον σοι Σ 10–11 καὶ τὸ ἄλλο ὀλίγον ὄπισθεν] ἄνωθεν A 11 ὄπισθεν] ἔμπροσθεν M 11 πρόσεχε να] βλέπε A 11 μὴ] μηδὲν MB 12 δ'] δύο A τέσσαρα B 13 με τὴν ἀντίδοτον τὴν λεγομένην] ἤγουν τὴν A 13 με] καὶ M 13 καὶ] om. A 13 τρώγῃ] δὲ add. A 14 σκόρδα] σκόροδα Σ 14 ἑψημένα] ὄρμ L^3 in marg. ψημένα M ἑψημέν.. A 15 δὲ] om. MB 15 διὰ] τὰ add. A 15 ὑγρὰ] τὰ εἶναι add. M ὅπου ἔνεν add. A κείμενα B 15 εἶναι] om. Σ 15–16 καὶ δι' ἐκεῖνα τὰ ὑγρὰ] om. AB 16 δι' ἐκεῖνα] διακῖνα L 16 ἀνορεκτεῖ] ἀνορεκτοῦσι MA 16 τοιαύτην] om. A 16 ἤγουν] om. A 17 καλὰ] κατὰ MB om. A 17 πρωῒ] om. A 17 πικρὰν] τὴν πικρὰν κατὰ πρωῒ A ἄλλο. ἃς προσέχῃ δὲ μὴ τρώγῃ τίποτε παστὰ ἢ κρέας ἢ ὀψάρια μηδὲ ζωμία πίνῃ μηδὲ νερὰ μηδὲ ὄσπρια. ἀλλὰ ἃς τρώγῃ μίαν τὴν ἡμέραν καὶ μὴ δειπνᾷ.⌝ εἰ μὴ ἃς πίνῃ θερμὸν κρασὶν παλαιόν.

ἄλλο. ἔμπλαστρον περὶ ἀνορεξίας· ⌜κοπάνισον ῥόδα ξηρὰ καὶ ἀλόην καὶ κηκίδιν καὶ κοπανισμένα φοινίκια καὶ οἰνάνθην ὀλίγον καὶ μελίλωτα ξηρὰ κοπανισμένα καὶ μαστίχην τριμμένον καὶ κρόκον τριμμένον καὶ ἀψινθίου σπόρον τριμμένον καὶ ψιλὸν ποιήσας καὶ αὐγοῦ κρόκον καὶ πίσσαν ὀλίγην καὶ πολλὴν ποίησον ἔμπλαστρον καὶ θέτε το εἰς τὸν στόμαχον.⌝

154 [127] ρνδʹ ⌜κυνώδης ὄρεξις ἔνι, ὅταν ὀρέγεται ἄνθρωπος πολλὰ πλήρης να φάγῃ καὶ τρώγῃ καὶ πάλιν πεινᾷ συνεχῶς καὶ πάλιν τρώγει ὥσπερ οἱ σκύλλοι τρώγουν καὶ πάντα ὄρεξιν ἔχει.⌝ οὕτως καὶ ἡ ὄρεξις τοῦ ἀνθρώπου ὁμοία ἐστὶ τοῦ κυνὸς καὶ διὰ ⊳om. A τοῦτο λέγεται κυνώδης ὄρεξις ἀπὸ τοῦ μετέχειν τῇ ὀρέξει τοῦ κυνός. ⌜αὐτὸ δὲ τὸ πάθος γίνεται διὰ ψυχρὸν χυμὸν⌝ καὶ ὀξώδει ἐν τῷ στομάχῳ. ⌜ὀφελεῖ δὲ τοὺς τοιούτους ἵνα πίνωσιν ὕδωρ καὶ γάρος μικτὸν καὶ τὸ λεγόμενον διὰ τριῶν πεπερέων,⌝ ὅπερ σε εἶπον τὴν σκευασίαν ὀπίσω ὅπου γράφει ὅτι διὰ χυμοὺς παχέους ἀνορεκτεῖν καὶ διοσπολιτικὸν καὶ αὐτὸ εἶπα σε τὸ πῶς σκευάζεται. ἃς τρώγῃ καὶ τὸ βραδὺ κυδωνάτον καὶ ἃς πίνῃ ἐπάνω θερμὸν κρασὶν παλαιὸν

---

17 ὡσεὶ] om. A 1 ἃς προσέχῃ δὲ μὴ τρώγῃ τίποτε παστὰ] παστὰ μὴ τρώγῃ A 1 προσέχῃ] βλέπῃ M 1 δὲ] om. M 1 τίποτε] .... L 1 κρέας] κρέη M κρέα B 2 ζωμία] ....μία L 2 νερὰ μηδὲ ὄσπρια] μηδὲ ὀψάρια χλωρὰ μηδὲ νερὰ πίνει μηδὲ ὄσπρια A 4 ἔμπλαστρον περὶ ἀνορεξίας] om. B 5 κηκίδιν] κικίδια B 5 κοπανισμένα] om. Σ 5 οἰνάνθην] οἰνάθην M 6–7 καὶ κρόκον τριμμένον] om. Σ 7 τριμμένον] τρίψας AB 7 καὶ] om. M 7 ποιήσας] om. M 8 πίσσαν] ...σσαν L 8 καὶ πολλὴν] πολλὰ καὶ Σ 9 θέτε] om. M 9 το] om. MA 9 στόμαχον] ἐπάνω add. A 10 κυνώδης] πρὸς ὄρεξιν κυνώδη· κυνώδης M περὶ κ..ους ὀρέξεως· κυνώδης B 10 ὄρεξις] ὀρέξεως B 10 ὀρέγεται ἄνθρωπος] transp. MA 11 πλήρης] πολλὰ MB om. A 12 τρώγει] καὶ πάλιν πεινᾷ add. A 12 τρώγουν] συνεχῶς B 12 ὄρεξιν ἔχει] ἔχουν ὄρεξιν A 12 ἔχει] ἔωσι M ἔχουσιν B 13 ὁμοία ἐστὶ] ὁμοιάζει A 13 κυνὸς καὶ διὰ] κυνο......ὰ L κυνὸς A 14 ἀπὸ] ..τὸ L 14 κυνός] σκύλλου M 15 ψυχρὸν χυμὸν] ψυχρῶν χυμῶν MB 15 ὀξώδει] ὀξίδα M ὀξίδιν B 17 σε] σοι M om. B 18 ὀπίσω] ὀπίσθεν M 18 γράφει ὅτι] γρά... ὅτι L γράφει MB 18 χυμοὺς παχέους] παχεῖς χυμοὺς MB 19 εἶπα] εἶπον MB 19 σε] σοι B 19 τὸ πῶς] το..σι L πῶς B 19 σκευάζεται] καὶ add. MB

καὶ μὴ δειπνᾷ.[om. A◁]

155 [128] [om. A]ρνε΄ ὅταν ἄνθρωπος ἔχει τὸν λεγόμενον παρὰ τῶν ἰατρῶν λυγμόν, [▷om. B]παρὰ δὲ τῶν χωρικῶν τοῦτο καλεῖται λούξικας. τινὲς δὲ λέγουσιν αὐτὸ κλόξον.[om. B◁] ποίησον δὲ θεραπείαν τοιαύτην· ⌜ἤγουν πήγανον βράσον με τὸ κρασὶν καὶ ἄς πίῃ.⌝ 2 ἄλλο. καλαμίνθην ποταμογείτωνα βραστὸν με τὸ κρασὶν καὶ ἄς το πίνῃ ἤ ⌜καστόριον τριμμένον μετὰ ὄξους ὀλίγου καὶ οἴνου καὶ ἄς το πίνῃ ἀλλὰ μηδὲν βράσῃ.⌝

156 [om. A]ρνϛ΄ ὅταν ὁ ἄνθρωπος ἔχῃ ἀπεψίαν, σημεῖον ἔχει τοῦτο, ὅτι ἐρεύγεται ὡς ἂν ὀξύδια. ποίησον θεραπείαν τοιαύτην· ἤγουν γάρος καὶ νερὸν ὀλίγον χλίανέ το καὶ ἄς το πίνῃ νηστικὸς καὶ ἀφ' ὅτου φάγῃ. 2 ἄλλο. ἂς τρώγῃ δὲ τὸ βραδὺ κυδωνάτον καὶ ἂς πίνῃ ἐπάνω θερμὸν κρασὶν παλαιὸν καὶ μὴ τρώγῃ ἄλλο τίποτε. 3 ἄλλο. ἂς τρώγῃ καὶ τὸ λεγόμενον διοσπολιτικὸν καὶ διὰ τριῶν πεπερέων καὶ τὸ διὰ καλαμίνθης οὕτως ὡς τὸ κυδωνάτον. αὐτὰ δὲ ἔγραψα σοι τὸ πῶς σκευάζονται ἐκεῖ ὅπου ἀνορεκτοῦσιν διὰ παχεὶς χυμοὺς ὀπίσω ὀλίγον. 4 ἄλλο. ἂς τρώγῃ δὲ καὶ τὸ λεγόμενον διὰ κιτρίου τὸ βραδὺ καὶ ἂς πίνῃ θερμὸν κρασὶν παλαιὸν καὶ μὴ δειπνᾷ. 5 ἄλλο. ἂς κοιμᾶται δὲ με τὸν ἄρρωστον αὐτὸν παιδίον μικρὸν πλὴν να ἔνι καλόσαρκον καὶ λιπαρὸν καὶ παχὺ τὸ παιδίον, ἵνα θερμαίνῃ τὸν ἄρρωστον. εἰ δὲ παιδίον οὐκ ἔνι, ἂς θέτε σκυλλόπουλα λιπαρὰ καὶ παχέα καὶ ἂς κοιμᾶται μετ' ἐκεῖνα. εἰ δὲ σκυλλόπουλα ταῦτα οὐκ εἰσί, ἂς κοιμᾶται ὁ ἄρρωστος μετὰ κορασέαν παρθένον

---

20 βραδὺ] τὸ add. MB 2 ὅταν ἄνθρωπος ἔχει] transp. M κλόξον add. B 2–3 παρὰ τῶν ἰατρῶν] om. B 3 τοῦτο] τὸ πάθος M 4 τινὲς δὲ] scripsi .νες δὲ L τινὲς M 4 κλόξον] κλόγξον M 4 δὲ] om. M 5 βράσον με τὸ] scripsi βρα....... L βρασμένον με τὸ MB 5 καὶ] om. M 5 ἄς] το add. B 5 πίῃ] πίνῃ MB 5–6 καλαμίνθην ποταμογείτωνα βραστὸν] ἢ κύμινον τριμμένον πολλὰ καὶ ἕνωσον MB 5–6 καλαμίνθην] scripsi καλαμιν...... L 7 καὶ] τοῦ add. M 9 ἔχει τοῦτο] ἔνι MB 10 ὀξύδια] ὀξείδιν MB 11–12 καὶ ἀφ' ὅτου φάγῃ] om. M 12 τὸ] κατὰ M 13 ἐπάνω θερμὸν] ἐπάνω ...μὸν L ἐπάνωθεν θερμὸν M 13 μὴ τρώγῃ ἄλλο τίποτε] transp. B 13 ἄλλο] α.. L 14 καὶ] τὸ add. MB 15 ὡς] ὥσπερ καὶ M 16 σοι] σ... L 16 σκευάζονται] κατασκευάζονται B 17 ὀπίσω ὀλίγον] ὀπίσω ὀλ... L ὄπισθεν ὀλίγον M 18 κιτρίου] κι L 18 κρασὶν παλαιὸν] ......μὸν L 19 κοιμᾶται] κοιμᾶ.... L 22 λιπαρὰ καὶ παχέα] λιπαρ..αχέα L 23 ταῦτα] τοιαῦτα M om. B

γυναῖκα πλὴν ἐκείνη να ἔνι παχέα καὶ λιπαρὴ καθὼς οἱ πάλαι τοῦτο ἐποίουσαν. συνήθειαν γὰρ τὸ ἤχασιν. εἰ δὲ παρθένος γυνὴ οὐκ ἔνι, κοιμᾶται καὶ μετὰ γαμημένην γυναῖκα. ἀλλὰ ἂς προσέχῃ καλὰ ὅσον δύναται να μὴ γαμήσῃ ἢ τὴν παρθένον τὴν ἄλλην καὶ αὐξήσῃ ἡ ἀπεψία. εἰ δὲ ἔνι ἄρρωστος γυναῖκα ἂς κοιμᾶται μετ' ἐκείνην παιδίον ἢ παρθένος γυνὴ ἢ σκυλλόπουλα πλὴν παχέα. τὸν αὐτὸν τρόπον καὶ ἡ παρθένος γυνὴ ἂς ἔνι παχέα καὶ λιπαρὴ πολλὰ ὥσπερ καὶ τὸ παιδίον οἷον ἐὰν ἔνι ἢ ἄρσεν ἢ θῆλυ να εἶναι λιπαρά. 6 εἰ δὲ θερμότης, ἡ ἀπεψία ἐγένετο ἐν τῷ κοίτει τῆς γαστρός. σημεῖον ἔνι ῥεύγεται ὡς ἂν τζίκνα καὶ καίεται ἀπέσω πολλὰ πλὴν οὐ κενώνει. ἂς τρώγῃ δὲ ὁ τοιοῦτος ἤτοι πρᾶγμα ψυχρὸν οἷον ὀπωρικὰ καὶ ἄγουρα καὶ ὥριμα καὶ ἀντύβια καὶ μαιούλια με τὸ ὀξείδιν τὸ δριμὺ καὶ πηκτὴν καὶ ὀξύγαλα πολλὰ ὄξυνον παχύ. ἀντὶ οἴνου ἂς πίνῃ νερὸν με τὸ ὀξείδιν τὸ δριμύ.

157 [129] ρνζ′ ὅταν ὁ ἄνθρωπος κενώνῃ χολὴν καὶ ἄνω καὶ κάτω μετὰ ἀνάγκης ποίησον θεραπείαν τοιαύτην· ἤγουν ⌜κοπάνισε τὰς

23 κοιμᾶται] ...μᾶται L 1 γυναῖκα] om. B 1 ἐκείνη να ἔνι] transp. B 1 ἔνι] ἔ.. L 1 λιπαρὴ] λιπαρὰ MB 1 πάλαι] παλαιοὶ MB 2 ἐποίουσαν] ἐποίησαν MB 2 συνήθειαν] συνήθειας M 2 ἤχασιν] .λασὶν L 2 εἰ] ἢ MB 3 κοιμᾶται] ἂς κοιμᾶται MB 3 μετὰ] με B 3 γαμημένην] om. B 3 γυναῖκα] scripsi γυναι... L γυναῖκα πλὴν να ἔνι παχέα M γυναικὸς φθαρμένην B 4 μὴ γαμήσῃ] μηδὲν μιάνη M μηδὲν συνουσιάσει B 4 ἢ] με B 4 παρθένον] παρθένιον ἢ B 4 καὶ] μᾶλλον add. B 5 εἰ] ... L 5 ἔνι] ὁ add. B 5 γυναῖκα] γυνὴ B 5 μετ' ἐκείνην] μετὰ τὴν B 6 γυνὴ ἢ] ...... L ἢ πάρθενος γυνὴ ἢ B 6 πλὴν] δὲ add. MB 6 παχέα] καὶ λιπαρὰ πολλὰ add. MB 6–7 τὸν αὐτὸν τρόπον] scripsi τὸν αὐ... τρόπον L τὸ αὐτὸ M om. B 7 καὶ ἡ παρθένος] om. B 7 γυνὴ] om. MB 7 ἂς ἔνι παχέα καὶ λιπαρὴ πολλὰ ὥσπερ] om. M ὥσπερ B 8 οἷον ἐὰν ἔνι] om. M ἔνι B 8 ἄρσεν] ἀρσενικὸν M 8 θῆλυ] θυλικὸν M 8 εἶναι] ἔνι B 8 εἰ] ἡ LB 9 θερμότης] ἀπὸ θερμότητος B 9 κοίτει] κύτη M 9 σημεῖον ἔνι] om. B 10 ῥεύγεται] ὅτι ἐρεύγεται M 10 τζίκνα] τζύκναν L 10 καίεται] καὶ add. M 10 οὐ κενώνει] οὐκ ἑνώσῃ MB 11 δὲ] om. B 11 οἷον ὀπωρικὰ καὶ] om. B 12 ἄγουρα] ἄωρα MB 12 ἀντύβια] ἰντύβια MB 12 καὶ μαιούλια] καιρούλια L καὶ ἰούλια M 12 με τὸ ὀξείδιν τὸ δριμὺ] μετὰ ὄξους δρυμαίου M 13 ὄξυνον παχύ] ὄξυνα παχέα MB 13 ἀντὶ] δὲ add. MB 15 ὁ ἄνθρωπος] om. M 15 κενώνῃ] κενώσῃ AB 15 χολὴν καὶ] om. MA καὶ B 15 κάτω] χολὴν add. Σ 16 ἀνάγκης] ἀναπυησ.. L 16 τοιαύτην] om. A

λεγομένας ψαλίδας παρὰ ἰδιωτῶν. παρὰ δὲ τῶν ἰατρῶν ἕλικας ▷om. L ἀμπέλων καὶ τὸν ζωμὸν ἐκείνων ἄς τον πίνῃ. 2 ἄλλο. ῥόδα ξηρὰ βράσον με τὸ κρασὶν τὸ αὐστηρὸν βάλε καὶ οἰνάνθην καὶ ἄς το πίνῃ. 3 ἄλλο. ▷om. B κοπάνισον ῥοΐδιν ὄξυνον καὶ τὸ ζωμὶν ἄς πίνῃ. om. B◁ 4 ἄλλο. ἂς πίνῃ δὲ καὶ αὐστηρὸν κρασὶν καὶ ἂς τρώγῃ καὶ ψίχας με τὸ κρασὶν τὸ αὐστηρόν.⌝ 5 ἄλλο. ▷om. A ἂς τρώγῃ καὶ σεμίδαλιν με τὸ κρασὶν τὸ αὐστηρὸν καὶ ἄς το βράζῃ. om. A◁ 6 ἄλλο. ἂς τρώγῃ δὲ καὶ ὀπωρικά οἷον τὰ λεγόμενα μέσπιλα καὶ κυδώνια ἄωρα καὶ ἀπίδια καὶ ῥοΐδια ὄξυνα. 7 ἄλλο. ⌜ἄντλει δὲ καὶ τοὺς πόδας θερμὸν νερὸν τὸ βραδὺ καὶ ἂς ἔνι ζεστά. 8 ἄλλο. τῶν σταφίδων τὰ κοκκία κοπάνισον καὶ ἄς πίνῃ με τὸ κρασὶν τὸ αὐστηρόν.⌝ 9 ἄλλο. ▷om. A βάτου καρπὸν ἤτοι τὰ κοκκία τοῦ ἄωρα ξηρὰ τρίψον καὶ πότισον μετὰ αὐστηροῦ οἴνου. om. A◁ 10 ἄλλο. βάλε ὀξείδιν καὶ κρασὶν καὶ δάφνης φύλλα καὶ μυρσίνης καὶ σχίνου φύλλα καὶ κηκίδιν τζακισμένον καὶ φλούδια ῥοϊδίων τὰ λεγόμενα παρὰ τῶν ἰατρῶν σίδια καὶ ἂς βράσουν καὶ πυρίᾳ τὸν στόμαχον τέσσαρες φοραῖς τὴν ἡμέραν κατὰ φορὰν δὲ ὡσεὶ πανία δέκα τὰ αὐτὰ δὲ πυρίᾳ καὶ τὴν ἕδραν ἀποκάτω. 11 ἄλλο. ⌜τὴν αὐτὴν δὲ θεραπείαν ὅταν συμβῇ τὸν ἄνθρωπον ἀπὸ ψύξεως ἤτοι ἀπὸ κρυότητος καὶ τρέχει ἡ κοιλία του πολλὰ ἂς ἔνι δὲ ζεστὰ ὁ τοιοῦτος καὶ κρύον μὴ

---

16 κοπάνισε] κοπάνισον Σ 1 ψαλίδας παρὰ ἰδιωτῶν. παρὰ δὲ τῶν ἰατρῶν] om. A transp. B 1 παρὰ] τῶν add. M 1 ἕλικας] scripsi ἐλλ... L ἕλιγκας MB ἐλίγγας τῶν A 2 ζωμὸν] ζουμὶν B 3 βάλε] καὶ βάλε A 3 καὶ] τὸ add. A 4 ῥοΐδιν] ῥόδιν A 4 τὸ ζωμὶν] τὸν ζωμὸν ἐκεῖνον A 4 ἄς] τον add. A 5 ἂς πίνῃ δὲ καὶ αὐστηρὸν κρασὶν καὶ ἂς τρώγῃ] om. A 5 καὶ] om. B 6 καὶ] ἂς βρέχῃ A 6 ψίχας] ψιχέας ἄρτου B 6 αὐστηρόν] καὶ ἂς τρώγῃ ἂς πίῃ καὶ κρασὶν αὐστηρὸν add. A 7 καὶ] πλὴν B 7 το] om. B 8 δὲ] om. B 8 ὀπωρικά] ὑπορικὰ A 8 μέσπιλα] μούσπουλα A 8 καὶ] om. A 10 θερμὸν νερὸν] transp. A 10–11 τῶν σταφίδων τὰ κοκκία κοπάνισον] transp. B 11 ἄς] τα add. AB 11–12 τὸ αὐστηρόν] om. A 12 καρπὸν] κοπάνισον add. B 12 ἤτοι] ἤγουν B 13 τρίψον] om. B 15–16 τὰ λεγόμενα παρὰ τῶν ἰατρῶν σίδια] om. A 16 σίδια] σίδιν B 16 πυρίᾳ] πυρίαζε AB 16–17 τέσσαρες] τέσσαρεις A 17 φοραῖς] φορὲς L φορὰς B 18 πυρίᾳ] πυρίαζε AB 18 ἀποκάτω] φοραῖς τέσσαρεις add. A ὑποκάτω B 18 δὲ] om. AB 18 θεραπείαν] ποίησον καὶ add. A ποίησον καὶ τὰ αὐτὰ add. B 19 ἤτοι ἀπὸ κρυότητος] καὶ κρυώσει ἢ περάσει ποταμὸν ἢ περιπατεῖ ὅλην τὴν ἡμέραν νερὰ καὶ κρυάνει AB

πίνῃ μηδὲ κρύα τρώγῃ καὶ ἐὰν οὐ πυρέσσῃ λοῦέ τον.⌝om. L◁

158 [130] ρνη′ ▷om. L⌜πρὸς φλεγμονὴν ἐντέρων· ὅταν τὸν ἄνθρωπον συμβῇ καὶ φλεγμαίνουσι τὰ ἔντερά του ἢ ξηρὰ κόπρος να διάβῃ καὶ πόρον οὐχ εὑρίσκει καὶ ἀναβαίνει πάλιν ἐξοπίσω ἡ κόπρος καὶ ξερᾷ την ὁ ἄνθρωπος om. L◁ καλεῖται τοῦτο τὸ πάθος παρὰ τῶν ἰατρῶν ἰλεός. ὅπερ τὸ νόσημα τοῦτο ἔνι θανάσιμον. εἴς τινας ἐὰν γένηται φλεγμονὴ τῶν ἐντέρων. εἰ δὲ δι' ἔμφραξιν κόπρου γένηται τὸ πάθος τοῦτο, σωτήριον ἔνι. καὶ ἐὰν ἔνι παιδίον ὁ ἄρρωστος ἤγουν χρονῶν δεκατεσσάρων, ποίησον ἐμβροχὰς καὶ καταπλάσματα καὶ πυρίας. εἰ δὲ ἔνι μειράκιον ὁ ἄρρωστος ἤγουν χρονῶν εἴκοσι και δύο, ποίησον φλεβοτομίαν. εἰ δὲ ἔνι ὁ ἄρρωστος ἀκμάζων ἤτοι χρονῶν τεσσαράκοντα δύο, ποίησον καὶ εἰς αὐτὸν φλεβοτομίαν καὶ κρασὶν μὴ πίνωσι ἂς εἶναι καὶ ζεστά.⌝

159 [131] ρνθ′ ⌜ὅταν οἱ νεφροὶ φλεγμαίνουσι καὶ γίνεται πόνος καὶ βάρος εἰς τὸν τόπον ἐκεῖνον, ὅπου ἔνι ἡ φλεγονὴ τῶν νεφρῶν καὶ πυρετοῦσιν οἱ τοιοῦτοι καὶ ξεροῦσι χολήν· τοὺς τοιούτους φλεβοτόμησον. 2 ἄλλο. κοπάνισον τὰ κοκκία τῆς τήλης καὶ τοῦ πηγάνου καὶ τοῦ ἀνήθου τὸν σπόρον καὶ μάλαξον μετὰ ὕδατος καὶ θὲς εἰς τὸν τόπον τῶν νεφρῶν. 3 ἄλλο. τὸν σπόρον τοῦ πέπονος τριμμένον ἄς το πίνῃ μικτὸν με τὸ μέλι καὶ με τὸ κρασίν. ἐπάνω δὲ τοὺς νεφροὺς πυρίαζε κρασὶν καλὸν χλίον με τὸ πανίν. 4 ἄλλο. ἔμπλαστρον δὲ τό τε διὰ χυλοῦ λεγόμενον παρὰ τῶν ἰατρῶν ἔμπλαστρον

---

20 πολλὰ] om. AB 1 οὐ] om. A 2 πρὸς φλεγμονὴν ἐντέρων] om. A 2 φλεγμονὴν] τῶν add. B 2–3 ὅταν τὸν ἄνθρωπον συμβῇ καὶ φλεγμαίνουσι τὰ ἔντερά του ἢ ξηρὰ] ἤγουν ὅταν B 3 κόπρος] παλαιὰ ἐμφραχθῇ εἰς τὰ ἔντερα καὶ ἔρχεται ἄλλη κόπρος add. AB 5 ἄνθρωπος] καὶ add. A 5 παρὰ τῶν] .... L 6 ἰλεός] εἰλεὸς MB 6 τοῦτο] ..... L 6 εἴς τινας] om. A 7 φλεγμονὴ] διὰ φλεγμονὴν AB 7–8 γένηται τὸ πάθος] om. A 8 καὶ ἐὰν] ἐὰν οὖν B 10 πυρίας] πυρία A 10 μειράκιον ὁ ἄρρωστος] transp. B 10 ὁ ἄρρωστος] om. MA 11 ποίησον] καὶ εἰς αὐτὸν add. A 11 ἀκμάζων] ἐκμάζων M 12 καὶ] om. MA 13 εἶναι καὶ] ἔνι δὲ A 14 ὅταν] πρὸς φλεγμονὴν νεφρῶν· ὅταν M πρὸς φλεγμονὴν νεφρῶν B 14 οἱ νεφροὶ φλεγμαίνουσι] om. B 14 καὶ] om. Σ 16 καὶ] ... L 16 τοιοῦτοι] τοι M 16 χολήν] ...... L 17 κοπάνισον] κο...... L 17 τὰ] om. Σ 19 τὸν σπόρον τοῦ πέπονος] transp. M 20 το] τὸν MA 20 μικτὸν] om. A 20 μέλι καὶ με τὸ κρασίν] κρασὶν καὶ τὸ μέλι M κρασὶν καὶ τὸ μέλι ὀλίγον A μέλι τὸ ὀλίγον καὶ τὸ κρασὶν B 20 δὲ] εἰς add. Σ 21 καλὸν] om B 22 δὲ τό τε] τὸ M δὲ τὸ AB

Μνάσεον θέτε καὶ αὐτὸ ἁπλωμένον εἰς τὸ πανὶν εἰς τοὺς νεφρούς, ἔνθα ἔνι ὁ πόνος. 5 ἄλλο. ἀνάλυσον κηρὸν καὶ βάλε ῥοδέλαιον καὶ αὐγοῦ κρόκον καὶ μάλαξον αὐτὰ καὶ θέτε ἐπάνω τοὺς νεφροὺς με τὸ πανίν.⌝

160 [132] ρξʹ ⌜διαβήτης λέγεται, ὅτε ἠνίκα πίνει κατουρεῖ, καὶ τοῦτο ποιεῖ τὸ συνεχῶς καὶ κατὰ ὥραν· τοὺς τοιούτους πότισον πολυγόνου χυλὸν ἢ τὸ λεγόμενον ῥάσδον.⌝ βράσε τὸ με τὸ κρασὶν τὸ παχὺ τὸ μαῦρον καὶ ἄς το πίνῃ. 2 ἄλλο. ▷L⌜βράσον φοινίκια καὶ τὸν ζωμὸν τῶν τοιούτων. ἄς το πίνῃ.⌝L◁ 3 ἄλλο. ▷om. Bβράσον μυρσίνης φύλλα με τὸ κρασὶν καὶ ἄς το πίνῃ.om. B◁ 4 ἄλλο. ⌜ἐπάνω δὲ εἰς τοὺς νεφροὺς θέτε κριθάλευρον βρασμένον με τὸ νερόν. εἶτα ἀφ' ὅτου βράσῃ πάλιν βάλε ὄξος καὶ ῥοδέλαιον καὶ ἔψησον εἶτα βάλε ἀμπέλου ψαλίδας κοπανισμένα καὶ φύλλα ἀμπέλου κοπανισμένα καὶ ἀμάραντα κοπανισμένα καὶ θέτε τα εἰς τοὺς νεφρούς.⌝

161 [133] ρξαʹ ⌜ῥαγάδες λέγονται παρὰ τῶν ἰατρῶν, ὅταν τὰ ὀρχίδια κατασχισθῶσιν ἢ ἡ ἕδρα· ὠφελεῖ δὲ αὐτὰς ῥητίνη τριμμένη καὶ μαλαγμένη με τὸ ῥοδέλαιον καὶ ἄλειφε τοὺς τόπους τοὺς κατασχισμένους πλὴν τὴν ῥητίνην φρύξε την εἰς τὸν ἥλιον. εἶτα τρίψε την. αὐγοῦ κροκὸν θέτε ἐπάνω εἰς τὰς πληγὰς ἢ ὄπιον με τὸν κρόκον.

---

22 παρὰ τῶν ἰατρῶν] καὶ A 1 αὐτὸ] αὐτὰ Σ 1 ἁπλωμένον] ἁπλωμένα Σ 1 εἰς τὸ] με τὸ MB om. A 1 πανὶν] om. A 2 ἀνάλυσον] ἀνάλυσε MB 2 βάλε] τὸ λεγόμενον add. MB 3 θέτε] θές τα M τα add. AB 3 τοὺς] εἰς τοὺς MA 5 διαβήτης] πρὸς πάθος λεγόμενον διαβήτην· διαβήτης M διαβήτης A πρὸς πάθος τὸ λεγόμενον διαβτ' διαβήτης δὲ B 5 λέγεται] λέγ.... L 5 ὅτε ἠνίκα] οὔτε ονικα L 6 ποιεῖ] εἰ L 6 κατὰ] καθ' AB 7 πολυγόνου] πολυγόνα M 7 ἢ τὸ] ... L 7 ῥάσδον] οὐράσδον M 7 τὸ με τὸ κρασὶν] ......σιν L 8 πίνῃ] ...... L 11 θέτε] θὲς M 12 βράσῃ πάλιν βάλε ὄξος καὶ ῥοδέλαιον καὶ ἔψησον εἶτα βάλε] scripsi βράσῃ LMA πάλιν βάλε ὄξος καὶ ῥοδέλαιον καὶ ἔψησον εἶτα ἀφ' ὅτου βράσῃ in marg. M πάλιν βάλε ὄξος καὶ ῥοδέλαιον καὶ ἔψησον εἶτα βάλε ἀφ' ὅτου βράσῃ B 13 ἀμπέλου] βάλε ἀμπέλου A 13 ψαλίδας] ἐλίγγας A 13 κοπανισμένα] κοπάνισον M om. A κοπανισμένας B 13 ἀμπέλου] om. A 13 κοπανισμένα] κοπανισμένας LM 13–14 καὶ ἀμάραντα κοπανισμένα] om. A 14 θέτε] θὲς M 14 τα] om. A 15 ῥαγάδες] πρὸς ῥαγάδες ὀρχιδίων M 15 παρὰ τῶν ἰατρῶν] om. MA 16 ἡ] om. M 16 αὐτὰς] αὐτοὺς A 16 τριμμένη] om. M 17 ἄλειφε] τοὺς πόνους add. B 17–18 κατασχισμένους] κατεσχιμένους A 18 φρύξε] τρίψε A 18 την] πρῶτον θὲς A 18 τρίψε την] τρίψον αὐτὴν M τρίψον A

2 ἄλλο. κίσσου φύλλα ξηρὰ πάσον τὴν πληγήν. 3 ἄλλο. ⊳om. Aτὴν στάκτην τῶν φύλλων τοῦ κισσοῦ μίξον με τὸ ῥοδέλαιον καὶ βάλε εἰς μολύβδινον ἀγγεῖον καὶ τάραξον αὐτὰ εἶτα ἄλειφε.om. A⊲ πρῶτον δὲ ἀποσφούγγιζε τὰς πληγάς. οἶνον ἀντὶ νεροῦ πλὴν τὸ κρασὶν ἂς ἔνι χλίον. εἶτα πάσον ῥόδα ξηρὰ τριμμένα.⌝ 4 ἄλλο. ψιμμίθιν καὶ λιθάργυρον τριμμένον πάσον τὰς πληγάς. ⌜ἐὰν δὲ φλεγμαίνῃ τὸ αἰδοῖον, κοπάνισον τρυφερὰ φύλλα ἀμπέλου καὶ μίξον με τὸ ψιμμίθιν καὶ θὲς ἐπάνω εἰς τὴν φλεγμονήν. 5 ἄλλο. ἄντλει δὲ τὸ αἰδοῖον πρῶτον νερὸν κρύον θαλάσσιον.⌝ εἰ δὲ οὐκ ἔνι θαλάσσιον, βάλε ἅλας καὶ νερόν. εἶτα κοπάνισον ἀμάραντα καὶ μίξε το με τὸ ψιμμίθιον καὶ κρόκον αὐγοῦ καὶ θέτε τα εἰς τὴν φλεγμονὴν τοῦ αἰδοίου. αἰδοῖον δὲ λέγεται διὰ τὸ αἰσχυνόμεθα ἀλλήλοις ταῦτα δεικνύναι.

162 [134] ρξβʹ ὅπου δὲ καταπίῃ βδέλλας ἀπροσδοκήτως, ⌜πότισον αὐτὸν ἅλμην με τὸ νερόν. 2 ἄλλο. νίτρον τριμμένον μίξε το με

19 θέτε] θὲς MA 1 φύλλα ξηρὰ πάσον] φυ... .... L τὰ φύλλα ξηρὰ κοπάνισον καὶ πάσον A φύλλα ξηράνας πάσον B 1 πληγήν] κοπανιστὰ δὲ τὰ φύλλα add. M 2 τῶν φύλλων] ....λων L 2 μίξον] μίξε την B 3 εἶτα ἄλειφε] om. M 4 νεροῦ] τοῦ ὕδατος Σ 4–5 τὸ κρασὶν ἂς ἔνι χλίον] transp. MA 5 χλίον] κρίον B 5 ῥόδα ξηρὰ] transp. M 5 ψιμμίθιν] ψιμμίθιον MA 6 τριμμένον] τριμμένα MA 6–7 τὸ αἰδοῖον] ἡ ψωλὴ μόνη Σ 7 ἀμπέλου] ἀμπελίου M 7 μίξον] αὐτὰ add. MA 7–8 ψιμμίθιν] ψιμήθ.. L ψιμμίθιον MA 8 θὲς ἐπάνω] θές τα M θέτε τα ἐπάνω A θέτε το B 8 φλεγμονήν] τῆς ψωλῆς add. M τῆς λέγοντος add. B 8 δὲ] πρῶτον add. Σ 8 τὸ αἰδοῖον] om. A 9 πρῶτον] om. Σ 9 ἔνι] ἔχεις MA ἔχι B 10 μίξε το] ἕνωσον αὐτὰ M μίξον αὐτὰ AB 10–11 ψιμμίθιον] ψιμίνθιον A ψιμμίθιν B 11 θέτε τα] θὲς αὐτὰ M θέτε το B 11 τὴν φλεγμονὴν τοῦ αἰδοίου] αἰδοῖου τὴν φλεγμονὴν ἤτοι τῆς ψωλῆς M τὴν φλεγμονὴν τοῦ αἰδοίου ἤτοι τῆς ψωλῆς A ἢ τῆς λέγοντος add. B 12 δὲ λέγεται] δέονται B 12 διὰ τὸ] ὅταν M ὅτι A ἐπηδεὶ B 12 ἀλλήλοις] ἀλλήλους B 12 ταῦτα δεικνύναι] ....κνύναι L δεικνύναι B 13 ὅπου] πρὸς καταπίοντας βδέλλας· ἐὰν M πρὸς ἀβδέλλας ἐὰν A πρὸς καταπίοντας ἀβδέλας B 13 δὲ] τις MA om. B 13 καταπίῃ] om. B 13 βδέλλας] scripsi ...... L βδέλλαν M ἀβδέλλας A om. B 13 ἀπροσδοκήτως] om. AB

6 ἐὰν] It appears as if this could be the beginning of a new chapter in L, but since a part of the margin is missing this cannot be stated with certainty. 6–7 τὸ αἰδοῖον] In the margins of L a fragmentary scholion by the first hand in red ink can be found; obviously, the word αἰδοῖον was not generally understood, and it had to be explained by a vernacular synonym. The text reads ..δοῖον ... λέ.. ται .. ψο... The Σ group of the manuscripts has the synonym as well.

τὸ νερὸν καὶ ἄς το κρατῇ εἰς τὸ στόμα του. καὶ ἄς το κινῇ τὸ νερὸν ἐκεῖνο, ἔνθεν καὶ ἔνθεν εἰς τὸ στόμα του. ἂς τρώγῃ δὲ καὶ σκόρδα.⌝ 3 ἄλλο. κοπάνισον κόριδας καὶ μίξε τας με τὸ ὀξείδιν τὸ δριμὺ καὶ ἄς τας πίῃ.

163 [134] ρξγ´ ἐὰν φάγει τις ἀμανιτάρια πολλὰ χλωρὰ καὶ ὡς ἂν πνίγεται· οὕτως ἀνάγκαζον ἵνα ξερᾷ πλὴν ποτίσε τον ἔλαιον καὶ να ξεράσῃ. 2 ἄλλο. ⌜κλήματος στάκτην πότισον με τὸ ὀξείδιν τὸ πολύ. καὶ τὸ κρασὶν ἤτοι δύο τὸ ὀξείδιν καὶ ἕνα τὸ κρασίν. 3 ἄλλο. τρίψον τὸ λεγόμενον νίτρον καὶ πότισον καὶ αὐτὸ ὡς ἐπότισας τὴν στάκτην τοῦ κλήματος. 4 ἄλλο. ⊳om. Aαὐγὰ ὀρνιθερὰ ἔψησον εἰς ὀξείδιν καὶ κρασὶν καὶ ἄς τα τρώγῃ ἐκζεστά.⌝ 5 ἄλλο. ἀψίνθιον βράσουν με τὸ κρασὶν καὶ ἄς το πίνῃ.om. A⊲ 6 ἄλλο. ⊳om. MAμελισσόφυλλον καὶ νίτρον τριμμένον βράσον με τὸ μέλι. καὶ βάλε καὶ νερὸν ὀλίγον καὶ ἄς το πίνῃ.om. MA⊲ 7 ἄλλο. ⊳om. L⌜πήγανον βράσον με τὸ ἔλαιον βάλε καὶ κρασὶν καὶ ἄς πίνῃ.om. L⊲ 8 ἄλλο. ⊳om. Bβράσε ὀρίγανον με τὸ κρασὶν καὶ μέλι καὶ ἄς το πίνῃ.om. B⊲ 9 ἄλλο. ὅσον δὲ δύναται, ἂς ξερᾷ,⌝ ἵνα τὰ ἐκβάλῃ καὶ λοῦέ τον καὶ συνεχῶς.

164 [135] ρξδ´ ⌜ἐὰν δὲ τυρωθῇ τὸ γάλα εἰς τὴν κοιλίαν τοῦ ἀνθρώπου· πότισον αὐτὸν ποταμογείτονος φύλλα ξηρὰ με τὸ νερὸν

---

14 μίξε το] ἕνωσον M μίξον A μίξε B 1–2 τὸ νερὸν ἐκεῖνο] om. A 2 καὶ] .. L 2 εἰς τὸ στόμα του] om. A καὶ add. B 2 δὲ] om. B 2 καὶ] ... L τὰ add. MB 3 σκόρδα] σκόροδα AB 3 κόριδας] σκόρυδας B 3 μίξε τας] ἕνωσον M 4 τας] το Σ 4 πίῃ] πίνῃ AB 5–6 ἐὰν φάγει τις ἀμανιτάρια πολλὰ χλωρὰ καὶ ὡς ἂν πνίγεται] transp. B 5 φάγει τις] transp. M 5 πολλὰ] om. MA 6 οὕτως] τοῦτον Σ 6 ἀνάγκαζον] ἀνάγκαζε MB ἀνάγκασον A 6 ξερᾷ] ξηρὰ MB 6 ποτίσε] πότισον MA 6 τον] αὐτὸν M om. A 6 καὶ] μόνον καὶ Σ 7 πότισον] αὐτὸν add. A 8 ἕνα] ἓν B 8 τρίψον] δὲ add. M 9 τὸ] om. A 9 λεγόμενον] om. A 9 καὶ αὐτὸ] om. A ἑαυτὸν B 10 ὀρνιθερὰ] ὄρνιθος M 10 εἰς ὀξείδιν] .... L 11 τα] om. B 11 ἀψίνθιον] ψ̣...θ̣ι̣ο̣ν̣ L 11 βράσουν] βράσον MB 12 το] om. M 12 μελισσόφυλλον] φελισσόφυλλα B 13 νίτρον] κίτρον L 13 καὶ] om. B 14 πήγανον] ἀπήγανον A 14 βράσον] πάλιν B 14 ἔλαιον] καὶ add. A 15 καὶ] τὸ μέλι καὶ add. B 15 ἄς] το add. A 15 βράσε] βράσον MA 15 ὀρίγανον] ὀριγάνην M ὀρύγανιν A 16 καὶ] τὸ add. MA 16 δὲ] om. A 17 ξερᾷ] ξεράσῃ A 17 τὰ] om. MB 17 ἐκβάλῃ] αὐτὰ add. MB 17 συνεχῶς] ἐὰν δύναται καθ’ ἣν ἡμέραν add. A 18 δὲ] om. Σ 18 εἰς τὴν κοιλίαν] ἐν τῇ κοιλίᾳ B

καὶ χλωρὰ κοπάνισον καὶ τὸν ζωμόν τους πότισον αὐτόν. 2 ἄλλο. βράσον θύμον με τὸ κρασὶν καὶ ἄς το πίνῃ.⌝ 3 ἄλλο. πυτία οἷα καὶ ἂν ἔνι πότισον με τὸ ὄξος πλὴν τριμμένη καλά.

165 ρξε΄ ὅταν πέσουν τὰ μαλία τῆς κεφαλῆς ὅλα, ποίησον θεραπείαν τοιαύτην· ἤγουν ἀδίαντον κοπάνισον καὶ μίξον με τὸ λάδανον καὶ ποίησέ το ὡς ἂν πήττας καὶ θέτε το εἰς τὴν κεφαλὴν ἐπάνω ὅλην. 2 ἄλλο. ▷om. A τοῦ ἀσφοδέλου καῦσέ την καὶ τὴν στάκτην αὐτῆς μάλαξον με τὸ χρηστέλαιον καὶ ἄλειφε τὴν κεφαλήν. 3 ἄλλο. καλάμου χλωροῦ τὸ φλοῦδιν κοπάνισον καὶ μάλαξον με τὸ ὄξος καὶ θέτε τα εἰς τὴν κεφαλήν του.om. A◁ 4 ἄλλο. πέπερι τριμμένον μίξον με τὸ κρασὶν καὶ τὸ ὄξος. καὶ ἄλειφε τὴν κεφαλὴν ἅπασαν. 5 ἄλλο. ἀγριοσυκέας φύλλα καὶ συκέας ἡμέρου κοπάνισον καὶ μάλαξον αὐτὸ με τὸ μέλι. καὶ θέτε τα εἰς τὴν κεφαλήν. 6 ἄλλο. ἐχίνου δέρμα καῦσον καὶ τὴν στάκτην ἐκείνην μίξε την με τὸ ὄξος καὶ τὸ χρηστέλαιον καὶ ἄλειφε τὴν κεφαλήν. 7 ἄλλο. ▷M ἐχῖνον θαλάσσιον καύσε τον οὕτως με τὸ ὄστρακόν του καὶ τὴν στάκτην μίξον με τὸ ὄξος καὶ τὸ χρηστέλαιον καὶ ἄλειφε τὴν κεφαλήν.M◁ 8

18–19 τοῦ ἀνθρώπου] om. B 1 καὶ] ἢ καὶ M ἢ B 1 τὸν ζωμόν τους] om. AB 1 τους] τῶν M 1 πότισον] πότι... L 1 αὐτόν] τὸν ζωμὸν A τὸν ζωμόν τους B 2 πυτία] scripsi πιτύα LM πυτάι A πίτυρα B 2–3 καὶ ἂν] om. M καὶ A 3 ἔνι] ἐν... L 3 με τὸ ὄξος] μετὰ ὄξους B 3 τριμμένη] τριμέν... L τριμμένην MB 4 ὅταν] ἐὰν A 4 πέσουν] πέσωσιν MB 4 τὰ μαλία] αἱ τρίχες M 4 ὅλα] om. MB 5 τοιαύτην] om. A 5 ἤγουν] ἤγουν Σ 5 ἀδίαντον κοπάνισον] transp. B 5 μίξον] ἕνωσον M 5 με τὸ] με τὸν MA 6 ποίησέ το] ποίησον αὐτὸ MA ποίησον αὐτὰ B 6 ὡς ἂν] ὡσεὶ A 6 θέτε] θές B 6 το] om. M τας B 6 κεφαλὴν ἐπάνω] scripsi .... ἐπάνω L κεφαλήν του MA κεφαλὴν B 7 τοῦ] τῆς M 7 ἀσφοδέλου] scripsi ἀσφοδελ...... L ἀσφοδελέας τὴν ῥίζαν M ἀσφοδέλου τὴν ῥίζαν B 7 καῦσέ] καῦσον M 7 την] om. M 9 φλοῦδιν] φλούδιον MB 9 με τὸ ὄξος] μετὰ ὄξους B 10 θέτε τα] θὲς M θές το B 10 του] ὅλην add. M ὅλην B 11 τὴν κεφαλὴν ἅπασαν] transp. B 11 ἅπασαν] om. M 12 συκέας ἡμέρου] transp. B 13 αὐτὸ] αὐτὰ Σ 13 θέτε τα] θὲς M 13 κεφαλήν] ὅλην add. MB του ὅλην add. A 14 ἐχίνου] scripsi σχοίνου L ἀγχήνου τὸ λεγόμενον παρὰ τῶν δυσικῶν σκαντζόχοιρον τοῦτο τὸ M ἐχίνου δέρμα τὸ λεγόμενον παρὰ τῶν δυσικῶν σκατζοχοίρου τοῦ ἔχοντος ὡσὰν σουγλία τούτου τὸ A σχίνου δέρμα τὸ λεγόμενον παρὰ τῶν δυσικῶν σκατζόχοιρος τοῦ ἔχοντος ὡσὰν σουφλία παρόξυντα τοῦτο τὸ B 14 ἐκείνην] ἐκείνου A 14 μίξε] μίξος M μίξον A 14 την] om. MA ἄλλο. ▷[om. L]πτελέας ῥίζαν κοπάνισον καλὰ καὶ μίξον με τὸ ἀξούγγιν τῆς ἄρκου καὶ ἄλειφε τὴν κεφαλήν.[om. L]◁ 9 ἄλλο. ἀγριοσταπίδαν τριμμένην καὶ τεάφην καὶ ἀρσενίκιν σχιστὸν καὶ μίξον μετὰ ὄξους καὶ ἄλειφε. 10 ἄλλο. ▷[om. A]ψιμμίθιον καὶ λιθάργυρον τριμμένον μίξον μετὰ ὄξους καὶ ἐλαίου καὶ ἄλειφε.[om. A]◁ 11 ἄλλο. ἀσβέστιν πλύνε με τὸ νερὸν καὶ χύσε τὸ νερὸν καὶ βάλε ἔλαιον παλαιὸν καὶ ἄλειφε τὴν κεφαλὴν. 12 ἄλλο. ▷[om. A]ξύρισε τὸ κεφάλιν του καλὰ ὅλον καὶ βάλε ὄξος καὶ χήνειον ἀξούγγιν λυτὸν καὶ ἄλειφε.[om. A]◁ 13 ἄλλο. ἀνάλυσε ἀξούγγιν αἴγιον καὶ βάλε κεδρέαν καὶ ὄξος καὶ ἄλειφε τὴν κεφαλὴν γαληνούτζικα. 14 ἄλλο. ▷[om. A]αὐγέλαιον τριμμένον καὶ μαστίχην με τὸ ὄξος ἄλειφε τὴν κεφαλήν.[om. A]◁ 15 ἄλλο. ▷[om. B]ξύρισον τὸ κεφάλιν του καλὰ ὅλον καὶ βάλε ὄξος καὶ χήνειον ἀξούγγιν λυτὸν καὶ ἄλειφε τὴν κεφαλήν.[om. B]◁ 16 ἄλλο. ὑδράργυρον καὶ πτύσμα ἀνθρώπου καὶ ναρδέλαιον ἄλειφε τὴν κεφαλήν. 17 ἄλλο. ἀξούγγιν λυτὸν ἀλώπεκος μίξον μετὰ ὄξους καὶ κρασίου καὶ ἄλειφε τὴν κεφαλήν. 18 ἄλλο. σευτλόφυλλα κοπάνισον καὶ τὸν ζωμὸν τούτων μίξον με τὸ ὄξος καὶ τὸ χρηστέλαιον καὶ ἄλειφε τὴν κεφαλήν. 19 ἄλλο. ▷[om. AB]Ποντικοῦ κόπρα ξηρὰ τριμμένα μίξον με τὸ ὄξος καὶ τὸ χρηστέλαιον καὶ ἄλειφε τὴν κεφαλήν.[om. AB]◁ 20

---

15 ἐχῖνον] scripsi ἀγχήνουν M 2 ἄρκου] ἄρκτου B 2 ἀγριοσταπίδαν] ἀγριοσταφίδαν Σ 3 τεάφην] τεάφιν Σ 3 καὶ] om. AB 3–4 μετὰ ὄξους] με τὸ ὄξος MA 4 ἄλειφε] τὴν κεφαλήν add. MA 4 ψιμμίθιον] ψιμμίθιν B 5 μετὰ ὄξους] με τὸ ὀξείδιν MB 5 ἐλαίου] τὸ ἔλαιον MB 5 καὶ] om. B 5 ἄλειφε] τὴν κεφαλήν add. M 6 καὶ χύσε τὸ νερὸν] om. B 6 καὶ βάλε ἔλαιον παλαιὸν] ....αι παλαιόν L καὶ βάλε ἔλαιον M 7 κεφαλὴν] κε..... L 7 ξύρισε] ξύρισον MB 7 τὸ κεφάλιν] τὴν κεφαλὴν B 8 ὅλον] ὅλην B 8 χήνειον ἀξούγγιν] χιναξούγγιν B 8 ἄλειφε] τὴν κεφαλήν add. M 9 ἀνάλυσε] ἀνάλυε A 9 αἴγιον] αἴγειον Σ 9 ὄξος] καὶ ἕνωσον add. M καὶ μίξε τα add. AB 10 γαληνούτζικα] scripsi γαλινούτζικα LM om. A 10–11 τριμμένον καὶ] καὶ τριμμένον M 11 μαστίχην] μαστίχιν M 11 με τὸ] μετὰ B 11 ὄξος] καὶ add. M ὄξους B 13–14 ὑδράργυρον] ὑδάργυρον B 14 κεφαλήν] ὅλην add. M om. A 15 μετὰ ὄξους] με τὸ ὄξος MA με τὸ ὀξείδιν B 15 καὶ κρασίου] καὶ τὸ κρασὶν Σ 16 σευτλόφυλλα] σεύτλου φύλλα Σ 17 τούτων] τῶν MA τους B 17 μίξον] ἕνωσον M 17 ὄξος καὶ τὸ] om. B 17 καὶ] om. M 18 τὴν κεφαλήν] om. A

ἄλλο. ▷om. A κοπάνισον ῥίζαν σελίνου καὶ μίξον αὐτὴν μετὰ ὄξους καὶ ἄλειφε τὴν κεφαλήν.om. A◁ 21 ἄλλο. ▷om. B σκουρέαν σιδήρου βάλε καὶ ὄξος καὶ χρηστέλαιον καὶ ἂς ποιήσῃ νυχθήμερον καὶ ἄλειφε τὴν κεφαλήν.om. B◁ 22 ἄλλο. ▷om. L ὑοσκυάμου τὰ φύλλα με τὸ ὄξος καταπλασσόμενα πλὴν πρῶτον κοπάνισέ τα καὶ ῥητίνην τριμμένην με τὸ ἀξούγγιν τοῦ γεράνου καὶ ἐλάφου μυελὸν καὶ ὄξος μίξον τὰ πάντα καλῶς ἄλειφε.om. L◁ 23 ἄλλο. ▷om. A λάδανον μάλαξον με τὸ κρασὶν καὶ με τὸ μυρσινέλαιον καὶ ἄλειφε τὴν κεφαλήν.om. A◁ 24 ἄλλο. ▷AB σχίνου βλαστάρια τρυφερὰ κοπάνισον καὶ μάλαξον πάντα με τὸ ῥοδέλαιον καὶ τὸ κρασὶν καὶ θέτε τα εἰς τὴν κεφαλήν.AB◁

166 ρξϛ´ εἰς πιτυρίδα κεφαλῆς· ▷om. M κρίνου ῥίζαν κοπάνισον καὶ ἄλειφε τὴν κεφαλήνom. M◁ 2 ἄλλο. ▷M συκαμινεάς τὴν ῥίζαν κοπάνισον καὶ βάλε ὄξος καὶ μάλαξον καὶ ἄλειφε τὴν κεφαλήνM◁. 3 ἄλλο. ἔστι βοτάνιν τὸ λεγόμενον στρούθιον. τούτου τὴν ῥίζαν κοπάνισον καὶ μίξας μετὰ ὄξους ἄλειφε. 4 ἄλλο. λάδανον μάλαξον μετὰ ὄξους καὶ ἄλειφε. 5 ἄλλο. ἀγριοκαρδάμου σπέρμα ἢ τὰ φύλλα κοπάνισον καὶ μίξας μετὰ ὄξους ἄλειφε. 6 ἄλλο. κολίανδρον τρίψον καὶ πηγάνου σπέρμα ἢ τὰ φύλλα βάλε δὲ καὶ ὄξος καὶ μάλαξον αὐτὰ

---

18 κόπρα] κόπρια M 1 κοπάνισον ῥίζαν σελίνου] transp. A 1 μίξον αὐτὴν] μίξον M μίξε την B 1 μετὰ ὄξους] με τὸ ὄξος M τοξείδιν B 2–3 σκουρέαν σιδήρου βάλε καὶ] scripsi σκουρέαν ...... L σκωρέσιδήρου βάλε M σκωρέαν σιδήρου βάλε καὶ A 3 ἂς ποιήσῃ] ἀ...... L 4 με τὸ] βάλε B 5 καὶ] βάλε add. A om. B 6 ὄξος] καὶ add. B 6–7 μίξον τὰ πάντα καλῶς] καὶ μίξον αὐτὰ A 7 ἄλειφε] τὴν κεφαλὴν add. B 7 λάδανον] .αδανον L πάδανον M 8 μυρσινέλαιον] μυρνέλαιον M 9 πάντα] αὐτὰ B 11 πιτυρίδα] scripsi πυτυρίδα L πιτυρίδαν M πητυρίδας B 11 κρίνου ῥίζαν] κρίνα ἤτοι τὴν ῥίνα B 12 συκαμινεάς] κυκλαμινέας B 14 ἔστι] ἔνι M 14 βοτάνιν] βότανον MA 15 μίξας μετὰ ὄξους ἄλειφε] βάλε ὄξος καὶ μάλαξον καὶ ἄλειφε τὴν κεφαλήν M τὴν κεφαλὴν add. B 15 μετὰ] με τὸ A 15 ὄξους] ὄξος AB 15–16 μετὰ ὄξους καὶ ἄλειφε] om. M τὴν κεφαλὴν add. B 17 μίξας μετὰ ὄξους] καὶ μετὰ ὄξους ἔνωσον Σ 17 ἄλειφε] τὴν κεφαλὴν add. MB 17 κολίανδρον] κολίαντρον AB 17 τρίψον] τρίψας Σ 18 τὰ] om. Σ 18 φύλλα] πλὴν ἀγρίου μίξον καὶ λάδανον μετ᾽ αὐτὰ καὶ add. M πλὴν ἀγρίου καὶ λάδανον μίξας add. A πλὴν ἄγρια καὶ λάδανον B 18 βάλε] καὶ B 18 δὲ καὶ] om. Σ 18 μάλαξον] μίξε B

---

1 κοπάνισον ῥίζαν σελίνου] This paragraph is at the end of the chapter in A.
7 λάδανον] The following two paragraphs are reversed in B.

καὶ ἄλειφε. 7 ἄλλο. ⊳om. B κισσοῦ φύλλα κοπάνισον καὶ τὸν ζωμὸν τούτων ἄλειφε τὴν κεφαλήν.om. B⊲ 8 ἄλλο. ⊳om. Σ τοῦ σεύτλου τὴν ῥίζαν κοπάνισον καὶ τὸν ζωμὸν τούτου ἄλειφε τὴν κεφαλήν.om. Σ⊲ 9 ἄλλο. ⊳om. LB πήγανον ἄγριον κοπάνισον καὶ τρίψον καὶ λιθάργυρον κοπάνισον καὶ ἀγριοσταφίδα καὶ βάλε ὄξος καὶ μυρσινέλαιον ὅσον ἀρκεῖ καὶ ἕνωσον ταῦτα καὶ ἄλειφε τὴν κεφαλήν.om. LB⊲ 10 ἄλλο. ⊳om. L τεάφιν τριμμένον μίξον μετὰ μυρσινέλαιον καὶ ἄλειφε. 11 ἄλλο. σεύτλου τὴν ῥίζαν κοπάνισον καὶ τὸν ζωμὸν τούτου ἄλειφε τὴν κεφαλήν.om. L⊲ 12 ἄλλο. λιθάργυρον τριμμένον καὶ ἅλας καὶ οἶνον καὶ ἔλαιον μίξας ἄλειφε τὴν κεφαλήν. 13 ἄλλο. βάλε ἀγριαγγουρέας ῥίζαν καὶ λουπηνάρια καὶ βράσε τα με τὸ νερὸν καὶ ἃς λούηται με τὸ νερὸν ἐκεῖνο. 14 ἄλλο. ⊳om. L τήλην βράσον καὶ κοπάνισον καὶ σεύτλου φύλλα καὶ τὸν ζωμὸν τούτων μίξον με τὸ νερὸν τῆς τήλης καὶ ἄλειφε τὴν κεφαλήν.om. L⊲ 15 ἄλλο. τὴν στάκτην τοῦ κλήματος μάλαξον με τὸ νερὸν καὶ ἄλειφε.

167 [136] ρξζʹ ⌜ἀλφοὶ λέγονται, ὅταν ὁ ἄνθρωπος εὐγάνει εἰς τὸ πρόσωπον ὡς ἂν φακῆς κοκκία.⌝ καὶ εἰς λειχῆνας καὶ εἰς λέπρας ὠφελεῖ· ⌜δενδρολίβανον κοπάνισον καὶ μίξον με τὴν κεδρέαν καὶ

---

18 αὐτὰ] αὐτὰ καλὰ M om. AB 1 ἄλειφε] τὴν κεφαλὴν add. MA 1 κισσοῦ] τὰ add. M 4 τρίψον] μίξον A 5 κοπάνισον] τριμμένον κοπάνισον A 6 ἕνωσον ταῦτα] μίξε τα ὅλα A 7 τριμμένον] τριμμένην A 7 μίξον] τρίψον B 7 μετὰ] με τὸ AB 7 ἄλειφε] τὴν κεφαλὴν AB τοῦ add. M 8 τὴν] om. A 8 τούτου] ταύτης A αὐτῆς B 9 κεφαλήν] om. A 9 λιθάργυρον τριμμένον] λιθαρ............ L 9 ἅλας] ἔλαιον A 10 οἶνον] ἅλας A 10 ἔλαιον μίξας] ἐλαι............ L ἔλαιον M οἶνον A 10 τὴν κεφαλήν] om. MA 10–11 βάλε ἀγριαγγουρέας] βα...γριαγγουρέας L 11 λουπηνάρια] λυπηνάρια Σ 12 με τὸ] τὸ AB 13 καὶ] μίξον add. A 13 τούτων μίξον] τοῦ σεύτλου A 13–14 με τὸ νερὸν τῆς τήλης] με τῆς τήλης A 14 τὴν κεφαλήν] om. A 14 τὴν στάκτην] scripsi τὴν στα.την L καῦσον κλῆμα καὶ εἰς τὴν στάκτην M καῦσον κλῆμα καὶ τὴν στάκτην A καῦσον κλῖμα καὶ εἰς τὴν καῦσιν B 15 μάλαξον με τὸ νερὸν καὶ] βάλε νερὸν καὶ τὸ νερὸν ἐκεῖνο μετὰ τὴν στάκτην MB βάλε με τὸ νερὸν καὶ A 15 ἄλειφε] τὴν κεφαλὴν add. A 16 ἀλφοὶ λέγονται] πρὸς ἀλφοὺς M ἀλφοὶ λέγονται A 16–17 ὁ ἄνθρωπος εὐγάνει εἰς τὸ πρόσωπον] εἰς τὸ πρόσωπον τοῦ ἀνθρώπου εὐγένουσιν M εἰς τοῦ ἀνθρώποῦ τὸ πρόσωπον εὐγένει A 17–18 καὶ εἰς λειχῆνας καὶ εἰς λέπρας ὠφελεῖ] ὃκαὶ παρὰ τῶν ἰατρῶν ἀλφοὶ ὀνομάζονται. ποίησον θεραπείαν τοιαύτην. ἤγουν M 17 εἰς] om. B 18 δενδρολίβανον κοπάνισον] transp. M 18 κοπάνισον] κοπανισμένον A 18 καὶ μίξον] om. A

ἄλειφε πλὴν βάλε καὶ χρηστέλαιον. 2 ἄλλο. τεάφην τριμμένην καὶ λιθάργυρον καὶ ἀγριοσταπίδα καὶ χρηστέλαιον μίξον καὶ ἄλειφε τὴν κεφαλήν. 3 ἄλλο. δᾳδὶν λιπαρὸν βράσον με τὸ χρηστέλαιον καὶ ἄλειφε. πλὴν βάλε καὶ λιθάργυρον καὶ ὄξος. ἄλειφε δὲ ἐν λουετρῷ.¬ 4 ἄλλο. μάρμαρον τριμμένον ψιλὸν πολλὰ μίξον με τὸ ἄσπρον τοῦ αὐγοῦ καὶ ἄλειφε. 5 ἄλλο. ψιμμίθιον καὶ κόπρα ξηρὰ χελιδωνίου καὶ ἄσπρον αὐγοῦ μίξον καὶ ἄλειφε. 6 ἄλλο. ▷om. AB κρίνου ῥίζαν κοπάνισον καὶ μίξας μετὰ ὄξους ἄλειφε. om. AB◁ 7 ἄλλο. ▷om. L συκαμινέαν ῥίζαν κοπάνισον καὶ μάλαξον αὐτὴν μετὰ ὄξους καὶ ἄλειφε. om. L◁ 8 ἄλλο. ἀγριοσυκέας τὰ φύλλα ἢ καὶ ἡμέρου συκέας κοπάνισον καὶ μάλαξον ταῦτα μετὰ ὄξους καὶ θέτε τα εἰς τὸ πάθος. 9 ἄλλο. om. A▷⌐ ἀγριοσταπίδαν τριμμένην καὶ τεάφην τριμμένην μίξον μετὰ ὄξους καὶ ἄλειφε.¬ om. A◁ 10 ἄλλο. ποταμογείτονος φύλλα κοπάνισον καὶ τρίψον καὶ τεάφην καὶ βάλε ὄξος καὶ μάλαξον καὶ ἄλειφε. 11 ἄλλο. ▷om. L καρπὶν κοπάνισον καὶ νίτρον καὶ βάλε ὄξος καὶ ἄλειφε om. L◁.

---

18 καὶ] τὸ χρηστέλαιον καὶ add. Σ 1 πλὴν βάλε καὶ χρηστέλαιον] om. AB ὀφελεῖ δὲ τὸ βότανον ἐ τοῦτο ἤγουν τὸ δέδρολίβανον καὶ εἰς λειχήνας καὶ εἰς λέπρας M 1 τεάφην] τεάφιν Σ 1 τριμμένην] τριμμένον MB 2 ἀγριοσταπίδα] ἀγριοσταφίδα MB ἀγριοσταφίδαν A 3 τὴν κεφαλήν] om. AB 3 λιπαρὸν] λειπαρὸν M 3 χρηστέλαιον] καλὰ καὶ τὸ ἔλαιον αὐτὸ μίξον μετὰ τριμμένου λιθαργύρου καὶ ὄξος add. M καλὰ καὶ τὸ ἔλεος αὐτὸ τρίψον λιθάργυρον καὶ μίξον καὶ ὄξος add. A 4 πλὴν] om. B 4 βάλε καὶ λιθάργυρον καὶ ὄξος] om. Σ 4 λιθάργυρον] scripsi ...θάργυρον L 4 ἄλειφε δὲ] om. MA 4 λουετρῷ] λουτρῷ MB τῷ λουτρῷ A 5 μίξον με τὸ] καὶ B 5 τοῦ] om. B 6 καὶ] om. B 6 κόπρα ξηρὰ χελιδωνίου] scripsi κοπρ.........δωνίου L κόπρια ξηρὰ χενδρώνος M κόπρια χελιδόνος ξηρὰ A κόπρια ξηρὰ χελιδόνος B 7 μίξον καὶ ἄλειφε] μι........λειφε L μίξας ἄλειφε AB 7–8 κοπάνισον καὶ] scripsi κοπάνισον .. L om. M 8 μίξας μετὰ ὄξους] καὶ ὄξος μίξας M 8 ἄλειφε] τὴν κεφαλὴν add. M 8–9 συκαμινέαν ῥίζαν κοπάνισον] κοπάνισον τὴν ῥίζαν τῆς συκαμινέας B 9 μετὰ ὄξους] ὄξος A 9 ἄλειφε] θέτε την B 10 τὰ] om. B 10 καὶ] om. B 10 ἡμέρου συκέας] ἡμεροσυκέας M ἡμέρου A συκέας B 11 ταῦτα] αὐτὰ B 11 μετὰ ὄξους] με τὸ ὄξος MA 11 θέτε] θὲς M 12 ἀγριοσταπίδαν] ἀγριοσταφίδαν M 12 τεάφην] τεάφιν MB 12 τριμμένην] τριμμένον MB 12 μίξον] μίξας M 14 τεάφην] τεάφιν Σ 14 ἄλειφε] scripsi ἀ...φε L ἔπειτα ἄλειφε M 15 καρπὶν] καρπὴν B

---

6 ἄλλο] The following two paragraphs are reversed in M.

168 ρξη´ ἄφθα ἔνι, ὅταν τοῦ ἀνθρώπου τὸ στόμα φουσκώσῃ ἀπέσω καὶ τὰ οὖλη· κράτει εἰς τὸ στόμα ῥοδόσταγμα. 2 ἄλλο. βερονίκην ἔψησον σὺν οἴνῳ καὶ ἄς το κρατῇ εἰς τὸ στόμα του. 3 ἄλλο. ἐλαίας μαύρης τὰς ἐλαίας ἆς μασσᾶται. καὶ ἄς τα κρατῇ εἰς τὴν γλῶσσαν του. 4 ἄλλο. φλούδια ῥοϊδίνων βράσον με τὸ κρασὶν καὶ κράτει εἰς τὸ στόμα. 5 ἄλλο. ▷om. MB ἀγριοσταπίδαν ἔψησον μετὰ ὄξους.om. MB◁ 6 ἄλλο. ▷L†....... ἄς τον κρατῇ εἰς τὸ στόμα του.L◁ 7 ἄλλο. βερονίκην καὶ ῥοδόσταγμα καὶ μέλι μίξον καὶ ἆς κρατῇ εἰς τὸ στόμα. 8 ἄλλο. ▷om. Lἀγριελαίας φύλλα τρυφερὰ ἆς μασσᾶται καὶ τὸν ζωμὸν τῶν φύλλων τῆς ἐλαίας ἆς ἀλείφηται.om. L◁ 9 ἄλλο. ἀγριελαίας φύλλα βράσον μετὰ ὄξους καὶ τὸν ζωμὸν ἆς κρατῇ εἰς τὸ στόμα αὐτοῦ. 10 ἄλλο. ▷Lβάτου φύλλα τρυφερὰ ἆς μασσᾶται.L◁ 11 ἄλλο. βάτου φύλλα τρυφερὰ ἆς βράσουν δὲ αὐτὰ με τὸ ὄξος καὶ ἆς κρατῇ τὸ ὄξος εἰς τὸ στόμα του. 12 ἄλλο. ▷om. Lἔνι ἄκανθα καὶ λέγεται τρίβολον καὶ ποιεῖ ἀκάνθια τετράγονα στρογγύλα. ταῦτα τὰ φύλλα τῆς ἀκάνθου βράσον με τὸ ῥοδόσταγμα καὶ ἄλειφε τὰ οὖλη.om. L◁

169 ρξθ´ περὶ ἀμβλυοπίας· πήγανον χλωρὸν κοπάνισον καὶ μάλαθρον καὶ τὸν χυλὸν τούτων μάλαξον με τὸ μέλι καὶ ἄλειφε τοὺς

---

1 ἄφθα] πρὸς ἄφθας· ἀφθα M πρὸς ἄφθας τὰς ἐν τῷ στόματι B 1 τοῦ ἀνθρώπου] om. A 1 στόμα] τοῦ ἀνθρώπου add. A 1 φουσκώσῃ] φουσκώνη M φουσκώνει A 2 εἰς τὸ στόμα ῥοδόσταγμα] transp. A 2 βερονίκην] βερωνίκην A 4 μαύρης] μαύρας Σ 4 τὰς ἐλαίας] om. AB 4 τα] τας AB 4 τὴν] τὸ AB 4 γλῶσσαν] στόμα AB 5 του] ἤτοι εἰς τὴν γλῶσσαν A καὶ εἰς τὴν γλῶτταν του add. B 5 φλούδια] φλούδιον M 5 ῥοϊδίνων] ῥοϊδίου M ῥοδίων A ῥοϊδίων B 5 κρασὶν] καὶ τὸν ζωμὸν add. MA 5 καὶ] om. A 6 στόμα] του add. MB 6 ἀγριοσταπίδαν] ἀγριοσταφίδας A 6 μετὰ ὄξους] με τὸ ὄξος καὶ κράτει εἰς τὸ στόμα A 7–8 βερονίκην καὶ] β........ L βερωνίκην καὶ A 8 μίξον] μίξ.. L μίξε B 8 ἆς κρατῇ] .... L ἄς το κρατῇ AB 8 στόμα] του add. MB 9–10 τὸν ζωμὸν τῶν φύλλων τῆς ἐλαίας ἆς ἀλείφηται] ἆς ἀλείφηται καὶ τὸν ζωμὸν τῶν φύλλων μετὰ ὄξους B 10 ἀγριελαίας] τῆς ἀγριελαίας τὰ Σ 11 βράσον] βράσουν B 11 τὸν ζωμὸν] om. Σ 11 κρατῇ] τὸ ὄξος add. MB τὸ κρατεῖ τὸ ὄξος A 11 αὐτοῦ] του M ομ. B 12 μασσᾶται] ἆς βράσουν δὲ αὐτὰ με τὸ ὄξος καὶ ἆς κρατῇ τὸ ὄξος εἰς τὸ στόμαν του B 14 λέγεται] λέγουνται B 15 τρίβολον] τριβόλην AB 15 τετράγονα] τρίγωνα A 15 ταῦτα] ταύτης B 16 ἀκάνθου] ἀκάνθης AB 16 βράσον] βράσε A 17 περὶ] πρὸς Σ 17 ἀμβλυοπίας] ἀμβλυωπίαν ποίησον θερπείαν A ποίησον θεραπείαν τοιαύτην add. M ἀμβλυωπίαν· ἀμβλιωπίας θερπείαν ἐχει ταύτην B

ὀφθαλμοὺς ἔξωθεν. 2 ἄλλο. ▷om. LBδενδρολίβανον καῦσον καὶ τὴν στάκτην μίξον με τὸ μέλι καὶ ἄλειφε τοὺς ὀφθαλμοὺς ἔξω.om. LB◁ 3 ἄλλο. κισσὸν μαῦρον κοπάνισον καὶ τὸν ζωμὸν τούτου βάλε εἰς τὴν μύτην του. 4 ἄλλο. ▷om. Lβεττονικῆς ζωμὸν στάξον ἀπέσω εἰς τοὺς ὀφθαλμούς.om. L◁ 5 ἄλλο. χήνας χολὴν ἄλειφε ἔξω τοὺς ὀφθαλμούς. 6 ἄλλο. ▷om. Lλιθάργυρον τριμμένον μίξον με τὸ ὄξος καὶ τὸ χρηστέλαιον καὶ ἄλειφε ἔξω τοὺς ὀφθαλμούς.om. L◁ 7 ἄλλο. ἅλας καππαδοκικὸν τριμμένον μίξον με τὸν χυλὸν τοῦ μαλάθρου καὶ μέλι ἀκάπνιον βάλε καὶ μίξον καὶ ἄλειφε ἔξω τοὺς ὀφθαλμούς. 8 ἄλλο. M▷ κοπάνισον μάλαθρον καὶ τὸν ζωμὸν τούτου ἄλειφε ἔξω τοὺς ὀφθαλμούς.◁M 9 ἄλλο. ▷MAκοπάνισον τὴν ῥίζαν τῆς χελιδονέας καὶ μίξον με τὸ μέλι τὸ ἀκάπνιν καὶ ἄλειφε ἔξω τοὺς ὀφθαλμούς.MA◁

170 [137] ρο΄ ⌜τοὺς πόνους τῶν ἀρμῶν καὶ τὰς σκληρότητας μαλακίζει· ἀμμωνιακὸν βάλε νυχθήμερον εἰς ὄξος δριμὺ καὶ ὅταν γένηται ἁπαλόν, μάλαξον τοῦτο με τὸ μέλι καὶ θέτε ἐπάνω εἰς τὸν τόπον τῶν ἀρμῶν. 2 ἄλλο. ▷om. Lτρίψον πύρεθρον ψιλὸν καὶ πέπερι καὶ βάλε χρηστέλαιον καὶ ἕνωσον καὶ ἄλειφε τοὺς πόνους.om. L◁⌝ 3 ἄλλο. ποταμογείτονος φύλλα χλωρὰ ἕψησον με τὸ νερὸν εἶτα κοπάνισον ταῦτα καὶ θέτε τα εἰς τὴν σκληρότητα καὶ εἰς τὸν πόνον τῶν ἀρμῶν. 4 ἄλλο. ⌜κεδρέαν μίξον με τὸ χρηστέλαιον καὶ ἄλειφε

17–18 μάλαθρον] μάλαθρ... L 1 ἔξωθεν] ἔξω MA om. B 3 τούτου] αὐτοῦ A 4 του] om. Σ 4 βεττονικῆς] scripsi βετωνίκης A βετονικῆς MB 4 ἀπέσω] ἔσω A 5 χήνας χολὴν] θύνας χολὴν MA χήνας χολὴν B 6 μίξον] μίξε το A 6 ὄξος] χυλὸν μετὰ ὄξους B 7 τὸ χρηστέλαιον καὶ] om. A καὶ χρηστελαίου B 7 ἔξω τοὺς ὀφθαλμούς] om. B 8 καππαδοκικὸν] καππαδόκιον A 8 τριμμένον] om. A 8 μαλάθρου] βάλε add. B 9 ἀκάπνιον] ἀκάπνιν καὶ AB 9 βάλε] om. B 9 μίξον] μίξε τα A μίξας αὐτὸ B 9 καὶ] om. B 13 τοὺς] πρὸς MB 13 πόνους τῶν ἀρμῶν] πόνον ἀρμῶν M ἀρμοπονίας B 13 καὶ τὰς σκληρότητας] om. A καὶ σκληρότητος· τοὺς πόνους τῶν ἀρμῶν καὶ τὰς σκληρότητας B 14 μαλακίζει] om. M 14 ἀμμωνιακὸν] om. A ὅπερ add. B 14 νυχθήμερον] om. A 14 καὶ] ἀμμωνιακὸν καὶ ἂς ποιήσῃ νυχθήμερον καὶ A om. B 15 τοῦτο] om. A 15 με τὸ μέλι] om. B 15 θέτε] θὲς M το add. AB 15 ἐπάνω] om. B 16 τόπον] πόνον MA 16 ἀρμῶν] ἄνωθεν add. B 16 καὶ] ζιγγίβερι ψιλὸν καὶ add. B 17 καὶ] ὅλα καὶ A ὅλα ταῦτα καὶ B 17 τοὺς πόνους] ἄρθρα B 18 ποταμογείτονος] ποταμ....... L 18 με τὸ νερὸν εἶτα] μετ........τα L 19 θέτε τα] θέτε τ.... L θὲς M 19 πόνον] τόπον B τὸν πόνον. 5 ἄλλο. τὴν ῥίζαν τοῦ κισσοῦ καῦσον καὶ τὴν στάκτην μίξον με τὸν κρόκον τοῦ αὐγοῦ καὶ θέτε εἰς τὸν πόνον.⌝ 6 ἄλλο. ἐπάνω εἰς τὸν πόνον θὲς τριμμένον καὶ ψιλούτζικον πολλὰ πέπερι. καὶ ἐπάνω εἰς τὸ πέπερι θὲς ὕσκαν καὶ καῦσε τὸν τόπον ἔνθα ἔνι ὁ πόνος. 7 ἄλλο. ▷om. LA⌜ὑοσκυάμου σπέρμα κοπάνισον καὶ ψιλὸν ποιήσας βάλε καὶ τριμμένον πολλὰ ψιλὸν νίτρον καὶ πίε το.om. LA◁ 8 ἄλλο. ▷om. ABἄλειφε δὲ τὸν τόπον ἔνθα ἔνι ὁ πόνος τῶν ἀρμῶνom. AB◁ ▷om. Bἐλαφίου ἀξούγγιν ἢ ὄρνιθος ἢ ταύρου ἢ ἄρκου.⌝om. B◁

171 [138] ροα′ ⌜ἀπόστημα σκορπίζει καὶ μαλακίζει· ὑοσκυάμου φύλλα κοπανισμένα καὶ μάλαξον αὐτὰ με τὸ ἀξούγγιν τὸ χοίρειον καὶ θέτε το ἐπάνω εἰς τὸ ἀπόστημα. 2 ἄλλο. σπαθοβότανον κοπάνισον καὶ θέτε τὸ ἐπάνω εἰς τὸν τόπον ἔνθα ἔνι ἡ σκληρότης.⌝ 3 ἄλλο. λαπάτου ἀγρίου ἢ καὶ ἡμέρου τὰ φύλλα κοπάνισον καὶ βράσον καὶ θέτε τὰ χλία εἰς τὸν τόπον. 4 ἄλλο. λινόσπορον κοπάνισον καὶ ψιλὸν ποιήσας βράσον με τὸ νερόν. εἶτα ▷om. MAβάλε χοίρειον ἀξούγγιν καὶ μάλαξον τοῦτο καὶ θέτε τὸ ἐπάνω εἰς τὸ ἀπόστημα. 5 ἄλλο. ⌜ἀγριομολόγχης τὰ φύλλα βράσον με τὸ νερὸν εἶταom. MA◁ κοπάνισον καὶ θέτε τα ἐπάνω εἰς τὸν τόπον.⌝ 6 ἄλλο. αἶρας ἀλεύριν βράσον με τὸ νερὸν τὸ γλυκύ, εἶτα βάλε χοίρειον ἀξούγγιν καὶ θέτε το εἰς τὴν πληγὴν ἐπάνω. 7 ἄλλο. ⌜κρίθινον ἀλεύριν καὶ λινόσπορον ψιλὸν βράσον με τὸ νερόν. ▷om. Bεἶτα βάλε χοίρειον

20 ἀρμῶν] ἄρθρων MB 1 τὸν] om. A 1 πόνον] τόπον MB om. A 1 τὴν στάκτην] om. A 2 θέτε] θὲς M τὴν ἐπάνω add. A τὰ add. B 3 ἐπάνω εἰς τὸν πόνον] om. A 3 καὶ] om. A 4 πέπερι] πιπέρην A αὐτὸ B 4 ὕσκαν] ἴσκαν A 4 καῦσε] καῦσον Σ 4 τόπον] πόνον B 5 πόνος] τόπος M 6 ψιλὸν] om. B 6 καὶ] βάλε καὶ νερὸν add. B 7 δὲ] om. M 8 ἐλαφίου] ἐλάφειον M ἐλάφου A 8 ἄρκου] ἄλειφε τὸν πόνον τῶν ἀρμῶν add. A 9 ἀπόστημα] πρὸς ἀπόστημα ML πρὸς ἀπόστημα· ἀπόστημα B 9 σκορπίζει καὶ μαλακίζει] om. M ταῦτα add. B 10 καὶ] om. M 10 αὐτὰ] om. M 10 με τὸ] μετὰ A 10 τὸ] om. A 11 θέτε το] θὲς M θέτε τα A 11–12 σπαθοβότανον κοπάνισον] σπαθοβόη L παθοβότανον κοπάνισον M 12 θέτε τὸ] θὲς M 12 τόπον ἔνθα] ....θα L 13 λαπάτου] λαπάθου M λαπάθου AB 13 ἀγρίου ἢ καὶ] scripsi ἀ.... καὶ L ἀγρίου ἢ MA 13–14 βράσον] πρῶτον add. Σ 18 θέτε] θὲς M 18 ἐπάνω] om. A 18 εἰς τὸν τόπον] om. B 18 αἶρας] ἦρας L 18 ἀλεύριν] ἄλευρον M 20 θέτε] θὲς B 20 το] τα A ἐπάνω add. B 20 ἐπάνω] ἐπ..νω L om. B

1 τὴν] This paragraph is in B in a different location.

ἀξούγγιν, χήνειον ὄρνιθος,om. B◃ ἐλάφου μυελόν, ταύρου ἀξούγγιν, παρδάλεος ▹om. Aἄρκτου μόσχου λεόντος κόρακος ἀλώπεκοςom. A◃ τὰ πάντα λυτὰ καὶ βάλε τὰ καὶ μάλαξον καὶ θές τα ἐπάνω με τὸ πανίν.⌝ 8 ἄλλο. κόπρον περιστερᾶς καὶ ὄξος βάλε καὶ μάλαξον ταῦτα καὶ θέτε τα ἐπάνω με τὸ πανὶν εἰς τὴν πληγήν.

172 [139] ροβ′ ⌜ἀπορυφὰς καὶ ἐκβατὰ μαλακίζει· καρπὶν κοπανισμένον ψιλόν. μίξε το με τὴν πρόπολιν καὶ θέτε τὸ ἐπάνω. 2 ἄλλο. κριθάλευρον βράσον με τὸν χυλὸν του σευτλοφύλλου καὶ μαλάξας καλὰ θέτε τα.⌝ 3 ἄλλο. ▹om. Aφλόμον βοτάνην κοπάνισον τὰ φύλλα καὶ θέτε τὰ ἐπάνω.om. A◃ 4 ἄλλο. ▹om. M⌜κρομμύδιν ἕψησον εἰς τὰ καρβούνια καὶ κοπάνισον καὶ θέτε τὸ ἐπάνω.⌝om. M◃ 5 ἄλλο. ▹om. Aκρομμύδιν κοπάνισον καὶ ποιήσας ψιλὸν κατάπλασε.om. A◃ 6 ἄλλο. λαπάτου ῥίζαν κοπάνισον καὶ μάλαξον με τὸ ἀξούγγιν τὸ χήνειον καὶ τὸ χοίρειον καὶ θέτε τα ἐπάνω. 7 ἄλλο. ▹om. L⌜πίσσαν τρίψον καὶ πάσσε την ἐπάνω.⌝om. L◃

173 [140] ρογ′ ὅταν δάκῃ τινὰ ὄφις ἢ ἀσπίδα ποίησον θεραπείαν τοιαύτην· ⌜κοπάνισον φύλλα μηλέας καὶ τὸν ζωμὸν τούτων πίε

20 ἀλεύριν] ἄλευρον A 1 ἀξούγγιν] ἢ add. MA 1 χήνειον] ἢ add. MA om. B 1 ὄρνιθος] ἢ add. A 1 μυελόν] ἢ add. A 2 παρδάλεος] ἀλώπεκος add. M λέοντος ἀλώπεκος add. A 2 ἄρκτου] ἄκρου M 2 ἀλώπεκος] om. M 3 τὰ] ταῦτα M βάλε τὰ B 3 καὶ βάλε τὰ καὶ] om. MA καὶ B 3 καὶ θές τα] .... L καὶ θέτε τὰ A καὶ οὕτως θέτε τὰ B 4 περιστερᾶς] ....στερᾶς L 5 ταῦτα] αὐτὰ A 5 θέτε] θὲς M 5 με τὸ πανὶν εἰς τὴν πληγήν] transp. B 5 τὴν πληγήν] τὸν τόπον A 6 ἀπορυφὰς] ἀπορυφὰς L πρὸς ἀπορυφὰς MB 6 καὶ ἐκβατὰ μαλακίζει] om. M καὶ εὐγατὰ μαλακίζει A ταῦτα add. B 6–7 κοπανισμένον] κοπανισμέναν M 7 μίξε το] μίξον M αὐτὸ add. A μίξας B 7 καὶ] om. B 7 τὸ] τα A om. B 8 με τὸν] με τὸ MB 8 χυλὸν] χλίον M 8 σευτλοφύλλου] σεύτλου A 9 καλὰ] καὶ M 9 τα] τα ἐπάνω εἰς τὴν ἀπορυφὴν A 9 φλόμον βοτάνην] φλομοβότανον M φλῶμος βοτάνη ἐστὶ οὗ B 10 τὰ] πλὴν τὰ M 10 τὰ] om. B 10 κρομμύδιν] scripsi .μύδην L κρεμίδιν A κομμίδιν B 11 καρβούνια] κάρβουνα A κάρβουνον B 12 κρομμύδιν] scripsi .ρομύδην L κομίδιν MB 12 ποιήσας ψιλὸν] transp. MB 13 λαπάτου] λαπάθου Σ 14 χήνειον] χοίρειον A 14 καὶ τὸ χοίρειον] om. M καὶ τὸ χήνειον A 14 θέτε] θὲς M 14 τα] το B 15 τρίψον] τρίψας AB 15 καὶ πάσσε] κατάπλασσε A θέτε B 15 την] αὐτὴν add. B 16 ὄφις ἢ] ἡ MA om. B 16 ἀσπίδα] ἀσπὶς AB 16–17 ποίησον θεραπείαν τοιαύτην] ποίησον θε........ την ἤγουν L om. MA 17 μηλέας καὶ] μη....... L μηλαίας καὶ A 17 τούτων] τοῦτον A τοὺς B

μετὰ οἴνου. κοπάνισον καὶ τὰ φύλλα καὶ θές τα εἰς τὴν πληγήν. 2 ἄλλο. μυρσίνης φύλλα κοπάνισον καὶ θέτε τα εἰς τὴν πληγὴν ἐπάνω.⌝ 3 ἄλλο. ἔγχελυν ζῶντα σχίσε το καὶ θὲς το εἰς τὴν πληγήν. 4 ἄλλο. ⌜ἡδύοσμον κοπάνισον καὶ πότισον μετὰ θερμοῦ νεροῦ. 5 ἄλλο. ▷om. A κόστου φύλλα κοπάνισον καὶ θέτε τα εἰς τὴν πληγήν. om. A◁ 6 ἄλλο. χαλβάνιν τρῖψον καὶ μίξον με τὸ χρηστέλαιον καὶ ἄλειφε τὸν τόπον.⌝ 7 ἄλλο. δρακοντίας τὴν ῥίζαν κοπάνισον καὶ τὰ φύλλα καὶ μίξον μετὰ ἀξουγγίου ἐλαφίου καὶ ἄλειφε. 8 ἄλλο. ▷om. AB ⌜ἀξούγγιον ἄρκου λύσας ἄλειφε τὴν πληγήν.⌝ om. AB◁

174 ροδʹ ὅταν ἄνθρωπος ἔχῃ ἀπεψίαν· πότισον αὐτὸν γλησχούνιν ξηρὰν τριμμένην με τὸ κρασίν. 2 ἄλλο. πεπέρι τριμμένον καὶ μακεδονησίου σπέρμα κοπανίσας καὶ τρίψας τὰ πάντα ὁμοῦ, πότισον με τὸ γάρος ▷om. L ἐντὸς τοῦ λουτροῦ, ὅταν γένηται καλά. πρόσεχε δὲ μὴ λουσθῇ ὅταν κενώνῃ τὴν ἀπεψίαν. 3 ἄλλο. κινάμωμον τριμμένον καὶ πηγάνου σπέρμα τριμμένον πότισον ἅμα με τὸ νερὸν τὸ χλίον. 4 ἄλλο. ▷om. AB ὕσσωπον κοπάνισον καὶ πότισον με τὸ νερόν. om. AB◁ om. L◁

175 om. A ροεʹ ▷om. L θεραπεία εἰς βῆχας· δέρμα δαύκου κοπάνισον

17 πίε] πότισον MB μίξον A 1 μετὰ οἴνου] μετ. οἰ.... L με τὸν οἶνον A σὺν οἴνῳ B 1 κοπάνισον καὶ τὰ φύλλα καὶ] καὶ τὰ φύλλα κοπανισμένα M om. A 1 κοπάνισον] d'e add. B 1 θές τα] καὶ θέτε A καὶ θέτε τα B 2 μυρσίνης φύλλα] μυρσινόφυλλα A 2 θέτε] θὲς M 2–3 εἰς τὴν πληγὴν ἐπάνω] om. A 3 ἐπάνω] om. M 3 ἔγχελυν] scripsi ἀχέλιν L ἐγχέλην M χελῶναν A ἐσχέλην B 3 ζῶντα] ζωντανὴν A ζωντανὸν B 3 σχίσε το] σχίσον M σχίσε την A χίσε B 3 θὲς] θέτε Σ 3 το] τὴν A 4 θερμοῦ] καὶ add. B 5 νεροῦ] ὕδατος A 5 θέτε] θὲς M 6 μίξον] ἕνωσον M 7 τὸν τόπον] om. B 7 δρακοντίας] ἔστι βοτάνη ἡ λεγομένη δρακοντία ταύτης Σ 7 κοπάνισον] κοπανισμένην A 8 μίξον] ἕνωσον M μίξε την A 8 μετὰ] με A 8 ἀξουγγίου ἐλαφίου] ἀξούγγιν ἐλάφειον M ἀξούγγιν ἐλάφου A 8 ἄλειφε] τὴν πληγὴν add. B 9 ἀξούγγιον] ἀξούγγιν M 9 ἄρκου λύσας] ἄρκου M 10 ὅταν] ὁ add. A 10 ἄνθρωπος ἔχῃ] ἔχῃ ὁ ἄνθρωπος MB 10 γλησχούνιν] βλησκούνιν M 11 ξηρὰν] ξηρὸν Σ 11 τριμμένην] τριμμένον Σ 11 καὶ] ξυλοζιζίβερι τριμμένον καὶ add. MB ξυλοζίγγιβερ τριμμένον καὶ add. A 11–12 μακεδονησίου] μακεδονι..σίου L μακαιδονοσίου M μακαιδοσίου B 12 τὰ] om. Σ 12 ὁμοῦ] om. A 13 γάρος] γάρου B 13 ὅταν] δὲ add. A 13 πρόσεχε] πρόσεξον A 13 δὲ] om. A 14 λουσθῇ] λουθῇ AB 15 πηγάνου σπέρμα] πεπέρι AB 15 τριμμένον] καὶ πηγάνου σπέρμα τιμμένον add. AB 18 δέρμα] σπέρμα B

καὶ πότισον μετὰ οἴνου. 2 ἄλλο. ▷om. Bβδέλλιον κοπάνισον καὶ πότισον μετὰ οἴνου.om. B◁ 3 ἄλλο. ἔστι βοτάνη ἡ λεγομένη δρακοντία. ταύτης τὴν ῥίζαν ἂς τρώγῃ ὠμὴν καὶ ἑψημένην. 4 ἄλλο. τὴν δὲ ὠμὴν ῥίζαν τῆς δρακοντίας ἂς τρώγῃ με τὸ μέλι. 5 ἄλλο. πηγάνου τὰ φύλλα κοπάνισον καὶ τὸν ζωμὸν πότισον. 6 ἄλλο. ῥέον Ποντικὸν κοπάνισον καὶ πότισον μετὰ ὕδατος. 7 ἄλλο. ἔστι βότανον τὸ λεγόμενον πολυγόνιον τοῦτο κοπάνισον καὶ τὸν ζωμὸν πότισον. 8 ἄλλο. ἀψίνθιον κοπάνισον χλωρὸν καὶ τὸν ζωμὸν πότισον. 9 ἄλλο. ἀδίαντον βράσον με τὸ νερὸν καὶ πότισον με τὸ κρασίν. 10 ἄλλο. ἀκάνθης ἄσπρης τὴν ῥίζαν κοπάνισον καὶ πότισον μετὰ ὕδατος. 11 ἄλλο. ἀμμωνιακὸν κοπάνισον καὶ ψιλὸν ποιήσας πότισον μετὰ μέλιτος ἢ μετὰ γυμνοκρίθου χυλόν. 12 ἄλλο. ἀνδράνχνην τὴν λεγομένην γλιστρίδαν ἕψησον με τὸ ὄξος καὶ τὸ κρασὶν καὶ ἄς την τρώγῃ. 13 ἄλλο. πρόπολιν θὲς εἰς ζῶντα κάρβουνα καὶ κάπνιζέ τον. 14 ἄλλο. ῥητίνην χλωρὴν ποίησον κοκκία ὅσον ἐρεβίνθια καὶ ἄς τα καταπίνῃ με τὸ χλίον τὸ νερόν. 15 ἄλλο. τὸ λεγόμενον γλυκοκάλαμον κάπνιζέ τον ὑποκάτω. 16 ἄλλο. ἀρσενίκιν καὶ χαλβάνιν θὲς εἰς τὰ κάρβουνα ἁπτόμενα καὶ κάπνιζέ τον καὶ τὸν καπνὸν ἄς τον καταπίνῃ. 17 ἄλλο. θὲς εἰς τὰ κάρβουνα ἁπτόμενα πίσσαν καὶ λίβανον καὶ στύρακα καὶ κάπνιζε. 18 ἄλλο. βαλσάμου καρπὸν κοπάνισον καὶ ψιλὸν ποιήσας πότισον μετὰ ὕδατος. 19 ἄλλο. κυπαρίσσου χλωρὰ βλαστάρια ἂς τρώγῃ μετὰ ὄξους. 20 ἄλλο. δενδρολίβανον βράσον μετὰ οἴνου καὶ πότιζε κατὰ πρωΐ. 21 ἄλλο. ▷om. Bμαλάθρου ῥίζαν βράσον με τὸ κρασὶν καὶ ἄς το πίνῃ.om. B◁ 22

18 δαύκου] δαυκίου B 3 ταύτης] ταύτην B 3 τὴν δὲ] πλὴν τὴν B 5 φύλλα] χλωρὰ B 5 ζωμὸν] αὐτὸν B 7 τὸ λεγόμενον] om. B 7 πολυγόνιον] πολιγόνιν καλούμενον B 7 τοῦτο κοπάνισον] transp. B 7–8 τὸν ζωμὸν πότισον] transp. B 8 κοπάνισον χλωρὸν] transp. B 8 τὸν ζωμὸν πότισον] transp. B 9 πότισον] πότιζε B 10 καὶ πότισον] om. B 11–12 πότισον] ἢ add. B 12 μετὰ μέλιτος] με τὸ μέλι B 12 μετὰ γυμνοκρίθου χυλόν] με τὸν χυλὸν τοῦ γυμνοκρίθου B 13 γλιστρίδαν] ταύτην add. B 13 με τὸ ὄξος] μετὰ ὄξους B 14 την] om. B 14 πρόπολιν] κρόπολιν B 14 ζῶντα κάρβουνα] καρβούνια ζωντανὰ B 15 τον] το B 15 ὅσον] ἴσον B 16 τὸ] om. B 18 θὲς] θέτε B 18 τὰ] om. B 18–19 τὸν καπνὸν ἄς τον καταπίνῃ] ἂς σπεύδῃ να καταπίνῃ τὸν καπνὸν B 19 τὰ κάρβουνα] καρβούνια B 19 πίσσαν] πίσαν B 20 κάπνιζε] τον add. B 21 πότισον] πότιζε B 21–22 κυπαρίσσου χλωρὰ βλαστάρια ἂς τρώγῃ] transp. B

ἄλλο. μαλά[om. L◁]θρου σπόρον κοπάνισον καὶ πότισον μετὰ οἴνου. 23 ἄλλο. τοῦ πεύκου τὰ φύλλα καὶ τὰ κλωνάρια τὰ τρυφερὰ βράσον μετὰ οἶνον καὶ πότισον. 24 ἄλλο. ὑοσκυάμου σπέρμα κοπάνισον καὶ πότισον με τὸ νερόν. 25 ἄλλο. ὕσσωπον καὶ πήγανον βράσον με τὸ μέλι καὶ τὸ κρασὶν καὶ πότισον. 26 ἄλλο. [▷om. L]ῥίζαν φλόμου βράσον μετὰ οἶνον καὶ πότισον.[om. L◁] 27 ἄλλο. στύραξ λιπαρός ποίησον αὐτὸν ἶσα κοκκία ἐρεβίνθου καὶ ἄς τα καταπίνῃ με τὸ κρασίν. 28 ἀλλο. χαλβάνιν ποίησον κοκκία καὶ με τὸ ἄσπρον τοῦ αὐγοῦ ἄς τα καταπίνῃ.

176 [141] [om. A]ρος′ τὸ γάλα τῶν γυναικῶν καταβάζει καὶ τὰ καταμήνια ὁρμᾷ κάτω· ⌜λυγαίας τὸν σπόρον κοπάνισον καὶ πότισον μετὰ οἴνου. 2 ἄλλο. τοῦ ἀνήθου τὴν ῥίζαν βράσον καὶ ἄς την πίνῃ.⌝ 3 ἄλλο. [▷L]τοῦ ἀνήθου τὸν σπόρον βράσον με τὸ νερὸν καὶ πότισον μαράθρου ῥίζαν. καὶ τὰ φύλλα καὶ τὸν σπόρον † ....νου καὶ πότισον.[L◁] 4 ἄλλο. [▷MB]τὴν κορυφὴν τοῦ ἀνήθου βράσον ἢ τὸν σπόρον με τὸ νερὸν πότισον.[MB◁] 5 ἄλλο. [▷om. L]⌜μαλάθρου σπόρον τὴν ῥίζαν καὶ τὰ φύλλα βράσον μετὰ οἶνον καὶ πότισον.[om. L◁] 6 ἀλλο. μελάνθιον κοπάνισον καὶ πότισον μετὰ οἴνου.⌝ 7 ἄλλο. βασιλικόσπορον κοπάνισον καὶ πότισον με τὸ νερόν. κράμβον ἃς ἑψηθῇ δεύτερον καὶ ἄς την τρώῃ. 8 ἄλλο. ῥάμνου φύλλα καὶ τὰ ἁπαλὰ βλαστάρια ἑψ-

---

23 πότιζε] αὐτὸν add. B 1 πότισον μετὰ οἴνου] ποτ.........νου L 2 τοῦ] τῆς MB 2 πεύκου] πεύκης B 2 τὰ κλωνάρια τὰ] τ............α L τὰ κλώνια τὰ B 3 σπέρμα] σπόρον M 4 με τὸ] μετὰ MB 5 πότισον] με τὸ νερὸν add. B 5 φλόμου] φλώμου B 6 οἶνον] οἴνου B 6 στύραξ λιπαρός] στύρακα λιπαρὸν MB 7 ἶσα] scripsi .σα L ἶσον M ἶσα με B 7 ἐρεβίνθου] ἐρεβινθίου M ἐρεβίνθια B 8 χαλβάνιν] χαλβάνην M 8 κοκκία] ὡς κοκκία M 8–9 με τὸ ἄσπρον τοῦ αὐγοῦ ἄς τα καταπίνῃ] transp. B 10 τὸ] πρὸς M πρὸς τὸ B 10 γάλα τῶν γυναικῶν καταβάζει καὶ] καταβηβάσαι γάλα γυναικῶν B 10 τὰ] om. MB 10–11 καταμήνια] τῶν γυναικῶν add. M 11 ὁρμᾷ κάτω] om. B 11 τὸν] om. B 11 κοπάνισον καὶ] om. M 14 μαράθρου] scripsi ἀράθρου L 14–15 πότισον] scripsi πότισ... L 15 βράσον ἢ τὸν σπόρον] transp. B 16 με τὸ] μετὰ B 16 νερὸν] καὶ add. B 17 οἶνον] οἴνου B 17–18 μελάνθιον κοπάνισον] μέλανθ........ L 18 βασιλικόσπορον] scripsi βασιλικ.σπορον L βασιλικοῦ σπόρον MB 19 με τὸ] μετὰ M 19 κράμβον ἃς ἑψηθῇ] ἕψησον με τὴν κράμβην B 19 κράμβον] κράμβη M 20 τρώῃ] τρώγῃ MB 20 ῥάμνου] ἔστιν ἄκανθα ἡ λεγομένη ῥάμνος ταύτης τὰ MB 20 ἁπαλὰ] τρυφερὰ MB

ηθέντα ἅς τα τρώη. 9 ἄλλο. ⌜ἅλμην θαλασσίαν ἄντλει συνεχῶς.⌝ 10 ἄλλο. βρωμοβότανον κοπάνισον καὶ τὸν ζωμόν του ἄλειφε τὰ βυζία τῆς γυναικὸς εἶτα πάσσε ἐπάνω ψιμμίθιν. 11 ἄλλο. ▷om. L ὑοσκυάμου φύλλα κοπάνισον καὶ τὸν ζωμὸν τούτων ἄλειφε τὰ βυζία.om. L◁ 12 ἄλλο. ⌜ὄξος καὶ νερὸν καὶ ἅλας μίξον, εἶτα βάλε ψιμμίθιον καὶ λιθάργυρον τριμμένον καὶ μίξας ὅλα ἄλειφε τὰ βυζία.⌝

177 [142] ροζʹ ⌜γυναῖκα ὅπου οὐδὲν δύναται να γενήσει συντόμως ἔπαρον καβούρους ποταμοῦ καὶ θὲς αὐτοὺς εἰς ζῶντα κάρβουνα καὶ κάπνισον τὴν⌝ καὶ εἰς μίαν θέλει γενήσειν.

178 [143] ροηʹ ⌜γλώσσης ἀνδρύτητα μαλακίζει· ἤγουν βράσον μέλι καὶ ἐπάνω κράτει πανὶν ψιλὸν ἔχων ἀπέσω λινόκουκον καὶ ἣν ἱδρώσει τὸ πανὶν ἀπὸ τὸν ἀτμὸν τοῦ μέλιτος τότε τρίβε τὴν γλῶσσαν.⌝ 2 ἄλλο. ἡδύοσμον χλωρὸν κοπάνισον καὶ ψιλὸν ποιήσας τρίβε τὴν γλῶσσαν. 3 ἄλλο. βράσον ἰσχάδια με τὸ νερὸν καὶ τὸ νερὸν τῶν ἰσχάδων, ἅς το κρατῇ εἰς τὸ στόμα του. 4 ἄλλο. ▷om. B ⌜κοπάνισον ἡδύοσμον χλωρὸν καὶ ἔκβαλε τὸν ζωμὸν καὶ μίξον ῥοδέλαιον καὶ χαμαίμηλον καὶ ἄλειφε τὴν γλῶσσαν.⌝om. B◁

284.20–285.1 ἐψηθέντα] ἔψησον καὶ M 1 τα] om. B 1 τρώη] τρώγη MB 1 ἅλμην] ἅλμη M 2 βρωμοβότανον] τὸ λεγόμενον βρωμοβότανον MB 2 του] τούτου MB 3 εἶτα] scripsi ..τα L καὶ ἀφότου τὰ ἀλείψας M καὶ ἀφ' ὅτου ἀλείψης B 3 πάσσε] πάσον MB 3 ψιμμίθιν] ψιμμίθιον MB 3 ὑοσκυάμου] τὰ add. B 4 τούτων] om. B 5 νερὸν] ..ρὸν L 5 ἅλας] ἅλμην B 5 μίξον] ἕνωσον M 5 ψιμμίθιον] ψιμμίθιν B 6 τριμμένον] τριμμένα M 7 γυναῖκα ὅπου οὐδὲν δύναται να γενήσει συντόμως] πρὸς γυναῖκα μὴ δυναμένην γεννήσαι συντόμως M γυναῖκα ὅπου οὐδὲν δύναται γεννήσαι συντόμως A πρὸς γυναῖκα τὴν μὴ δυναμένην γεννήσαι· γυναῖκα τὴν μὴ δυναμένην γεννῆσαι συντόμον B 7 γυναῖκα] scripsi γυναικ... L 7 συντόμως] συντόμ.. L 8 καβούρους] τοῦ add. A 8 αὐτοὺς] τους M τούτους AB 8 κάρβουνα] κάρβονα B 9 τὴν] αὐτὴν Σ 9 εἰς μίαν θέλει γενήσειν] εὐθὺς γεννᾶ M εὐθὺς θέλει γεννῆσαι B 9 θέλει] θέλη A 9 γενήσειν] γενῆσαι A 10 γλώσσης ἀνδρύτητα μαλακίζει] πρὸς γλῶσσαν ἔχουσαν ἀνδρότητα M περὶ γλώσσης ἀδρότητα. τὴν ἀδρότητα τῆς γλώσση μαλακίζει A πρὸς ἀδρώτητα γλώσσης· γλώσσης ἀδρότητα μαλακίζει ταῦτα B 10 ἤγουν] om. Σ 11 ἀπέσω] scripsi ἀπέ...ω L om. Σ 11 λινόκουκον] λινόκου.... L λινόκοκκον M 12 ἣν ἱδρώσει τὸ πανὶν] scripsi ηνιδ.........ην L ἡνίκα ὑδρώσει τὸ πανὶν M ἡνίκα ἱδρώση τὸ πανὶν AB 12 τὸν ἀτμὸν] τοῦ ἀτμοῦ Σ 12 μέλιτος τότε τρίβε] μέλι........βε L 14 ἰσχάδια] ἰσχάδας Σ 15 τὸ στόμα του] εἰς τὴν γλῶσσαν A 16 ζωμὸν] τοῦ add. MA 17 μίξον] με τὸ add. M 17 καὶ] τὸ add. M

5 ἄλλο. ▷Bκοπάνισον ἡδύοσμον χλωρὸν καὶ ἄλειφε τὴν κεφαλήν.B◁ 6 ἄλλο. ▷om. Aἡδύοσμον χλωρὸν βράσον με τὸ νερὸν καὶ κρασὶν ὀλίγον καὶ τὸν ζωμὶν ἐκεῖνο κράτει εἰς τὸ στόμα.om. A◁ 7 ἄλλο. ⌜ἐλαίας φύλλα μασοῦ, ἐπικράτει εἰς τὴν γλῶσσαν.⌝

179 ροθ′ θεραπεία εἰς δυσουρίαν· ἀψίνθιον βράσον με τὸ νερὸν καὶ ἀφ' ὅτου βράσῃ, μίξον με τὸ κρασὶν καὶ ἄς το πίνῃ. 2 ἄλλο. ▷om. Bἀσπαράγγια βράσον με τὸ νερὸν καὶ ἄς το πίνῃ.om. B◁ 3 ἄλλο. ▷LAσκολοπένδριον βράσον με τὸ νερὸν καὶ ἄς το πίνῃ.LA◁ 4 ἄλλο. κορίδας κοπάνισον καλὰ καὶ ἄς τας θέσῃ ἔμπροσθεν εἰς τὴν τρύπαν τοῦ αἰδοίου. 5 ἄλλο. ▷Bσπαράγγια βράσον με τὸ νερὸν καὶ ἄς το πίνῃ.B◁ 6 ἄλλο. ▷MBἄγριον σταφύλην καὶ ἥμερον τὰ κοκκία τῶν κοπάνισον καὶ πότισον μετὰ οἴνου.MB◁ 7 ἄλλο. ▷om. Aβοτάνιν τριφύλλιν τούτου τὸν σπόρον κοπάνισον καὶ πότισον μετὰ οἴνου.om. A◁ 8 ἄλλο. χαλικολίθαρον βράσον μετὰ οἴνου καὶ πότισον. καλεῖται παρὰ τῶν ἰατρῶν πυρίτης λίθος. 9 ἄλλο. λιθάριν τὸ εὑρίσκεται εἰς τοὺς καινουργίους σπόγγους τρίψον τοῦτο ἀκριβῶς καὶ μετὰ οἴνου πότισον. 10 ἄλλο. om. ▷Lδενδρολίβανον βράσον μετὰ οἴνου καὶ πότισον αὐτόν. 11 ἄλλο. κιχωρίων ῥίζας τὰ φλούδια τούτων καὶ σελίνου ῥίζας καὶ μακεδονησίου ῥίζας καὶ μαλάθρου ῥίζας καὶ

17 χαμαίμηλον] χαμεμηλέλαιον MA 2–3 νερὸν καὶ κρασὶν ὀλίγον] κρασὶν καὶ τὸ νερὸν M 3 τὸν ζωμὶν] τὸ βράσμα M 4 ἐπικράτει] καὶ κράτει M 4 γλῶσσαν] γλῶτταν M σου add. B 5 θεραπεία εἰς] πρὸς B 5 δυσουρίαν] θεραπεύει add. B 5 βράσον] βρά... L βραστὸν B 6 μίξον με τὸ κρασὶν] ἕνωσον με τὸ νερὸν M πάλιν συγκέρασέ το με τὸ νερὸν A μίξον με τὸ νερὸν καὶ κρασὶν B 7 ἀσπαράγγια] σπαράγγια A 8 νερὸν καὶ ἄς] .... L 9 καλὰ καὶ ἄς τας θέσῃ] ...... τὰς θέσῃ L καλὰ καὶ θὲς αὐτὰς M καλὰ καὶ ἄς τα θέσῃ B 9 ἔμπροσθεν] om. M 9 τὴν τρύπαν] τ....υπαν L τὴν ὀπὴν A 10 τοῦ] τῆς Σ 10 αἰδοίου] ψωλῆς Σ 10 βράσον] scripsi βράσον add. B 11 ἄγριον σταφύλην καὶ ἥμερον] ἀγρίου σταφηλίου ἢ ἡμέρου B 12–13 βοτάνιν τριφύλλιν τούτου τὸν σπόρον] ἄγριον σταφύλην καὶ ἥμερον τὰ κοκκία τῶν M 14 χαλικολίθαρον] καλικολίθαρον AB ὃ καλεῖται παρὰ τῶν ἰατρῶν πυρίτης λίθος τοῦτο B 14 καλεῖται] δὲ add. MA om. B 15 παρὰ τῶν ἰατρῶν πυρίτης λίθος] om. B 15 τὸ] ὅπου A ὅπερ B 16 τοὺς] om. AB 16 καινουργίους σπόγγους] transp. MA καινούργιον σπόγγον B 16 τοῦτο] om. M 16–17 καὶ μετὰ οἴνου πότισον] scripsi κα..τα οἴνου πότισον L transp. Σ 18 αὐτόν] om. AB 18 κιχωρίων] ιχωρίων L βάλε κιχωρίων B 18 τούτων] om. B 19 καὶ μακεδονησίου ῥίζας καὶ μαλάθρου ῥίζας] om. B 19 ῥίζας] ῥίζαν A

ἰντυβίου ῥίζαν καὶ ἀνήθου καὶ πηγάνου ἀγρίου καὶ ἡμέρου βάλε καὶ ἂς βράσουν με τὸ νερὸν καὶ τὸν ζωμὸν ἐκείνων ἂς πίνῃ. σὺν τούτοις βάλε καὶ ῥεφανίου ῥίζαν καὶ ζοχίου.[om. L◁]

180 ρπ´ εἰς δυσεντερίαν· ἀμάραντον κοπάνισον καὶ τὸν ζωμὸν πότισον. 2 ἄλλο. σμύρναν ὅσον φάβα τρίψον καὶ πότισον μετὰ οἴνου. 3 ἄλλο. λυγαίας φύλλα βράσον μετὰ οἴνου καὶ ἂς πίνῃ. 4 ἄλλο. [▷L]βάτου φύλλα καὶ τὰ βλαστάρια βράσον με τὸ κρασὶν καὶ ἄς το πίνῃ.[L◁] 5 ἄλλο. ἔστι βοτάνιν τὸ λεγόμενον χαμαίπιτυς. ταύτην βράσε μετὰ οἴνου καὶ ἄς το πίνῃ κατὰ πρωῒ με τὸ κρασίν.

181 ρπα´ ὅπου πονεῖ τὸ ὠτίον του καὶ οὐδὲν ἀκούει, ποίησον θεραπείαν τοιαύτην· ἤγουν κοπάνισον καρύδια καὶ τὸ ἔλαιον τούτων ἢ τῶν γλυκέων ἀμυγδάλων. χλίανε καὶ στάξε εἰς τὸ ὠτίον. 2 ἄλλο. βαλσαμέλαιον ἔνσταξε εἰς τὸ ὠτίον. 3 ἄλλο. λαγωοῦ χολὴν με τὸ γάλα τὸ γυναικεῖον στάξον εἰς τὸ ὠτίον. 4 ἄλλο. κόρακος χολὴν μίξον με τὸ μέλι τὸ καλὸν τὸ καινούργιον καὶ στάξον εἰς τὸ ὠτίον.

---

19 ῥίζας] ῥίζαν A 1 ἰντυβίου] ἰδιβίου A ἰντιβίων B 1 ῥίζαν] om. B 1 βάλε] καὶ ῥεφανίου ῥίζαν καὶ ζοχίου A om. B 2 καὶ] om. A 2 τὸν ζωμὸν] om. AB 2 ἐκείνων] om. AB 2–3 σὺν τούτοις βάλε καὶ ῥεφανίου ῥίζαν καὶ ζοχίου] om. A 3 βάλε] om. B 4 εἰς] πρὸς MB 4 δυσεντερίαν] δυσεντεριας M 4 ἀμάραντον] μάραντα M ἀμάραντα A εἰς δυσεντερίαν ὠφελεῖ ταῦτα B 4 κοπάνισον] ἀμάραντα add. B 4–5 τὸν ζωμὸν πότισον] transp. B 5 φάβα τρίψον] φάβατ...ψε L 6 φύλλα] καὶ βλαστάρια add. B 6 μετὰ οἴνου] om. B 6 ἂς πίνῃ] πότισον A 6 πίνῃ] τὸ π... L 7 βράσον] scripsi βρά..... L 8 βοτάνιν] βότανον M βοτάνη AB 8 χαμαίπιτυς] scripsi χαμαιχίστιν L χαμεπίτης M χαμαιπύτης A χαμαιχίστη B 8 ταύτην] τοῦτο A 9 βράσε] βράσον Σ 9 το] την B 9 με τὸ κρασίν] om. M νηστικὸς A 10 ὅπου] πρὸς ὅτινος M πρὸς πόνον ὠτίου· εἰς τὸν B 10 πονεῖ τὸ ὠτίον] πον.... ..ων L πονοῦν τὰ ὠτία ἢ τὸ ὠτίον B 10 του] αὐτοῦ B 10 οὐδὲν ἀκούει] οὐ ο..ακούει L οὐ δύναται ἀκούειν B 10–11 ποίησον θεραπείαν] ποίησ......... L 11 τοιαύτην] om. A 11 ἤγουν] om. AB 11 καρύδια καὶ τὸ] καρι........ L καρύδια ἢ ἀμύγδαλα γλυκέα καὶ τὸ A 11–12 ἢ τῶν γλυκέων ἀμυγδάλων] om. A 12 ἀμυγδάλων] τὸ ἔλαιον add. MB 12 χλίανε καὶ στάξε] βάσταζε M στάζε A καὶ στάξον B 12 ὠτίον] οὖς M 13 βαλσαμέλαιον] ὠφελεῖ καὶ βαλσαμέλαιον B 13 ἔνσταξε] στάξον M ἐνσταζόμενον B 13 ὠτίον] οὖς B 14 ὠτίον] οὖς B 15 μίξον] ἕνωσον M 15 μέλι τὸ καλὸν] transp. B 15 καινούργιον] σύνωρον A

---

13 λαγωοῦ] The following two recipes are reversed and in different positions in MAB.

5 ἄλλο. ⊳om. B κρομμυδίου χυλὸν χλίον στάξον εἰς τὸ ὠτίον. om. B⊲ 6 ἄλλο. τὴν ῥίζαν τοῦ ἀσφοδέλου κοπάνισον καὶ τὸν ζωμὸν ταύτης στάξον εἰς τὸ ὠτίον.

182 ρπβ΄ ἐὰν τῆς γυναικὸς τὰ καταμήνια κρατηθῶσιν, ποίησον θεραπείαν τοιαύτην· ἤγουν ἀγαρικὸν ὅσον κοκκία ϛ᾽ ἂς πίνῃ κατὰ πρωῒ με τὸ νερόν. 2 ἄλλο. ⊳om. B ἀδίαντον βράσον με τὸ νερὸν καὶ ἄς το πίνῃ κατὰ πρωΐ. om. B⊲ 3 ἄλλο. τζουκνίδας φύλλα κοπάνισον καὶ σμύρναν κοπάνισον καὶ ἀνάμιξον καὶ δός τα ἄλλην γυναίκα καὶ ἄς τα βάλῃ εἰς τὴν μήτραν αὐτῆς ἢ ἄς τα θέσῃ ἔξω † .....ωμα αὐτῆς. 4 ἄλλο. ἄσαρ κοπάνισον καὶ ᾽ το πίνῃ με τὸ μέλι. 5 ἄλλο. ⊳om. MA ῥίζαν ἀσφοδέλου κοπάνισον καὶ τὸν ζωμὸν ταύτης ἄς το πίνῃ με τὸ κρασίν. om. MA⊲

183 [144] ρπγ΄ ⌜ἐὰν συμβῇ τὴν γυναῖκα τὴν ἐγγαστρομένην καὶ ἀποθάνῃ τὸ παιδίον ἀπέσω εἰς τὴν μήτραν αὐτῆς,⌝ ποίησον θεραπείαν τοιαύτην, ὅπως να τὸ ἐκβάλῃς· ⌜πότισέ την ἀψίνθιον βρασμένον με τὸ κρασίν.⌝ 2 ἄλλο. ⊳om. A κοπάνισον ἀψίνθιον καὶ μάλαξον

15 ὠτίον] οὖς Σ 1 χλίον] χλί... L 2 τὴν ῥίζαν τοῦ ἀσφοδέλου κοπάνισον] transp. B 2 τοῦ ἀσφοδέλου] ἀσφοδελέας M τῆς ἀσφοδελαίας A 3 ὠτίον] οὖς AB 4 τῆς γυναικὸς τὰ καταμήνια κρατηθῶσιν] transp. B 4 τῆς] om. M 5 τοιαύτην] ὅπως πάλιν να κενώνῃ ἡ γυνὴ τὰ τοιοῦτα ὡς εἶχεν συνήθεειαν add. M ὅπως πάλιν να κενώνει ὡς εἶχε συνήθειαν add. A ὅπως κενώσῃ πάλιν ἡ γυνὴ ταῦτα ὡς ἔχει συνήθεια add. B 5 ἤγουν] om. A 5 ὅσον] ὡσεὶ B 6 πρωῒ] νηστικὴ add. M νῆστις add. A 6 με τὸ νερόν] om. B 6 με τὸ νερὸν] om. A 7 το] τον MA 7 πρωῒ] νηστικός add. M νῆστις με τὸ νερὸν add. A 8 καὶ σμύρναν κοπάνισον καὶ ἀνάμιξον] om. MA 8 κοπάνισον] om. B 8 τα] το A 9 τα] το AB 9 εἰς] scripsi ... L ἀπέσω εἰς MB ἔσω εἰς A 9 αὐτῆς] της Σ 9 ἄς τα θέσῃ] om. A 9 ἔξω] ἔξωθεν A 9 † ......ωμα] πλησίον τῆς μήτρας M om. A εἰς τὸ μορίον B 10 αὐτῆς] καὶ οὕτως θέλωσιν κατέρχεσθεν τὰ καταμήνια αὐτῆς ὡς εἶχεν συνήθειαν πρότερον add. M καὶ οὕτως θέλουν κινήσειν τὰ καταμήνια ὡς ἔθος A ἔσω καὶ οὕτως θέλωσι κατέρχεσθαι κατὰ τὸ συνηθὲς add. B 10 κοπάνισον καὶ ᾽] κοπαν....... L 11 το] om. B 13 ἐὰν] δὲ add. A πρὸς B 13 συμβῇ τὴν] om. B 13 ἐγγαστρομένην] ἐγγαστωμένην A ἐὰν συμβῇ add. B 13 καὶ] να B 14 παιδίον] αὐτῆς add. MA βρέφος B 14 ἀπέσω] om. B 14 εἰς τὴν] ἐν τῇ B 14 μήτραν αὐτῆς] μήτραν τῆς A κοιλίᾳ αὐτῆς οὕτως δεῖ ποιεῖν B 14–15 θεραπείαν τοιαύτην] om. A 15 να] om. A 15 ἐκβάλῃς] ἐκβάλῃ θεραπείαν τοιαύτην A 15 πότισέ την] scripsi πότισ... L ἤγουν πότιζε αὐτὴν M πότιζε αὐτὴν AB 16 κρασίν] om. A

με τὸ μέλι καὶ θές το μέσῳ εἰς τὸ σῶμα αὐτῆς.[om. A◁] 3 ἄλλο. δαυκὶν να γένηται εἰς ἀλλότριον τόπον χωρὶς ἵνα τὸ σπείρωσίν του τὸ σπέρμα, πότισον μετὰ οἴνου. 4 ἄλλο. [om. A▷]ἀγριοελαίας δάκρυον κοπάνισον καὶ πότισον μετὰ οἴνου.[om. A◁] 5 ἄλλο. δρακοντίας τὸ σπέρμα κοπάνισον καὶ πότισον με τὸ μέλι καὶ τὸ κρασὶν τὸ πρωΐ. 6 ἄλλο. ⌜καππάρεως φλοῖον ἢ τῆς ῥίζης, βράσον μετὰ οἴνου καὶ πότισον. 7 ἄλλο. κινάμωμον ψιλὸν καὶ σμύρναν ψιλὴν πότισον.⌝ 8 ἄλλο. κυκλαμίνου ῥίζαν βράσον με τὸ νερὸν καὶ ἄς το πίνῃ. 9 ἄλλο. ⌜μακεδονησίου σπόρον κοπάνισον καὶ πότισον με τὸ νερόν.⌝ 10 ἄλλο. ἔστι βοτάνιν τὸ λεγόμενον τριφύλλιν. τούτου τὰ φύλλα καὶ τὰ ἄνθη καὶ τὸν σπόρον βράσον μετὰ νεροῦ καὶ πότισον. 11 ἄλλο. ⌜ὀρίγανον βράσον με τὸ νερὸν καὶ ἄς το πίνῃ.⌝ 12 ἄλλο. [▷om. AB]τῆς πτελέας τὴν ῥίζαν βράσον με τὸ μέλι καὶ τὸ κρασὶν καὶ ἄς το πίνῃ.[om. AB◁] 13 ἄλλο. ἔστι βοτάνη ἡ λεγομένη πρασία. ταύτην βράσον μετὰ νεροῦ καὶ ἄς το πίνῃ.

184 [145] ρπδʹ ⌜ἐὰν δάκῃ τινὰ ὄφις ἢ ἔχιδνα·⌝ κοπάνισον ῥίζαν ἀσφοδέλου καὶ πότισον βρασμένην με τὸ κρασίν. 2 ἄλλο. τὴν ῥίζαν τοῦ ἀσφοδέλου κοπάνισον καλὰ καὶ θέτε ἐπάνω εἰς τὴν πληγήν. 3 ἄλλο. ⌜καστόριον τριμμένον πότισον μετὰ οἴνου. 4 ἄλλο. κρί-

16 μάλαξον] μάλαξέ το M 1 μέσῳ] μέσα M εἰς τὸ αἰδοῖον αὐτῆς μέσα B 1 εἰς τὸ σῶμα αὐτῆς] πλησίον τῆς μήτρας ὕγουν μέσα εἰς τὴν γένναν τῆς M om. B 1 δαυκὶν] δαυκὴν L φακὶν B 2 χωρὶς] om. B 2 ἵνα] νὰ Σ 2 του] τούτου Σ 3 σπέρμα] δέρμα M 3 ἀγριοελαίας] scripsi ἀ..οελαίας L ἀγριλαίας M ἀγρυελαίας B 4 δρακοντίας] ἔστι βοτάνη ἡ λεγομένη δρακοντία ταύτης Σ 6 καππάρεως φλοῖον] τὸ φλίον τῆς καππάρεως B 6 φλοῖον] ὁ φλοιὸς M ἡ φλοιὸς A 6 ἢ] om. A 7 κινάμωμον] κι...μωμ... L 7 καὶ σμύρναν] ....μην L σμίρνα M καὶ σμύρνα B 7 ψιλὴν] ψιλῆς B 7 πότισον] με τὸ νερὸν add. Σ 8 κυκλαμίνου] ......αλαμίνου L καλαμίου B 8 νερὸν καὶ] ..... L 9 μακεδονησίου] μακεδονίου M 9 σπόρον] ῥίζαν ἢ σπόρον A 10 βοτάνιν] βότανον M βοτάνη A 11 μετὰ νεροῦ] με τὸ νερὸν Σ 11 πότισον] ἂς πίνῃ M ἄς το πίνῃ AB 12 το] om. A 13 πτελέας] scripsi πτε.έας L παλαίας M 13 τὴν ῥίζαν] scripsi τηρίζαν L τῆς ῥίζαν M 14 πρασία] πρασεῖα A καὶ add. M 15 μετὰ νεροῦ] με τὸ νερὸν Σ 15 το] om. M 16 ἔχιδνα] ποίησον θεραπείαν τοιαύτην add. M ποίησον οὕτως add. B 17 πότισον βρασμένην] βράσον M πότιζον πλὴν βράσε ταύτην A αὐτὴν πότιζον βρασμένην B 17 κρασίν] καὶ πότισον add. M 17–18 τὴν ῥίζαν τοῦ ἀσφοδέλου] ἀσφοδελέαν M ἀσφολδελαίαν A τοῦ αὐτοῦ τὴν ῥίζαν B 18 καλὰ] με τὸ κρασὶν M

νου φύλλα κοπάνισον καὶ θέτε τα εἰς τὴν πληγὴν ἐπάνω. 5 ἄλλο. ξυλοκέρατα κοπάνισον καλὰ καὶ πότισον μετὰ οἴνου.⌝

185 ρπε´ τὰ λεγόμενα παρὰ τῶν ἰδιωτῶν ἐρμίγγια, παρὰ δὲ τῶν ἰατρῶν ἕλμινθας θεραπεύει καὶ ἐκβάλλει αὐτάς· ἤγουν κοπάνισον ἀμαράντα καὶ τὸν ζωμὸν τούτων πότιζον μετὰ οἴνου. 2 ἄλλο. ἀψίνθιον καὶ χολὴν ταυρίαν πότισον πλὴν τὸ ἀψίνθιον, βράσον πρῶτον με τὸ νερὸν εἶτα ἑνώσας πότιζε. 3 ἄλλο. κοπάνισον λουπηνάρια πικρὰ καὶ τὸ ἀλεύριν τούτων μίξον με τὸ μέλι καὶ ἄς το τρώῃ. 4 ἄλλο. θρύμβον βράσον με τὸ νερὸν καὶ ἄς το πίνῃ. 5 ἄλλο. ▷Bκρίμβον βράσον με τὸ νερὸν καὶ ἄς το πίνῃ.B◁ 6 ἄλλο. κράμβης σπέρμα κοπάνισον καὶ ἄς το πίνει με τὸ νερόν. 7 ἄλλο. ῥίζαν συκαμινέας βράσον με τὸ νερὸν καὶ ἄς το πίνῃ τὸ πρωΐ. 8 ἄλλο. ▷om. Lῥίζαν πτελέας βράσον με τὸ μέλι καὶ πότισον πλὴν δὲ βάλε καὶ κρασίν.om. L◁ 9 ἄλλο. πολυποδίου ῥίζαν βρεχέ την εἰς τὸ ἕψημα καὶ τρῶγέ την. 10 ἄλλο. ▷om. Aὕσσωπον ἕψησον με τὸ νερὸν καὶ τὸ κρασὶν καὶ ἄς το πίνῃ κατὰ πρωΐ.om. A◁ 11 ἄλλο. ῥεπανίου σπόρον κοπάνισον καὶ πότισέ τον με τὸ νερόν. 12 ἄλλο. σκόροδα ἆς τρώγῃ.

---

18 θέτε ἐπάνω] θές την M θέτε A θέτε την B 1 φύλλα] ῥίζαν M 1 θέτε] θὲς M 1 ἐπάνω] om. MA ἀπάνω B 2 ξυλοκέρατα] ...... L 2 πότισον μετὰ] π.......ετὰ L 3 τὰ] πρὸς ἕλμινθας· τὰ MB 3 ἰδιωτῶν] ἰατρῶν M 3 ἐρμίγγια] ἐρμύγγια L ἑλμίνθια M 3 παρὰ δὲ τῶν] om. B 4 ἰατρῶν] χορικῶν M om. B 4 ἕλμινθας] ἐρμίγγια M ἕλμιθες A om. B 4 θεραπεύει καὶ ἐκβάλλει αὐτάς] θεράπευε οὕτως M 4 αὐτάς] αὐτὰ τοῦτο B 4 ἤγουν] om. B 5 τούτων] om. Σ 5 πότιζον] πότισον Σ 6 ταυρίαν] ταυρείαν M 6 πρῶτον] om. Σ 7 εἶτα] καὶ ἔπειτα M καὶ αὖθις A καὶ τότε B 7 ἑνώσας πότιζε] ἕνωσον τὴν χολὴν καὶ ἄς την πίνῃ M μίξον τὴν χολὴν καὶ ἄς την πίνῃ AB 7 λουπηνάρια] λυπηνάρια MA 8 με τὸ] τὴν B 8 μέλι] ....λι L χολὴν B 8 το] om. M την πίνῃ ἤγουν σὺν μέλιτι B 8 τρώῃ] νηστικὸς add. A om. B 9 θρύμβον] θύμον MA ζύμον B 9 το] om. MA 11 το] om. M την B 12 το] om. MA 12 τὸ] κατὰ B 13 βράσον] βράσε A 13–14 πότισον πλὴν δὲ βάλε καὶ κρασίν] καὶ κρασὶν καὶ ἄς το πίνῃ ὠφέλιμον B 13 πότισον] πίνε A 14 την] om. Σ 15 τρῶγέ την] ἆς τρώγῃ MB τρῶγε A ἐστὶ ὠφέλιμον add. B 15 ὕσσωπον] πολλυποδίου ῥίζαν M τοῦ αὐτοῦ τὴν ῥίζαν B 15 ἕψησον] βράσον MB 15–16 καὶ τὸ κρασὶν] om. MB 16 το] τὴν MB 16 πίνῃ] καπίνῃ B 16 ῥεπανίου] ῥεφανίου A 17 καὶ πότισέ τον] .... L καὶ πότισον AB 17 με τὸ] με M 17 σκόροδα ἆς τρώγῃ] ἆς τρώγῃ δὲ καὶ σκόροδα MA transp. B

186 ρπϛʹ ὅταν ὁ ἄνθρωπος πονεῖ κἂν εἰς τὸ λεγόμενον ἰσχίον· πότισον τριμμένον ἀμμωνιακόν. 2 ἄλλο. βάλε ἀφ' ἑσπέρας ἀμμωνιακὸν εἰς ὄξος καὶ ἂς βράσῃ. καὶ ὅταν μαλακίσῃ ἅπλωνέ το εἰς τὸ πανίν. καὶ θέτε ἐπάνω εἰς τὸν τόπον ἔνθα πονεῖ. 3 ἄλλο. ▷om. A ἄσαρ βράσον με τὸ νερὸν καὶ ἄς το πίνῃ. 4 ἄλλο. βράσον ἄσαρ καὶ τὸ μέλι καὶ τὸ κρασὶν καὶ πότισον με τὸ νερόν. om. A◁ 5 ἄλλο. ▷om. L ὕσσωπον ἔψησον με τὸ μέλι καὶ τὸ νερόν. ἄς το πίνῃ κατὰ πρωΐ. om. L◁ 6 ἄλλο. τὴν ῥίζαν τοῦ ἀσπαράγγου βράσον με τὸ κρασὶν καὶ ἄς το πίνῃ. 7 ἄλλο. ▷om. B τὸ σπέρμα τῆς καππάρεως βράσον με τὸ κρασὶν καὶ ἄς το πίνῃ. om. B◁ 8 ἄλλο. καππάρεως τῆς ῥίζης τὴν φλοῦδαν, βράσον με τὸ κρασὶν καὶ πότισον. 9 ἄλλο. κενταύριον βράσον με τὸ κρασὶν καὶ ἄς τον δόσουν κλυστήριν. 10 ἄλλο. πήγανον βράσον με τὸ ἔλαιον †... καὶ ἄς το πίνῃ.

187 [146] ρπζʹ ⌜ὅπου γένηται τὸ πρόσωπόν του κίτρινον, ὅπερ καλεῖται παρὰ τῶν ἰατρῶν ἴκτερος, παρὰ δὲ τῶν χωρικῶν χρυσιασμός, ποίησον δὲ θεραπείαν τοιαύτην· βράσον ἀδίαντον με τὸ

1 ὅταν ὁ ἄνθρωπος πονεῖ κἂν εἰς τὸ λεγόμενον ἰσχίον] πρὸς ὀδύνην ἰσχίου ἤτοι περινεφρίου M πρὸς ὀδύνην ἰσχίου B 1 ὅταν ὁ ἄνθρωπος] scripsi ὅ........ L ὅπου A 2 πότισον] scripsi πο...σον L πότισον αὐτὸν Σ 2 τριμμένον] τὸ λεγόμενον Σ 2–3 ἀμμωνιακὸν] om. A 3 εἰς] τὸ add. A 3 ὄξος] ἀμμωνιακὸν add. A 3 εἰς] ..ς L 4 θέτε] το add. Σ 4 ἔνθα] ὁποῦ Σ 5 με τὸ] με... L 5 το] om. M 5 καὶ τὸ] με τὸ MB 6 πότισον] ἄς την πίνῃ B 6 με τὸ] με M μετὰ B 6 νερόν] νεροῦ B 7 νερόν] κρασὶν B 7 ἄς το πίνῃ] καὶ ἂς πίνῃ AB 8 τὴν ῥίζαν τοῦ ἀσπαράγγου] transp. B 8 το] την M om. A 10 το] om. MA 10 καππάρεως τῆς ῥίζης] τὴν ῥίζαν τῆς καππάρεως A 10 καππάρεως] τῆς καππάρεως MB 10 τὴν φλοῦδαν] τὸ φλοῦδιν MB om. A 10–11 βράσον με τὸ κρασὶν καὶ πότισον] πότισον βρασμένον με τὸ κρασίν M πότισον βρασμένον B 11 κενταύριον] πενταύριον M ενταύριον B 11 κρασὶν] νερὸν Σ 12 δόσουν] δώσουν MA 13 †...] om. Σ 13 το] om. AB 14 ὅπου γένηται τὸ πρόσωπόν του κίτρινον] ὅπου γένηται τὸ πρό ......τρινον L πρὸς πρόσωπον ὅταν γένηται κίτρινον MB 15–16 καλεῖται παρὰ τῶν ἰατρῶν ἴκτερος, παρὰ δὲ τῶν χωρικῶν χρυσιασμός] καλοῦσι παρὰ τῶν ἰδιωτῶν χρυσιασμὸν παρὰ δὲ τῶν ἰατρῶν ἴκτερος τὸ πάθος καλεῖται M χρυσιασμὸς παρὰ δὲ ἰατρῶν ἴκτερος A παρὰ μὲν τῶν ἰατρῶν καλεῖται ἔκτερος τὸ τοιοῦτον πάθος παρὰ δὲ τῶν ἰδιωτῶν χρυσιασμὸς B 15 ἰατρῶν] scripsi ...... L 15–16 χρυσιασμός] scripsi ....σμός L 16 δὲ] om. Σ 16 τοιαύτην] om. A

5 βράσον] This recipe has a different location in MB. 11 κενταύριον] The following two paragraphs are reversed in B.

κρασὶν καὶ ἄς το πίνῃ. 2 ἄλλο. ἀλόην τριμμένην πότισον με τὸ νερόν.⌝ 3 ἄλλο. ῥίζαν ἀσφοδέλου βράσον με τὸ κρασὶν καὶ πότιζε καὶ τὰ βλαστάρια ταύτης ἄς τα τρώῃ ἀντὶ λαχάνων. 4 ἄλλο. ⌜ἀψινθίου κορυφὴν βράσον με τὸ νερὸν ἢ με τὸ κρασὶν καὶ πότισον. 5 ἄλλο. ▷[om. A]σκολοπένδριν βράσον με τὸ νερὸν καὶ ἄς το πίνῃ.[om. A◁]⌝ 6 ἄλλο. ⌜χρυσολαχάνου σπόρον κοπάνισον καὶ πότισον ἢ με τὸ νερὸν ἢ με τὸ τυρόγαλον.⌝ 7 ἄλλο. ⌜τὴν ῥίζαν τοῦ λαπάτου βράσον με τὸ κρασὶν καὶ πότισον.⌝ 8 ἄλλο. βράσον χαμαίμηλα καὶ τὸν ζωμὸν τούτων πότισον. 9 ἄλλο. τὴν ῥίζαν τῆς χελιδονέας βράσον μετὰ οἴνου καὶ πότισον. 10 ἄλλο. τὴν ῥίζαν τῆς κυκλαμινέας βράσον με τὸ μέλι καὶ τὸ νερὸν καὶ πότισον.

188 ρπη′ περὶ κεφαλαλγίας· τὰ φύλλα τῆς λυγαίας κοπάνισον καὶ τὸν σπόρον ταύτης καὶ μίξον ῥοδέλαιον καὶ ἄλειφε τὴν κεφαλήν. 2 ἄλλο. ▷[L]ἀμάραντα κοπάνισον καὶ τὸν ζωμὸν τούτων μίξον με τὸ ῥοδέλαιον καὶ ἄλειφε τὴν κεφαλήν.[L◁] 3 ἄλλο. ἀλόην τριμμένην μίξον με τὸ ῥοδέλαιον καὶ ὄξος καὶ ἄλειφε τὸ μέτωπον καὶ τοὺς μήλιγγας. 4 ἄλλο. ἀμύγδαλα πικρὰ κοπάνισον καλὰ καὶ μίξον ὄξος

---

16 βράσον] ἤγουν βράσον Σ 1 καὶ] .. L 1 το] om. M 2 ῥίζαν] τὴν ῥίζαν τοῦ A 2 ἀσφοδέλου] ἀσφοδίλλου M 2 με τὸ] με.. L 2 κρασὶν] νερὸν B 2–3 πότιζε] ἄς πίνῃ M ἄς το πίνῃ A ἄς την πίνῃ B 3 ταύτης] τούτου τὰ τρυφερὰ ἡνίκα αὐξήσουν MB τοῦ τὰ τρυφερὰ ἡνίκα αὐξήσουν A 3 τα] om. B 3 τρώῃ] τρώγῃ Σ 3 λαχάνων] λαχάνου B 4 ἀψινθίου] ἀψ....θίου L ψινθίου M ἀψίνθιον B 4 κορυφὴν] κορυφὰς MA 4 ἢ] καὶ A 4 κρασὶν] ...σὶν L 5 σκολοπένδριν] σκολοπένδριον MB 5 βράσον] βρ.... L 5 με τὸ] μετὰ M 7 ἢ] om. M 7 νερὸν] νερ... L ἢ με τὸ κρασὶν ἢ add. B 7 ἢ] πολλάκις add. B 8 λαπάτου] scripsi λαπ... L λαπάθου M λαπάθου A 8 πότισον] ἄς πίνῃ MB ἄς το πίνῃ A 8–9 χαμαίμηλα] με τὸ νερὸν add. A 9 τούτων] om. A 9 πότισον] ἄς πίνῃ A 9 τὴν] τῆς M 10 χελιδονέας] χελιδονίας A 10 βράσον] βρα.... L 10 πότισον] πότιζον M πότιζε A 10 τὴν] τῆς M 11 κυκλαμινέας] κυκλαμεναίας M κυκλαμηναίας A 11 βράσον με τὸ μέλι καὶ τὸ νερὸν καὶ πότισον] πότισον με τὸ νερὸν καὶ τὸ μέλι βρασμένον M πότισον με τὸ νερὸν καὶ τὸ μέλι βρασμένα A πότισον με τὸ νερὸν καὶ τὸ μέλι βρασμένην B 12 περὶ] πρὸς AB 12 κεφαλαλγίας] κεφαλαλγίαν MB 12 τὰ] ... L θεράπευε δὲ οὕτω. τὰ A οὕτως ποίει ὠφέλειμων. τὰ B 12–13 κοπάνισον καὶ τὸν σπόρον] κοπάνισ....... L 13 ταύτης] αὐτῆς B 13 καὶ] om. A 13 μίξον] ἕνωσον με τὸ M με τὸ add. A μίξας B 13 καὶ ἄλειφε] ....ε L ἄλειφε B 14 καὶ] scripsi ..ι L 16 καὶ] τὸ add. Σ 16 ὄξος] ὀξείδιν B 17 μήλιγγας] μήνιγγας MA 17 ἀμύγδαλα] ..γδαλα L

καὶ ῥοδέλαιον καὶ ἄλειφε τὸ μέτωπον καὶ τοὺς μήλιγγας. 5 ἄλλο. [▷om. LB]ἀμύγδαλα πικρὰ κοπάνισον καὶ μίξον ὄξος καὶ ῥοδέλαιον καὶ ἄλειφε τὸ μέτωπον καὶ τοὺς μήνιγγας.[om. LB◁] 6 ἄλλο. κοπάνισον φύλλα χλωρὰ ἀμπελῶνος καὶ [▷om. A]τὰ λεγόμενα παρὰ τῶν ἰδιωτῶν κυκλίδια, παρὰ δὲ τῶν ἰατρῶν[om. A◁] ἕλικες ἀμπέλου. ταῦτας κοπάνισον καὶ μίξον καὶ τὰ φύλλα. καὶ θέτε τα εἰς τὸ μέρος τῆς κεφαλῆς ἔνθα ἔνι ὁ πόνος. 7 ἄλλο. φύλλα χλωρὰ κράμβης κοπάνισον καὶ τὸν ζωμὸν τούτων βάλε εἰς τὴν μύτην πλὴν χλίον. 8 ἄλλο. ἡδύοσμον χλωρὸν κοπάνισον καὶ θέτε εἰς τὸ μέτωπον. 9 ἄλλο. ἔστι βοτάνη ἡ λεγομένη κόνυζα. ταύτης τὰ φύλλα κοπάνισον καὶ μίξον με τὸ χρηστέλαιον καὶ ἄλειφε τὴν κεφαλήν. θέτε δὲ καὶ τὰ φύλλα ταύτης. 10 ἄλλο. [▷om. B]ψύλλεον κοπάνισον καὶ μίξον ῥοδέλαιον καὶ ὄξος καὶ ἄλειφε τὴν κεφαλήν.[om. B◁] 11 ἄλλο. [▷om. A]ἔστι βοτάνιν τὸ λεγόμενον βρωμοβότανον. καλεῖται δὲ παρὰ τῶν ἰατρῶν στρύχνος. τοῦτο κοπάνισον καὶ τὸν ζωμὸν μίξον με τὸ χρηστέλαιον καὶ ἄλειφε τὴν κεφαλήν.[om. A◁] 12 ἄλλο. [▷B]μελάνθιν κοπάνισον καὶ μίξον χρηστέλαιον καὶ ὄξος καὶ ἄλειφε τὴν κεφαλήν.[B◁] 13 ἄλλο. σινάπιν κοπάνισον καὶ μάλαξον με τὸ μέλι καὶ θέτε εἰς τὸ μέτωπον. 14 ἄλλο.

292.17–293.1 ὄξος καὶ ῥοδέλαιον] καὶ ἔλαιον M ὄξος καὶ ἔλαιον A ὄξος B 1 τὸ μέτωπον καὶ] om. Σ 1 μήλιγγας] μίνιγγας M μήνιγγας A 2 καὶ μίξον] καλὰ καὶ μίξον A 2 ῥοδέλαιον] ἔλαιον A 3 τὸ μέτωπον καὶ] om. A 3–4 κοπάνισον φύλλα χλωρὰ ἀμπελῶνος] ἀμπελόφυλλα χλωρὰ κοπάνισον B 4 ἀμπελῶνος] ἀμπέλου A 4 τὰ λεγόμενα] τὰς καλουμένας B 5 κυκλίδια] κηλλίδας ἢ ψαλλίδας M κικλίδια ἢ ψαλίδας B 5 ἕλικες] scripsi ἕλλικες L ἕλλιγγας MA ἕλικας B 5 ταῦτας] ταῦτα M om. A 5–6 κοπάνισον καὶ] κοπανίσας A 6 μίξον] ἕνωσον M μίξε τα A 6 καὶ τὰ] μετὰ A 6 θέτε] θὲς M 6 τα] om. B 6 εἰς] ... L om. B 6 τὸ μέρος τῆς κεφαλῆς] om. B 7 πόνος] εἰς τὸ μέρος ἐκεῖνο τῆς κεφαλῆς B 7 χλωρὰ] om. Σ 7 κράμβης] χλωρὰ add. AB 7 κοπάνισον] χλωρὰ add. M 8 εἰς] ... L 8 μύτην] ῥίναν M 9 θέτε] το add. MB τον add. A 9 μέτωπον] με..... L 9 ἔστι] ἐ.... L στι M 9–10 βοτάνη ἡ] βοταν... L βοτάνη τὸ B 10 λεγομένη κόνυζα. ταύτης] scripsi λεγόμεν.....αύτης L κόρυζα. ταύτης M λεγομένη κόνιζα. ταύτης A τὸ λεγόμενον κόνιζον ἢ κόνηζα. ταύτης B 10 μίξον] ...... L 11 ταύτης] scripsi ταυ...L 12 μίξον] ἕνωσον M 12 καὶ] τὸ add. M 13 βοτάνιν] βότανον M βοτάνη AB 13 τὸ λεγόμενον] ἡ λεγομένη A 14 καλεῖται δὲ παρὰ] παρὰ δὲ Σ 14 στρύχνος] στ..χνος L λέγεται στρήχνος M 15 μίξον] ...ξον L 18 θέτε] θὲς M το add. B

13 ἔστι] This paragraph is located at the end of the chapter in manuscript A.

πήγανον ἥμερον ἕψησον με τὸ ὄξους καὶ τὸ ῥοδέλαιον καὶ ἄλειφε τὴν κεφαλὴν καὶ τὸ μέτωπον καὶ τοὺς μήλιγγας.

189 [147] ρπθ′ ὅπου τρέχῃ ἡ κοιλία του ποίησον θεραπείαν τοιαύτην ὅπως ἱσταθῇ· ⌜ῥίζαν βάτου τὰ βλαστάρια βάλε καὶ κοπάνισε καὶ τὸν ζωμὸν τούτων πότισον. 2 ἄλλο. προυν⊳om. Lέας ῥίζαν κοπάνισον καὶ ἕψησον με τὸ κρασὶν καὶ ἂς πίνῃ. 3 ἄλλο. ⊳om. Aπαλιουρέας ῥίζαν καὶ τὰ φύλλα βράσον μετο κρασὶν καὶ ἂς πίνῃ.om. A⊲ 4 ἄλλο. μυρσίνης φύλλα βράσον με τὸ κρασὶν καὶ ἰάτρευε κλυστήριν. 5 ἄλλο. ⊳om. Aσχίνου τὸ σπέρμα ἕψησον με τὸ κρασὶν καὶ ἰάτρευε κλυστήριν.om. A⊲ 6 ἄλλο. τὸ σπέρμα τοῦ σχίνου καὶ τὰ τρυφερὰ φύλλα βράσον με τὸ νερὸν καὶ ἄς το πίνῃ. 7 ἄλλο. κολοκύνθιν ἕψησον με τὸ κρασὶν καὶ ἄς το πίνῃ με τὸ κρασίν. 8 ἄλλο. κυπαρίσσου τρυφερὰ βλαστάρια βράσον με τὸ κρασὶν καὶ ἂς πίνῃ.⌝ 9 ἄλλο. ⊳Bσεύκτου τὰ φύλλα βράσον με τὸ κρασὶν καὶ ἄς τα πίνῃ.B⊲ 10 ἄλλο. ἔστι βότανον τὸ λεγόμενον ἀρνόγλωσσον, τούτου τὸν σπόρον κοπάνισον καὶ πότισον μετὰ ὕδατος.om. L⊲

190 [148] om. Lρϙ′ πρὸς ἐὰν δάκῃ τινα σκύλλος ποίησον θεραπείαν τοιαύτην· ἤγουν ⌜κοπάνισον μελισσόφυλλα καὶ πότισον μετὰ οἴνου. 2 ἄλλο. ποταμογείτονος φύλλα κοπάνισον καὶ μίξον με τὴν στάκτην καὶ θὲς εἰς τὴν πληγήν.⌝ 3 ἄλλο. ⊳Aσκόροδα κοπάνισον καὶ θέτε τα ἐπάνω.A⊲ 4 ἄλλο. ⌜σκόροδα κοπάνισον καὶ πότισον με τὸ νερόν.⌝ ⊳om. Bἂς τρώγῃ δὲ καὶ σκόροδα.om. B⊲ 5 ἄλλο. ⌜πήγανον

---

18 μέτωπον] μετω.... L 1 ἥμερον ἕψησον] om. B 1 με τὸ ὄξους] μετὰ ὄξος M με τὸ ὄξος A μετὰ ὄξους B 1 καὶ τὸ] ... τὸ L καὶ MB 1 ῥοδέλαιον] ῥοδελαίου B 2 τὴν κεφαλὴν καὶ] om. Σ 2 τὸ μέτωπον καὶ] om. A 2 μήλιγγας] μίνιγγας MB μήνιγγας καὶ τὸ μέτωπον A 3 ὅπου τρέχῃ ἡ κοιλία του] πρὸς κοιλίαν ὅπου ῥέῃ πολλά M 3–4 θεραπείαν τοιαύτην] οὕτως B 4 ὅπως] ἵνα A 4 ἱσταθῇ] να σταθῇ MB σταθῇ A 4 βάτου] scripsi βατ.. L βάτου καὶ Σ 4 τὰ] φύλλα καὶ τὰ B 4 κοπάνισε] κοπάνισον Σ 4 καὶ] ... L 6 πίνῃ] τὸ πίνῃ AB 6 παλιουρέας] scripsi ἀλλιουρέαν M παλουρέας B 7 ἂς] τὸ add. B 8 ἰάτρευε] τὸν add. A 9 σπέρμα] καὶ τὰ φύλλα τρίψας καὶ B 10 τὸ σπέρμα τοῦ σχίνου] σχίνου τὰ φύλλα A 10 σχίνου] αὐτοῦ B 11 φύλλα] βλαστάρια καὶ τὸ σπέρμα A 11 βράσον] ἕψησον A 11 νερὸν] κρασὶν A 11 ἄς το πίνῃ] καὶ ἰάτρευε τὸ κλυστήριν A 11 κολοκύνθιν] κολοκύνθην A 12 το] om. B 12 με τὸ κρασίν] τὸ πρωῒ AB 13 ἂς] τα add. A 15 βότανον] βοτάνη A βοτάνιν B 17 πρὸς] om. A 17 τινα] ὁ add. B 17 σκύλλος] οὕτως ποίησον B 18 τοιαύτην] om. A 20 θὲς] θέτε τα AB

κοπάνισον καὶ μίξον αὐτὸ με τὸ χρηστέλαιον καὶ θὲς εἰς τὴν πληγήν. 6 ἄλλο. ἀγριοσυκέας φύλλα κοπάνισον τρυφερὰ καὶ μάλαξον με τὸ μέλι καὶ θές τα εἰς τὴν πληγήν. 7 ἄλλο. [▷om. A]πολύγονον κοπάνισον καὶ τὸν χυλὸν πότισον μετὰ οἴνου.[⌝om. A◁]

191 [om. B]ρϙα′ [▷om. L]πρὸς συνάχην ποίησον θεραπείαν τοιαύτην· σκύλλου κόπρον ξηρὰν λευκὴν τρίψον καὶ μίξον με τὸ μέλι[om. L◁] καὶ θέτε τα εἰς τὸν τράχηλον. 2 ἄλλο. [▷L]κεδρέαν θέτε εἰς τὸν τράχηλον. [L◁] 3 ἄλλο. μέλι καὶ νερὸν χλίανε καὶ ἄς το κρατῇ εἰς τὸ στόμα του. 4 ἄλλο. πίτυρα βράσον με τὸ νερὸν καὶ ἕψημα εἶτα σακέλλισον καὶ τὸν ζωμὸν ἄς τον βάνῃ εἰς τὸ στόμα του καὶ ἂς τον κράτῃ πολλὴν ὥραν.

192 ρϙβ′ τὸν λεγόμενον λούξικαν, παρὰ δὲ τῶν ἰατρῶν λυγμὸς, θεράπευε οὕτως· κοπάνισον ἡδύοσμον χλωρὸν καὶ τὸν ζωμὸν τούτου μίξον με τὸν χυλὸν τοῦ ῥοϊδίου τοῦ ὀξύνου καὶ πότισον. 2 ἄλλο. ἔστι βοτάνιν τὸ λεγόμενον σκολόπενδρα. ταύτην βράσον με τὸ νερὸν καὶ πότισον. 3 ἄλλο. ἀριστολόχια κοπάνισον καὶ πότισον μετὰ θερμοῦ ὕδατος. 4 ἄλλο. [▷om. A]ἔστι βοτάνιν ἡ λεγομένη ἄλυσσος. ταύτην βράσον με τὸ νερὸν καὶ πότισον.[om. A◁]

---

21 πότισον] θέτε τα B 1 θὲς] θέτε το AB 2 μάλαξον] μίξον A 3 με τὸ μέλι] σὺν μέλιτι B 3 θές] θέτε AB 3 τὴν] om. B 3 πληγήν] ἢ τοῦ σκύλλου ἢ τῆς κολοπένδρας ἢ τῆς λεγομένης γαλῆς. ζῶον γὰρ αἶνει τὸ τοιοῦτον ὥσπερ καλοῦμενος αἴλουρος. καὶ ἀναβαίνει εἰς δένδρα ἐπάνω add. M 4 μετὰ οἴνου] καὶ με τὸ κρασὶν ἂς τρώγῃ καὶ σκόρωδα B 5 τοιαύτην] om. A 7 καὶ θέτε τα] .....έθε L 7–8 κεδρέαν θέτε εἰς τὸν τράχηλον] scripsi κεδρ..........ὸν τράχηλον L 9 του] ἢ μόνον χλίανε add. M om. A 9 καὶ] τὸ add. M om. A 9 ἕψημα] om. A 9 εἶτα] om. M 10 σακέλλισον] σακέλισον MA 10 ζωμὸν] αὐτῶν add. MA 10 ἄς τον βάνῃ] βάλε MA 11 τον] τὸ M 11 πολλὴν ὥραν] transp. A 12 τὸν] πρὸς τὸν B 12 λούξικαν] λούτζικαν B 12 παρὰ δὲ τῶν ἰατρῶν] om. B 12–13 λυγμὸς] λυγμὸν MA ἥως κλόξον B 13 θεράπευε] θεραπεύεται δὲ M θεράπευσον A om. B 13 οὕτως] om. B 13 χλωρὸν] om. B 14 ῥοϊδίου τοῦ ὀξύνου] transp. B 14 ῥοϊδίου] ῥοδίου M 15 βοτάνιν] βοτάνη MA 15 τὸ λεγόμενον] ἡ λεγομένη Σ 15 σκολόπενδρα] σκολοπενδρία A 15–16 με τὸ νερὸν] σὺν τῷ ὕδωρ B 16 ἀριστολόχια] ἀριστολογχίαν MB ἀριστολόχιαν A 17 βοτάνιν] βοτάνη M 18 ἄλυσσος] ἄλυσμον L ἄλυσος M 18 με τὸ νερὸν] om. M μετὰ οἴνου B

---

3 πληγήν] The additional text in M is probably a scholion that was copied in by mistake.

193 [149] ρϞγʹ ὅταν λυσσιάρης σκύλλος δάκῃ τινά, ποίησον θεραπείαν τοιαύτην· ἤγουν ⌈ἔστι βοτάνιν ἡ λεγομένη γεντιανή. ταύτης τὴν ῥίζαν βράσον μετὰ οἴνου καὶ πότισον. 2 ἄλλο. καβούρους ποταμίους καῦσον εἰς ἁπτόμενα κάρβουνα καὶ τὴν στάκτην τῶν καβούρων πότισον μετὰ οἴνου. 3 ἄλλο. ▷om. L τὸν λεγόμενον λίθον αἱματίθην τοῦτον εἰς ἀκόνην τρίψον καὶ ἐκείνας τρίμματα τὰ ἔκβαλε. πότισον μετὰ οἴνου. 4 ἄλλο. βδέλλιον σὺν οἴνῳ πινόμενον.⌉om. L◁

194 ρϞδʹ τὰ λιθάρια τὰ εἶναι εἰς τὴν κύστιν καὶ εἰς τοὺς νεφρούς· ταῦτα δὲ ἐκβάλλει δάκρυον ἀμπέλου, πότισον δὲ τοῦτο μετὰ οἴνου. 2 ἄλλο. τὴν ῥίζαν τοῦ λαπάτου βράσον μετὰ οἴνου καὶ πότισον. 3 ἄλλο. ἀμυγδαλέας δάκρυον πότισον μετὰ ὄξους καὶ μετὰ ἑψήματος. 4 ἄλλο. τὴν ῥίζαν τοῦ ἀσπαράγγου κοπάνισον καὶ τὸν ζωμὸν πότισον. 5 ἄλλο. σκολοπενδρίου τὴν ῥίζαν καὶ τὰ φύλλα κοπάνισον καὶ πότισον. 6 ἄλλο. καππάρεως τῆς ῥίζης τὸ φλοῦδιν κοπάνισον καὶ πότισον μετὰ οἴνου. 7 ἄλλο. ▷LB γλυκόριζον κοπάνισον και τὸν ζωμὸν πότισον μετὰ οἴνου.LB◁ 8 ἄλλο. ▷Σ στρουθίου τὴν ῥίζαν κοπάνισον καὶ τὸν ζωμὸν πότισον μετὰ οἴνου. Σ◁ 9 ἄλλο.

1 λυσσιάρης σκύλλος δάκῃ τινά] transp. B 1 λυσσιάρης σκύλλος] ....-κύλος L λυσιάρις σκύλος M 1 τινά] τὸν ἄνθρωπον MA 2 τοιαύτην] om. A 2 ἤγουν] om. A 2 βοτάνιν] βότανον M βοτάνη AB 2 ταύτης] ταύτην B 3 ῥίζαν] κοπάνισον καὶ add. M 3 καβούρους] καβάρους M 3–4 ποταμίους] ποταμίους add. L 4 ἁπτόμενα] om. MB 4 κάρβουνα] καρβούνια B 5 οἴνου] ὕδατος B 5–6 τὸν λεγόμενον λίθον αἱματίθην τοῦτον εἰς ἀκόνην τρίψον] transp. B 5–6 αἱματίθην] αἱματίτην A 6 ἀκόνην] ἀκόνιν B 6 ἐκείνας] ἐκεῖνα τὰ A τὰ B 6 τὰ] om. A ἐκεῖνα B 6 ἔκβαλε] καὶ add. A ἃς ἐκβάλλῃ B 7 βδέλλιον] ὠφελεῖ βδέλλιον B 7 πινόμενον] πότισον A 8 τὰ] πρὸς MB 8 τὰ] ἅπερ AB 8 εἶναι] εἰσὶν MA 8 κύστιν] κύστην A 9 ταῦτα δὲ] ... L om. MA 9 ἐκβάλλει] ἐκβάλλονται δὲ οὕτως M ἐκβάλλονται οὕτως AB 9 δὲ τοῦτο] om. Σ 10 λαπάτου] λαπάθου M λαπάθου AB 11 καὶ μετὰ] ἢ M ἢ μετὰ B 11–12 ἑψήματος] ὑψήματος M ἑψήματα B 12 ἀσπαράγγου] σπαραγγίου Σ 12–13 τὸν ζωμὸν πότισον] transp. B 13 τὴν ῥίζαν καὶ] om. Σ 14 καὶ πότισον] ὁμοίως καὶ τὴν ῥίζαν καὶ πότισον τὸν ζωμόν M ὁμοίως καὶ τὴν ῥίζαν του καὶ πότισον τὸν ζωμὸν A τὸν ζωμὸν add. B 14 καππάρεως τῆς ῥίζης τὸ φλοῦδιν] transp. B 14 καππάρεως τῆς ῥίζης] τὴν ῥίζαν τῆς καππάρεως Σ 14 τὸ φλοῦδιν] scripsi τὸ ..ου δην L καὶ τὸ φλοῦδιν Σ 15 μετὰ οἴνου] μετ....... L 15 γλυκόριζον] .....λυκορίζην L 16 τὸν ζωμὸν πότισον μετὰ οἴνου] transp. B 17 τὸν ζωμὸν πότισον μετὰ οἴνου] transp. B

τὴν ῥίζαν τῆς χαμαιδάφνης κοπάνισον καὶ τὸν ζωμὸν πότισον μετὰ οἴνου.

195 ρϘε΄ θεραπεία εἰς λειχῆνας καὶ λέπρας· ἔστι βοτάνιν ἡ λεγομένη βάτραχος. ταύτης τὰ φύλλα κοπάνισον καὶ θέτε τὰ ἐκεῖ. 2 ἄλλο. ἐλαίας μαύρης δάκρυον βάλε ἐν ὄξει. καὶ ἂς ἀποβραχῇ νυχθήμερον καὶ τότε ἄλειφε. 3 ἄλλο. τὰ λεγόμενα κογχύλια καῦσον καὶ τὴν στάκτην μίξον με τὸ ὄξος καὶ ἄλειφε. 4 ἄλλο. ▷om. B δαμασκηνέας φύλλα κοπάνισον καὶ μίξας μετὰ ὄξους ἄλειφε. om. B◁ 5 ἄλλο. ▷om. AB δαμασκηνέας δάκρυον ἂς βραχῇ εἰς ὄξος δριμὺ εἶτα ἄλειφε. om. AB◁ 6 ἄλλο. ▷LB τὴν ῥίζαν τῆς δρακοντίας κοπάνισον καὶ μίξας μετὰ ὄξους ἄλειφε LB◁ 7 ἄλλο. ▷om. L καρπὶν κοπάνισον καὶ δενδρολιβάνου φύλλα καὶ βάλε ἔλαιον καὶ ὑγροπίσσιν καὶ ὄξος καὶ μίξας ἄλειφε. om. L◁ 8 ἄλλο. ▷om. B κεδρέαν ἕψησον μετὰ ὄξους καὶ ἄλειφε. om. B◁ 9 ἄλλο. ▷om. MA μελάνθιν κοπάνισον καὶ μίξας μετὰ ὄξους ἄλειφε. om. MA◁ 10 ἄλλο. ἀγριοκαρδάμου ῥίζαν κοπάνισον καὶ ἄλειφε μετὰ ὄξους. 11 ἄλλο. τρίψον τεάφην καὶ ῥητίνην καὶ νίτρον καὶ βάλε ὄξος καὶ χρηστέλαιον καὶ ἄλειφε. τοῦτο ὠφελεῖ καὶ εἰς κνησμάραν μορίων.

196 ρϘϛ΄ ἐὰν τὰ βυζία τῆς γυναικὸς ἢ τοῦ ἀνδρὸς ἔχωσι σκληρότητας· κοπάνισον λυγαίας φύλλα καὶ μίξον μετὰ βουτύρου καὶ θέτε με τὸ πανὶν ἐπάνω. 2 ἄλλο. τὴν ῥίζαν τοῦ ἀσφοδέλου καὶ τὰ φύλλα

---

17 τὸν ζωμὸν] om. MA 1 χαμαιδάφνης] καμαιδάφνης B 3 θεραπεία εἰς] πρὸς B 3 βοτάνιν] βοτάνη Σ 4 τὰ] τε M 4 ἐκεῖ] εἰς τὸν τόπον ἐκεῖνον M 5 ἐν ὄξει] εἰς τὸ ὀξείδιν A 6 καὶ τότε] εἶτα A 6 ἄλειφε] τὰς λειχῆνας καὶ τὰς λέπρας add. MB τὰς λυχίνας καὶ τὰς λέπρας add. A 6 κογχύλια] κοχίλια B 7 με τὸ ὄξος] μετὰ ὄξους Σ 8 φύλλα] καὶ δάκρυον add. A 8 μίξας] μίξον MA 8 ὄξους] καὶ add. MA 9 δάκρυον] φύλλα M 9 ἂς βραχῇ εἰς ὄξος δριμὺ εἶτα] κοπάνισον καὶ μίξον μετὰ ὄξους καὶ M 9 βραχῇ] βραχὺ L 10 δρακοντίας] βοτάνη add. B 11 ἄλειφε] .... L 12 δενδρολιβάνου φύλλα] δενδρολίβανον B 13 ὄξους] ὄξος M 14 μίξας] μίξον B 15 κοπάνισον] κοπανι... L 16 ἄλειφε μετὰ ὄξους] μίξας ὄξος ἄλειφε A τρανσπ. B 16 τεάφην] τεάφιν Σ 16 ῥητίνην] ῥητίνην M 17 βάλε] βαλὼν B 17 τοῦτο] τὸ αὐτὸ Σ 18 κνησμάραν] κνισμάρας M κνησμάρας A ξησμάραν B 18 μορίων] ὀρχιδίων Σ 19 ἐὰν τὰ] πρὸς B 19 τῆς γυναικὸς ἢ τοῦ ἀνδρὸς] transp. M ἢ B 19–20 σκληρότητας] σκληρώτητα M σκληρότητα A κἂν τε ἀνδρῶν ἄν τε γυναικῶν εἰσὶν ποίησον θεραπείαν τοιαύτην B 20 μετὰ βουτύρου] με τὸν βούτυρον A 20 θέτε] θές τα M τα add. A το add. B

καὶ τὰ ἄνθη, ἔψησον καὶ ἄς τα τρώῃ μετὰ ὄξους. 3 ἄλλο. φύλλα ἀμπέλου κοπάνισον καὶ μίξον μετὰ βουτύρου καὶ θέτε τα ἐπάνω. 4 ἄλλο. ▷om. LA τὴν ῥίζαν τῆς ἀλθαίας κοπάνισον καλὰ καὶ θέτε αὐτὴν ἐπάνω.om. LA◁ 5 ἄλλο. ἡδύοσμον χλωρὸν κοπάνισον καὶ θέτε ἐπάνω.

197 ρϙζʹ βυζία γυναικῶν ἢ ὀρχίδια παιδὸς ἐμποδίζει διὰ τὸ να μηδὴν γενῶσι † .. μεγάλα τοῦτο ἤγουν· φύλλα ὑοσκυάμου κοπάνισον καὶ θέτε τὰ ἐπάνω. 2 ἄλλο. ▷om. B φύλλα μανδραγόρας κοπάνισον καὶ θέτε τὰ ἐπάνω. 3 ἄλλο. λιθάργυρον ψιλὸν ποίησε καὶ ψιμμίθιον καὶ ὄξος μίξας ἄλειφε. 4 ἄλλο. φύλλα σεύτλου βράσον με τὸ νερόν. καὶ ὕστερον κοπάνισον καὶ θέτε ἐπάνω.om. B◁ 5 ἄλλο. κοπάνισον κύμινον ψιλὸν καὶ μίξον ἄλευρον σίτου ψιλὸν καὶ μέλι καὶ μαλάξας κατάπλασσε. 6 ἄλλο. δενδρολίβανον κοπάνισον καὶ μίξον ῥοδέλαιον καὶ θέτε ἐπάνω. 7 ἄλλο. λινόσπορον κοπάνισον καὶ ποίησον ψιλὸν καὶ τήλην καὶ μίξον ῥοδέλαιον. ▷om. B μάλαξον καὶ θέτε ἐπάνω. 8 ἄλλο. μολοχίου φύλλα καὶ λαπάτου κοπάνισον καὶ μίξας μετὰ ῥοδελαίου κατάπλασσε.om. B◁ 9 ἄλλο. ἀλεύριν φαβάτινον μίξον με τὸ μέλι καὶ θέτε ἐπάνω εἰς τὰ βυζία.

---

21 με τὸ πανὶν ἐπάνω] transp. B 1 τρώῃ] τρώγῃ Σ 1 μετὰ ὄξους] με τὸ ὀξείδιν A σὺν ὄξους B 1–2 φύλλα ἀμπέλου κοπάνισον] transp. B 2 βουτύρου] βουτήρου M 2 τα] om. A 3 αὐτὴν] την B 4 θέτε] τε add. M την B 4 ἐπάνω] εἰς τὴν σκληρότητα add. A 5 βυζία γυναικῶν] πρὸς βυζία γυναικὸς MB βυζία γυναικὸς A 5 ἢ] καὶ M εἴτε καὶ B 5 παιδὸς ἐμποδίζει διὰ τὸ] scripsi παιδ........ ποδίζει διὰ τὸ L παιδὸς M παιδὸς ἐμποδίζει A παιδίων B 5–6 να μηδὴν] μὴ A ἵνα μὴ B 6 γενῶσι] γένουσιν M γενέσθαι A γένονται B 6 † ..] om. Σ 6 τοῦτο ἤγουν] om. MA ποίησον οὕτως B 6 φύλλα ὑοσκυάμου] transp. MA 7 τὰ] τε M om. A 8 τὰ] om. MA 8 ποίησε] ποίησον MA 9 ψιμμίθιον] ψιμίνθιν A 9 μίξας] om. A 9 ἄλειφε] τὸν τόπον add. A 10 ὕστερον] om. A 11 ψιλὸν] καὶ ψιλὸν ποιήσας B 11 καὶ] om. B 11 ἄλευρον] με ἀλεύριν M καὶ ἀλεύριν A με τὸ ἄλευριν τοῦ B 11 ψιλὸν] om. AB 12 κατάπλασσε] κατάπλαττε B 13 μίξον] ἕνωσον με τὸ M με τὸ add. AB 13 θέτε] θές τα M τὰ add. B 13 ἐπάνω] εἰς τὸν τόπον add. B 14 τήλην] ὁμοίως add. B 14 ῥοδέλαιον] κατάπλασσε add. B 14 μάλαξον] καὶ μάλαξον M καὶ μαλάξας A 15 καὶ] om. A 15 θέτε] τα add. M 15 μολοχίου] μολογχίου M 15 λαπάτου] λαπάθου MA 16 μίξας] μίξον M 16 μετὰ ῥοδελαίου] scripsi ......δελαίου L με τὸ ῥοδέλαιον M ῥοδέλαιον A 17 φαβάτινον] βάτινον B 17 θέτε] θές τα M το add. AB

198 [150] ρϙη′ ⌜ὅταν δάκῃ τινὰ μελίσσιν ἢ σφῆκα, ποίησον θεραπείαν τοιαύτην· μολοχίου ἀγρίου τὰ φύλλα, κοπάνισον καὶ θές τα ἐν τῷ δήγματι. 2 ἄλλο. ▷Mμολογχίου ἡμέρου τὰ φύλλα κοπάνισον καὶ θέτε τα ἐκεῖ εἰς τὸν τόπον.M◁ 3 ἄλλο. ▷om. Lδάφνης φύλλα τρυφερὰ κοπάνισον καὶ θές τα ἐπάνω.om. L◁⌝ 4 ἄλλο. μελισσοφύλλου φύλλα κοπάνισον καὶ θές τα εἰς τὴν πληγήν. 5 ἄλλο. ▷L⌜πήγανον χλωρὸν κοπάνισον καὶ θές το εἰς τὴν πληγήν.⌝L◁

199 [151] ρϙθ′ ὅπου βρωμῆ ἡ μύτη του, ποίησον θεραπείαν τοιαύτην. τὰ ⌜φύλλα τοῦ κίσσου κοπάνισον καὶ τὸν χυλὸν χλίανε καὶ στάξε εἰς τὴν μύτην. 2 ἄλλο. ▷om. MBποταμογείτονος τὰ φύλλα κοπάνισον καὶ τὸν χυλὸν χλίανε. καὶ στάξε εἰς τὴν μύτην.om. MB◁ 3 ἄλλο. ὀρίγανον χλωρὸν κοπάνισον καὶ τὸν χυλὸν στάξον εἰς τὴν μύτην. 4 ἄλλο. μελάνθιν κοπάνισον καὶ μίξον χρηστέλαιον καὶ ἄλειφε τὴν μύτην ἔξω. καὶ στάξε καὶ ἀπέσω.⌝ 5 ἄλλο. σμύρναν καὶ μίξον με τὸ γάλα τὸ γυναικεῖον καὶ στάξον εἰς τὴν μύτην.

200 [152] σ′ ὅταν δείρουν τι καὶ ποιήσει πληγάς, ποίησον θεραπείαν τοιαύτην· ▷om. B⌜ἤγουν τὰ βλαστάρια τοῦ πλατάνου κοπάνισον καὶ μάλαξον με τὸ ἀξούγγιν τὸ χοίρειον καὶ θέτε εἰς τὰς πληγάς.om. B◁ 2 ἄλλο. ▷om. Lτὴν λεγομένην λημνισφραγίδα κοπάνι-

1 ὅταν] πρὸς ἐὰν B 1 μελίσσιν] μελίσσην L μέλισσα B 1–2 ποίησον θεραπείαν τοιαύτην] om. A 1–2 θεραπείαν τοιαύτην] οὕτως B 2 μολοχίου] ἤγουν μολογχίου M 2 θές] θέτε AB 3 δήγματι] ὁμοίως καὶ τοῦ ἡμέρου add. A ἢ καὶ ἡμέρου μολοχίου κοπανίσας θέτε ἐπάνω B 3 ἡμέρου] scripsi ἡμέρου add. M 5 θές] θέτε B 5 μελισσοφύλλου] μελισσοφύλλου M μελισσοβοτάνου τὰ A μελισσόφυλλα B 6 φύλλα] om. B 6 καὶ θές] θέτε B 6 τα] om. MB 6 εἰς τὴν πληγήν] ἐπάνω AB 8 του] ἀνθρώπου add. MB 8–9 τοιαύτην] om. A 9 τὰ] ἤγουν Σ 9 φύλλα τοῦ κίσσου κοπάνισον] transp. A 9 χλίανε] χλιάνας B 10 καὶ] om. B 10 στάξε] στάξον MB 10 μύτην] ῥίνα M 11 στάξε] στάξον A 12 χυλὸν] της add. B 13 μύτην] ῥίνα MA 13 μίξον] με τὸ add. M στάξον με τὸ B 14 τὴν] κεφαλὴν καὶ τὴν add. B 14 μύτην] ῥίνα M 14 στάξε] βάλε MB om. A 14 καὶ] om. A 14 ἀπέσω] ἔσω A 14 σμύρναν] σμύρνα B 14 καὶ] τρίψον καὶ Σ 15 στάξον] ..ον L 15 τὴν μύτην] τὰ ῥωθώνια Σ 16 ὅταν] πρὸς ἐὰν B 16 δείρουν] δέρωσι M δήρωσι AB 16 τι] τινα M τινας AB 16 ποιήσει] ποιήσῃ M ποιήσωσι A 16–17 ποίησον θεραπείαν] om. B 17 τοιαύτην] om. AB 17 τὰ] om. A 18 θέτε] θὲ τὰ M θές τας A 19 τὴν] κοπάνισον τὴν B 19 λημνισφραγίδα] scripsi λημίσφαγίδα M λιμνίαν σφραγίδα AB

σον καὶ τρίψον καὶ μάλαξον με τὸ νερὸν καὶ ἄλειφε.[om. L◁] 3 ἄλλο. βῶλον τρίψον καὶ μάλαξον με τὸ νερὸν καὶ θέτε εἰς τὰς πληγάς. 4 ἄλλο. ἡδύοσμον χλωρὸν κοπάνισον καὶ μάλαξον με τὸ ἀξούγγιν τὸ χοίρειον καὶ ἄλειφε.⌉ 5 ἄλλο. [▷om. A]φύλλα σχίνου τρυφερὰ κοπάνισον καὶ θέτε τα ἐπάνω.[om. A◁] 6 ἄλλο. [▷om. M]βάτου φύλλα τρυφερὰ κοπάνισον καὶ θέτε εἰς τὴν πληγήν[om. M◁].

201 [om. Σ]σα′ περὶ ἀσυλληψίας· γαλεκοτὴν δίσχισον αὐτὴν καὶ ἔπαρον μετὰ τῆς ἀριστερᾶς σου χειρὸς μήτε ἡλίου ὄντος μήτε σελήνης. καὶ δῆσον εἰς λινοῦν πανίον καὶ φόρει καὶ να θαυμάσῃς. 2 ἄλλο. ὑοσκυάμου σπέρμα † ..... μετὰ γάλακτος ἱππίου, ἔμβαλον εἰς δέρμα ἐλαφινὸν καὶ ἂς δήσῃ αὐτὸ εἰς τὸν βραχίωνα καὶ να θαυμάσῃς. 3 ἄλλο. σιδερίτην βοτάνιν βαλὼν εἰς ἐλάφιον δέρμα δὸς αὐτὴν καὶ ἂς φορῇ.

202 σβ′ ὅπου ξερᾷ πολλὰ εἴπερ θέλῃ ἵνα μηδὲν ξερᾷ, ποίησον θεραπείαν τοιάυτην· ἡδύοσμον χλωρὸν κοπάνισον καὶ τὸν χυλὸν πότισον με τὸ ὄξος καὶ τὸ κρασίν. 2 ἄλλο. κοπάνισον ῥοΐδιον ὄξυνον καὶ τὸν χυλὸν τούτου καὶ ἄς τον πίνῃ. 3 ἄλλο. ψαλίδας ἀμπέλου κοπάνισον καὶ τὸν χυλὸν πότισον. 4 ἄλλο. φρύξον φακὴν καὶ ἄς την τρώῃ με τὸ μέλι. 5 ἄλλο. [▷om. B]ὀρόβιν φρυγμένον με τὸ μέλι ἂς τρώῃ.[om. B◁]

299.19–300.1 κοπάνισον καὶ τρίψον] om. B 1 ἄλειφε] τὰς πληγὰς add. AB 2 βῶλον] τὸν λεγόμενον βῶλον Σ 2 μάλαξον] μίξον Σ 2 θέτε εἰς τὰς πληγάς] ἄλειφε MB ἄλειφε τὰς πληγὰς A 4 ἄλειφε] θές το M θέτε ἐπάνω A θέτε το B 5 θέτε] θὲς M 5 ἐπάνω] εἰς τὰς πληγάς add. M ὡσαύτως καὶ τῆς add. B 6 εἰς τὴν πληγήν] ἐπάνω A εἴληφε τέρμα ἑκατὸν τῆς βιβλίου. ἀπέντευθεν ἄρχεται ἑκατοντὰς ἡ τρίτη add. B 7 δίσχισον] scripsi δύσχισον L 8 σελήνης] scripsi σελίνης L 10 ὑοσκυάμου] scripsi ὑοσκιλάμου L 10 δέρμα] scripsi .έρμα L 14 ὅπου] εἴ τις A ὁπόταν τις B 14 πολλὰ εἴπερ θέλῃ ἵνα μηδὲν ξερᾷ] εἴτι φάγῃ M 14 εἴπερ θέλῃ ἵνα] ἐὰν θέλῃς να AB 14–15 ποίησον θεραπείαν τοιάυτην] ἔπαρον A ἤγουν add. M 15 κοπάνισον] και κοπάνισέ τον A 16 με τὸ ὄξος καὶ τὸ κρασίν] με τὸ κρασὶν καὶ με τὸ ὀξείδιν B 16 καὶ τὸ] ἢ με τὸ A 17 καὶ τὸν χυλὸν τούτου καὶ ἄς τον πίνῃ] καὶ ἂς πίνῃ τὸν χυλὸν ὀλίγον ὀλίγον A ὀλίγον ὀλίγον add. B 17 τούτου καὶ] om. M του B 18 τὸν χυλὸν πότισον] transp. B 18 φρύξον] φρύξε A 19 τρώῃ] τρώγῃ Σ 19 ὀρόβιν] ῥόβιν L 19 φρυγμένον] φρυμένον M φρυγμένον A 19–20 με τὸ μέλι ἂς τρώῃ] ἂς τρώγῃ με τὸ μέλι MA

203 [153] ϛγ′ εἰς νεφέλιν ὀφθαλμῶν ποίησον θεραπείαν τοιαύτην· ⌜δρακοντίαν κοπάνισον καὶ τὸν χυλὸν αὐτῆς ἄλειφε ἔξω τοὺς ὀφθαλμούς.⌝ 2 ἄλλο. κινάμωμον τρίψον καλὰ καὶ μίξον με τὸ γάλα τὸ γυναικεῖον. καὶ ἄλειφε ἐκτὸς τῶν ὀφθαλμῶν. 3 ἄλλο. ⌜τριφύλλιν κοπάνισον καὶ τὸν ζωμὸν μίξον με τὸ μέλι καὶ ἄλειφε ἐκτὸς τῶν ὀφθαλμῶν.⌝ 4 ἄλλο. ἀμμωνιακὸν τρίψον καὶ μάλαξον με τὸ γάλα τὸ γυναικεῖον καὶ ἄλειφε τοὺς ὀφθαλμοὺς ἔξωθεν. 5 ἄλλο. ▷om. L⌜σμύρναν τρίψον καὶ μίξον με μέλι καινούργιον καὶ ἄλειφε τοὺς ὀφθαλμούς.⌝om. L◁ 6 ἄλλο. ▷Aκοπάνισον μαλάθρου φύλλα καὶ τὸν ζωμὸν βάλε εἰς χαλκὸν ἀγγεῖον καὶ ἂς ἔνι ἡμέρας τρεῖς εἶτα ἔπαρον ἐξ αὐτοῦ καὶ ἄλειφε τοὺς ὀφθαλμοὺς κατὰ ὥραν καὶ πρωΐ. 7 ἄλλο. βάτου φύλλα κοπάνισον καὶ τὸν ζωμὸν ἄλειφε τοὺς ὀφθαλμούς.A◁ 8 ἄλλο. ▷Bἀμμωνιακὸν μάλαξον με τὸ γάλα τὸ γυναικεῖον καὶ ἄλειφέ τους ἔξω τοὺς ὀφθαμούς.B◁

204 ϛδ′ εἰς πόνον ὀφθαλμῶν· κοπάνισον ἀγριελαίας φύλλα καὶ τὸν χυλὸν ἄλειφε τοὺς ὀφθαλμούς. 2 ἄλλο. ▷om. Bἀμάραντα κοπάνισον καὶ θέτε τα ἐπάνω εἰς τοὺς ὀφθαλμούς.om. B◁ 3 ἄλλο. ▷om. Aγλιστρίδας φύλλα κοπάνισον καὶ μίξον μετὰ πίτυρα θέτε εἰς τοὺς ὀφθαλμούς. 4 ἄλλο. κοπάνισον ἀψίνθιον καὶ μάλαξον με τὸ μέλι. καὶ θέτε ἐπάνω εἰς τοὺς ὀφθαλμούς. 5 ἄλλο. ἀμμωνιακὸν τρίψον καὶ ἑνώσας μετὰ γάλακτος γυναικείου ἄλειφε τοὺς ὀφθαλμούς. 6

---

1 εἰς νεφέλιν ὀφθαλμῶν] πρὸς ὀφθαλμοὺς ἔχοντας νεφέλας MB ὅπου ἐχωσιν οἱ ὀφθαλμοὶ νεφέλην A 1 ποίησον] οὕτως πράττε. ποίησον B 1 τοιαύτην] om. A 2 δρακοντίαν κοπάνισον] ἤγουν κοπάνισον τὴν λεγομένην δρακοντίαν MA τὴν λεγομένην δρακοντίαν κοπάνισον B 2 ἔξω] om. M 3 τρίψον] τρίψας MA 3 καλὰ] ψιλὸν B 3 καὶ] om. MA 3 μίξον] τρίψον A 3 με τὸ] μετ. L 3 γάλα] .. λα L 4 καὶ ἄλειφε ἐκτὸς τῶν ὀφθαλμῶν] scripsi ἄλειφ. .... τῶν ὀφθαλμῶν L ἄλειφε τοὺς ὀφθαλμοὺς ἔξω M ἄλειφε ἔξω τοὺς ὀ... A ἄλειφε ἔξω τοὺς ὀφθαμοὺς B 4–5 τριφύλλιν] τὸ λεγόμενον τρυφύλλιον MA 5–6 ἐκτὸς τῶν ὀφθαλμῶν] τοὺς ὀφθαλμούς MB ἔξω τοὺς ὀφθαλμούς A 6 τρίψον καὶ] om. Σ 7 τοὺς ὀφθαλμοὺς ἔξωθεν] ἔξω τοὺς ὀφθαλμοὺς AB 7 ἔξωθεν] ἔξω M 8 καὶ] om. B 8 με μέλι] μέλι A με τὸ μέλι B 9 τοὺς] ἔξω τοὺς A 15 εἰς πόνον ὀφθαλμῶν] πρὸς ὀδύνην ὀφθαλμῶν MB εἴ τις πονεῖ τοὺς ὀφθαλμούς του A 15 ἀγριελαίας] ἐλαίας M 15 φύλλα] πλὴν ἀγρίας add. M 17 εἰς] om. M 18 θέτε] καὶ θέτε M καὶ θέτε τα B 19 μάλαξον] μίξον M 20 θέτε] τε add. M το add. B 20 ἐπάνω εἰς τοὺς ὀφθαλμούς] om. B

ἄλλο. κοπάνισον μαράθρου φύλλα καὶ τὸν ζωμὸν βάλε εἰς χαλκὸν ἀγγεῖον καὶ ἆς ἔνι ἡμέρας τρεῖς. εἶτα ἔπαρον ἐξ αὐτοῦ καὶ ἄλειφε τοὺς ὀφθαλμοὺς καθὴν ὥραν καὶ τὸ πρωΐ. 7 ἄλλο. βάτου φύλλα τρυφερὰ κοπάνισον ▷om. B καὶ θέτε τα ἐπάνω εἰς τοὺς ὀφθαλμούς. 8 ἄλλο. κυκλάμινον κοπάνισον om. B◁ καὶ τὸν χυλὸν με τὸ μέλι, ἄλειφε τοὺς ὀφθαλμούς. om. A◁ 9 ἄλλο. δενδρολίβανον κοπάνισον καὶ μάλαξον με τὸ μέλι καὶ τὸ γάλα τὸ γυναικεῖον καὶ ἄλειφε τοὺς ὀφθαλμούς. 10 ἄλλο. ▷om. L μανδραγόραν κοπάνισον καὶ τὸν χυλὸν μάλαξον μετὰ πίτυρα ἢ με τὸ κριθάλευρον καὶ θὲς εἰς τοὺς ὀφθαλμούς. om. L◁ 11 ἄλλο. χλωρὸν κολοκύνθιν ξύσε καὶ τὰ ξύσματα κοπάνισον καὶ τὸν ζωμὸν μίξον μετὰ πίτυρα. καὶ μάλαξον καὶ θέτε ἐπάνω εἰς τοὺς ὀφθαλμούς. 12 ἄλλο. σέλινον κοπάνισον καὶ τὸν ζωμὸν τούτου μίξον μετὰ πίτυρα καὶ θέτε εἰς τοὺς ὀφθαλμούς. 13 ἄλλο. ▷om. L κοπάνισον σμύρναν καὶ μίξον με τὸ μέλι καὶ ἄλειφε τοὺς ὀφθαλμούς. 14 ἄλλο. ὑοσκυάμου φύλλα κοπάνισον καὶ μίξον μετὰ πίτυρα καὶ θέτε τα εἰς τοὺς ὀφθαλμούς. om. L◁ 15 ἄλλο. ῥόδα χλωρὰ κοπάνισον καὶ τὸν χυλὸν τούτων μίξον με τὸ γάλα τὸ γυναικεῖον καὶ ἄλειφε. 16 ἄλλο. πήγανον κοπάνισον καὶ τὸν χυλὸν μίξον μετὰ πίτυρα καὶ θέτε εἰς τοὺς ὀφθαλμούς. 17 ἄλλο. ἀμάραντα κοπάνισον καὶ τὸν χυλὸν μίξον με τὸ μέλι καὶ ἄλειφε τοὺς ὀφθαλ-

---

21 μετὰ γάλακτος γυναικείου] γάλα γυναικεῖον M μετὰ γάλατος γυναικεῖον B 1 κοπάνισον μαράθρου φύλλα] transp. B 1 μαράθρου] scripsi ..αράθρου L μαλάθρου MB 1–2 εἰς χαλκὸν ἀγγεῖον] .... L 2 ἡμέρας τρεῖς] ἡ .... ρεῖς L 2 εἶτα] om. B 2 ἔπαρον] ἔπαρε M 2 αὐτοῦ] αὐτ. L 2 ἄλειφε] ἄλει.. L 3 καθὴν] κατὰ MB 3 τὸ] om. M 4 ἐπάνω] om. M 5 χυλὸν] μίξον add. M 5 με τὸ μέλι] om. A 5 μέλι] καὶ add. MB 6–7 κοπάνισον καὶ μάλαξον] om. A 7 μέλι] κοπάνισον καὶ βάλε add. A 7 τὸ] om. A 8 μανδραγόραν] μανδραγόρα A μανδραγούραν B 9 θὲς] θέτε B 9 εἰς] τὸ ἐπάνω εἰς A 10 χλωρὸν κολοκύνθιν ξύσε] transp. B 10 ξύσε] ξύσον M το add. A 10 ξύσματα] ὀξύσματα M 11 καὶ μάλαξον] om. A καὶ μαλάξας B 11 καὶ] om. B 11 θέτε] θές τα M το add. A 12 ἐπάνω εἰς τοὺς ὀφθαλμούς] transp. MA 13 τούτου] τοῦ A 13 θέτε] θές τα M τα add. A το add. B 14 κοπάνισον σμύρναν] transp. A 15 ὑοσκυάμου φύλλα] transp. A 16 εἰς τοὺς ὀφθαλμούς] om. A 18 ἄλειφε] τοὺς ὀφθαλμοὺς add. B 18 κοπάνισον] χλωρὸν add. M χλωρὸν κοπανισμένον A 18 καὶ τὸν χυλὸν] om. MA 19 πίτυρα] .... ρα L 19 θέτε] θές τα M τα add. A 20 χυλὸν] τούτων add. A

μούς. 18 ἄλλο. ▷om. MA ἡδύοσμον κοπάνισον καὶ τὸν χυλὸν μίξον με τὸ γάλα τὸ γυναικεῖον καὶ ἄλειφε. om. MA◁ 19 ἄλλο. βάτου φύλλα τρυφερὰ κοπάνισον καὶ τὸν χυλὸν μίξον με τὸ μέλι καὶ γυναικεῖον γάλα καὶ ἄλειφε. 20 ἄλλο. ▷om. L ψύλλεον κοπάνισον καὶ μίξον με τὸ μέλι καὶ τὸ γάλα τὸ γυναικεῖον καὶ ἄλειφε τοὺς ὀφθαλμούς. om. L◁ 21 ἄλλο. ▷M ἐλαίας φύλλα τρυφερὰ κοπάνισον καὶ μίξον με τὸ μέλι καὶ τὸ γάλα τὸ γυανικεῖον καὶ ἄλειφε τοὺς ὀφθαλμούς. M◁ 22 ἄλλο. καλάμου χλωρὰ φύλλα κοπάνισον καὶ τὸν χυλὸν ἄλειφε τοὺς ὀφθαλμούς. τὰ δὲ κοπανισμένα φύλλα θέτε ἐπάνω.

205 [154] σε′ εἰς πόνον ὀδόντων ποίησον θεραπείαν τοιαύτην. ⌜καρπὶν μαῦρον βράσον μετὰ ὄξους καὶ ἃς κρατῇ τὸ ὄξος ἔνθα πονεῖ.⌝ 2 ἄλλο. τὴν ῥίζαν τοῦ ἀσπαράγου, βράσον με τὸ ὄξος καὶ κράτει εἰς τὸ στόμα. 3 ἄλλο. τὴν ῥίζαν τοῦ ἀσφοδέλου κοπάνισον καὶ τὸν χυλὸν στάξον εἰς τὸ ῥουθούνιν τῆς μερέας ἧς πονεῖ. 4 ἄλλο. ▷Σ ἀγριοσταφίδας ἕψησον μετὰ ὄξους καὶ μέλιτος καὶ ἃς κρατῇ εἰς τὸ στόμα του. 5 ἄλλο. ⌜τῆς πλατάνης τὰ βλαστάρια τὰ τρυφερὰ βράσον με τὸ ὄξους καὶ κράτει εἰς τὸ στόμα. 6 ἄλλο. Σ◁ βάτου φύλλα βράσον με τὸ ὄξος καὶ κράτει⌝ εἰς τὸ στόμα.

302.20–303.1 τοὺς ὀφθαλμούς] om. M 1 ἡδύοσμον κοπάνισον] transp. B 3 κοπάνισον καὶ] om. AB 3 τὸν χυλὸν] om. Σ 3–4 γυναικεῖον γάλα] τὸ γάλα τὸ γυναικεῖον MB καὶ με τὸ γάλα τὸ γυναικεῖον A 4 ἄλειφε] τοὺς ὀφθαλμούς add. AB 4 μίξον] μίξας B 4–5 με τὸ μέλι καὶ τὸ] om. AB 5 τὸ] om. A 5 καὶ] om. B 5 τοὺς ὀφθαλμούς] om. A 8 καλάμου χλωρὰ φύλλα κοπάνισον] transp. B 8 χλωρὰ] χλωροῦ τὰ Σ 8 καὶ] καλὰ καὶ MA 9 τὰ δὲ] καὶ τὰ Σ 9 κοπανισμένα φύλλα] transp. MB φύλλα A 9 θέτε] θὲς MA τα add. B 9 ἐπάνω] κοπανισμένας add. A 10 εἰς...τοιαύτην] εἴ τις πονεῖ τοὺς ὀδόντας A 10 ποίησον θεραπείαν τοιαύτην.] om. B 11 καρπὶν] κραμπὴν L ἤγουν καρπὶν M 11 μαῦρον] om. MA 11 καὶ] τὸ ζουμὶν add. M 11 ἃς] om. B το add. A 11 τὸ ὄξος ἔνθα] εἰς τὸ μέρος ἐκεῖνο ὅπου Σ 12 με τὸ ὄξος] μετὰ ὄξους Σ 14 χυλὸν] ζωμὸν B 14 ῥουθούνιν τῆς μερέας ἧς πονεῖ] ῥ..θούνιντῆς μερέας ἧς πονεῖ L δεξιὸν ῥωθώνιν ἐὰν τὸ δεξιὸν μέρος πάσχει, εἰ δὲ τὸ ἀριστερὸν μέρος τῶν ὀδόντων πάσχῃ, βάλε εἰς τὸ ἀριστερὸν ῥωθώνιν M δεξιὸν ῥοθώνιν A δεξιὸν ῥοθώνιον ἐὰν πάσχει τὸ δεξιὸν μέρος τῶν ὀδόντων εἰ δὲ τὸ ἀριστερὸν πάσχει βάλλε τὸ εἰς τὸ ἀριστερὸν ῥοθούνιον B 15 ἀγριοσταφίδας] γριοσταφίδας M ἀγριοσταφίδα A 15 ἃς] το add. AB 16 του] om. A 16 βλαστάρια τὰ] om. B 17 τρυφερὰ] φύλλα καὶ βλαστάρια add. B 17 με τὸ] μετὰ A 17 καὶ] ἃς τα A 18 βάτου] .... L 18 με τὸ ὄξος] μετὰ ὄξους Σ 18 εἰς τὸ στόμα] om. A

206 [155] ϛ′ ὅταν οἱ ὀδόντες σείονται, ποίησον θεραπείαν τοιαύτην· ἂς δήσῃ εἰς τὸν τράχηλον τοῦ ἀγριοκαρδάμου ῥίζαν. 2 ἄλλο. ⊳Σ⌜βράσον ὕσσωπον μετὰ ὄξους καὶ ἂς κρατῇ εἰς τὸ στόμα του.⌝Σ⊲ 3 ἄλλο. συκαμινέας φλοῦδιν ἕψησον μετὰ ὄξους καὶ κράτει εἰς τὸ στόμα. 4 ἄλλο. τὴν ῥίζαν τοῦ κυπέρου ἄς την μασσᾶται. 5 ἄλλο. ⊳Σἔστι δένδρον τὸ λεγόμενον μελία, τούτου τὰ φύλλα κοπάνισον καὶ ἄς τα κρατῇ εἰς τὸ στόμα του.Σ⊲ 6 ἄλλο. ⊳LB⌜καρπὸν καππάρεως ἂς βράσῃ μετὰ ὄξους καὶ ἂς κρατῇ εἰς τὸ στόμα.LB⊲ ἄλλο. 7 καππάρεως τῆς ῥίζης τὸ φλοῦδιν ἂς βράσῃ με τὸ ὄξος καὶ ἂς κρατῇ εἰς τὸ στόμα. 8 ἄλλο. ῥοΐδων φύλλα βράσον με τὸ ὄξος. καὶ ἂς κρατῇ⌝ εἰς τὸ στόμα. 9 ἄλλο. φύλλα ἀγριοελαίας ἕψησον μετὰ ὄξους καὶ κρατῇ εἰς τὸ στόμα.

207 [156] ϛζ′ ὅταν ὁ ἄνθρωπος καύσῃ μέρος τι τοῦ σώματος αὐτοῦ· ⌜κοπάνισον ἀρνόγλωσσον καὶ θές το ἐπάνω⌝ εἰς τὴν καῦσιν. 2 ἄλλο. ἀμάραντα κοπάνισον θέτε ἐπάνω. 3 ἄλλο. ⌜συκαμινέας φύλλα ἕψησον μετὰ ἐλαίου. εἶτα κόψας τὰ φύλλα μικρά, θέτε ἐπάνω⌝.

---

1 ὅταν οἱ ὀδόντες σείονται] πρὸς σειομέμους ὀδόντας MB 1 σείονται] σειο.. L 1–2 τοιαύτην] om. A 2 δήσῃ] δείσῃ L δήσῃ M δύσει B δήσει A 2 εἰς] om. M 2 τράχηλον] σφόνδηλον A 3 ἂς] το add. AB 3 του] om. A 4 φλοῦδιν] φλούδια M φλούδα B 4 ὄξους] ὄξος M 5 τὴν] ἂς μασσᾶται τὴν B 5 ἄς την μασσᾶται] om. B 7 του] om. AB 8–9 καππάρεως τῆς ῥίζης τὸ φλοῦδιν] τὸν φλοῦν τῆς ῥίζης τῆς καππάρεως M τὸ φλοῦδιν τῆς ῥίζης τῆς καππάρεως A ὡς αὐτὸς καὶ τὸ φλοῦδιν τῆς ῥίζης τῆς καππάρεως B καππάρεως τῆς ῥίζης τὸ φλοῦδιν L 9 ἂς βράσῃ] βράσον MA 9 με τὸ ὄξος] μετὰ ὄξους B 9 ἂς] το add. A om. B 10 στόμα] του add. M 10 ῥοΐδων] ῥοϊδίων A τὸ ῥοϊδίων B 10 βράσον] κοπάνισον καὶ βράσον M 10 με τὸ] μετὰ AB 10 ὄξος] ὄξους AB 10 ἂς] om. A το add. M 11 εἰς τὸ στόμα] om. A 11 στόμα] του add. M 11 ἀγριοελαίας] ἀγριελαίας MB 11 ἕψησον] βράσον A 11 μετὰ ὄξους] με τὸ ὄξος M 11 καὶ] ἂς add. MA ἄς τα A 12 στόμα] του add. M 13 ὁ ἄνθρωπος] καύσῃ ὁ ἄνθρωπος B 13 καύσῃ] om. B 13 μέρος τι] transp. MA τιποτας μέλος B 13 τοῦ σώματος] om. B 13–14 αὐτοῦ] om. M του AB 14 θές] θέτε A 14 καῦσιν] καύσην B 15 ἀμάραντα] μάραντον B 15 κοπάνισον] κοπ..... L 15 θέτε] καὶ θές τα MA καὶ θέτε το B 16 φύλλα ἕψησον] φυ.....ψησον L 16 ἐλαίου] ἐλαία B 16 εἶτα κόψας] καὶ κατακόψας ὕστερον MB 16 μικρά] μικρὰ add. MB 16 θέτε] θές τα Σ 17 ἐπάνω⌝] εἰς τὸν τόπον ἔνθα ἐκάη add. M εἰς τὸν τόπον A εἰς τὸν καυθέντα τόπον add. B

208 [157] ση΄ ποδαλγικοὺς ὠφελεῖ ῥίζα βολβοῦ. ταύτην κοπάνισον καὶ θέτε ἔνθα ἔνι ἡ ὀδύνη. 2 ἄλλο. ⌜κρίνου ῥίζαν κοπάνισον καὶ θέτε ἔνθα ἔνι ἡ ὀδύνη⌝. 3 ἄλλο. ▷Σῥίζαν γογγυλίδος κοπάνισον καὶ ἐπίθες ταύτην ἔνθα ἔνι ἡ ὀδύνη. 4 ἄλλο. ⌜τὸν σπόρον τῆς γογγυλίδος βράσον με τὸ νερὸν καὶ τὸ νερὸν ἐκεῖνο ἄντλει τοὺς πόδας.Σ◁ 5 ἄλλο. ῥεπανίου ῥίζαν κοπάνισον καὶ θέτε εἰς τὸν τόπον. 6 ἄλλο. ▷MBσπέρμα ῥεπανίου βράσον με τὸ νερὸν καὶ αὐτὸ τὸ νερὸν ἄντλει τοὺς πόδας.⌝MB◁ 7 ἄλλο. φύλλα κράμβης κοπάνισον καὶ μίξον μετὰ σιτίνου ἀλεύρου κατάπλασσε. 8 ἄλλο. ▷Σῥίζαν κυκλαμίνου κοπάνισον καὶ θὲς ταύτην ἔνθα ἔνι ὁ πόνος. 9 ἄλλο. ῥίζαν βοτάνης τῆς λεγομένης βρυωνίας ταύτην κοπάνισον καὶ ἐπίθες εἰς τὸν πόνον.Σ◁

209 [158] σθ΄ ⌜ὅταν οἱ ἁρμοὶ τῶν ποδῶν ἢ τῶν χειρῶν γένωνται σκληροὶ⌝ καὶ ὡς ἂν πῶροι, ποίησον θεραπείαν τοιαύτην. ⌜βάλε ἀμμωνιακὸν εἰς ὄξος καὶ ἂς ποιήσῃ δύο νυχθημέρα καὶ ἂς βραχῇ, εἶτα μάλαξον αὐτὸ με τὸ ὑγροπίσσιν, καὶ θέτε τὸ ἐπάνω εἰς τὸν

1 ποδαλγικοὺς ὠφελεῖ] πρὸς ποδαλγηκούς M πρὸς ποδαλγικοὺς ὠφελεῖ B 1 ῥίζα βολβοῦ.] τὴν ῥίζαν τοῦ κλοβοῦ M 1 ταύτην] om. M 2 θέτε] θές την ἐπάνω M θέτε την ἐπάνω AB 2 ὀδύνη] τῶν ποδῶν add. Σ 3 θέτε] θές την MB θὲς ταύτην B 3 ἔνθα ἔνι ἡ] ἐπάνω A 3 ὀδύνη] τῶν ποδῶν add. MB 3 ῥίζαν γογγυλίδος] τὴν ῥίζαν τῆς γογγιλίδος B 4 ἐπίθες] θὲς B 4 ταύτην] ἐπάνω A τὴν B 4 ὀδύνη] ὀδύνη τῶν ποδῶν B 4–5 τῆς γογγυλίδος] ταύτης B 5 τὸ νερὸν] μετ' B 6 κοπάνισον] ὁμοίως καὶ σπέρμα βράσον με τὸ νερὸν καὶ μετ' αὐτὸ τὸ νερὸν ἄντλει τοὺς πόδας A 6–7 καὶ θέτε εἰς τὸν τόπον] om. A 6 θέτε] ἐπίθες ταύτην MB 6–7 τόπον] πόνον MB 7 σπέρμα] δέρμα M 7–8 αὐτὸ τὸ νερὸν] μετὰ τοιούτου ὕδατος B 9 μετὰ σιτίνου ἀλεύρου] με τοῦ σίτου ἄλευρον M μετὰ σίτωνος ἀλεύρου B 9 κατάπλασσε] κατάπλασε L καὶ κατάπλασσε AB 10 θὲς ταύτην] ἐπίθες ταύτην AB 10 ἔνι] om. A 10 ῥίζαν] τὴν ῥίζαν τῆς B 11 βρυωνίας] βρυονίας B 11 ταύτην] om. AB 11 ἐπίθες] θὲς A 12 πόνον] τόπον B 13 ἁρμοὶ] ἤτοι τὰ ἄρθρα add. M ἢ τὰ ἄρθρα add. A 13 τῶν ποδῶν ἢ τῶν χειρῶν] τῶν χειρῶν καὶ τῶν ποδῶν εἶτα δακτύλια B 14 σκληροὶ] σκηροὶ M σκληρὰ AB 14 καὶ] om. MA 14 ὡς ἂν πῶροι] om. AB 14 πῶροι] πόροι LM 14 ποίησον θεραπείαν τοιαύτην] καὶ ποιοῦσιν ὡσὰν πόρους εἶτα ἄρθρα ἐπάνω ποίησον οὕτως B 14 τοιαύτην] om. A 14 βάλε] ἤγουν βάλε MA 15 ἂς ποιήσῃ] ποίησε B 15 δύο νυχθημέρα] νηχθήμερα δύο M νυχθήμερον B 15 καὶ ἂς βραχῇ] om. M βράσ. B 16 αὐτὸ] αὐτὰ M αὐτὸν B 16 θέτε τὸ] θὲς M

πόνον, ἔνθα ἔνι ἡ σκληρότης. 2 ἄλλο. θαλάσσιον νερὸν ἄντλει τοὺς πόδας συνεχῶς. 3 ἄλλο. ▷Στρίψον ἅλας καὶ τὸ λεγόμενον νίτρον καὶ μίξον μετὰ χρηστελαίου καὶ ἄλειφε συνεχῶς τοὺς πόνους τῶν ἄρθρων.⌝Σ◁

210 [159] σι′ πρὸς τό τινα ποιήσεις να κρατηθῇ τὸ αἷμα· ⌜ἀδίαντον κοπάνισον, τὸν χυλὸν τούτου, πότισον μετὰ οἴνου. 2 ἄλλο. ἡδύοσμον χλωρὸν κοπάνισον καὶ τὸν χυλὸν μίξον με τὸ ὄξος καὶ τὸ κρασὶν καὶ πότισον. 3 ἄλλο. ἀμυγδάλων δάκρυον σὺν οἴνῳ πότιζε. 4 ἄλλο. βάτου φύλλα τρυφερὰ κοπάνισον καὶ θές τα ἔνθα ῥέει τὸ αἷμα⌝. 5 ἄλλο. ▷Βπήγανον χλωρὸν κοπάνισον καὶ μετὰ χρηστελαίου μαλάξας θὲς ἐκεῖ ἔνθα ῥέει τὸ αἷμα.Β◁ 6 ἄλλο. ⌜ἡδύοσμον τρίψον καὶ θέτε εἰς τὴν μύτην του⌝. 7 ἄλλο. ποταμογείτονος φύλλα χλωρὰ κοπανισθέντα καὶ ἐπιθέμενα εἰς τὴν μύτην, ἵστησι τὴν κένωσιν τοῦ αἵματος. 8 ἄλλο. θεῖον ἄπυρον θὲς εἰς τὴν μύτην.

211 [160] σια′ ὅταν ὁ στομαχὸς ἔχῃ ἀτονίαν ἤτοι ἀδυναμίαν καὶ καίεται ἀπέσω πολλά, ⌜κοπάνισον πολυγόνου τὰ φύλλα καὶ θέτε εἰς τὸν στόμαχον. 2 ἄλλο. ▷ΜΒῥέον Ποντικὸν σὺν ὕδατι πινόμενον.⌝ΜΒ◁

305.16–306.1 εἰς τὸν πόνον] om. A 1 πόνον] τόπον MB 1 ἡ σκληρότης] ἡ σκηρότης M ἡ σ̣κ̣λ̣ηρότης L 1 νερὸν] ζεστὸν add. Σ 1 ἄντλει] ἄν... .....L 2 συνεχῶς] om. A 2 τρίψον] ῥίψον L τρίψας B 2 καὶ] om. B 3 καὶ] om. B 3 πόνους] τόπους B ὀδυνηροὺς πόνους M 5 πρὸς τό τινα ποιήσεις να κρατηθῇ τὸ αἷμα] cf. Index L αἱμορρα.....αν ἐὰν κρατεῖ L πρὸς αἱμορραγίαν ὅπως να σταθῇ τὸ αἷμα B 6 τὸν] καὶ τὸν Σ 6 οἴνου] καὶ ἔστι ὠφέλιμον add. B 7 χλωρὸν κοπάνισον] transp. MA 7 με τὸ ὄξος] μετὰ ὄξους M μετ' ὀξείδιν B 7–8 τὸ κρασὶν] οἴνου A 8 πότισον] ἂς πίνῃ M ἄς το πίνῃ AB 8 ἀμυγδάλων] ἀμυγδαλέας MA ἀμύγδαλα B 8 δάκρυον] τὸ δάκρυον αὐτῶν B 8 πότιζε] πινόμενον MB πιννώμενον A 9 θές] θέτε AB 9 τα] om. M 9 ἔνθα] ἐκεῖ ὅπου Σ 11 τρίψον] τριμμένον Σ 12 καὶ θέτε] ἂς θέσῃ Σ 12 μύτην] ῥίναν M 12–13 ποταμογείτονος... μύτην] om. B 13 ἐπιθέμενα] ἐπιθέντα MA 13 μύτην] ῥίνα M 13 ἵστησι] ἵστισε L ἵστισιν M 14 θεῖον] θῆον B 14 ἄπυρον] scripsi ἄπειρον L ἄπυρον ἤτοι τεάφιν τριμμένον MB ἄπειρον ἤτοι τεάφιν τριμμένην A 14 θὲς] θέτε το B 14 μύτην] ῥίναν του M 15 ὁ στομαχὸς ἔχῃ] transp. B 15 ἀτονίαν ἤτοι ἀδυναμίαν] scripsi ἀτονίαν ἤτοι ἀδυναμείαν M ἀτονίαν A ἀδυναμίαν B 15–16 καὶ καίεται ἀπέσω πολλά] om. B 16 ἀπέσω] ἔσω A 16 πολυγόνου] scripsi πολι... L πολλύγονον A πολύγονον MB 16 τὰ φύλλα καὶ θέτε] καὶ τὰ φύλλα θέτε M καὶ τὰ φύλλα θέτε τα A καὶ τὰ φύλλα καὶ θέτε το B 17 στόμαχον] στο........ L 17 ῥέον Ποντικὸν] scripsi ῥέου ποτικοῦ M νεὸν ποντ̣ικ̣ὸ̣ν B

212 [160] ϛιβ´ ἡ κοιλία ὑπάγει, ἵστησι ταύτην· βερονίκιν καὶ ψιλὸν ποιήσας, πότισον μετὰ οἴνου. ἄλλο. 2 φύλλα ἀμπέλου κοπάνισον καὶ ψαλίδας καὶ τὸν χυλὸν πότισον. ἄλλο. 3 ⌜τῆς γεντιανῆς τὴν ῥίζαν κοπάνισον καὶ πότισον μετὰ οἴνου.⌝

213 [160] σιγ´ στόμαχος ἐὰν ἔχῃ ζέσιν ἀπέσω· ⌜γλιστρίδος φύλλα κοπάνισον μίξον μετὰ πίτυρα καὶ θέτε εἰς τὸν στόμαχον.⌝ 2 ἄλλο. ἀψίνθιον βράσον με τὸ σησαμέλαιον καὶ πότισον μετὰ οἴνου. 3 ἄλλο. μαρούλια ἄπλυτα τρῶγε. 4 ἄλλο. ⌜δάφνης φύλλα κοπάνισον καὶ τὸν χυλὸν πότισον.⌝ 5 ἄλλο. ▷MA γλυκόριζον κοπάνισον καὶ τὸν ζωμὸν πότιζε χλίον.MA◁ 6 ἄλλο. ⌜μακεδονησίου σπόρον κοπάνισον καὶ ποτισον με τὸ κρασίν. 7 ἄλλο. ζοχὶν ἥμερον ἢ καὶ ἄγριον ἂς τρώγῃ.⌝ 8 ἄλλο. ▷om. A τὸ αὐτὸ βράσον μετὰ ὕδατος καὶ πότισον.om. A◁ 9 ἄλλο. ⌜χαμαίμηλα βράσον με τὸ νερὸν καὶ πότισον.⌝ 10 ἄλλο. ἀριστολοχίας ῥίζαν κοπάνισον καὶ θέτε εἰς τον στόμαχον. 11 ἄλλο. ▷L ἀσπαράγγια βράσον καὶ πότισον τὸν ζωμὸν.L◁

214 σιδ´ θεραπεία εἰς σπληνικοὺς καὶ πρησμένους· λυγαίας σπόρον

---

1 ἡ κοιλία ὑπάγει, ἵστησι ταύτην] πρὸς ἐὰν ὑπάγῃ ἡ κοιλία πολλὰ M ἐὰν ὑπάγει ἡ κοιλία καὶ θέλης ἵνα τὴν στήσης A πρὸς πολλὰ ἡ κοιλία του ἵστισι ταύτην B 1 βερονίκιν] φονίκην L φερονίκην M βερωνίκην A βερονίκην B βερονίκιν scripsi 1 καὶ] om. A 1–2 ψιλὸν ποιήσας] τοῦτο ψιλὸν MA ψιλὸν B 3 ψαλίδας] ἕλιγκας M 3 τῆς] τὴν B 3 γεντιανῆς] γεντιλιανῆς B 5 στόμαχος ἐὰν ἔχῃ ζέσιν ἀπέσω] πρὸς ἐὰν ἔχει ζέσιν ὁ στόμαχος. ὠφελεῖ δὲ τοῦτον M πρὸς ἐὰν ἔχει ὁ στόμαχος ζέσιν B 5 γλιστρίδος] γλυτρίδα, ἤτοι ἔπαρον τὰ M κοπάνισον τῆς γλιστρίδος τὰ B 5 φύλλα] φύλλα καὶ MB om. A 6 κοπάνισον] κοπάνισον τὰ φύλλα αὐτῆς A om. B 6 μίξον] καὶ μίξον M 6 θέτε] θὲς M θέτε τα AB 7 σησαμέλαιον] συσαμέλαιον L σισαμέλαιον A σησαμαίλαιον B 8 μαρούλια] μαρούλλια M μαρούλλην A ὠφελεῖ καὶ μαρούλιν B 8 ἄπλυτα] ὠφελεῖ A ἄπλυτον B 8 τρῶγε] ἐσθιώμενα M ἐσθιώμενον A ἐσθιόμενον B 9 καὶ] ... L 9 χυλὸν] ζωμὸν M 9 πότισον] χυλὸν τοῦτον πότιζε χλίον B 10 ζωμὸν] χυλὸν A 11 με τὸ κρασίν] μετὰ οἴνου B 11 ζοχὶν] ζοχὴν L ζόχον MA ἢ add. B 11–12 ἢ καὶ ἄγριον] ἄγριον L καὶ ἄγριον MA ἢ ἄγριον B 12 τὸ αὐτὸ] ζόχον ἥμερον καὶ ἄγριον M ἢ B 12 μετὰ] ..τὰ L 13 χαμαίμηλα βράσον] χαμεμη ..... ράσον L 13 καὶ] om. A 14 ἀριστολοχίας] ἀ..ρόχης L τὴν λεγομένην ἀριστολοχίαν ταύτης τὴν MA om. B 14 ῥίζαν] τὴν ῥίζαν τῆς λεγομένης ἀριστολογχίας B 14 θέτε] θές την M θέτε την B 15 πότισον] ..τισον L

κοπάνσιον καὶ πότισον με τὸ κρασίν. 2 ἄλλο. ▷Σσκολοπένδρου τὴν ῥίζαν καὶ τὰ φύλλα κοπάνισον καὶ θές τα εἰς τὴν σπλῆνα.Σ◁ 3 ἄλλο. ▷om. LBσκολοπενδρίου τὴν ῥίζαν καὶ τὰ φύλλα βράσον με τὸ νερὸν καὶ πότισον.om. LB◁ 4 ἄλλο. ἀσπάραγγια κοπάνισον καὶ τὸν χυλὸν πότισον. 5 ἄλλο. ▷om. Aἀγρίας βράσον με τὸ νερὸν καὶ τὸν ζωμὸν πότισον.om. A◁ 6 ἄλλο. ἀμμωνιακὸν οὐγγίαν μίαν καὶ ἄς βραχῇ εἰς κρασὶν νυχθήμερα τρία καὶ τὸ κρασὶν αὐτὸ πότισον. 7 ἄλλο. ▷Σβράσον ἀψίνθιον καὶ σῦκα καὶ νερὸν καὶ τὸν ζωμὸν αὐτῶν βράσον καὶ μάλαξον ἀλεύριν αἶρας καὶ ποίησον κατάπλασμα καὶ θὲς εἰς τὴν σπλῆνα.Σ◁ 8 ἄλλο. ▷om. Aκαππάρεως τῆς ῥίζης τὸ φλοῦδιν κοπάνισον με τὸ κρασὶν καὶ πότισον. 9 ἄλλο. φύλλα κισσοῦ τρυφερὰ κοπάνισον καὶ θέτε τα εἰς τὴν σπλῆνα.om. A◁ 10 ἄλλο. φύλλα τρυφερὰ κισσοῦ βράσον μετὰ οἴνου καὶ πότισον. 11 ἄλλο. ▷Σἀγριοκάρδαμον κοπάνισον καὶ θὲς εἰς τὸν σπλήνα ἕως ὥρας τρεῖς.Σ◁ 12 ἄλλο. ἰτέας φύλλα καὶ τὸν καρπὸν καὶ τὸ φλοῦδιν τῆς ῥίζης βράσον με τὸ νερὸν καὶ πότισον.

---

17 θεραπεία εἰς σπληνικοὺς καὶ πρησμένους] πρὸς θεραπείαν σπληνὸς B 1 με τὸ κρασίν] μετὰ οἴνου A 1–2 σκολοπένδρου τὴν ῥίζαν καὶ τὰ φύλλα κοπάνισον] transp. B 1 σκολοπένδρου] σκολοπενδρίου A 2 καὶ τὰ φύλλα] om. A 2 θές τα] θέτε την A om. B 2 σπλῆνα] πλίνα L θέται add. B 3 βράσον] κοπάνισον ἢ καὶ ἀκοπάνιστα βράσον A 4 καὶ] τὸ κρασὶν καὶ add. A 4 ἀσπάραγγια] ἀ.αράγγια L σπαράγγια AB 4–5 τὸν χυλὸν πότισον] transp. B 5 ἀγρίας] ἀγρίαν B 5 καὶ] ... L 5–6 τὸν ζωμὸν πότισον] transp. B 6 οὐγγίαν] ο... L 6 καὶ] om. AB 6 βραχῇ] βραχὺ L βραχὴ MB 7 εἰς κρασὶν νυχθήμερα τρία] scripsi εἰς κρασὶν νυχθη..ρα τρία L τρία νυχθήμερα εἰς τὸ κρασὶν MA νυχθήμερον τρία εἰς τὸ κρασὶν B 7 καὶ τὸ] εἶτα πότισον αὐτὸ τὸ A om. B 7 κρασὶν αὐτὸ] κρασὶν A om. B 7 πότισον] πο...σον L 8 νερὸν] νίτρον καὶ νερὸν AB 8 τὸν ζωμὸν αὐτῶν βράσον] βράσας A 8 αὐτῶν βράσον] αὐτὸν βάλε B 9 καὶ μάλαξον] βάλε A 9 ἀλεύριν αἶρας] αἶρας ἀλεύριν B 9 ἀλεύριν] ἀλείριν M ἄλευρον A 9 θὲς] θέτε ἐπάνω A θέτε το B 10 σπλῆνα] πλίνα MB 10 καππάρεως τῆς ῥίζης] ῥίζης καππάρεως MB 10 φλοῦδιν] φλο.... L 10 κοπάνισον] καὶ βράσον add. MB 11 πότισον] πότι... L ἄς το πίνη B 11 φύλλα κισσοῦ τρυφερὰ] transp. MB 12 θέτε τα] θὲς M θέτε τα B 12–13 φύλλα τρυφερὰ κισσοῦ] φύλλα τρυφερὰ ....του L τρυφερὰ φύλλα κισσοῦ MA ἢ B 14 θὲς] το add. A θέτε το B 14 τὸν] om. B 14 ἰτέας] ... L αἰταίας M ἐταίας A 15 καὶ τὸν καρπὸν] om. A 15 φλοῦδιν] φλού....... L φλούδιον M φλοῦδι B 16 πότισον] ....ιον L πότιζε A

215 [161] σιε΄ ὅταν ὁ ἄνθρωπος ἔχῃ στρόφους ποίησον θεραπείαν τοιαύτην· ▷LBβράσον ἀγρίαν με τὸ νερὸν καὶ πότισον.LB◁ 2 ἄλλο. ▷Σ⌜τοῦ ἀνήθου τοῦ ξηροῦ τὴν κορυφὴν καὶ τὸ σπέρμα πότισον με τὸ νερόν.Σ◁ 3 ἄλλο. ▷MBπήγανον βράσον με τὸ νερὸν καὶ πότισον.MB◁ 4 ἄλλο. ▷Σσάμψυχον κοπάνισον καὶ τὸν χυλὸν βράσον με τὸ κρασὶν καὶ πότισον.Σ◁ 5 ἄλλο. κύμινον βράσον με τὸ ἔλαιον καὶ τὸ ἔλαιον ἐκεῖνο ἰάτρευε⌝ κλυστήριν. 6 ἄλλο. ▷om. Lπλατυκύμινον κοπάνισον καὶ πότισον μετὰ ὕδατος.om. L◁ 7 ἄλλο. ⌜καστόριον τρίψον καὶ πότισον με τὸ κρασίν. 8 ἄλλο. ▷om. Bναρθηκίου τὸν σπόρον⌝ κοπάνισον καὶ πότισον με τὸ κρασίν.om. B◁ 9 ἄλλο. πήγανον καὶ κύμινον καὶ ἄνισον πότισον με τὸ νερὸν καὶ τὸ μέλι.

216 [162] σιϛ΄ ἐὰν δάκῃ τινὰ ὁ σκόρπιος· ⌜τρίψον τὸν καρπὸν τοῦ ἀσφοδέλου καὶ πότισον μετὰ οἴνου. 2 ἄλλο. ▷Σσάμψυχον κοπάνισον καὶ μίξον με τὸ ὄξος καὶ τὸ ἅλας τριμμένον καὶ μαλάξας θὲς εἰς τὴν πληγήν.⌝Σ◁ 3 ἄλλο. ῥίζαν κυπέρου κοπάνισον καὶ πότισον μετὰ οἴνου. 4 ἄλλο. ⌜μελισσόφυλλα κοπάνισον καὶ πότισον μετὰ οἴνου.⌝

217 [163] σιζ΄ ἐὰν συμβῇ εἰς πληγὴν καὶ γένωνται σκωλήκια, ποίησον θεραπείαν τοιαύτην ὅπως ἐκβάλῃς ταῦτα· ⌜κοπάνισον τὰ

---

1 ὅταν] πρὸς ὅταν M πρὸς B 1 ὁ ἄνθρωπος ἔχῃ] transp. M om. B 1 στρόφους] στρόφον M στρόφον ἀνθρώπου B 1–2 ποίησον θεραπείαν τοιαύτην] scripsi om. MA ...ησον θεραπείαν τοιαύτην L οίησον θεραπείαν τοιαύτην ἤγουν B 2 ἀγρίαν] ἀγρία B 3 σπέρμα] τούτου add. A 3–4 πότισον με τὸ νερόν] βράσον με τὸ νερὸν καὶ πότισον A 5 βράσον] πότισον B 6 καὶ πότισον] om. B 6 κύμινον] κιμ.... L κυκλάμινον A 6 τὸ ἔλαιον] μετ' αὐτοῦ τοῦ ἐλαίου B 7 ἐκεῖνο] αὐτὸ MA om. B 7 πλατυκύμινον κοπάνισον] transp. B 9 τὸν] om. A 10–11 πήγανον καὶ κύμινον καὶ ἄνισον] κύμινον καὶ ἄνισον καὶ πήγανον B 11 ἄνισον] ἄνηθον MA 11 μέλι] ... B 12 ἐὰν] πρὸς ἐὰν M 12 τινὰ] τ̣ι̣ν̣ὰ̣ L τινὰν M textum AB 12 τρίψον] ὠφελεῖ δὲ ὁ καρπὸς τοῦ ἀσφοδέλου ἤτοι τρίψον M ἔπαρον A ὠφελεῖ δὲ ὁ καρπὸς τοῦ ἀσφοδέλου ὃν τρίψας B 12–13 τὸν καρπὸν τοῦ ἀσφοδέλου] om. MB 13 καὶ] καὶ τρίψε τον καὶ A om. B 14 ὄξος] ἅλας B 14 ἅλας] ὄξος B 14 τριμμένον] om. B 14 θὲς] θέτε AB 16 μελισσόφυλλα] μελισσόφιλον M μελισσόφυλλον AB 16–17 πότισον μετὰ οἴνου] θές το εἰς τὴν πληγὴν Σ 18 πληγὴν] τὴν πληγὴν A 18 γένωνται] γένον.... L γίνωνται AB 18 σκωλήκια] σκουλίκια L 19 ποίησον θεραπείαν τοιαύτην ὅπως ἐκβάλῃς ταῦτα] om. A 19 θεραπείαν τοιαύτην] οὕτως B 19 ὅπως] om. B

φύλλα τῆς πτελέας καὶ θέτε ἐπάνω. 2 ἄλλο. ▷om. Aἀψίνθιον χλωρὸν κοπάνισον καὶ θέτε ἐπάνω εἰς τὴν πληγήν.⌝om. A◁ 3 ἄλλο. ⌜κεδρέαν ἄλειψον⌝ καὶ ἐκβαίνουσιν. 4 ἄλλο. ▷Σ⌜τὸ αὐτὸ ποίησον καὶ ὅταν εἰς τὸ ὠτίον ἀνθρώπου ὦσι σκώληκες.⌝Σ◁.

218 [164] σιη΄ ἐὰν δάκῃ τινὰ ἡ θαλασσία τρυγών, ποίησον θεραπείαν τοιαύτην. ⌜ἐλελίσφακον βράσον μετὰ οἴνου καὶ πότισον. 2 ἄλλο. φύλλα τοῦ ἐλελισφάκου κοπάνισον καὶ θέτε εἰς τὴν πληγήν.⌝ 3 ἄλλο. ▷om. Aἀσφοδέλου σπέρμα κοπάνισον καὶ πότισον μετὰ οἴνου.om. A◁ 5 ἄλλο. ▷MBφύλλα ἀσφοδέλου κοπάνισον καὶ θές τα εἰς τὴν πληγήν.MB◁

219 [165] σιθ΄ εἰς σπληναρίους καὶ πρησμένους ⌜σάμψυχον βράσον με τὸ νερόν, εἶτα βάλε καὶ κρασίν καὶ πότισον.⌝ 2 ἄλλο. κόπρον βοείαν ξηρὰν πότισον μετὰ οἴνου. 3 ἄλλο. ▷Lρίζαν κυπέρου βράσον με τὸ νερόν. εἶτα βάλε κρασὶν καὶ πότισον.L◁ 4 ἄλλο. ▷Σ⌜καππάρεως φλοῦδιν κοπάνισον καὶ πότισον μετὰ οἴνου.⌝Σ◁ 5 ἄλλο. ἀσάρου ῥίζαν κοπάνισον καὶ πότισον μετὰ οἴνου. 6 ἄλλο. MA▷ἀκόρου ῥίζαν κοπάνισον καὶ πότισον μετὰ οἴνου.MA◁ 7 ἄλλο. ⌜ῥίζαν συκέας κοπανισον καὶ μίξον με τὸ ἀλεύριν τὸ κρίθινον καὶ θές το εἰς τὴν σπλῆναν πλὴν πρῶτον ἔψησον τὸ κριθάλευρον με τὸ νερόν.⌝ 8 ἄλλο.

---

19 ἐκβάλης ταῦτα] scripsi ἐκβάλης ταυ...... L ἐκβληθῶσιν ἤτοι M om. B 1 πτελέας] ἐπτελαίας καλὰ B 1 θέτε] θές τα MB τα add. A 1 ἐπάνω] καὶ ἐκβαίνουσι add. A 2 κοπάνισον καὶ] om. M 2 θέτε] θὲς MB 2 εἰς τὴν πληγήν] om. MB 3 ἐκβαίνουσιν] ἐκβαίνωσιν L ἐκβαίνουσι A ἐκβλη^θ B 3 καὶ] om. A 4 τὸ ὠτίον] τὰ ὠτία A 4 ὦσι] εἰσὶ AB 5 τινὰ] ἄνθρωπον AB 5 ἡ θαλασσία τρυγών] τρυγὼν θαλασσία A θαλασσιεσὰ τρυγὼν B 5–6 ποίησον θεραπείαν] om. B 6 τοιαύτην] om. AB 6 ἐλελίσφακον] ἤγουν τὸ λεγόμενον ἐλελίσφακον ὅπερ ἡ κυράμας ἦθκος ἔβρασεν μετ' ἐκεῖνον νερὸν καὶ ἔλουσεν τὸν αὐθέντην καὶ θν ἡμῶν ἰν χρ τοῦτο οὖν M ἔπαρον ἐλελίσφακον καὶ A τὸ λεγόμενον ἐλιλήφασκον B 6 μετὰ οἴνου] με τὸ κρασὶν Σ 7 φύλλα τοῦ ἐλελισφάκου] τοῦ αὐτοῦ τὰ φύλλα B 7 ἐλελισφάκου] ἐλελυφάσκου L ἀλισφάσκου M 7 θέτε] θές τα Σ 9 φύλλα ἀσφοδέλου] τοῦ αὐτοῦ τὰ φύλλα B 11 σάμψυχον] σύμψυχον L 12 εἶτα] καὶ Σ 12 καὶ] ὕστερον Σ 12–13 κόπρον βοείαν ξηρὰν] κόπρον ξηρὰν βοείαν MA κόπρον ξηράνας B 13 μετὰ οἴνου] ἤγουν βόειον μετὰ ὄνου B 15 ἀσάρου] σάρου M 17 ῥίζαν] ῥίζας B 17–18 συκέας κοπανισον] συ..έα.......πάνισον L συκῆς κοπάνισον M 18 κρίθινον] κριθιν.. L 18 θές] θέτε B 18 το εἰς τὴν] scripsi το ... L εἰς τὸν MA εἰς τὴν B 19 σπλῆναν] σπλῆνα AB 19 πρῶτον] πρῶτα B

τὸ λιθάριν τὸ εὑρίσκεται εἰς τὴν κεφαλὴν τοῦ βατράχου ἔπαρον καὶ δῆσον εἰς τὴν κεφαλὴν τοῦ πρησμένου καὶ ὠφελεῖ τὰ μέγιστα.

220 [165] ϛκʹ εἰς φλεγμονὴν θεραπεία· ⌜κοπάνισον LB▹γληχούνιν μετὰ πίτυρα καὶ θέτε εἰς τὴν σπλῆνα.LB◃ 2 ἄλλο. λουπηνάρια πικρὰ ἄλεσον, καὶ τὸ ἀλεύριν τούτων μάλαξον μετὰ πίτυρα καὶ τὸ νερὸν καὶ θέτε ἐπάνω.⌝ 3 ἄλλο. ▹LMδάφνης φύλλα χλωρὰ κοπάνισον καὶ μίξον με τὰς ψίχας τοῦ ἄρτου καὶ θέτε ἐπάνω.LM◃ 4 ἄλλο. ▹B⌜κράμβης φύλλα κοπάνισον καὶ μίξον μετὰ πίτυρα καὶ μάλαξον καὶ θέτε τα ἐπάνω.B◃ 5 ἄλλο. ▹Σκολοκύνθης χλωρῆς τὰ φύλλα κοπάνισον καὶ θές τα ἐπάνω.⌝Σ◃ ἄλλο. 6 ▹om. A κυπέρου τὰ φύλλα κοπάνισον καὶ θέτε ἐπάνω.om. A◃ ἄλλο. 7 ▹Στῆς πλατάνης τὰ τρυφερὰ φύλλα κοπάνισον καὶ ἔψησον μετὰ οἴνου καὶ θές ἐπάνω.Σ◃ ἄλλο. 8 ▹L⌜κολοκύνθης ξηρῆς τὰ ξύσματα κοπάνισον καὶ θέτε τὰ ἐπάνω.⌝L◃ 9 ἄλλο. φύλλα σεύτλου βράσον καὶ κοπάνισον καὶ θές τα ἐπάνω εἰς τὴν φλεγμονήν. 10 ἄλλο. ▹om. L⌜λινόσπορον κοπάνισον καὶ τήλης σπόρον μίξον, βάλε καὶ νερὸν καὶ ἂς βράσουν καὶ ὕστερον βάλε χήνειον ἀξούγγιν, χοίρειον, ὀρνίθειον, ἀλώπεκος, ἄρκτου, ἐλαφίου

---

19 ἔψησον] ἔψησ.. L 1 λιθάριν] λιθάρην A 1 τὸ] ὅπου MA ὅπερ B 1 εὑρίσκεται] εὑρίσκετε M 1 τὴν κεφαλὴν] τὸ κεφάλιν MA 1 ἔπαρον καὶ] om. B 2 δῆσον] δεῖσον L 2 τὴν κεφαλὴν] τὸν τράχηλον Σ 2 ὠφελεῖ τὰ μέγιστα] ὠφελεθήσεται B 2 τὰ μέγιστα] πολλὰ A 3 φλεγμονὴν] φλεγμονὰς M σπληνὸς add. B 3 θεραπεία] scripsi θεραπείαν L θεράπευε δὲ οὕτως M ποίησον θεραπείαν A om. B 3 κοπάνισον] om. A 3 γληχούνιν] scripsi γληχούνην LB 4–5 λουπηνάρια πικρὰ ἄλεσον] transp. B 4 λουπηνάρια] λυπηνάρια MA 5 ἄλεσον] om. M 5 μετὰ πίτυρα καὶ τὸ νερὸν] om. B 6 θέτε] θές τα M θέτε τα AB 6 ἐπάνω] μετὰ ψυχίας ἄρτου ἢ μετὰ πίτυρα add. B 7 ψίχας] ψιχίας M 7 καὶ] ἢ μετὰ πίτυρα καὶ M 7 θέτε] θὲς M 9 κολοκύνθης χλωρῆς] κολοκύνθης χλορὶς A κολοκύνθης ~~ξερὶς~~ χλωρὶς B 9 φύλλα] ξύσματα AB 10 θές] θέτε A 10 κυπέρου τὰ] κυπέρου MB 11 θέτε] θές τα M τα add. B 11 πλατάνης] πλατάνου B 11 τρυφερὰ] ἁπαλὰ A χλωρὰ B 12 θές] θέτε τα AB 12–13 κολοκύνθης] κολοκίνθης L 14 καὶ θές] κ...... L 15 λινόσπορον κοπάνισον] κοπάνισον λινόσπερμα B 15 λινόσπορον] λινόσπερμα AB 15–16 τήλης σπόρον μίξον] μίξας τήλεος σπόρον B 16 βάλε καὶ] καὶ βάλε A 17 χήνειον] χοίρειον B 17 χοίρειον, ὀρνίθειον] καὶ ὀρνίθειον καὶ χοίρειον A καὶ χείνιον ὀρνίθιον B 17 ἀλώπεκος] καὶ ἀλώπεκος A ἀλλώπεκος B 17 ἄρκτου] om. A 311.17–312.1 ἐλαφίου μυελόν] om. B

---

9 ἄλλο] This entry and the following one are reversed in B.

μυελόν, ταύρου ἀξούγγιν, μόσχου ἀξούγγιν. τὰ πάντα λύσας καὶ μάλαξον καὶ ἐπίθες ἐπάνω τὴν φλεγμονήν.⌝ λύσε δὲ καὶ ἀξούγγιν λέοντος.[om. L◁]

221 [166] σκα΄ πρὸς φωνὴν ἀποκοπεῖσαν· ἀποκαθαίρει ⌜σκόροδα ὠμὰ καὶ ἑψημένα ἂς τρώγῃ. 2 ἄλλο. χαλβάνιν ποίησον ἴσα ἐρεβινθίου κοκκία. καὶ δίδε αὐτὸ κατὰ βραδὺ πέντε πέντε καὶ ἄς τα καταπίνῃ καὶ ἐπάνω ἂς πίνῃ θερμὸν κρασὶν καὶ τίποτε μὴ τρώγῃ.⌝

222 σκβ΄ φθείρας ἀπόλυσι· κίσσου δάκρυον μίξον με τὸ ὄξος καὶ χρίε τὴν κεφαλήν. 2 ἄλλο. ῥοδάφνην βράσον με τὸ νερὸν καὶ θερμὸν λοῦε τὴν κεφαλήν. 3 ἄλλο. ἀγριοσταφίδα κοπάνισον καὶ μίξον με τὸ χρηστέλαιον καὶ ἄλειφε τὴν κεφαλήν.

223 [167] σκγ΄ πρὸς τὰς λεγομένας χοιράδας· ⌜κοπάνισον ἀρνόγλωσσον καὶ θέτε ἐπάνω. 2 ἄλλο. [▷A]κοπάνισον ἀδίαντον καὶ θὲς ἐπάνω.[A◁] 3 ἄλλο. λουπηναρίων ἀλεύριν μίξον με τὸ ὄξος. καὶ θέτε ἐπάνω εἰς τὴν πληγήν. 4 ἄλλο. [▷om. A]κοπάνισον λάπαθον καὶ

---

17 ἐλαφίου] ἐλάφειον A 1 ταύρου ἀξούγγιν, μόσχου ἀξούγγιν] om. A ταύρου μόσχου λέοντος καὶ μυελὸν ἐλάφιον B 1 τὰ] καὶ A 1 καὶ] om. A 2 μάλαξον] μαλάξας B 2 καὶ] om. B 2 ἐπίθες] θέτε A 2 τὴν] εἰς τὴν B 2–3 λύσε δὲ καὶ ἀξούγγιν λέοντος] καὶ ταύρου ἀξούγγιν A om. B 4 πρὸς φωνὴν ἀποκοπεῖσαν] πρὸς ἐὰν κρατηθῇ ἡ φωνὴ τοῦ ἀνθρώπου B 4 ἀποκοπεῖσαν] ἀπὸ κοπῆσαν L ἀποκοπίσαν M 4 ἀποκαθαίρει] scripsi ἀπὸ ...θαίρει L ποίησον θεραπείαν τοιαύτην· ἤτοι ἂς τρώγῃ M om. B τηνικαῦτα καθαίρει A 4 σκόροδα] κόρδα M ἂς τρώγῃ add. B 5 ὠμὰ] ἂς τρώγῃ add. A 5 ἂς τρώγῃ] ἂς ...ει L om. MA 5 ἴσα] ἶσον MA 5–6 ἐρεβινθίου] scripsi ἐρεβιν.... L ῥεβινθίου MA 6 αὐτὸ] αὐτῷ M αὐτὸν A 6 τα] om. A 7 θερμὸν] καὶ add. M 7 καὶ] om. A 7 μὴ] ἄλλο μὴ MA 7 τρώγῃ] μήτε πολλὰ μήτε ὀλίγα add. A 8 φθείρας ἀπόλυσι] πρὸς φθείρας ἐν τῇ κεφαλῇ M πρὸς ἐν τῇ κεφαλῇ φθείρας B 8 μίξον] μίξας AB 8 με τὸ] μετὰ MB 9 καὶ χρίε] χρίεται B 9 ῥοδάφνην] scripsi ῥο..δάφνην L τὴν λεγομένην ῥοδάφνην Σ 10 θερμὸν] χλίον M τὸ θερμὸν A om. B 10 ἀγριοσταφίδα] ἀγριοσταπιδ.. L ἀγριοσταφίδαν MA ἀγριοσταφίδα B 11 χρηστέλαιον] χρηστελ... L 11 καὶ] om. B 12 πρὸς τὰς λεγομένας χοιράδας] scripsi τὰς λεγομέ... χοιράδας L πρὸς χοιράδας τὰ λεγόμενα χελωνικὰ M τὰς λεγομένας χοιράδας ὠφελεῖ A πρὸς χειράδας καὶ τὰ λεγόμενα χελιδονικὰ B 12–13 κοπάνισον ἀρνόγλωσσον] ἀρνόγλωσσον κοπανισμένον Σ 13 θέτε] θὲς M θέτε το AB 14 λουπηναρίων] λυπηνάρια ποίησον MA 14 μίξον] καὶ μίξον MA 14 με τὸ] μετὰ Σ 14 ὄξος] ὄξους AB 14–15 θέτε] θέτε τα M θέτε το AB 15 εἰς τὴν πληγήν] om. A

μίξε με τὸ ὄξος καὶ θέτε ἐπάνω.[om. A◁⌉] 5 ἄλλο. κοπάνισον λινόσπορον καὶ θές το ἐπάνω. 6 ἄλλο. [▷MB]φύλλεον κοπάνισον καὶ θές το ἐπάνω με τὸ ὄξος.[MB◁] 7 ἄλλο. [▷Σ⌈]κοπάνισον τὸ λεγόμενον βρωμοβότανον καὶ τὸν χυλὸν τούτου μίξον μετὰ πίτυρα καὶ θέτε τα ἐπάνω. 8 ἄλλο. κυκλαμινέαν ῥίζαν κοπάνισον καὶ θὲς ἐπάνω.[⌉Σ◁] 9 ἄλλο. [▷ om. B]δρακοντίας ῥίζαν κοπάνισον καὶ θέτε ἐπάνω. 10 ἄλλο. ῥίζαν δρακοντείαν κοπάνισον καὶ τὸν χυλὸν πότισον.[om. B ◁] 11 ἄλλο. [▷B]πότισον δὲ καὶ τὸν χυλὸν τῆς τοιαύτης ῥίζης.[B◁]

224 σκδʹ θεραπεία εἰς ψώρας· λουπηνάρια πικρὰ βράσον μετὰ ὕδατος καὶ τὸ νερὸν ἐκεῖνο χλιαρὸν ἄντλει τὰς ψώρας. 2 ἄλλο. κεδρέαν ἕψησον μετὰ ὄξους καὶ ἄλειφε ἐν βαλανείῳ. 3 ἄλλο. καρδίαν ῥοδάφνης καὶ μυρσίνης τηγάνισε μετὰ χρηστελαίου καὶ ἄλειφε. 4 ἄλλο. φύλλα κοπάνισον κισσοῦ καὶ βράσον με τὸ νερὸν καὶ πότισον. 5 ἄλλο. πήγανον βράσον μετὰ ὕδατος καὶ ἄντλει εἰς τὸ λουτρόν. 6 ἄλλο. ἀγριοσταπίδα κοπάνισον καὶ σὺν ἐλαίῳ ἄλειφε.

225 σκεʹ τὸν πόνον τῶν ὠτίων θεραπεύει· ἀμυγδαλέλαιον χλιαρόν, στάξον εἰς τὸ οὖς. 2 ἄλλο. [▷LB]καριδίων ἔλαιον στάξον εἰς τὸ οὖς.[om. LB◁] 3 ἄλλο. [▷ om. A]δαφνέλαιον χλιαρὸν στάξον εἰς τὸ οὖς. 4 ἄλλο. βαλσαμέλαιον στάξον εἰς τὸ οὖς.[om. A◁] 5 ἄλλο. [▷om. B]δρακοντίας τὸν σπόρον κοπάνισον καὶ ἐκβάλας τὸν ζωμὸν μίξον μετὰ χρηστελαίου καὶ ἄλειφε. 6 ἄλλο. ἡδύοσμον κοπάνισον καὶ τὸν

15 λάπαθον] λάπατον L καὶ λάπαθον M 1 μίξε] μίξον M μίξας B 1 με τὸ ὄξος] μετὰ ὄξους B 1 καὶ] ..... L om. B 1 θέτε] θὲς M 2 θές] θέτε B 2 το] τον MB om. A 2 καὶ] μίξας με τὸ ὄξος add. B 2 θές το] θέτε τον B 3 με τὸ ὄξος] om. B 4 θέτε τα] θέτε A 5 ἐπάνω] om. B 5 κυκλαμινέαν] scripsi κυκλαμιναίαν M κυκλαμηναίαν A κυκλαμίνου B 6–7 θέτε ἐπάνω. 10 ἄλλο. ῥίζαν δρακοντείαν κοπάνισον καὶ] om. M 6 θέτε] την add. A 6–7 ἄλλο. ῥίζαν δρακοντείαν κοπάνισον] om. A 9 λουπηνάρια] λυπηνάρια M 9 μετὰ] με τὸ A om. B 10 ὕδατος] ὕδωρ MA om. B 10 νερὸν] ὕδωρ B 10 ἐκεῖνο] τοῦτο B 10 χλιαρὸν] χλιερὸν B 12 καὶ] ἢ M 12 τηγάνισε] τυγάνισον M τηγάνισον AB 13 κοπάνισον] om. Σ 13 καὶ] om. Σ 13–14 πότισον] ἄντλη τὰς ψωράς M ἄντλει A ἄντλυε B 14 πήγανον] ὀρίγανον Σ 14–15 εἰς τὸ λουτρόν] ἐν λουτρῷ A εἰς λουτρὸν B 15 ἀγριοσταπίδα κοπάνισον] transp. B 15 ἀγριοσταπίδα] ἀγριοσταφίδαν M ἀγριοσταφίδα AB 15 σὺν ἐλαίῳ] συνελ.... L 16 τὸν] πρὸς M 16 τῶν ὠτίων] ὠτίου M ὠτίων B 16 θεραπεύει] ......ευει L om. MB 17 στάξον] στάξ̣ο̣ς L σταζόμενον A 17 οὖς] ὠτίον M 17 καριδίων ἔλαιον] καρυδέλαιον B 17 στάξον] ...ξον L 20 μίξον] ἕνωσον M

ζωμὸν μίξον με τὸ μέλι καὶ στάξον εἰς τὸ ὠτίον. om. B◁ 7 ἄλλο. ▷Σκολοκυνθίου χλωροῦ τὰ ξύσματα κοπάνισον καὶ τὸν ζωμὸν ἔκβαλε καὶ ἕνωσον με τὸ ῥοδέλαιον καὶ στάξον εἰς τὸ οὗς.Σ◁ 8 ἄλλο. κοπάνισον κρομμύδιν καὶ τὸν ζωμὸν χλίον στάξον εἰς τὸ οὗς. 9 ἄλλο. ▷Bπήγανον χλωρὸν ἕψησον μετ' ἐλαίου καὶ χλιάνας στάξον εἰς τὸ οὗς.B◁ 10 ἄλλο. ▷Σῥεφανέλαιον στάξον εἰς τὸ οὗς.Σ◁ 11 ἄλλο. ▷ om. Aκοπάνισον κύμινον καὶ βάλε καὶ ἔλαιον καὶ ἕψησον καὶ χλιάνας στάξον εἰς τὸ οὗς.om. A◁ 12 ἄλλο. τὴν ῥίζαν τοῦ ἀσφοδέλου κοπάνισον καὶ τὸν χυλὸν μίξον μετὰ σμύρνης τριμμένης καὶ δενδρολιβάνου καὶ στάξον εἰς τὸ ὠτίον.

226 [168] σκϛʹ ἂν θέλης να μὴ διψᾷ τίς, ποίησον θεραπείαν τοιαύτην· ⌜ἐλαίαν κολυμβάδα κράτει εἰς τὴν γλῶσσαν σου. 2 ἄλλο. αὐγὸν ὄρνιθος ῥόφα κατὰ πρωῒ νηστικὸς ὠμόν.⌝ 3 ἄλλο. ▷Σ⌜μαρουλλίου σπόρον καὶ γλυκόριζον βράσον με τὸ νερὸν καὶ ἂς κρυώσῃ καὶ δίδε τὸ καὶ ἂς πίνῃ.⌝Σ◁ 4 ἄλλο. φοινίκια καὶ ξυλοκέρατα καὶ γλυκύριζον καὶ νερὸν βράσον καὶ κρυάνας δὸς πιεῖν.

227 [168] σκζʹ ἀλφοὶ λέγονται, ὅτε ὁ ἄνθρωπος εὐγάνῃ εἰς τὸ πρόσωπον ὡς ἂν φακὴν μαύρην. ποίησον δὲ θεραπείαν τοιαύτην·

---

21 χρηστελαίου] χρ..στελαίου L χρηστέλαιον M 1 μίξον] ἕνωσον M 1 εἰς τὸ ὠτίον] ἐν τῷ ὠτίῳ A 2–3 ἔκβαλε] ἐκβαλὼν B 3 καὶ] om. B 3 ἕνωσον] μίξον AB 4 τὸν] τοῦτον τὸν B 6 ῥεφανέλαιον] ῥεπανέλαιον B 6 οὗς] ὠτίον A 7 κύμινον] κυκλάμινον MB 7 βάλε] om. B 7 καὶ] om. MB 7 ἔλαιον καὶ ἕψησον] ἕψησον μετ' ἐλαίου B 8 καὶ] εἶτα M 9 σμύρνης] μυρσίνης A 9 τριμμένης] om. B 10 ὠτίον] οὗς B 11 διψᾷ τίς] δίψας B 11–12 ποίησον θεραπείαν τοιαύτην] ἔπαρον A om. B 12 ἐλαίαν κολυμβάδα] ἐλαίας κολυμβάδας M ἐλαίαν κολυμβάδαν καὶ A ἐλαίαν κολυμβάδαν B 12 σου] om. A 13 ὄρνιθος] ὀρνιθίτικον AB 13 ῥόφα] ὠμὸν ῥόφα B 13 κατὰ πρωῒ νηστικὸς] νηστικὸς κατὰ πρωῒ B 13 ὠμόν] om. Σ 13–14 μαρουλλίου] μαιουλίου M 14 γλυκόριζον] γλυκορίζου A 14–15 ἂς κρυώσῃ καὶ δίδε τὸ καὶ] om. B 15 πίνῃ] τον πίνῃ κρύον B 15 καὶ] om. A 15 ξυλοκέρατα] ξυλοκαίρατα B 16 καὶ νερὸν βράσον] βράσον με τὸ νερὸν AB 16 κρυάνας] κριάνας L ἂς κρυώσῃ A 16 δὸς πιεῖν] δίδε τὸν καὶ ἂς πίνῃ M καὶ δίδε τὸν καὶ ἂς πίνῃ ἀντὶ νεροῦ A δίδου ποιεῖν αὐτὸ B 17 ἀλφοὶ] πρὸς ἀλφούς. ἀλφοὶ M πρὸς ἐκβαλόντας εἰς τὸ πρόσωπον ἀλφοὺς B 17 ὅτε] ὅταν A ἔτι B 17–18 ὁ ἄνθρωπος εὐγάνῃ εἰς τὸ πρόσωπον] om. B 17 ἄνθρωπος] ἄν.... L 17 εὐγάνῃ] εὐγάλῃ A 18 πρόσωπον] του add. MA 18 μαύρην] μαῦριν L 18 ποίησον] καὶ ποίησον A om. B 18 δὲ] εἰς αὐτοὺς M om. AB 18 θεραπείαν] om. B

καρπὶν ἄσπρον τρίψας καὶ μίξας ὄξος ἄλειφε. 2 ἄλλο. ▷om. B καρπὶν μαῦρον τριψας καὶ μίξας ἔλαιον καὶ ὄξος ἄλειφε ἢ ἐν ἡλίῳ ἢ ἐν πυρί. om. B◁ 3 ἄλλο. τεάφην τριμμένην καὶ ὄξος ἄλειφε. 4 ἄλλο. ▷Σ λιθάργυρον τριμμένον καὶ τεάφην καὶ ἀγριοσταφίδα τριμμένην μίξας μετὰ ὄξους καὶ ἄλειφε. Σ◁

228 σκη′ ὅπου ἐκβάλωσιν εἰς τὸ πρόσωπον ὡς ἂν φακὴν ἄσπρην, ποίησον θερπείαν τοιαύτην. σμύρναν ξηρὰν τρίψον καὶ τεάφην καὶ μίξας μετὰ ὄξους, ἄλειφε. 2 ἄλλο. κηκιδίου τὴν καρδίαν καὶ ὑγροπίσσιν μίξας ἄλειφε.

229 σκθ′ ἀλωπεκία ἔνι, ὅταν τὰ μαλία τῆς κεφαλῆς ῥέουσιν. ποίησον θεραπείαν τοιαύτην· καλάμου φύλλα καῦσον καὶ μίξον με τὸ ὑγροπίσσιν καὶ ἀξούγγιν πρόβειον καὶ ξύρησε τὴν κεφαλὴν ὅλην καὶ ἄλειφε. 2 ἄλλο. ἀγριοσταπίδαν τριμμένην καὶ μαστίχην καὶ ὑδράργυρον καὶ ὄξος, μίξον ὅλα καὶ ἄλειφε τὴν κεφαλήν. 3 ἄλλο. ἀξούγγιν ἀλώπεκος ἀνάλυσε καὶ αἴγας βάλε πίσσαν καὶ ὄξος καὶ ἄλειφε.

---

18 τοιαύτην] om. AB 1 καρπὶν] ἤγουν καρπὶν MA 1 μίξας] μίξον B 1 ὄξος] μετὰ ὄξους MA με τὸ ὀξείδιν καὶ ἔλαιον καὶ ὄξος καὶ B 1 ἄλειφε] ἄλειφε ἢ ἐν ἡλίῳ ἢ ἐν πυρί B 2 τριψας καὶ μίξας] τρίψον καὶ μίξον MA 2 ἔλαιον] ἐλ...ον L 2 ἄλειφε] καὶ ἄλειφε M 2 ἢ] om. A 2–3 ἢ ἐν πυρί] ....ρι L ἢ ἐν πῦρ A 3 τεάφην] scripsi τεάφ.. L τεάφιν MA τιάφιν B 3 τριμμένην] τριμμένον MB 4 τεάφην] scripsi τέφιν M τεάφιν AB 5 καὶ] om. AB 6 ὅπου] πρὸς ὅπου M τινὲς A ὡσαύτως καὶ εἴτι B 6 ἐκβάλωσιν] εὐγάνουσιν M ἐκβάλλουσιν A ἐκβάλλει B 6 τὸ] om. B 6 πρόσωπον] τῶν add. A 6 φακὴν] φακὶν B 6 ἄσπρην] ἄσπρον L ἄσπριν M μέλαινα B 7 ποίησον θερπείαν τοιαύτην] om. Σ 7 τρίψον] τρίψας B 7 τεάφην] τεάφιν Σ 8 μίξας μετὰ ὄξους] ὄξος καὶ ἑνώσας M ὄξος μίξας A μίξας ταῦτα μετὰ ὄξους B 8 κηκιδίου] τὸ ἀπέσω ἤγουν add. M τὸ ἀπέσω τοῦ κηκιδίου ἤγουν B 8 καρδίαν] του add. M ἀπέσω add. A 8–9 καὶ ὑγροπίσσιν μίξας] μίξας με ὑγροπίσσιν M μίξας μετὰ ὑγροπίσσιν B 10 ἀλωπεκία] πρὸς ἀλωπεκίαν· ἀλωπεκία M πρὸς πάθος ἀλωπεκίαν λεγόμενον ὃς B 10 ἔνι] ἔστιν B 10 ῥέουσιν] πίπτουσιν M 10–11 ποίησον θεραπείαν τοιαύτην] ἔπαρον A ἤγουν add. MB 11 καῦσον] καὶ καύσας A 11 καὶ] om. A 11 μίξον] αὐτὰ add. Σ 12 ὅλην] om. MA οὕως B 13 καὶ] om. AB 13 ἀγριοσταπίδαν] ἀγριοσταφίδαν Σ 14 ὑδράργυρον] λιθάργυρον A 14 μίξον ὅλα] ἕνωσον M μίξας B 14 καὶ] om. B 14 τὴν κεφαλήν] om. A 15 ἀλώπεκος] ἀλώπεκος καὶ αἰγὸς AB 15 ἀνάλυσε] ἀνάλυσον M om. A 15 καὶ αἴγας] om. AB 15 βάλε] βάλε καὶ A καὶ βαλὼν B 15 καὶ] om. B

230 ϛλʹ πρὸς παρωτίδας θεραπεία· ἀξούγγιν μόσχου βόϊον μίξον καὶ ἀσβέστιν καὶ ἄλειφε. 2 ἄλλο. τήλης σπόρον κοπάνισον καὶ καρδαμόσπορον καὶ βάλε καὶ ἀσβέστιν καὶ ἄλειφε.

231 σλαʹ ὅπου κατουρῇ αἷμα ποίησον θεραπείαν τοιαύτην· ξυλοκέρατα ἔψησον, τρίψον καὶ ἄκορον καὶ μίξον αὐτὸ με τὸν ζωμὸν τῶν ξυλοκεράτων καὶ πότισον.

232 σλβʹ πρὸς βῆχας· ὕσσωπον καὶ γλησχούνιν καὶ ἰσχάδια βράσον με τὸ νερὸν καὶ πότισον. 2 ἄλλο. πηγάνου σπόρον τριμμένον καὶ κύμινον καὶ πέπερι μίξον μέλι καὶ πότισον. 3 ἄλλο. ▷om. L στύρακα ἴσον ὀρόβιν κοπάνισον καὶ πότιζε μετὰ νεροῦ ἕως ἡμέρας τρεῖς. om. L◁ 4 ἄλλο. ξυλοτζιτζίβερ τριμμένον καὶ κύμινον καὶ μέλι πότισον. 5 ἄλλο. ▷om. L σιναπίου σπόρον τριμμένον πολλὰ μετὰ νεροῦ χλιαροῦ πότισον. 6 ἄλλο. καρδάμου σπόρον λεῖον κοπάνισον καὶ πότισον μετὰ νεροῦ. om. L◁ 7 ἄλλο. κοπάνισον πηγάνου φύλλα καὶ τὸν χυλὸν μίξον με τὸ βούτυρον καὶ πότισον τὸ αὐτὸ καὶ εἰς στῆθος ἄλειφε. 8 ἄλλο. ▷om. L κοπάνισον τὴν βερονίκην καὶ τὸν χυλὸν αὐτῆς μίξον με τὸ μέλι καὶ πότισον. om. L◁ 9 ἄλλο.

1 πρὸς παρωτίδας θεραπεία] .ρὸς πρωτι..... ραπει..... L πρὸς παρωτίδας AB 1 ἀξούγγιν μόσχου βόϊον] scripsi μόσχου βόϊον L ἀξούγγιν μόσχου M ἀξούγγιν μόσχου βόειον A ἀξούγγιν μόσχου καὶ βοὸς B 2 καὶ] μὲ M 2 ἀσβέστιν] ἂς β..... L 2 τήλης] τ̣ήλεως B 3 καρδαμόσπορον] καρδάμου σπόρον B 3 καὶ] om. A 3 ἄλειφε] τὸν τόπον add. A 4 ὅπου] πρὸς ὅπου B 4 ποίησον θεραπείαν τοιαύτην] θεραπεία A 4 τοιαύτην] ἤγουν add. MB 4–5 ξυλοκέρατα] ξυλοκέρατον B 5 τρίψον] καὶ τρίψον MA 5 καὶ] τὸ λεγόμενον add. Σ 5 αὐτὸ με τὸν] om. M με τὸν A 6 καὶ] om. B 7 ὕσσωπον] ..σωπον L 7 γλησχούνιν] βλησκούνιν M χλισχούνην AB 7 ἰσχάδια] ἰσχάδας M ἰσχάδιν B 8 με τὸ νερὸν] μετὰ ὕδατος M μετὰ νεροῦ A 8–9 πηγάνου σπόρον τριμμένον καὶ κύμινον καὶ πέπερι] πέπερι τριμμένον καὶ κίμμινον καὶ σπόρον πηγάνου ὁμοίως τριμμένον B 9 κύμινον] κυκλάμινον A 9 μίξον] τριμμένον ἑνώσας ταῦτα με τὸ M τριμμένον μίξον A μετο add. B 9 μέλι] μέλιτος B 10 ὀρόβιν] ῥόβιν A 10 πότιζε] πότισον B 10 νεροῦ] ὕδατος B 11 ξυλοτζιτζίβερ] ζινκίβερ M τζιτζίβερ A 11 καὶ] μετὰ B 12 μέλι] μέλι ἑνώσας M μέλι ταῦτα A μέλιτος μίξας B 13 χλιαροῦ πότισον] transp. B 13 ἄλλο] ὁμοίως A 13 λεῖον] om. AB 13–14 κοπάνισον] om. A 14 καὶ πότισον μετὰ νεροῦ] om. A 15 χυλὸν] ζωμὸν A 15 με τὸ] μετ̣. L μετὰ B 15 βούτυρον] βουτύρου B 15–16 τὸ αὐτὸ καὶ εἰς στῆθος] om. A καὶ B 16 εἰς] τὸ M 16 ἄλειφε] εἰς τὸ στῆθος add. A τὸ στῆθος B

om. A▷γλησχούνιν βράσον με τὸ νερὸν καὶ πότισον με τὸ μέλι.om. A◁ 10 σησάμιν ἂς μασάται συνεχῶς. 11 ἄλλο. ῥητίνην τρίψον καὶ πότισον μετὰ οἴνου. 12 ἄλλο. ὕσσωπον καὶ γλησχούνιν καὶ τήλην βράσον με τὸ μέλι καὶ τὸ κρασὶν καὶ πότισον. 13 ἄλλο. ▷om. LBἄνηθον καὶ κύμινον καὶ πήγανον βράσον με τὸ κρασὶν καὶ πότισον. 14 ἄλλο. βλησκούνιν καὶ πήγανον καὶ βούτυρον ἕψησον με τὸ μέλι καὶ τὸ νερὸν καὶ πότισον ἐν τῷ λουτρῷ.om. LB◁ 15 ἄλλο. ▷om. MAφύλλα σελίνου βράσας με τὸ νερὸν καὶ πότισον.om. MA◁ 16 βούτυρον ἄλειφε τὸ στῆθος. 17 ἄλλο. ▷om. Aπηγάνου φύλλα καὶ κύμινον καὶ κρασὶν ὀλίγον ἕψησον καὶ ἄλειφε τὸ στῆθος. 18 ἄλλο. κύπεριν τριμμένην πολλὰ ψιλὴν ποίησον καὶ πότισον μετὰ οἴνου καλοῦ ἐν τῷ λουετρῷ.om. A◁

233 [169,170] om. Aςλγʹ ὅπου βρωμῇ τὸ στόμα του· ▷om. L⌜τρίψον πεπέρι κοκκία ιε' καὶ καῦσον ξυλοκέρατα καὶ τρίψον φλοῦδιν ῥοΐδων ἐξάγιον ἓν κοπάνισον καὶ κηκίδιον ἓν ἀτρύπητον καὶ μίξον μετὰ μέλιτος ἄλειφε τὸν τόπον ὅπου ἔνι σαπημένος.om. L◁ 2 ἄλλο. μύρον καλὸν ἄλειφε. 3 ἄλλο. ἀρσενίκιν ψιλὸν μίξον μετὰ μέλιτος. καὶ

16 βερονίκην] βετονίκην A 1 γλησχούνιν] βλησκούνιν M λισχούνην B 1 νερὸν] ...... L νερὸν MB 1 καὶ] τὸ .. add. B 1 με τὸ] μετ. L με τὸ M om. B 1 μέλι] om. B 2 σησάμιν ἂς μασάται συνεχῶς] transp. B 2 σησάμιν] ἠσάμιν M 2 ἂς μασάται] ......α τε L 3 γλησχούνιν] βλησκούνιν M γληχούνην A γλησχούνιν B 4 μέλι] νερὸν B 4 καὶ] μέλι καὶ add. B 6 βλησκούνιν] γληχούνην A λιχούνιν B 6 με τὸ] μετὰ B 6–7 μέλι καὶ τὸ νερὸν] transp. A μέλιτος καὶ ὕδατος B 7–8 φύλλα σελίνου] transp. B 9 τὸ] εἰς τὸ A 9 πηγάνου φύλλα καὶ κύμινον] transp. B 10 ἕψησον] ἑψήσας MB 10 καὶ] om. MB 11 ψιλὴν] om. B 11 ποίησον καὶ] om. B 11 οἴνου] κρασίου B 11 καλοῦ] μυροδάτου MB 11 ἐν] .... L 13 ὅπου] πρὸς ὅταν B 13 στόμα του] στόμα M ἀνθρώπου ποίησον οὕτως add. B 14 κοκκία] κηκίδια M 14 καὶ καῦσον] transp. B 14 τρίψον] καὶ add. B 15 κηκίδιον] κικίδιν B 15 ἀτρύπητον] ἀτρίπητον L 15 μίξον] μίξας B 16 μέλιτος] μέλι B 16 τὸν] σαπημένον add. B 16 ὅπου ἔνι σαπημένος] om. B 16–17 μύρον καλὸν ἄλειφε] transp. B 17 καλὸν] om. M 17 ἀρσενίκιν] πρὸς σῆψιν οὐλῶν· ἀρσενίκιν B

6 βλησκούνιν] This recipe and the following one are reversed in B. In A it forms the end of the chapter. 8 βούτυρον] The following recipes are reversed in B. 11–12 λουετρῷ] The following chapters are missing in A; however, the chapter headings can be found in the margin of the page. 16–17 μύρον καλὸν ἄλειφε] This recipe is at the end of the chapter in manuscript B. 17 ἀρσενίκιν] B starts a new chapter here.

ἄλειφε στυππεῖα καὶ τὸ στυππεῖον βάλε με τὸν δάκτυλόν σου εἰς τὸ στόμα ἔνθα ἡ σῆψις ἔνι. 4 ἄλλο. φλοῦδιν ἰτέας καῦσον καὶ χαρτὶν καὶ τὴν στάκτην των, βάλε καὶ πέπερι ψιλὸν καὶ μάλαξον μετὰ μέλιτος καὶ θέτε εἰς τὴν σῆψιν. 5 ἄλλο. ὑγράργυρον καὶ πέπερι ψιλὸν καὶ μέλι ἑνώσας, ἄλειφε τὸν τόπον. 6 ἄλλο. [⊳om. M]λυπηναρίων ἀλεύριν καὶ ἄρκου χολὴν μίξον καὶ θέτε εἰς τὸν τόπον.[⌝om. M⊲]

234 [171] [om. A]ςλδ΄ ⌜πρὸς βουβῶνας ἤτοι ἀπορυφᾶς· ἀσβέστιν ζῶντα καὶ μέλι καὶ σαπούνιν μάλαξον καὶ θὲς ἐπάνω.⌝

235 [172] [om. A]ςλε΄ ⌜πρὸς βιασμούς· ῥεπανίου σπόρον ψιλὸν καὶ κηκίδιν βράσον με τὸ κρασὶν καὶ πότισον. 2 ἄλλο. ῥοϊδίων φλούδια καὶ πίσσαν ξηρὰν κάπνιζε ὑποκάτω. 3 ἄλλο. ὀρίγανον βράσον με τὸ κρασὶν καὶ πότισον.⌝

236 [om. A]ςλϛ΄ ὅπου καταπίῃ βδέλλας· ὄξος ζεστὸν πότισον. 2 ἄλλο. κόριδας θὲς ἐπάνω τῶν καρβουνίων καὶ κάπνιζε εἰς τὸ στόμα. καὶ τὸν καπνὸν ἂς καταπίνῃ. 3 ἄλλο. βδέλλας καὶ ὄφεις ἐκβάλλει ἔλαιον ἄγουρον ποτιζόμενον. 4 ἄλλο. [⊳om. L]βούτυρον καὶ ὄξος μίξον. εἶτα πύρωσον σίδηρον καλὰ καὶ βάλε το ἀπέσω καὶ ἂς σβεσθῇ. εἶτα

---

17 μίξον] μίξας B 1 στυππεῖα] στυππεῖον M συπτίον B 1 τὸ στυππεῖον] om. MB 1 βάλε] ...ε L 1 σου] ἀπέσω add. B 1–2 εἰς τὸ στόμα] ἀπέσω M om. B 2 ἔνθα] scripsi..θα L ὅπου M ἔνι add. B 2 σῆψις] τῶν οὐλῶν add. M 2 ἔνι] om. B 3 χαρτὶν] χάρτην B 3 στάκτην των] scripsi στάκτου τῶν L στάκτην τούτων M στάκτην B 3 βάλε καὶ] με τὸ M μίξον με τὸ B 3 πέπερι] τὸ add. B 3–4 μάλαξον μετὰ μέλιτος] scripsi ..άξον μετὰ μέλιτ.. L τὸ μέλι M τὸ μέλι καὶ μαλάξας θέτε B 4 καὶ] om. MB 4 εἰς τὴν σῆψιν] scripsi ....ọυψιν L ἐπάνω M εἰς τὴν σῆτιν B 4 ὑγράργυρον] ὑδράργυρον M ὑδάργυρον B 5 ψιλὸν καὶ] ὑψη...... L 5 ἑνώσας] μίξας B 5 τὸν τόπον] ἐν τῷ τόπῳ MB 5 λυπηναρίων] λουπηναρίων B 8 μέλι καὶ σαπούνιν] transp. B 8 μάλαξον] μάλαξε M μαλάξας B 8 καὶ] om. MB 8 ἐπάνω] εἰς τὰς ἀπορυφὰς add. M 9 βιασμούς] τὸ πάθος τῶν βιασμῶν B 10 φλούδια] ξηρὰ ἢ χλωρὰ add. M φύλλα ἢ χλωρὰ ἢ ξηρὰ B 12 πότισον] ἂς το πίνῃ B 13 ὅπου καταπίῃ βδέλλας] scripsi ὅπου καταπίẹι ...λας L πρὸς εἴτις καταπίει βδέλλας M πρὸς ἐὰν καταπίῃ ἀβδέλλας B 13 ὄξος] τοξείδιν B 13 ζεστὸν πότισον] transp. B 14 θὲς] ... L 14 καρβουνίων] καρβούνων B 14 κάπνιζε εἰς τὸ στόμα] καὶ καπνὸν ἀπὸ τὸ λούτρων ἂς χασκήσῃ B 15 καὶ τὸν καπνὸν ἂς καταπίνῃ] transp. B 15 βδέλλας] πρὸς βδέλλας καὶ ὄφεις· βδέλλας B 16 ἄγουρον] ἄωρον MB 17 ἂς σβεσθῇ] scripsi ἀσβεσθῇ MB

---

15 βδέλλας] B has here a new chapter.

πότισον καὶ τὸν βούτυρον καὶ τὸ ὄξος.om. L◁

237 om. Aςλζʹ γάλα γυναικὸς να καταβάσῃ· κοπάνισον μελάνθιον καὶ πότισον μετὰ οἴνου ἐν βαλανίῳ. 2 ἄλλο. τήλης σπόρον κοπάνισον καὶ θέτε ἐπάνω εἰς τὰ βυζία. 3 ἄλλο. κρομμύδιν ἃς τρώῃ. 4 ἄλλο. ▷om. Bλουπηνάρια τὰ λεγόμενα θέρμια ἃς τρώῃ.om. B◁ 5 ἄλλο. μαράθρου ῥίζαν βράσον με τὸ κρασὶν καὶ πότισον. 6 ἄλλο. τζουκνίδας κλωνάρια βράσον με τὸ νερὸν καὶ πότισον. 7 ἄλλο. ἡδύοσμον κοπάνισον καὶ τὸ χυλὸν ἄλειφε τὰ βυζία.

238 [173] om. Aςληʹ ὅπου φαίνηται τὰ γόνατά του, ὅτι εἶναι κρύα ὡς ἂν πάγος· ⌜μίξον ἀσβέστιν ψιλὸν καὶ ὑγροπίσσιν καὶ μάλαξέ τα καλῶς καὶ ἄπλενέ το εἰς τὸ πανὶν καὶ θέτε ἐπάνω. 2 ἄλλο. δάφνης κοκκία κοπάνισον καὶ κοσκίνησον καὶ μάλαξον μετὰ θαλασσίου νεροῦ καὶ θέτε εἰς πανὶν καὶ θέτε ἔνθα ἡ ὀδύνη τῶν γονάτων ἐστὶν⌝ ▷Mπλὴν πρῶτον κοσκίνησον τὰ κοκκία τῆς δάφνης.M◁

239 σλθʹ πρὸς δυσουρίαν· ▷om. A▷om. Lκοπάνισον ῥίζαν μαιουλίου καὶ πότισον πρωῒ με τὸ νερόν.om. L◁ 2 ἄλλο. καρδάμου σπόρον κοπάνισον καὶ πότισον μετὰ οἴνου. 3 ἄλλο. ▷om. Lκορίδας κοπάνι-

---

17 εἶτα] καὶ τότε B 2 γάλα γυναικὸς να καταβάσῃ] scripsi γάλα ..... ἐὰν καταβάζει L πρὸς γάλα γυναικὸς καταβάσαι M πρὸς γάλα γυναικὸς ἐὰν θέλει να κατεβάσῃ ποίησον θεραπείαν τοιαύτην B 2 μελάνθιον] μ...ιον L μελάνθιν B 3 μετὰ οἴνου] om. B 3 ἐν] τῷ add. M 3 βαλανίῳ] μετὰ οἴνου add. B 3 τήλης] ψήλεως B 3–4 κοπάνισον καὶ] κοπανισμένον M 4 θέτε] θὲς M 4 κρομμύδιν] κρομῳ.. L 4 τρώῃ] τρωγῃ M τρώγῃ συνεχῶς B 5 λουπηνάρια] λυπηνάρια M 5 θέρμια] θ..μια L 5 τρώῃ] τρώγῃ M 6 μαράθρου] μαλάθρου B 6 καὶ] om. B 6–7 τζουκνίδας] τζουκνιδ.....ξας L ἓξ add. M 7 κλωνάρια] ἓξ add. B 7–8 ἡδύοσμον κοπάνισον] transp. B 8 τὸ χυλὸν] τὸν ζωμὸν MB 9 ὅπου] πρὸς ὅταν B 9 φαίνηται] scripsi ....ίνεται L φένεται M φαίνονται B 9 του] om. B 9 εἶναι] εἰσὶ B 10 ὡς] ...... L οἷον B 10 ἂν] τὸ B 10 πάγος] παθὸν B γ s.l. 10 μίξον ἀσβέστιν ψιλὸν] transp. M μίξας ἀσβέστι ψιλὸν B 10 καὶ] με τὸ M 10 καὶ] om. B 10 μάλαξέ τα] scripsi μάλαξ. .α L μάλαξον M μάλαξον αὐτὰ B 11 το] τα MB 11 τὸ] om. MB 11 πανὶν] π..... ... L 11 θέτε] τα add. MB 12 κοκκία] κου..... L 12 κοσκίνησον καὶ] scripsi κοσκινη... καὶ L om. MB 12 μάλαξον] καλῶς add. M 13 θέτε] το add. MB 13 εἰς πανὶν] .....ανὶν L εἰς τὸ πανὶν B 13 θέτε] ἐπίθες MB 13 ἡ ὀδύνη] ἡ ὀδυ L ἔνι ἡ ὀδύνη M 13 τῶν γονάτων ἐστὶν] scripsi τῶν γονάτ.. ... L καὶ ἡ ψύξις τῶν γονάτων MB καὶ ἡ ψύξις τῶν γονάτων ἐστὶν B 15 δυσουρίαν] ο... ......αν L θεραπεία add. A 15 μαιουλίου] μαϊλίου M 16 πότισον] τὸ add. B 17 κοπάνισον] κοπάν..... L

σον καὶ βάλε τας εἰς τὴν τρύπαν τῆς ψωλῆς καὶ εὐθὺς θέλει κατουρήσειν.[om. L◁] 4 ἄλλο. σελίνου ῥίζαν καὶ ῥεφανίου καὶ πράσου βράσον μετὰ οἴνου καὶ πότισον. 5 ἄλλο. [▷om. B]μαράθου ῥίζαν καὶ σελίνου βράσον με τὸ κρασὶν τὸ παλαιὸν τὸ καλὸν καὶ πότισον.[om. B◁] 6 ἄλλο. ἂς τρώῃ καβούρια καὶ ἂς πίνῃ καὶ τὸ ζωμὸν αὐτῶν. 7 ἄλλο. [▷om. L]ἄσαρ βλησκούνιν καὶ πέπερι πότισον μετὰ οἴνου εἰς τὸ λουτρὸν πλὴν πρῶτον ἂς βράσουν.[om. L◁] 8 ἄλλο. σελίνου σπόρον καὶ σκελίδας ἑπτὰ σκορόδου τρίψας ἅμα τὰ δύο, ἕψησον με τὸ κρασὶν καὶ πότισον. 9 ἄλλο. [▷om. L]κοπάνισον βεττονίκην καὶ τὸν χυλὸν αὐτῆς πότισον. [om. L◁om. A◁] 10 ἄλλο. [▷om. B]χαμαίμηλα βράσον με τὸ νερὸν καὶ πότισον.[om. B◁] 11 ἄλλο. δαφνέλαιον πότισον μετὰ οἴνου ἐν λουετρῷ. 12 ἄλλο. ἀσπαράγγου ῥίζαν καὶ ἰτέας βράσον με τὸ κρασὶν τὸ ἄσπρον καὶ πότισον. 13 ἄλλο. πεντάνευρον καὶ ἀδίαντον βράσον με τὸ νερὸν καὶ ἄς το πίνῃ. 14 ἄλλο. κόψον τοῦ λυσσομαμούδου τοὺς πόδας καὶ τὰ πτερὰ καὶ τὴν κεφαλὴν ζύγισέ τα ἁ σώζουν κοκκία τέσσαρα καὶ πότισον.

240 σμʹ πρὸς δυσεντερίαν· τήλης σπόρον φρυγμένον ἐν ἡλίῳ καὶ

319.17–320.1 κοπάνισον] τζακύσας B 1 τας] om. B 1 τρύπαν] ἔμπροσθεν add. B 1–2 κατουρήσειν] scripsi κατουρῆσιν M κατουρίσει B 2 σελίνου ῥίζαν] transp. B 2–3 βράσον μετὰ οἴνου] βράσον με... οἴνου L βάλε καὶ κρασὶν καὶ ἂς βράσουν M βράσε με τὸ κρασὶν τὸ καλὸν τὸ παλαιὸν τὸ μυροδάτον B 3 μαράθου ῥίζαν] μαράθρου ῥι.. L 4 βράσον] ἂς βράσουν M 4 με τὸ] με M 5 τρώῃ] τρώγῃ MB 5 καβούρια] καὶ βούρους B 5 τὸ] τὸν B 5 ζωμὸν] ζωμὶν M 6 βλησκούνιν] γληχούνι B 6 πέπερι] βράσον καὶ add. B 6–7 εἰς τὸ λουτρὸν] ἐν τῷ λουτρῷ B 7 πλὴν πρῶτον ἂς βράσουν] om. B 7 σελίνου σπόρον] σελινόσπορον B 7–8 καὶ σκελίδας] κ.. κελίδας L σκελίδας M καὶ σκόροδα B 8 σκορόδου] σκορόδ.. L σκόροδον M σκελίδας B 8 ἅμα τὰ δύο] transp. B 8 ἅμα] ..μα L 9 καὶ πότισον] ...ότισον L 9 βεττονίκην] scripsi βετονίκιν B βετονίκην M 10 με τὸ] ....ο L 11 δαφνέλαιον] δαφνελαι...... L 11 μετὰ οἴνου] με τὸ κρασὶν Σ 12 λουετρῷ] τῷ λουτρῷ Σ 12 ἀσπαράγγου] α.....ύγγου L ἀσπαραγγίου M σπράγγου B 12 βράσον με τὸ] βρά......ο L βράσον μετὰ B 13 κρασὶν] οἴνου B 13 τὸ ἄσπρον] om. Σ 13 πεντάνευρον] ...τάνευρον L κεντάνευρον M 14 ἄς το πίνῃ] πότισον MB ἂς πίνῃ A 14–15 κόψον τοῦ λυσσομαμούδου] ..ψο.. .... μούδου L κόψον τοῦ λυσσομουμούδου B 15 τὰ πτερὰ καὶ τὴν κεφαλὴν] transp. B 15 τὰ πτερὰ καὶ] ..... L 15 ζύγισέ] .....υγησε L ζυγήσε A 15–16 ἁ σώζουν] ......ζοὺν L σώσουν A νασό.ουν B 16 κοκκία] κουκκία B 17 τήλης σπόρον] τὸν σπόρον τῆς τήλεως B

φακὶν καὶ κηκίδιν καὶ κύμινον. τὰ πάντα κοπανίσας πότισον μετὰ οἴνου χλιαροῦ τὸ πρωῒ νῆστιν. 2 ἄλλο. τῆς βεττονικῆς τὸν χυλὸν πότισον μετὰ οἴνου καὶ νεροῦ χλιαροῦ τὸ πρωΐ.

241 σμα′ τὰ συνήθη τῶν γυναικῶν καὶ οὖρα καταβάζει καὶ παῖδας ἐκβάλλει ἀπὸ τῆς μήτρας καὶ σπλῆναν ὠφελεῖ· βότανον λεγόμενον χαμαίδρυον τοῦτο βράσον μετὰ οἴνου καὶ πότισον. 2 ἄλλο. χαλβάνιν θυμίαζε ὑποκάτω. 3 ἄλλο. ▷om. MA χαλβάνιν τρίψας θέτε ἔξω εἰς τὸ κρυπτὸν τῆς γυναικός om. MA◁. 2 ἄλλο. λουλάκιν καὶ μελάνθιν τρίψας om. L▷ μάλαξον μετὰ μέλιτος καὶ ποιήσας πανὶν ὡσὰν κόμπον βάλε ἀπέσω καὶ ῥάψε τα εἶτα δός τα ἄλλην γυναῖκα καὶ ἄς τα βάλλῃ ἀπέσω εἰς τὴν μήτραν με τὴν χεῖραν της. εἰς δὲ τὸν κόμπον δῆσε ῥάμμα καὶ ὅτε τὸ θέλης ἐκβάλλειν να ταυρίσῃς τὸ ῥάμμα καὶ να ἐκβάλῃς αὐτὸ, τοῦτο δὲ παρὰ τῶν ἰατρῶν καλεῖται πεσσός. 3 ἄλλο. ▷om. A χαλβάνην χλωρὴν ποίησον ὁμοίως ἐν πέσσῳ. om. A◁om. L◁

242 [174] om. L σμβ′ ⌜περὶ ἐσοχάδας καὶ ἐξοχάδας· ὀφιδίου δέρμα

15 περὶ] This chapter is listed in the pinax to L.

17 φρυγμένον] ..γμένον L ἔπαρον καὶ φρύξον B 1 φακὶν] φακὴν MA 1 κύμινον] καὶ add. B 1 τὰ πάντα κοπανίσας] transp. B 1 κοπανίσας] κοπάνισον M 2 τὸ] .. L κατὰ B 2 βεττονικῆς] scripsi βετονικῆς ΛMAB 3 καὶ νεροῦ] καὶ νερ.. L om. MA 3 χλιαροῦ] χ̣λ̣ι̣... L χλιεροῦ M χλίου B 3 τὸ] .. L 3 πρωῒ] νηστικὸν add. M νῆστιν add. A 4 τὰ συνήθη τῶν] πρὸς τὰ λεγόμενα συνήθεια M τὰ λεγόμενα συνήθη τῶν A πρὸς τὸ καταβάσαι τὰ λεγόμενα συνήθια τῶν B 4 τῶν] ... L 4 καταβάζει] καταβιβάζει A om. B 4 παῖδας] ....ας L παιδία Σ 5 τῆς] om. B 5 καὶ] ποίησον θεραπείαν τοιαύτην B 5 σπλῆναν] ..... L σπλῆνα A om. B 5 βότανον] τὸ βοτάνιν τὸ Σ 6 χαμαίδρυον] scripsi χαμα....... L χαμέδριον MA χαμαίδριον B 6 τοῦτο βράσον] πινόμενον βρεμένον B 6 καὶ] om. B 6 πότισον] πο....... L φελεῖ δὲ αὐτὸ καὶ τὴν σπλῆνα τὸ B 7 ἔξω] ὑποκάτω B 8 εἰς τὸ κρυπτὸν] εἰ .....μα L 8–9 λουλάκιν καὶ μελάνθιν τρίψας] δωλακ.......θιν τρίψ.. καὶ ψιλὸν L 8 λουλάκιν] λουλλάκιν A ζουλάκιν B 9 τρίψας] καὶ ψιλὸν ποιήσας B 9 ποιήσας πανὶν ὡσὰν κόμπον] om. B 10 βάλε] τα add. A τὸ add. B 10 ἀπέσω] εἰς πανὶν B 10 τα] ὡς ἂν κόμπον B 10 δός] ἂς B 10 τα] βάλλῃ B 10–11 καὶ ἄς τα βάλλῃ ἀπέσω] om. B 11 με τὴν χεῖραν] om. B 11 δῆσε] θῆσε A δέσον B 12 ῥάμμα] ῥάμα AB 12 καὶ ὅτε] ἵνα ὅταν B 12 τὸ] om. AB 12 θέλης] θέλει B 12 ἐκβάλλειν] να τὸ εὐγάλῃς A 12 ῥάμμα] ῥάμαν M ῥάμα AB 13 ἐκβάλῃς αὐτὸ] τὸ εὐγάλῃς A 13 παρὰ τῶν ἰατρῶν καλεῖται] καλεῖται παρὰ τῶν ἰατρῶν A 14 χαλβάνην] χαλβάνιν B 14 χλωρὴν] χλωρὸν B 15 περὶ] om. A πρὸς B 15 ἐξοχάδας] θεραπεύει add. A 15 ὀφιδίου] ὄφεως B

βρέξον εἰς τὸ χρηστέλαιον καὶ θές το ἐπάνω εἰς τὰ κάρβουνα τὰ ἁπτόμενα καὶ κάπνιζε,⌝ ὅπως ὁ καπνὸς ἀνέρχεται ἐκεῖ εἰς τὸν πάσχοντα τόπον. 2 ἄλλο. ⌜τὸ φλοῦδιν τῆς ῥίζης τῆς παλιουρέας βράσον με τὸ κρασὶν καὶ πότισον ἀπέσω εἰς τὸ λουτρόν. καὶ ἂς ἔνι νηστικὸς ἕως ὥρας τέσσαρις.⌝ 3 ἄλλο. λυπηναρίων ἀλεύριν κοσκινισμένον πάσον ἐπάνω πλὴν πρῶτον ἀποσφόγγισέ τας με τὸ κρασίν. 4 ἄλλο. δάφνης τὸ σπέρμα τῶν κοκκίων κοπάνισον καὶ ἔκβαλε τὸν ζωμὸν καὶ τὴν καρδίαν τῶν κοκκίων καὶ ποίησον αὐτὰ ψιλὰ καὶ πάσον ἐπάνω. 5 ἄλλο. ▷om. A⌜θύννας κεφαλὴν ἢ παλαμύδας παστῆς ἀπὸ τὰς δύο οἷαν εὕρης καῦσον καὶ τὴν στάκτην ταύτης πάσον ἐπάνω.⌝om. A◁ 6 ἄλλο. ξηρὸν ἵππουριν τὸ λεγόμενον πολυκόμπιν τρίψον καὶ πότισον μετὰ οἴνου. 7 ἄλλο. ⌜τὴν ῥίζαν τοῦ πολυκόμπου τηγάνισον με τὸ χρηστέλαιον καὶ ἄλειφε. 8 ἄλλο. ψιμμίθιον τρίψον καὶ μίξον αὐτὸ με τὸ γυναίκειον γάλα καὶ ἄλειφε. 9 ἄλλο. κοπάνισον ῥίζαν κρίνου καὶ μίξον μετὰ κριθάλευρον βάλε καὶ νερὸν καὶ ἕψησον καὶ ἄλωνέ τα εἰς πανὶν καὶ θέτε ἐπάνω.⌝ 10 ἄλλο. ἀειδάρας γάλα καὶ καταστατὸν τριμμένον μίξον καὶ ἄλειφε. 11 ἄλλο. ▷om. Aκοπάνισον κρόκον καὶ συρικὸν καὶ ἑνώσας ἄλειφε.om. A◁ 12 ἄλλο. συκαμινεάς σπόρον καὶ πεντάνευρον βράσον μετὰ ὕδατος καὶ πότισον. 13 ἄλλο.

15 δέρμα] εὑρῶν add. B 1 βρέξον] βρέξε το A 1 τὰ] om. B 1 τὰ] om. B 2 ἁπτόμενα] ἁπτούμενα A 2 κάπνιζε] κάπνισον A πλὴν ὑποκάτω τοῦ σχουνίου βάλε τὰ ἁπτόμενα κάρβουνα add. B 2 καπνὸς] ἀτμὸς AB 5 ἕως] μέχρις A μέχρι B 5 τέσσαρις] τεσσάρεις B 5 λυπηναρίων ἀλεύριν] λουπηνάρια ποίησον ἀλεύριν B 5 κοσκινισμένον] κοπανισμένον B 6 πάσον ἐπάνω πλὴν] om. AB 6 πρῶτον] καὶ ἐκεῖνον B 6 ἀποσφόγγισέ τας] ἀποσφογγίσας A ἀποσπόγγιζε τὸ B 6 κρασίν] εἶτα πάσε τὸ ἀλεύριν ἐπάνω εἰς αὐτὰς add. A εἶτα πάσσον τὸ αὐτὸν ἀλεύριν ἐπάνω B 7 δάφνης τὸ σπέρμα τῶν κοκκίων κοπάνισον] transp. B 8 τῶν κοκκίων] τοῦ κοκκίου B 8 ποίησον] ποιήσας B 8 ψιλὰ] ψιλὸν B 8 καὶ] om. B 9 θύννας] θύνας L θύας B 9 παλαμύδας] παλαμίδας B 9 τὰς] τῶν B 10 οἷαν] οἷον B 10 καῦσον] καῦσέ την B 10 ταύτης] τούτων B 11 ξηρὸν] ξηρὰν AB 11 ἵππουριν] ἤτοι add. B 12 πολυκόμπου] πολυκόμπιου AB 12 τηγάνισον] τηγανισέ την A 13 ψιμμίθιον] ψιμμίθιν B 13 τρίψον καὶ] ξηρὸν τριμμένον B 14 αὐτὸ] om. A 14 γυναίκειον γάλα] transp. B 14 γυναίκειον] τὸ add. AB 14–15 κοπάνισον ῥίζαν κρίνου] transp. B 15 μετὰ] με τὸ AB 15 βάλε καὶ] καὶ βαλὼν B 16 τα] το AB 16 εἰς] τὸ add. B 16 θέτε] το add. AB 16 ἐπάνω] om. B

⌜ἀποσφούγγισε πρῶτον τὸν τόπον με τὸ ὄξος, εἶτα ἔχε καλάμου φύλλα κεκαυμένα καὶ πάσον ἐπάνω.⌝ 14 ἄλλο. σκύλλου κόπρον ξηρὰν λεῖαν τριμμένην ἄντλει δὲ πρῶτον τὸν τόπον οἴνῳ εἶτα πάσον αὐτὴν ἐπάνω. 15 ἄλλο. ⌜καῦσον κρόκον αὐγοῦ καὶ πάσον ἐπάνω.⌝

243 [175] om. Lσμγʹ πρὸς εἴ τινος ἀνθρώπου ὁ κῶλος εὐγένει· ⌜ἔπαρον καταστατὸν τριμμένον καὶ ψιμμίθιον καὶ βάλε κρασὶν καὶ μάλαξον εἶτα ἄλειφε. 2 ἄλλο. μελίλωτα καὶ ῥόδα ξηρὰ καὶ τριμμένον ψιμμίθιον. βάλε καὶ κρασὶν καὶ βράσε τα εἶτα ἄλειφε τοὺς τόπους. 3 ἄλλο. ἀμανιτάριν ἕψησον μετὰ ὕδατος καὶ κοπάνισον αὐτὸ καὶ μίξον μετὰ βούτυρον καὶ μάλαξον καὶ ἄλειφε τὸν τόπον με τὸν δάκτυλόν σου. 4 ἄλλο. ἄλειψον πρότερον τὸν τόπον κρασὶν εἶτα ἔχε τριμμένον κέγχρου τὸ πίτυρον ψιλὸν κοσκινισμένον καὶ οὕτως πάσον ἐπάνω καὶ τότε πάλιν βάλε τὸν κῶλον του ἀπέσω ἀλειμμένον ὄντα μετὰ πίτυρα τῆς κέγχρου.⌝

244 om. Lσμδʹ ὅπου ἔχῃ πλατέας ἕλμινθας, ποίησον θεραπείαν τοιαύτην· καρδαμόσπορον κοπάνισον καὶ πεπέρι τριμμένον πότισον μετὰ νερὸν τὸ πρωῒ νῆστιν. 2 ἄλλο. κοπάνισον μελάνθιν καὶ πότι-

---

6 πρὸς] This chapter is listed in the pinax to L.

16–17 ἀειδάρας γάλα καὶ καταστατὸν τριμμένον μίξον] transp. B 1 ἀποσφούγγισε] ἀποσφούγγιζε B 1 πρῶτον] πρότερον B 1 με τὸ ὄξος] μετὰ ὄξους B 1–2 καλάμου φύλλα] καλαμόφυλλα B 2 πάσον] πάσε τα B 2 σκύλλου] τοῦ σκύλλου τὸ B 3 λεῖαν] om. B 3 ἄντλει δὲ πρῶτον τὸν τόπον οἴνῳ εἶτα] om. B 4 αὐτὴν] om. B 4 ἐπάνω] om. AB 4 πάσον] τοῦτον add. B 5 ἐπάνω] ἤστις δὲ ἔχεις ταύτας μηδόλως τρῶγε ἐρεβίνθια B 6 πρὸς] om. A 6 εἴ] εἂν B 6 ἀνθρώπου ὁ] om. B 8 ἄλειφε] τὸν κῶλον add. A 8–9 τριμμένον] om. A 9 ψιμμίθιον] ψιμμίνθιν B 9 εἶτα] καὶ A καὶ ἀπὸ τοῦτο B 11 μετὰ] om. AB 11 καὶ] om. A 11 μάλαξον] δὲ καὶ τὰ δύο add. A δὲ τὰ δύο B 12 πρότερον] πρῶτον A 13 τριμμένον] τριμμένα τῆς A τῆς add. B 13 τὸ πίτυρον] πίτυρα A 13 ψιλὸν] ψιλὰ A 13 κοσκινισμένον] κοσκινισμένα A 14 οὕτως] om. AB 14 πάσον] πάσσε τα A αὐτὰ add. B 14 ἐπάνω] ἀπάνω A 15 ὄντα] om. B 15 τῆς κέγχρου] om. AB 16 ὅπου] This chapter is listed in the pinax to L. 16 ἕλμινθας] σκώληκας ἤτοι ἕλμινθας A σκώληκας ἤτοι ἕλμινθας τὰς λεγομένας ἑρμίγγια B 16–17 τοιαύτην] ἤγουν add. B 17 καρδαμόσπορον κοπάνισον] transp. B 17 καὶ] μίξον καὶ add. B 17 τριμμένον] καὶ add. B 18 μετὰ] με τὸ A 18 νερὸν] ὕδατος B 18 τὸ πρωῒ] om. A

σον με τὸ κρασὶν τὸ ἄκρατον ἢ με τὸ γάρος. 3 ἄλλο. κοπάνισον χαλβάνιν καὶ ▷om. ABπότισον με τὸ κρασίν. 4 ἄλλο. χαλβάνιν κοπάνισον καὶom. AB◁ θὲς εἰς τὸν ὀμφαλὸν τοῦ ἀνθρώπου. 5 ἄλλο. κόστου σπόρον ἢ τὰ φύλλα κοπάνισον καὶ πότισον με τὸ νερόν. 6 ἄλλο. σκόροδον με τὸ χρηστέλαιον τὸ παλαιὸν ἃς τρώγῃ. 7 ἄλλο. κοπάνισον ἡδύοσμον χλωρὸν καὶ τὸν χυλὸν πότισον τὸ πρωΐ. 8 ἄλλο. κύμινον τριμμένον μίξον με τὴν χολὴν ταύρου καὶ θὲς εἰς τὸν ὀμφαλόν. 9 ἄλλο. λυπηναρίων ἀλεύριν μάλαξον με τὸ μέλι καὶ θὲς ἐπάνω. 10 ἄλλο. ▷om. Aκοπάνισον κολίανδρον ψιλὸν καὶ ἕνωσον με τὸ χρηστέλαιον καὶ θὲς εἰς τὸν ὀμφαλόν.om. A◁ 11 ἄλλο. ▷om. Bκοπάνισον καναβούριν καὶ πότισον με τὸ νερόν.om. B◁ 12 ἄλλο. κράμβης καυλία τρυφερὰ γουλία ἤτοι τὰς κορυφὰς βράσον με τὸ νερὸν τὸ καθαρὸν εἶτα ἔπαρον ἀπ' αὐτοῦ τοῦ νεροῦ ἐξάγιον ἓν καὶ ἅλας χοντρὸν ἐξάγιον ἓν καὶ τὸν σπόρον τῶν σκωλίκων θέλει εὐγάλην. 13 ἄλλο. λυπηναρίων ἀλεύριν καὶ ἀψίνθιον κοπάνισον καὶ βάλε δαφνέλαιον καὶ μαλάξας ἅμα ὅλα ἐπίθες ἐπὶ τὸν ὀμφαλόν.

245 [176] om. Lσμε′ εἰς τὰ ἐκβατὰ καὶ εἰς τὰς ἀπορυφάς· ⸢ὠφελεῖ σεύτλον καὶ τὸν χυλὸν μίξον με τὸ σίτινον ἀλεύριν καὶ μαλάξας ἐν τῷ ἅμα θὲς ἐπάνω. 2 ἄλλο. ▷Bἰξόν, ὑαλὶν τριμμένον καὶ σαπώνιν καὶ χρηστέλαιον καὶ μαλάξας θές τα ἐπάνω.B◁ 3 ἄλλο. κοπάνισον

18 νῆστιν] νηστικὸν B 1 γάρος] .... A 3 θὲς] το add. A 3 ὀμφαλὸν] ὀφαλὸν Σ 3 ἀνθρώπου] ἢ πότισον με τὸν κρασὶν A 4 κόστου σπόρον ἢ τὰ φύλλα] transp. B 5 σκόροδον με τὸ χρηστέλαιον τὸ παλαιὸν ἃς τρώγῃ] transp. B 5 τὸ παλαιὸν] om. A 6 τὸν χυλὸν πότισον] transp. B 7 χολὴν] τοῦ add. A 7 θὲς] θέτε την A 8 ὀμφαλόν] ὀφαλόν MA 9 θὲς ἐπάνω] θέτε τον εἰς τὸν ὀφ.. A 9 κολίανδρον] κολίαντρον καὶ B 9 ψιλὸν] ποιήσας add. B 9–10 καὶ ἕνωσον] μίξον B 10 θὲς] το add. B 10 ὀμφαλόν] ὀφαλόν M 12 καυλία] γουλία A 12 τρυφερὰ] ἤτοι τὰ add. B 12 γουλία] om. A 12 ἤτοι] om. B 13 εἶτα] καὶ AB 13 ἀπ'] ἐξ B 14 χοντρὸν] χονδρὸν A 14 καὶ] πότισον τὸν ἄνθρωπον add. B 14–15 τὸν σπόρον τῶν σκωλίκων θέλει εὐγάλην] ὅπως ἐκβάλῃς τὸν σπόρον τῶν σκωλίκων B 15 εὐγάλην] εὐγάλει A 15–16 λυπηναρίων ἀλεύριν καὶ ἀψίνθιον κοπάνισον καὶ βάλε δαφνέλαιον] transp. B 16 μαλάξας] μίξας A 16 ἐπίθες] θές τα A 16 ἐπὶ] εἰς A 17 εἰς] This chapter is listed in the pinax to L. 17 τὰ] om. B 17 εἰς] om. AB 17 ὠφελεῖ] om. B 18 σεύτλον] σεύκλον A σεύτλου B 18 καὶ] om. B 18 σίτινον ἀλεύριν] transp. B 18 μαλάξας] μάλαξον A 19 θὲς] καὶ τὰ δύο καὶ θέτε τα A τὰ δύο θέτε τα B

φύλλα χλωρὰ καλάμου καὶ τὸν χυλὸν μίξον με τὸ μολύβιν τὸ ψιλὸν πολλὰ ῥινισμένον καὶ μαλάξας θὲς ἐπάνω. 4 ἄλλο. καρπὶν μαῦρον τριμμένον μίξον με τὴν πρόπολιν καὶ μάλαξον καὶ θὲς ἐπάνω.⌉ 5 ἄλλο. ἀγριοπεριστερίου κόπρον τρίψας καὶ ἑνώσας καὶ μαλάξας με τὸ μέλι θὲς ἐπάνω. 6 ἄλλο. [▷B]χηναρίου κόπρον τρίψον καὶ ἑνώσας καὶ μαλάξας με τὸ μέλι θὲς ἐπάνω.[B◁] 7 ἄλλο. ⌈κοπάνισον ῥίζαν λαπάθου καὶ μάλαξον αὐτὴν μετὰ χοίρειον ἀξούγγιν καὶ θές την ἐπάνω.⌉

246 [om. L]ςμϛ′ ὅπου κατουρῇ τὴν νύκτα παρὰ τὸ θέλημά του· ποίησον θεραπείαν τοιαύτην ἤγουν· βράσον ποταμογείτονα με τὸ κρασὶν καὶ πότισον αὐτόν. 2 ἄλλο. σμύρναν τρίψον καὶ τὸν ζωμὸν τοῦ ποταμογείτονος με τὸ κρασὶν καὶ πότισον. 3 ἄλλο. ὄνυχας μαύρου χοίρου κεκαυμένας πότισον μετὰ οἴνου. 4 ἄλλο. κύστιν αἰγείαν καῦσον καὶ πότισον με τὸ ὄξος καὶ τὸ κρασίν. 5 ἄλλο. χηναρίου γλῶσσαν ἂς τρώγῃ. 6 ἄλλο. κοπάνισον δάφνης κοκκία καὶ πηγάνου σπόρον καὶ μίξον καὶ ἕνωσον με τὸ ῥοδέλαιον καὶ ἄλειψον τὸ αἰδοῖον τοῦ ἀνθρώπου. 7 ἄλλο. ὀρχίδια λαγωοῦ πότισον με τὸ κρασὶν τὸ καλόν. 8 ἄλλο. πηγάνου ἀγρίου σπόρον φρυγμένον εἰς ἥλιον ἂς τρώγῃ ἕως ἡμέρας τρεῖς. 9 ἄλλο. πετεινοῦ γούργουρον καῦσον καὶ τὴν στάκτην πότισον μετὰ οἴνου.

---

9 ὅπου] This chapter is listed in the pinax to L.

---

1 μολύβιν] scripsi μολύβην M μολίβην A μολίμβιν B 2 ῥινισμένον] ῥενισμένον B 2 καὶ μαλάξας] μάλαξον καὶ A μαλάξας B 3 μάλαξον καὶ θὲς] μαλάξας θέτε τα B 4 ἀγριοπεριστερίου] ἀγριοπεριστέρου B 4 καὶ ἑνώσας] om. B 7 μάλαξον] μαλάξας B 7 μετὰ] με τὸ AB 7 χοίρειον ἀξούγγιν] transp. B 7 καὶ] om. B 7 την] om. A 9 ὅπου] εἴτις A πρὸς ὅταν B 9 θέλημά του] θέλημα αὐτοῦ B 10 ἤγουν] om. AB 10 βράσον] om. B 10 ποταμογείτονα] ποταγείτωνα M ποταμογείτονος φύλλα βράσον B 11 αὐτόν] σὺν αὐτὸ B 11 καὶ] βάλε add. B 12 με τὸ] βάλε καὶ A καὶ B 12 καὶ] om. B 13 χοίρου] om. M 13 μετὰ οἴνου] om. B 13 κύστιν] ὕστην M 13 αἰγείαν] αἰγίαν B 14 ὄξος] ξείδιν B 14 ἄλλο] καὶ πότε A 14–15 χηναρίου γλῶσσαν ἂς τρώγῃ] transp. B 14 χηναρίου] χοιρίου B 16 καὶ ἕνωσον] om. AB 16 ἄλειψον] ἄλειφε B 17 ὀρχίδια] ὀρχίδιν B 18 τὸ καλόν] om. A 18 φρυγμένον εἰς ἥλιον] transp. B 19 ἕως] om. A 19 πετεινοῦ γούργουρον καῦσον] transp. B 19 γούργουρον] γούργουρα B

---

19 ἰξόν] The following three paragraphs are reversed in B.

247 [177] om. L ςμζ′ ⌜ὅπου δάκῃ τινὰ ἔχιδνα· ▷om. B πήγανον κοπάνισον καὶ τὸν χυλὸν πότισον. om. B◁ 2 ἄλλο. τὸ λεγόμενον παρὰ τῶν ἰατρῶν ἄμωμον τοῦτο τρίψον καὶ πότισον με τὸν χυλὸν τοῦ πηγάνου.⌝ 3 ἄλλο. σινάπιν καὶ καρδαμόσπορον τριμμένα μίξον με τὸ ὄξος καὶ ἄλειφε.

248 [179] om. L ςμη′ ⌜πρὸς ἡμίκρανον· γῆς ἔντερα ἤτοι σκωλήκια τὰ μακρέα ὅπου πιάνουν μετ' αὐτὰ οἱ ἄνθρωποι τὰ ὀψάρια ταῦτα φρύξον εἰς τὴν ἑστίαν καὶ πεπέρι κοκκία ιε' τριμμένα μίξον με τὸ ὄξος καὶ ἄλειφε τὴν κεφαλήν. 2 ἄλλο. ἀγριοπεριστερᾶς κόπρον τριμμένον βάλε ὄξος καὶ ἄλειφε. 3 ἄλλο. λυγαίας σπόρον ἕψησον με τὸ χρηστέλαιον καὶ κοπάνισον καὶ ἄλειφε τὸ μέτωπον καὶ τοὺς μήνιγγας. 4 ἄλλο. ἀσβέστιν ζῶντα μάλαξον με τὸ μέλι καὶ ἄλειφε.⌝ 5 ἄλλο. ▷om. A πηγάνου καρδίαν βάλε εἰς τὸ ὠτίον καὶ τὴν κεφαλὴν κατάσφιγγξε καὶ σκέπασον καλά. om. A◁ 6 ἄλλο. ⌜κοπάνισον ἀμύγδαλα πικρὰ καὶ μαστίχιν τριμμένον καὶ μίξον με τὸ ὄξος καὶ ἄλειφε τὸ μέτωπον.⌝ 7 ἄλλο. βλαστάρια τρυφερὰ κισσοῦ κοπάνισον καὶ τὸν χυλὸν ἕνωσον μετὰ ῥοδέλαιον καὶ τὸ ὄξος ἄλειφε. 8 ἄλλο. ▷om. A φύλλα ξηρὰ ῥοδάφνης θὲς εἰς τὰ καρβούνια τὰ ἁπτόμενα καὶ κάπνιζε εἰς τὰ ῥωθώνια. om. A◁

249 [180] om. L ςμθ′ ὅπου πονεῖ τὸ ἰσχίον του· ⌜βράσον ἐρεβίνθια

---

1 ὅπου] This chapter is listed in the pinax to L. 6 πρὸς] This chapter is listed in the pinax to L. 20 ὅπου] This chapter is listed in the pinax to L.

---

1 ἔχιδνα] ποίησον θεραπείαν add. A 2 τὸ] τρίψον τὸ B 3 τοῦτο τρίψον] om. B 4–5 με τὸ ὄξος] μετὰ ὄξους B 5 ἄλειφε] τὴν πληγὴν add. A 6 ἤτοι] ἤγουν A 6–7 σκωλήκια τὰ μακρέα] transp. B 7 τὰ μακρέα] om. A τὰ μικρὰ B 7 μετ' αὐτὰ οἱ ἄνθρωποι] om. B 7 ὀψάρια] μετ' αὐτῶν B 8 ἑστίαν] scripsi ἱστίαν Σ 8 πεπέρι κοκκία] κοκκία AB 8 ιε'] πέπερι add. AB 8 τριμμένα] om. B 8 μίξον] καὶ μίξας B 8–9 με τὸ ὄξος] ὄξους A μετὰ ὄξους B 9 καὶ] om. B 10 βάλε ὄξος καὶ] om. B 10 ἄλειφε] μετὰ ὄξους add. B 10 λυγαίας] scripsi λιγέας MB λιγαρέας A 11 με τὸ χρηστέλαιον] μετὰ χρηστελαίου A 12 ζῶντα] ἄβροχον A 14 κατάσφιγγξε] κατάσφιγξον B 15 μαστίχιν] μαστίχην A 15 με τὸ ὄξος] μετὰ ὄξους B 16 βλαστάρια τρυφερὰ κισσοῦ κοπάνισον] transp. A κοπάνισον βλαστάρια τρυφερὰ B 17 ἕνωσον] μίξον AB 17 μετὰ] με τὸ A 17 ἄλειφε] καὶ ἄλειφέ την A καὶ ἄλειφε B 18 καρβούνια] κάρβουνα B 19 κάπνιζε] τὸν add. B 19 ῥωθώνια] ῥουθούνια B 20 ὅπου] εἴ τις A πρὸς B 20 πονεῖ] om. B 20 τὸ] λεγόμενον add. A om. B 20 ἰσχίον] ἰσχίου B 20 του] ὠφελεῖ τοῦτον ἤγουν A om. B

με τὸ νερὸν καὶ ἃς τρώγῃ καὶ ἃς πίνῃ τὸν ζωμόν.⌝ 2 ἄλλο. ▷om. A τῆς λεγομένης ζέας τὰ φλούδια βράσον με τὸ νερὸν καὶ πότισον.om. A◁ 3 ἄλλο. ▷om. B⌜σινάπιν κοπάνισον καὶ πεπέρι καὶ μίξον με τὸ μέλι καὶ ἄλειφε.⌝om. B◁ 4 ἄλλο. κυκλαμίνου ῥίζαν ἢ δάφνης δῆσον εἰς τὸν τόπον. 5 ἄλλο. ▷om. B⌜ἀλόην τρίψον καὶ ῥητίνην καὶ πεπέρι καὶ θὲς ἐπάνω εἰς τὸν τόπον πλὴνom. B◁ πρότερον ἄλειψε τὸν τόπον με τὸ μέλι.⌝ 6 ἄλλο. φύλλα κράμβης καῦσον καὶ τὴν στάκτην μίξον με τὸ ἀξούγγιν τὸ χοίρειον καὶ θὲς ἐπάνω. 7 ἄλλο. χρηστέλαιον καὶ λιπαρὸν δᾳδὶν βράσον καὶ ἄλειφε τὸν τόπον. 8 ἄλλο. τὴν λεγομένην ἄφθαν ἄλειψον τὸν τόπον ἔνθα ἔνι ἡ ὀδύνη. 9 ἄλλο. ⌜ἑψημένα σκόροδα με τὸ χρηστέλαιον καὶ τὸ γάρος ἃς τρώγῃ.⌝ 10 ἄλλο. ▷om. A κοπάνισον βλησκούνιν καὶ πεπέρι καὶ μάλαξον με τὸ νερὸν καὶ θές τα ἐπάνω.om. A◁ 11 ἄλλο. ⌜βούτυρον μίξον με τὸ ἀνηθέλαιον καὶ ἄλειφε τὸν τόπον.⌝ 12 ἄλλο. κλῆμαν καῦσον καὶ τὴν στάκτην μίξον με τὸ ὑγροπίσσιν καὶ θὲς ἐπάνω. 13 ἄλλο. ▷B κοπάνισον τὸ λεγόμενον βρωμοβότανον καὶ τὸν τούτου χυλὸν μίξον μετὰ πίτυρα καὶ θές τα ἐπάνω.B◁ 14 ἄλλο. ▷om. A κυκλαμινου ῥίζαν κοπάνισον καὶ θές την ἐπάνω. κυκλαμίνου ῥίζαν κρέμασον εἰς τὸν τράχηλον αὐτοῦ.om. A◁ 15 ἄλλο. δρακοντίας ῥίζαν κοπάνισον καὶ θές την ἐπάνω. ῥίζαν δρακοντίας κοπάνισον καὶ τὸν χυλὸν πότισον.

250 om. L ςν΄ περὶ ἰκτερικοὺς ἤτοι ὅπου κυτρινιάσῃ τὸ πρόσωπόν τους· κοπάνισον ἀρσενίκιν καὶ ἕνωσον με τὸ κρόκον τοῦ αὐγοῦ καὶ πότισον ἀπέσω εἰς τὸ λουτρόν. 2 ἄλλο. κοπάνισον κενταύ-

---

22 περὶ] This chapter is listed in the pinax to L.

---

20 βράσον ἐρεβίνθια] transp. B 1 ἃς] τα add. B 1 πίνῃ] καὶ add. AB 3 με τὸ] om. A 5 ῥητίνην] ῥηθίνην A 6 πρότερον] δὲ add. B 6 ἄλειψε] ἄλειφε A 7 με τὸ] om. A 7 φύλλα] τὰ φύλλα τῆς B 7 στάκτην] τῶν φύλλων add. AB 8 θὲς] θέτε το AB 8 ἐπάνω] ἀπάνω A 9 τὸν τόπον] om. B 10 ἄλειψον] ἄλειφε B 11 ἑψημένα σκόροδα με τὸ χρηστέλαιον καὶ τὸ γάρος ἃς τρώγῃ] transp. B 11 καὶ τὸ γάρος] om. B 12 βλησκούνιν] γλησχούνιν B 14 τὸν τόπον] om. B 15 θὲς] θέτε το A το add. B 18 κυκλαμίνου] τὴν αὐτοῦ τὴν B 19 δρακοντίας] δρακοντίαν AB 20 θές] θέτε AB 20 ῥίζαν δρακοντίας κοπάνισον] om. A δρακοντίας βοτάνην ῥίζαν κοπάνισον B 20–21 τὸν χυλὸν πότισον] transp. B 22 ἤτοι] ἤγουν A om. B 22–23 ὅπου κυτρινιάσῃ τὸ πρόσωπόν τους] om. B 23 ἕνωσον με τὸ] μίξον AB

ριον καὶ πότισον με τὸ νερὸν εἰς τὸ λουτρόν. 3 ἄλλο. κοπάνισον μελάνθιν καὶ ζίγγιβερ καὶ βάλε ὄξος καὶ μάλαξον καὶ στάξον τὸ ὄξος ἐκεῖνο εἰς τὰ ῥωθώνιά του. 4 ἄλλο. κυκλαμίνου ῥίζαν βράσον με τὸ κρασὶν καὶ πότισον [▷om. B]χλίον. 5 ἄλλο. κοκκία ξυλοκεράτων βάλε εἰς κρασὶν καὶ ἂς εἶναι[om. B◁] νυχθήμερον, εἶτα πότιζε τὸ κρασίν. 6 ἄλλο. λαγωοῦ κόπρον τρίψον καὶ πότισον μετὰ οἴνου. 7 ἄλλο. κοπάνισον σελινόσπορον καὶ πότισον μετὰ οἴνου ἐν τῷ λουτρῷ. 8 ἄλλο. βλησκούνιν πότισον βρασμένον με τὸ κρασίν. 9 ἄλλο. [▷om. A]βράσον μάλαθρον καὶ ἄνηθον καὶ σέλινον με τὸ νερὸν καὶ πότισον.[om. A◁] 10 ἄλλο. [▷A]μόσχον καὶ ξυλάλαν καὶ λιβάνιν καὶ καστόριον καὶ ὕδωρ τῶν ἁγίων θεοφανίων καὶ λάδανον πότισον.[A◁]

251 [181] [om. L]σνα′ ⌜πρὸς λέπρας·⌝ κοπάνισον ⌜ῥίζαν ἀσφοδέλου⌝ καὶ ἔκβαλε τὸν ζωμὸν κοπάνισον καὶ καρδαμόσπορον καὶ αἶρας ἀλεύριν καὶ ἕνωσον ταῦτα καὶ μάλαξον καὶ βάλε ὄξος καὶ ἄλειφε τὰς λέπρας. 2 ἄλλο. [▷om. B]κοπάνισον ῥίζαν ἀσφοδέλου καὶ ἔκβαλε τὸν ζωμόν. τρίψον καὶ τεάφιν καὶ ἕνωσον καὶ ἄλειφε τὰς λέπρας.[om. B◁]

252 [182] [om. L]σνβ′ πρὸς ψώρας· κοπάνισον ἀγριοσταφίδα καὶ καρδάμου σπόρον καὶ βάλε ὄξος καὶ μάλαξον καὶ ἄλειφε τὰς ψώρας. 2 ἄλλο. κοπάνισον ῥίζαν κρίνου καὶ μίξας με τὸ μέλι ἄλειφε. 3 ἄλλο. κοπάνισον τεάφιν καὶ κόπρον αἰγὸς καὶ βάλε ὄξος καὶ ἄλειφε τὰς

---

12 πρὸς] This chapter is listed in the pinax to L. 18 πρὸς] This chapter is listed in the pinax to L.

---

23 τοῦ] om. AB 1 λουτρόν] λουτρῷ B 2 μελάνθιν] μαλάνθιον B 2 ζίγγιβερ] ζιγγίβερι A ζιζίβερ B 2 καὶ στάξον] om. B 2–3 τὸ ὄξος ἐκεῖνο] om. AB 3 εἰς] om. B 3 ῥωθώνιά] ἀπὸ τοῦ ὄξους add. B 3 του] om. AB 3 κυκλαμίνου] κυκαμίνου M σικαμίνου A 4 πότισον] ἂς ποιήσουν B 4–5 ξυλοκεράτων] ξυροκεράτων M 5 εἰς] εἰ τὸ A 5 εἶναι] ἕνε A 6 τὸ] με τὸ A 6 οἴνου] ἐν τῷ λουτρῷ add. B 7–8 ἐν τῷ λουτρῷ] εἰς τὸ λουτρὸν A 8 βλησκούνιν] γλήχωνα AB 8 πότισον βρασμένον με τὸ κρασίν] βράσον με τὸ κρασὶν καὶ πότισον B 12 λέπρας] ῥέπρας M 13 ἔκβαλε] ἐκβαλὼν B 14 ἕνωσον ταῦτα] μίξον ὅλα ἐν τῷ ἅμα AB 14 καὶ μάλαξον καὶ βάλε ὄξος] transp. B 14 βάλε] βαλὼν B 16–17 τὰς λέπρας] om. A 18 ψώρας] καὶ λέπρας add. B 18 ἀγριοσταφίδα] ἀγριοσταφίδας A 18–19 καρδάμου σπόρον] καρδαμόσπορον AB 19 βάλε] βαλὼν B 19 καὶ] om. B 20 μίξας] μίξον B 20 μέλι] καὶ add. B 21 καὶ βάλε] transp. B 21 καὶ] μίξας add. B

ψώρας. 4 ἄλλο. φύλλα ῥοδάφνης ἔψησον με τὸ χρηστέλαιον εἶτα ἔκβαλε τὰ φύλλα καὶ βάλε κηρὶν καὶ τεάφιν τριμμένον καὶ ἄλειφε εἰς τὸν ἥλιον τὰς ψώρας. 5 ἄλλο. βάλε ῥοδέλαιον καὶ λιθάργυρον τριμμένον καὶ ψιμμίθιον καὶ ὑγροπίσσιν καὶ ὄξος καὶ μίξον ταῦτα καὶ ἄλειφε εἰς τὸν ἥλιον.⌉

253 [183] om. Lςνγ′ ⌈πρὸς λειχήνας· ὑγροπίσσιν ἄλειψον τὰς λειχήνας. 2 ἄλλο. δαμασκηνέας δάκρυον ἀπόβρεχε νυχθήμερον εἰς ὄξος καὶ ἄλειφε. 3 ἄλλο. λυγαίας φύλλα κοπάνισον καὶ θὲς ἐπάνω ἢ μίξον μετὰ ὄξος. 4 ἄλλο. καππάρεως φύλλα τρίψας καὶ μίξας με τὸ ὄξος ἄλειφε.⌉ 5 ἄλλο. ▷om. Aκοπάνισον λίβανον καὶ μίξας με τὸ ὄξος ἄλειφε. 6 ἄλλο. ⌈τεάφιν τρίψας καὶ μίξας με τὸ ὄξος ἄλειφε.⌉ τέλος εἴληφε ἡ βίβλος αὕτη.om. A◁

---

6 πρὸς] This chapter is listed in the pinax to L.

328.21–329.1 τὰς ψώρας] om. A 1 ῥοδάφνης] ῥοδοδάφνης B 2 κηρὶν καὶ] κριὸν καὶ M om. A 2 τεάφιν] καὶ κηρίον add. A 3 εἰς τὸν ἥλιον] ἐν ἡλίῳ B 3 βάλε] κηρὶν καὶ B 4 μίξον] μίξας B 4 ταῦτα] ὅλα A om. B 5 καὶ] om. B 5 εἰς τὸν ἥλιον] τὰς ψώρας A om. B 6 λειχήνας] καὶ λέπρας add. B 6 ἄλειψον] ἄλειφε B 7 δάκρυον] ἤτοι κομίδι add. B 8 καὶ] μίξας ὄξος καὶ add. B 8–9 ἢ μίξον μετὰ ὄξος] om. B 9 ὄξος] ὄξους A 9 τρίψας] τρίψον B 9 καὶ μίξας] om. B 9–10 με τὸ] om. A σὺν B 11 ὄξος] καὶ add. B 11 τρίψας] τρίψον B 12 τέλος εἴληφε ἡ βίβλος αὕτη] om. B

APPENDIX

1. *Synoptic table of contents*

This table provides an overview of ω and $\aleph$, listing the headings of ω along with the corresponding chapter in $\aleph$. If a chapter is absent from $\aleph$, the heading is italicized. All data is simplified.

Subject	ω	$\aleph$	Mss. ω
On acute headache	1	1	LMAB
On headache and migraine	2	2	LMAB
On heat in the head	3	3	LMAB
On the eyes	4	5	LMAB
If the eyes are itching	5	4	LMAB
On flux from the eyes	6	5	LMAB
On nose bleed	7	6	LMAB
If someone spits blood	8	7	LMAB
On pain of the tongue	9	8	LMAB
If the bowels harden	10	9	LMAB
On diseases of the liver	11	10	LMAB
If the spleen hardens	12	11	LMAB
Vapour bath for kidney pain	13	12	LMAB
For anasarca	14	13	LMAB
For blocked ears	15	14	LMAB
On the pityriasis of the head	16	15	MAB
On headache	17	-	MAB
On pain in the eyes	18	16	MAB
On flux from the eyes	19	-	MAB
On pain in the eyes	20	-	AB
On the hyposphagma of the eye	21	17	MAB
On pus flowing from the ears	22	18	LMAB

On blood flowing from the ears	23	19	LMA
On worms in the ear	24	20	LMA
If something gets in the ear	25	21	LMA
When there is a noise in the ear	26	22,23	LMA
On the parotids, that is those which appear underneath the ears	27	24	LMA
On heavy nosebleeds	28	25	LMA
On bleaching of the teeth	29	26	LMA
If (mouth) lesions are bleeding	30	27	LMA
On rotting (mouth) lesions	31	28	LMA
On soreness of the lips	32	29	LMA
If the tongue is inflamed	33	30	LMA
When the mouth smells	34	31	LM
On swelling of the throat	35	32	LMA
If someone spits blood	36	33	LMA
On breast inflammation	37	34	LMA
If someone vomits whenever he eats	38	35	MA
On dysentery	39	36	LMA
On belly cramps	40	-	LMA
On colic	41	37	LMA
On worms in the belly	42	38	LMA
On kidney stones	43	39	LMA
On dysuria	44	-	LMA
On pus in the urine	45	40	LMA
On blood in the urine	46	41	LMA
On strangury	47	42	LMA
If someone wets the bed at night involuntarily	48	-	LMA
On hepatitis	49	43	LMA
On liver abscesses	50	44	LMA
If the liver harden like stone	51	44	LMA

On dropsy that is when the entire body is inflamed and swollen, the belly, feet and testicles	52	45	LMA
On inflammation and swelling of the spleen	53	46	LMA
If the spleen hardens	54	47	LMA
On jaundice	55	48	MA
On prolapse of the navel	56	49	MA
On the inflammation and swelling of the testicles	57	50	MA
On gonorrhea that is when his semen flows without intercourse involuntarily and without coveting women	58	51	MA
If someone has an involuntary erection	59	52	MA
On the inflammation of the anus	60	53	MA
On external hemorroids	61	54	MA
On the prolapse of the anus	62	-	MA
If the menses stop	63	55	MA
On sciatica, that is the so-called kidney area	64	56	MA
On soreness of the fingers	65	57	MA
For a patient when from lying on a bed for a long time his bones and his side get sore	66	58	MA
For wounds when they develop on the feet because he happens to have worn his sandals for a month or two	67	59	LMA
On calouses which occur in the hands and feet	68	60	MAB
On paronychia	69	61	MAB
On the pterygium of the nail, hand and foot	70	-	MAB
If the nails are smashed by a stone	71	-	MAB
If the nails are scaly, as with psoriasis	72	-	MAB
On leprosy	73	-	MAB

On psoriasis	74	64	LMAB
On leprosy	75	64	LMAB
On neuralgia	76	65	LMAB
On dropsy	77	65	LMAB
On eyes when they do not see at night	78	66	LMAB
On migraine	79	-	LMAB
On lichen	80	-	LMAB
On *aporufai* (cf. ω 234)	81	67	LMAB
If a wound gets black	82	68	LMAB
When the hair of the head falls out	83	69	LMAB
So that the hair of the head never falls out	84	-	LMAB
If you want to make the hair on the head grow	85	70	LMAB
If you want to make the hair black	86	71	LMAB
If you want to make the hair blond	87	-	LMAB
When there are a lot of lice on the head	88	72	LMAB
On the so called scurf that occurs in the head	89	73	LMAB
On alopecia and ophiasis	90	74	LMAB
On headache	91	75	LMAB
If the head appears to be inflated, it is a sign that he is suffering from pneuma	92	75	LMAB
If the skin on his head seems to be pricked with a needle and his head has a lot of heat	93	75	LMAB
If the skin on the head appears to be cold as ice and numb	94	76	LMAB
If someone has a headache caused by the stomach	95	77	LMAB
If someone drinks a lot of wine and has a headache because of this	96	78	LMAB
Ways to cleanse the head from phlegm	97	79,80	LMAB

If you want to cleanse the head from liquids	98	81	LMAB
If you want to sneeze a lot	99	82	LMAB
If you want to stop sneezing	100	83	LMAB
If a person's head aches because of the cold	101	84	LMAB
If a person's head aches because it was burnt by the sun a lot	102	84	LMAB
On catarrh	103	85	LMAB
If this happens because of the cold, the discharge is thick and cold	104	85	LMAB
If the hair of the eye falls out, do this to make them grow	105	86	LMAB
If you want to make the hair of the eyebrows black	106	87	LMAB
On hair, if it is double on the eyebrows, and you want it each to be single	107	88	LMAB
If the eyelid of a person has a so called stye or a hordeolum	108	89	LMAB
If the white of the eye gets very red and it appears as if the white of the eye is higher than the pupil	109	90	LMAB
Emphysema is a swelling of the eyelid	110	91	LMAB
On hardening of the eye	111	92	LMAB
On drying of the eye	112	93	LMAB
Ectropion is when the lower eyelid is turned towards the outside	113	94	LMAB
Dacryocystitis is when there develops an abscess close to the nose in the inner corner of the eye	114	95	LMAB
Therapy if the hair of the eyelid falls out	115	96	LMAB
If tears flow involuntarily from the inner corner of the eye	116	97	LMA
If the eye has a white (spot)	117	98	MA
On pterygium of the eye	118	99	MA

On blindness and dim-sightedness	119	100	LMAB
Dim-sightedness is when someone sees sometimes little, sometimes a bit less, sometimes not at all	120	100	LMAB
When the eye has a white (spot)	121	-	LMAB
If it gets black under the lower eyelid. This disease occurs mostly in women and in children	122	101	LMAB
If the lower eyelid gets red like blood or the colour of the place (gets) like yellow	123	101	LMAB
If there is pain from the cold	124	102	LMAB
If there is pain in the ear from a hot imbalance, as from sun, heat or severe exhaustion	125	103	LMAB
If there is a pain inside of the ear because of a hot imbalance	126	103	LMAB
If there is a pain in the ear because of thick humours	127	104	LMAB
If the ear is inflamed	128	105	LMAB
If you want to induce growth of the teeth of a child	129	106	LMAB
If there is a swelling under the tongue	130	107	LMAB
An aphthous ulcer is when the lips of a child are being corroded when it is breast-fed	131	108	LMAB
When the so called jaws are inflamed	132	109	LMAB
On cough	133	110	LMAB
For those who have a very hoarse voice	134	111	MAB
For those who are found to have dyspnea and are unable to breathe properly	135	112	MAB
For if someone spits a large mass of blood at once this is sign that a vein is ruptured	136	123	MAB
On pleuritis	137	112	MAB

On pneumonia. The disease pneumonia is when he coughs and spits a thick and yellow pus	138	113	LMAB
Empuema is called when pneumonia takes a long time and he coughs and spits something like pus and the sputum stinks	139	114	LMAB
When the disease that is called empuema results in an illness the locals call phthisis	140	115	LMAB
If someone has a headache and fever	141	115	LMAB
Phthisis is called when the lungs of a person produce pus and he spits pus and it stinks and he is wasting away	142	115	LMAB
When a person becomes at first irascible because of a fever and develops a fever because of that	143	116	MAB
On tertian fever	144	117	MAB
On quartan fever	145	118	MAB
On quotidian fever	146	119	MAB
On the disease which is called hectic fever	147	120	LMAB
On the cancerous disposition, or when the breasts of women or of men have the illness which is called cancer by the doctors	148	121	LMAB
Once this illness, the cancer itself, is healed and gone, and when you want her breasts to be immaculate so that they want develop another illness similar to this cancer or worse	149	121	LMAB
When their armpits smell and stink	150	122	LMAB
When a person vomits something resembling fried eggs, it is a sign that there is a humoural imbalance in his stomach	151	124	LMAB
For those who have no appetite because of frost, or because of cold that means when they get cold	152	125	LMA

When someone has a lot of fluid in the stomach and has no appetite because of the fluids	153	126	LMA
A dog-like appetite is when a person has the desire to eat more and he eats and he is continuously hungry and eats again like the dogs	154	127	LMA
When a person has hiccup	155	128	LM
When a person has dyspepsia, the sign is that he vomits something like vinegar	156	-	LMB
When a person excretes yellow bile from the upper and the lower (part of the alimentary tract) with discomfort	157	129	LMAB
For the inflammation of the intestines	158	130	MAB
When the kidneys are inflamed and there is pain and pressure in the place where the inflammation of the kidneys is, and those people have fever and vomit yellow bile	159	131	LMA
It is called diabetes when someone drinks while he urinates, and he does this continuously and every hour	160	132	LMA
Rhagades are called by the doctors, when there are fissures on the testicles or on the anus	161	133	LMA
If someone swallows leeches by mistake	162	-	LMA
If someone eats a lot of green mushrooms and somehow chokes	163	134	LMAB
If the milk in the stomach of a person turns to cheese	164	135	LMAB
When the entire head hair falls out	165	-	LMAB
On pityriasis of the head	166	-	LMAB
Alphoi are called when the person gets something resembling lentils on his face. It is also useful for the treatment of lichen and leprosy	167	136	LMAB

An aphthous ulcer is when the mouth of a person has tumours inside, and the lesions	168	-	LMAB
On dim-sightedness	169	-	LMAB
On pain of the joints. And it softens hardening	170	137	LMAB
(It) dissipates and softens an ulcer	171	138	LMAB
It softens *aporufai* (cf. ω 234)	172	139	LMAB
If someone is bitten by an asp	173	140	LMAB
If a person has indigestion	174	-	LMAB
Therapy for cough	175	-	MB
It induces lactation in women, and periods	176	141	LMB
(For) a women if she is entirely unable to conceive	177	142	LMAB
This softens hardening of the tongue	178	143	LMAB
Therapy for dysuria	179	-	LMAB
For dysentery	180	-	LMAB
If his ear hurts and he does not hear	181	-	LMAB
When the periods of women stop	182	-	LMAB
If it happens to a pregnant woman that the child inside her womb dies	183	144	LMAB
If someone is bitten by a snake or a viper	184	145	LMAB
(It) treats worms and expels them	185	-	LMAB
If a person is in pain and also for the so called sciatica	186	-	LMAB
When his face gets yellow, this is called icterus by the doctors and by the locals gold-colour	187	146	LMAB
On headache	188	-	LMAB
When there is a flow from the abdomen	189	147	LMAB
If someone is bitten by a dog	190	148	MAB
For a sore throat	191	-	MA
Treat the so called hiccup like this	192	-	LMAB

When someone is bitten by a rabid dog	193	149	LMAB
The stones which are in the bladder and in the kidneys	194	-	LMAB
Therapy for lichen and leprosy	195	-	LMAB
If the breasts of a woman or a man are hardened	196	-	LMAB
Impedes (the growth) of women's breasts and the testicles of children so that they will not get big	197	-	LMAB
If someone is bitten by a bee or a wasp	198	150	LMAB
If a person's nose stinks	199	151	LMAB
If something causes abrasion and leads to wounds	200	152	LMA
On infertility	201	-	L
If someone vomits a lot although he does not want to vomit	202	-	LMAB
On a cloud like spot the eye	203	153	LMA
On pain of the eye	204	-	LMAB
On toothache	205	154	LMAB
If the teeth are loose	206	155	LMAB
If a part of the body is burnt	207	156	LMAB
Podagra	208	157	LMAB
If the joints of the feet or the hands harden	209	158	LMAB
To stop someone bleeding	210	159	LMAB
If the stomach is exhausted, that means incapacitated, and burns internally	211	160	LMAB
If the bowels have diarrhoea, this stops it	212	160	LMAB
If the stomach is burning inside	213	160	LMAB
Therapy for those suffering from a spleen-disease and from a swelling	214	-	LMAB
If a persons has colics	215	161	LMAB
If someone is bitten by a scorpion	216	162	LMAB

If it happens that there are worms in a wound	217	163	LMAB
If someone is bitten by a sea roach	218	164	LMAB
On those suffering from a spleen disease and from a swelling	219	165	LMAB
On therapy of inflammation	220	165	LMAB
On the loss of voice	221	166	LMAB
(It) dissolves lice	222	-	LMAB
For the so called scrofula	223	167	LMAB
Therapy for psoriasis	224	-	LMAB
Therapy for pain in the ear	225	168	LMAB
If you want someone to stop being thirsty	226	-	LMAB
It is called alphos when the person gets something resembling a black lentil on his face	227	-	LMAB
If they develop something like a white lentil	228	-	LMAB
Alopecia is when the head hair falls out	229	-	LMAB
Therapy for the salivary glands	230	-	LMAB
If someone urinates blood	231	-	LMAB
On cough	232	-	LMAB
When his mouth smells	233	169,170	LMB
On swollen glands or *aporufai*	234	171	LMB
On belly cramps	235	172	LMB
If someone swallows leeches	236	-	LMB
To induce lactation of a woman	237	-	LMB
If a person's knees appear to be cold like ice	238	173	LMB
On dysuria	239	-	LMAB
On dysentery	240	-	LMAB
On how to induce periods in women and urination and it expels children from the womb and helps the spleen	241	-	LMAB
On internal and external hemorrhoids	242	174	MAB

On a prolapse of the colon	243	175	MAB
On flat worms	244	-	MAB
On swelling and *aporufai* (cf. ω 234)	245	176	MAB
If someone urinates at night involuntarily	246	-	MAB
If someone is bitten by a viper	247	177	MAB
On migraine	248	179	MAB
If someone has a pain in his hip	249	180	MAB
On jaundice, that is when the face is yellow	250	-	MAB
On leprosy	251	181	MAB
On psoriasis	252	182	MAB
On lichen	253	183	MAB

2. *Collation and List of Words*

This is a sample of a full collation of a chapter of ω. In the edition various trivial and recurrent mistakes were silently corrected. This was for instance the case with the variants of the verb 'drink'; accordingly the spelling of the word 'vinegar' was standardized throughout the edition. A comprehensive list of these words can be found on the following pages.

γ' εἰς ζέσιν κεφαλῆς· ῥοδέλαιο̣ βάλλε καὶ ὀξήδην καὶ χλιάνε τα. εἶ ἄλειφε τὸ κεφάλην τοῦ ὅλον. ζέσης δὲ λέγεται ὅταν φαίνεται τινὰ ὅτι κέεται ἡ κεφαλὴ αὐτοῦ καὶ ἐξάπτει. πλὴν πρῶτον φλεβοτόμησον αὐτὸν κρανιακὴν φλέβαν. εἶτα ἄλειφέ τον τὸ ὀξύδην. καὶ τὸ ῥοδέλαιον χλίον. καὶ κρασὶν μὴ πίνειν μηδὲ ἥλιος τὸν κρούει μηδὲ ἄνεμος μηδὲ κάπνος. μηδὲ δρυμέα τρώγει μηδὲ θερμὰ καὶ ἃς ἰατρεύεται καὶ κλυστήριν.

1 ῥοδέλαιο̣] ῥόδιον ἤτοι ῥοδόλαιον MAB 1 βάλλε] βάλλαι B 1 καὶ ὀξήδην] εἰς ὄξος M ὀξύδην A καὶ ὀξίδιν B 1 χλιάνε τα] χλιααvετα M 2 εἶ] καὶ MA ἔπητα B 2 ἄλειφε] ἄληφε B 2 τὸ κεφάλην] τὴν κεφαλὴν AB 2 ὅλον] ὅλην MAB 2 ζέσης] ζέσις MA ζέσεις B 2 ὅταν] ὅταν τὸν AB 2 φαίνεται] φένεται B 3 τινὰ] τὸν ἄνθρωπον M om. AB 3 κέεται] καίεται MAB 3 αὐτοῦ] τοῦ MAB 3 ἐξάπτει] ἐξάπτη B 3–4 φλεβοτόμησον αὐτὸν] φλεβοτόμησον τοῦτον M φλεβοτομησέτον A φλογοτόμηον B 4 φλέβαν] φλέφαν M 4 ἄλειφέ τον] ἄλειψον ἤγουν τὴν κεφαλήν του M ἄλειφέ την A ἄληφέτον B 4 τὸ] om. B 4 ὀξύδην] ὀξείδιν M ὀξίδιν B 5 κρασὶν] κρασὴν A 5 πίνειν] πίννη M πίννει A πίνην B 5 ἥλιος] ὁ ἥλιος AB 5 κρούει] κρούη MB 6 δρυμέα] δρυμαία M δριμμαῖα B 6 τρώγει] τρόγει B 6–7 ἃς ἰατρεύεται] ἰασιατρεύεται B 7 κλυστήριν] γλυ$^{\sigma\tau\rho}$ M κλυστήρην A κλειστήριν B

Words standardized to Classical spelling

ἀγγῶνος → ἀγκῶνος
αἰσθάννεται, ἐσθάνετε → αἰσθάνεται
ἄληφε → ἄλειφε
ἀλθέα → ἀλθαία
ἄλλας → ἄλας
ἄλυψον → ἄλειψον
ἀμίγδαλα → ἀμύγδαλα
ἀμυγδαλαίας → ἀμυγδαλέας
ἀμωνιακόν → ἀμμωνιακόν
ἀνιθέλαιον → ἀνηθέλαιον
ἄνιθον → ἄνηθον
ἀντίδωτον → ἀντίδοτον
ἄπειρον → ἄπυρον
ἀπήδια, ἀππίδια → ἀπίδια
ἄπλητα → ἄπλυτα
ἀπόστιμα → ἀπόστημα
ἀρώστου → ἀρρώστου
ἀσφοδέλλου → ἀσφοδέλου
ἀψίθιον → ἀψίνθιον
βάττος → βάτος
βδέλα → βδέλλα
βδέλιον → βδέλλιον
βλέβαν → φλέβα
βοήθειμα → βοήθημα
βόλον → βῶλον
βούτηρον → βούτυρον
βράσσον → βράσον
γάγγρινον → γάγγρηνον
Γαλινοῦ → Γαληνοῦ
γάλλα → γάλα
γλήχονος→ γλήχωνος
γλυσχρός → γλισχρός
γονοροία → γονορροία
γυναικία → γυναικεῖα
δεῖξις → δῆξις
διαβίτης → διαβήτης
δρακοντεία → δρακοντία
δρυμύ, δρυμή → δριμύ
δυπλαί → διπλαί
δῦσον → δῆσον
ἐγίλωψ → αἰγίλωψ
εἶστερον → ὕστερον
ἑκάστιν → ἑκάστην
ἐλέας → ἐλαίας
ἕλμιθας → ἕλμινθας
ἔμπραστρον → ἔμπλαστρον
ἕνη, ἕνει → ἕνι
ἐξήγισις → ἐξήγησις
ἐξοπήσω → ἐξοπίσω
ἐξωχάδας → ἐξοχάδας
ἔστη → ἔστι
εὐπατώριον → εὐπατόριον
ἕψιμα, ἕψιμαν → ἕψημα
ἕψησον → ἕψησον
ζέσεις, ζέσης → ζέσις
ζίγκιβερ, ζυγγίβερ, ζίγγιβιρ → ζίγγιβερ
ζομόν → ζωμόν
ζυμία → ζημία
ἦτα → εἶτα
ἤτι → ἤτοι
θεραπία → θεραπεία
θερμένοντα → θερμαίνοντα
θέττε → θέτε
θρίμβον → θρύμβον
θυριακή → θηριακή
ἰάτρεβε → ἰάτρευε
ἰδίοσμου → ἡδύοσμου
καθαρίδας → κανθαρίδας
καινούριον → καινούργιον
καπάρεος → καππάρεως
καστώριον → καστόριον
κατάρου → κατάρρου
κάχριος → κάχρυος
κεδραία → κεδρέα
κέεται → καίεται
κενούργιον → καινούργιον
κέχρου → κέγχρου
κηδονέας, κιδωνέας → κυδωνέας

κικιδίου → κηκιδίου
κίμινον → κύμινον
κλίματος → κλήματος
κνίθονται → κνήθονται
κοκία → κοκκία
κόλαν → κόλλαν
κολλυτικός → κολλητικός
κολοκίνθη → κολοκύνθη
κολοφόνια → κολοφώνια
κομάτια → κομμάτια
κόνιον → κώνειον
κονύδας → κονίδας
κοπάνησον → κοπάνισον
κορύδας → κορίδας
κοσκινήσας → κοσκινίσας
κοτύλλη → κοτύλη
κρέμαντε → κρέμανται
κριθάλεβρον → κριθάλευρον
κρίον → κρύον
κριτικοῦ → κρητικοῦ
κρόμια, κρόμυα → κρόμμυα
κρυώτης, κριότης → κρυότης
κυκλαμήνου → κυκλαμίνου
κυπαρίσου → κυπαρίσσου
κωλυκῆς → κωλικῆς
λαγοοῦ → λαγωοῦ
λαππάθου → λαπάθου
λειπαρός → λιπαρός
ληχήν, λιχήν, λιχίν → λειχήν
λιπαρώτητα → λιπαρότητα
λιτόν → λυτόν
λύα → λεῖα
λυγέας, λιγαίας → λυγαίας
λυμνία → λημνία
λυποθυμήση → λειποθυμήση
λωπαδίῳ → λοπαδίῳ
μαγνίτην → μαγνήτην
μάλλαξον → μάλαξον
μαλλία → μαλία
μάλλιστα → μάλιστα
μαρένεται → μαραίνεται
μασσάτε → μασσᾶται
μελαχολικόν → μελαγχολικόν
μέλενα → μέλαινα
μέτοπον → μέτωπον
μῆξον → μῖξον
μί → μή
μίλιγγας → μήλιγγας
μίνιγγας → μήνιγγας
μιρσυνέλαιον → μυρσινέλαιον
μίτη, μύττη,
μίτιν, μήτη → μύτη
μοιράκιον → μειράκιον
μῦλα → μῆλα
μυλέα → μηλέα
μύρμικος → μύρμηκος
μυρσύνη → μυρσίνη
ναρθικίου → ναρθηκίου
ξερά → ξηρά
ξεροφθαλμία → ξηροφθαλμία
ξηρᾷ → ξερᾷ
ξυρά → ξηρά
ξύρανον → ξήρανον
ὄγγος → ὄγκος
ὀξίν → ὀξύν
ὀρνίθιον → ὀρνίθειον
ὀρύγανον → ὀρίγανον
ὄρχεος → ὄρχεως
ὁσάν → ὡσάν
ὀτίον → ὠτίον
ὀφέλιμος → ὠφέλιμος
ὀφεληθῆ → ὠφεληθῆ
ὀφρίδια → ὀφρύδια
πάλαι → πάλιν
παλιουραία, παλλιουρέα → παλιουρέα
παστέλωσον → παστέλλωσον
πεπονθώτα → πεπονθότα
πέπωνος → πέπονος
πετρέους → πετραίους
πλείωνα → πλείονα
πλευρήτις →πλευρίτις
πνεῦμον → πνεύμων

ποίνη → πίνη
πολήν → πολλήν
πολλυαίμοις → πολυαίμοις
πολλύτριχον → πολύτριχον
πολλυχρόνιον → πολυχρόνιον
πολυγώνου → πολυγόνου
πίννει → πίνει
ποταμογείτωνος → ποταμογείτονος
πότησον, πότυσον → πότισον
πρισθῇ → πρησθῇ
πτελαίας → πτελέας
πυκρά → πικρά
πυκτήν → πηκτήν
πυρέσει → πυρέσσει
πύτυρα → πίτυρα
ῥινίματα → ρινίσματα
ῥιτίνη, ῥυτίνη,
ῥητήνη → ῥητίνη
ῥοδαίλαιον → ῥοδέλαιον
ῥοδακηνέα, ῥοδακηναία→ ῥοδακινέα
ῥοδόσταμα → ῥοδόσταγμα
ῥοφίσει, ῥοφύσει → ῥοφήσει
ῥύνησον → ῥίνισον
σακελίσας, σακκελίσας → σακελλίσας
σάμψιχον → σάμψυχον
σελήνου → σελίνου
σικέαν, σικίαν,
σικήαν → σικύαν
σκληρώτης → σκληρότης
σκολήκια → σκωλήκια
σκολίκων → σκωλήκων
σκολωπένδριον → σκολοπένδριον
σκολύκια, σκολίκια,
σκολήκια → σκωλήκια
σκόρωδα → σκόροδα
σκύλου → σκύλλου
σμίρνα → σμύρνα
σπλάχνα → σπλάγχνα
σπλίν → σπλήν
σταφήδας → σταφίδας
στείρακαν → στύρακα
στηβή, στοιβή → στιβή
στρογγίλον → στρογγύλον
στυπεία, στυπία → στυππεία
συκαμιναία, συκαμηναία → συκαμινεά
σύναπι → σίναπι
σύπετε → σήπεται
σύψις → σήψις
σφῖκες → σφῆκες
τείλης → τήλης
τειχοδέμονα → τειχοδαίμονα
τίλεως → τήλεως
τίλης → τήλης
τίν → τήν
τιούτους → τοιούτους
τριμένον → τριμμένον
τρίπα → τρύπα
τριτταῖος → τριταῖος
τριφερά → τρυφερά
τρυμένον → τριμμένον
τρυχῶν → τριχῶν
τρύψας → τρίψας
τυγάνιν → τηγάνιν
τυγάνισον, τυγάνησον,
τηγάνησον → τηγάνισον
τύλης → τήλης
ὑδύοσμον → ἡδύοσμον
ὑοσκιάμου → ὑοσκυάμου
ὕπαρ → ἧπαρ
ὕσωπον → ὕσσωπον
φένεται → φαίνεται
φθήρας → φθεῖρας
φθήσις → φθίσις
φλεβοτομηθίσει → φλεβοτομηθήσει
φλεγμένη → φλεγμαίνη
φλέφα → φλέβα
φλέφαρα → βλέφαρα
φοινίκεια, φοινίκοια → φοινίκια
φρίξας → φρύξας
φύλα → φύλλα
χαμαίμυλα → χαμαίμηλα
χαμεμηλέλαιον → χαμαιμηλέλαιον

χελιδώνων → χελιδόνων
χήναιον → χήνειον
χήνιον → χήνειον
χιλόν → χυλόν
χλορόν → χλωρόν
χλυαρόν → χλιαρόν
χλύον → χλίον
χοίνιον → χήνειον
χοίριον → χοίρειον
χοίραιον → χοίρεον
χορηκῶν → χωρικῶν
χορῆς → χωρίς
χρείε → χρίε
χυναρίου → χηναρίου
χύνειον → χήνειον
χωλή → χολή
ψίλλεον, ψίλαιον → ψύλλεον
ψύχας → ψίχας
ψυχίας → ψιχίας
ὡμιλείτω → ὁμιλείτω
ὠμωτριβές → ὠμοτριβές
ὤνυχας → ὄνυχας
ὤρεξις → ὄρεξις
ὥρημα, ὅριμα → ὥριμα
ὠτεῖον → ὠτίον
ὠφελία → ὠφελεία

Accusatives of classical words spelt with an additional -n have been converted back to their classical form

αἷμαν → αἷμα
βλέβαν → φλέβα
δέρμαν → δέρμα
θέλημαν → θέλημα
μέλιν → μέλι
πεπέριν → πεπέρι
ῥίναν → ῥίνα
ῥοδόσταμαν → ῥοδόσταγμα
σκληρότηταν → σκληρότητα
σπέρμαν → σπέρμα
σπλῆναν → σπλῆνα
στείρακαν → στύρακα
στόμαν → στόμα
σφραγίδαν → σφραγίδα
σῶμαν → σῶμα

Ambiguous words were adjusted to

ἀξούγγην, ἀξούγκην,
ἀξούγκιν → ἀξούγγιν
ἀλεύρην → ἀλεύριν
ἀλεύρειν → ἀλεύριν
ἀσβέστην → ἀσβέστιν
ἀκόνιν → ἀκόνην
ἀρσενίκην → ἀρσενίκιν
ἀψίνθην → ἀψίνθιν
βάλλε → βάλε
βαμβάκην, βαμπάκιν,
βαβάκιν → βαμβάκιν
βλυσκούνην → βλησκούνιν
τὸ βοτάνην → τὸ βοτάνιν
γουλήν → γουλίν
δαδήν → δᾳδίν
ἐλάδην → ἐλάδιν
ἐν τῷ ἅμα standardized form
ἑξαγίου τὸ ἥμισυ standardized form
ἐρεβίνθην → ἐρεβίνθιν
ἐφόλβιον → εὐφόλβιον
ζουμήν → ζουμίν
ζοχήν → ζοχίν
ζυμάρην → ζυμάριν
ζωμήν → ζωμίν
ἱππούρην → ἵππουριν f
καλάμην → καλάμιν
καρίδην → καρύδιν
τὸ κεφάλην → τὸ κεφάλιν
κικίδην, κηκίδην → κηκίδιν
κλυστήρην → κλυστήριν
κολοκίνθη → κολοκύνθη
κολοκύνθην → κολοκύνθιν
κομίδην → κομίδιν

κουκία → κοκκία
κρασήν, κρασί → κρασίν
κρεβάτην → κρεβάτιν
κριθάρην → κριθάριν
κρομύδην, κρομύδιν → κρομμύδιν
κυπερήν → κυπέριν
λαρδήν → λαρδίν
λιβάνην → λιβάνιν
λυπινάρια, λυμπινάρια,
ληπηνάρια, λιπηνάρια
→ λυπηνάρια
μαχαίρην, μαχέριν → μαχαίριν
μελάνθην → μελάνθιν
μέλην → μέλι
μολίβδην, μολύβδην → μολύβδιν
ξηράφην, ξυράφην → ξυράφιν
ὀξήδιν, ὀξίδιν,
ὀξύδην, ὀξήδην
ὀξίδην → ὀξείδιν
ὀρίζιν → ὀρύζην
πετζήν → πετζίν
πανήν → πανίν
πίσιν → πίσσιν
πολλυκόμπην,
πολυκόμπην → πολυκόμπιν
πουλίτζην → πουλίτζιν
πρινοκούκην → πρινοκούκιν
προζύμην → προζύμιν
ῥεφάνην → ῥεφάνιν
ῥουθούνην → ῥουθούνιν
σπογγάρην →σπογγάρην
συκότην → συκώτιν
σισάμην → σησάμιν
σαπούνην → σαπούνιν
σεμιγδάλι, σεμιδάλην,
σεμιδάλις → σεμιδάλιν
σμῖξον → μῖξον
σχοίνου, σχύνου → σχίνου
τηγάνην → τηγάνιν
τζουκάλην → τζουκάλιν
τριφύλλη, τριφύλλην,
τρυφίλην, τρυφύλην,
τρυφύλιν → τριφύλλιν
ὑγροπίσσην → ὑγροπίσσιν
φλοῦδην → φλοῦδιν
χαλβάνην → χαλβάνιν
χαρτήν → χαρτίν
χιλάριν, χυλάρην → χυλάριν
χριστέλαιον → χρηστέλαιον
χριστόν → χρηστόν
χρυσάφην → χρυσάφιν
ψιμίθιον, ψιμίθην
→ ψιμμίθιον, ψιμίθιν
ψιμίθιον, ψιμίθην
→ ψιμμίθιον, ψιμίθιν
ψιμμήθιν → ψιμμίθιν
ψωμήν → ψωμίν

Verb forms were adjusted to

δέναι, δέννε → δένε
καῦσαι → καῦσε
μῖξαι → μῖξε
πάσαι → πάσσε
πλύναι → πλύνε
στάξαι → στάξε
τάραξαι → τάραξε
φρύξαι → φρύξε

Modern Greek words were adjusted to

ἀγρυαγκουρέας → ἀγριαγγουρέας
ἀκροάγγουρα → ἀκροάγγουρα
ἀμανυτάρια → ἀμανιτάρια
ἀπέσσω → ἀπέσω
ἄπλωννε → ἄπλωνε
ἀσπαράγκου,
ἀσπαράγγου → ἀσπαράγγου
βαβάκιν, βαμπάκιν → βαμβάκιν
βιζύα, βιζία → βυζία
βρεμένον, βρεμαίνον → βρεμμένον
γλήχονος → γλήχωνος

γλισχούνιν, γλυσχούνιν → γλησχούνιν
γλύχων, γλίχων → γλήχων
γλυστρίδα → γλιστρίδα
δαμασκηναία → δαμασκηνέα
ἰντίβια → ἰντύβια
καλοκέριν → καλοκαίριν
καρίδιον → καρύδιον
κικοίδι → κηκίδιν
κόκινον → κόκκινον
κομάτια → κομμάτια
κρόκκον → κρόκον
μακεδονισίου,
μακεδονυσίου → μακεδονησίου
μαρούλλια → μαρούλια
μνάσεων → μνάσεον
μονόκιθρον → μονόκυθρον
ὄξινα → ὄξυνα
ὀστριδίου → ὀστρυδίου
πίτας, πίττας → πήττας
ῥεβίνθια → ἐρεβίνθια
ῥοφιτόν → ῥοφητόν
σύκοσε → σήκοσε
σκυλοκρόμυδον → σκυλλοκρόμμυδον
σπογκάριν → σπογγάριν
συνάπιν → σινάπιν
τζυκάλην, τζικάλιν → τζυκάλιν
τηλήσουν → τηλίσσουν
ὑψηλόν → ψιλόν
φλισκούνιν → φλησκούνιν
ψιμίνθιον → ψιμμίθιον

3. Pinax to L (in diplomatic transcription)

...ν̣ε̣φρο̣ὶ̣ λίθους
μ' πρ̇ὸς δυσουρί..
μα' πρ̣..... κατουροῦντα ἕ̣λκος
μβ' πρ̇ὸς κατουροῦντα α̣ἷ̣.α̣
μγ' πρὸς στραγγουρίαν, ὅπου οὖρεῖ καὶ στάζει
μδ' πρὸς κατουροῦντας .α̣ροῦχα
με' πρὸς φλεγμονὴν ἥπατος ἤτοι συκοτήν
μϛ' ὅταν γένηται ἀπόστη$^{\mu}$ εἰς τὸ̣ .παρ
μζ' πρὸς ἧπαρ ὅταν γένηται σκληρὸν ὡς ἂν πέτρα δυνατή
μη' ἐὰν πρισθῇ̣ ἡ σπλῆνα
μθ' ἐὰν γένηται ἡ σπλῖνα σκληρά
ν' πρὸς ἴκτερον τὸ λέγουσιν οἱ χωρικοὶ χρυσιασμόν
να' ὅταν ὁ ὀμφαλὸς τοῦ ἀνθρώπου εὔγει
νβ' πρὸς φλεγμονὴν ὀρχιδίων
νγ' πρὸς γο̣ν̣ο̣ρίαν
νδ' πρὸς πριασμὸν ἤτοι ὅταν κ̣.....ει δ. νατα πα.ὰ θέλημα
νε' πρὸς φλεγμονὴν δακτύλου
νϛ' πρὸς ἐξοχάδας ρὸς ἐξοχάδας
νζ' ὅτ.. τὰ σ̣υ̣ν̣ήθ̣. τῆς γυνικὸς κρατων..αι
νη' περὶ .σχυάδ̇ος ἤτοι τὸ λεγόμεν.. .σχυο.
νθ' ὅταν οἱ δάκτυλ.ι κατὰ σχ.σθῶσιν
ξ' πρὸς ἀρρω.. ὅταν ἀπὸ τὸ πο̇λὺν κειμέν... εἰς τὸ .ραββάτην καὶ πληγωθῶσι τὰ κόλα καὶ τὰ πλεῦρα αὐτοῦ
ξα' πρὸς πληγὰς, ὅπου γίνονται εἰς τὰ ποδάρια ἀπὸ τῆς ὑποδήσε̣ω̣ς
ξβ' πρὸς ἥλους, ἃ λέγωσιν αἱ ἰδιῶται κότζια
ξγ' πρὸς παρονυχίδας
ξδ' πρὸς πτερύγιον ὄνυχος ἢ χειρὸς ἢ ποδός.
ξε' πρὸς ὀνύχια, ὅταν συνθλασθῶσιν ἀπὸ λίθου
ξϛ' πρὸς ὀνύχια, ὅταν ἔχωσιν αἷμα
ξζ' πρὸς ὄνυχας λεπρούς
ξη' πρὸς κελεφίασιν
ξθ' πρὸς ψώρας
ο' προς λέπρας
οα' πρὸς πόνον νεύρων
οβ' πρὸς συκάμινον
ογ' πρὸς ὀφθαλμούς, ὅταν οὐ βλέπουσι τὴν νύκτα
οδ' πρὸς πόνον ἡμικρανίου
οε' πρὸς λιχίνας

1 ...ν̣ε̣φρο̣ὶ̣] L 108   22 ξ'] L 109

οϛ' πρὸς ἀπορυφάς
πρὸς οὐλὴν πληγῆς ἐὰν γένηται μαύρη
πρὸς τὸ ὅταν πίπτουσιν αἱ τρίχαι τῆς κεφαλῆς
ρκβ' πρὸς ἀποφωνῆς βραχνιάζουσι
ρκγ' πρὸς ὅσοι εὑρετῶσι οἵτε γέροντες οἵτε νέοι καὶ ἔχωσι δυσπνοίαν, ὅτι οὐ δύνανται ἵνα ἀναπνέουσι
ρκδ' ὅταν πτύῃ τις αἷμα πολὺν καὶ εἰς μίαν
ρκ..' πρὸς πλευρίτιδα
πρὸς περιπνευμονίαν
πρὸς ἐμπτύημα
πρὸς ὅπου μετὰ πέσῃ ἀπὸ τοῦ ἐμπτυήματος καὶ ἔλθῃ εἰς πάθος τὸ λεγόμενον φθῆσιν
πρὸς φθῆσιν
ὅταν ἀπὸ πυρετοῦ πρῶτον θυμωθῇ καὶ οὕτως πυρέξῃ
πρὸς τριταῖον
πρὸς τεταρταῖον
πρὸς αὐθημερινὸν
ρλβ' πρὸς νόσημαν τὸ λεγόμενον πυρετικήν
ρλγ' πρὸς βυζία ἔχοντα καρκῖνον πάθος
ρλδ' πρὸς τὸ ἂν θέλῃς ἵνα ἔνι τὰ βυζία καλὰ πάντοτε
ρλε' πρὸς μασχάλας, ὅταν βρωμοῦσι
ρλϛ' πρὸς ὅταν ἐρεύγεται
ὡς ἂν αὐγὰ τυγάνου
ρλζ' πρὸς ὅσοι διὰ ψ.ξιν ἀνορεκτοῦσιν
ρλη' πρὸς ὅσοι διὰ πολλὰ ὑγρὰ τὰ εἶναι εἰς τὸν στόμαχ.. ἀνορεκτοῦσιν
ρλθ' πρὸς ὄρεξιν κυνῶδ..
.μ' ὅταν ἔχῃ ὁ ἄνθρωπος λουξίκαν
ρμα' ὅταν ἔχῃ ὁ ἄνθρωπος ἀπεψίαν
..β' ὅταν κενων.. καὶ ἄνω καὶ κάτω χολήν
..γ' πρὸς φλεγμονὴν ἐντέρων
..δ' πρὸς φλεγμονὴν νεφρῶν
..ε' πρὸς πάθος τὸ λεγόμενον διαβίτην
...' πρὸς ῥαγάδας ὀρχ.δίων
ρμζ' ὅταν καταπίῃ βδέλλας
ρμη' ὅταν πνίγηται τις ἀπὸ τὰ μαγ.τάρια τὰ χλωρά
ρμθ' ὅταν τυρωθῇ τὸ γάλα ἐν τῇ κοιλίᾳ
ρν' πρὸς τὸ ἐὰν πέσουν τὰ μαλία τῆς κεφαλῆς
ρνα' πρὸς πυτυρίδα κεφαλῆς
ρνβ' πρὸς ἀλφούς, ὅταν τὸ πρόσωπον τοῦ ἀνθρώπου εὐγάνει ὡς ἂν φακήν
...' πρὸς ἀφθ.. ἐν τῷ στόματι

4 ρκβ'] L 110 23 ὡς] L 111

πρὸς ἀβλυοπίαν
ρνγ' πρὸς πόνους ἁρμῶν καὶ σκληρότητας
ρνδ' πρὸς ἀπόστημα
...' πρὸς ἀπορυφάς
ρνϛ' ὅταν δάκη τινὰ ἀσπίδα
ρν.' πρὸς ἀπεψίαν ἀνθρώπου
ρν.' πρὸς βῆχας
ρ..' πρὸς τὸ καταβάσαι γάλα γυναικῶν
ρ..' πρὸς γυναῖκα τὴν μὴ δυναμένην γεννῆσαι
πρὸς γλῶσσαν ἔχουσαν ἀνερύτητα
πρὸς δυσουρίαν
πρὸς δυσεντερίαν
ὅταν πονεῖ το ὠτίον καὶ οὐδὲν ἀκούει
ὅταν κρατηθῶσι τῆς γυναικὸς τὰ συνήθη
ὅταν ἀποθάνῃ τὸ παιδίον ἐντὸς τῆς γυναικός
ὅταν δάκη τινὰ ὄφις ἢ ἔχιδνα
πρὸς ἑλμίνθας
ρξ.' πρὸς ὀδύνην ἰσχύου
ρξϛ' πρὸς πρόσωπον ὅταν γένηται κίτρινον
ρξζ' πρὸς κεφαλαλγίας
ρξη' πρὸς κοιλίαν ὅταν ῥέει πολλά
ρξθ' ὅταν δάκη τινὰ σκύλος
ρο' πρὸς συάγχην
ροα' πρὸς λούξικαν
ραβ' ὅταν δάκη τινὰ λυσιάρου σκύλος
ρογ' πρὸς λιθ..ρια τὰ εἶναι εἰς τὴν κύστιν καὶ εἰς τοὺς νεφρούς
ροδ' πρὸς λιχίνας καὶ λέπρ..
.οε' πρὸς σκληρίαν βυζίων
πρὸς βυζία καὶ ὀρχιδία νὰ μηδὲν γένωσι μεγάλα
ὅταν δάκη τινὰ μέλισσα ἢ σφίκα
ὅταν βρωμεῖ ἡ μύτη τοῦ ἀνθρώπου
ὅταν ἀπὸ δαρμοῦ ποιήσει τ.. πληγάς
περὶ ἀσυλληψίας
ὅταν ξερᾶ εἴτι φάγη
ὅταν οἱ ὀφθαλμοὶ ἔχωσι νεφέλας
..β' πρὸς ὀδύνηνη ὀφθ.μων
ρπγ' πρὸς σειομένους ὀδόντας
ρπδ' πρὸς τὸ ἐὰν κα.ψη ἄνθρωπος τι τῶν μελῶν αὐτοῦ
.πε' πρὸς ποδαλγικούς
ρπϛ' ὅταν οἱ ἁρμοὶ τῶν δακτύλων γένονται σκληροὶ καὶ ποιήσουν ὡς ἂν

1 πίαν] L 112 25 δάκη] L 113

πόρους, ἐπάνω εἰς τὰ ἄρθρα
ρπζ’ πρὸς τὸ τινὰ ποιήσεις νὰ κρατηθῇ τὸ αἷμα.
ρπη’ ὅταν ὁ στόμαχος ἔχ.
πρὸς τὸ νὰ μὴ πέσουν ποτὲ αἱ τρίχες
πρὸς αὔξησιν τριχῶν κεφαλῆς
πρὸς τὸ νὰ γένονται μαῦραι τρίχαι
οζ’ πρὸς τὸ νὰ ποιήσει ξανθὰς τρίχας
οη’ πρὸς φθεῖρας τὰς ἐν τῇ κεφαλῇ
οθ’ πρὸς τὰ λεγόμενα γλυκέα ἤτοι λυκοκεφαλίαν
π’ πρὸς ἀλωπεκίαν καὶ ὀφίασιν
πα’ πρὸς τὸ ἐὰν αἰσθάνεται ὁ ἄνθρωπος ἐν τῇ κεφαλῇ
πβ’ πρὸς τὸ ἐὰν φαίνεται τὸν ἄνθρωπον ὅτι τὸ δέρμα τῆς κεφαλῆς αὐτοῦ ἀποταυρίζεται
πγ’ πρὸς τὸ ἐὰν φαίνεται αὐτὸν ὅτι κανεις κεντα τὸ δέρμα τῆς κεφαλῆς αὐτοῦ
π.’ πρὸς κεφαλαλγίαν
πε’ πρὸς τὸ ἐὰν πονεῖ τις ἀπὸ τὸ κρασὶν
πϛ’ πρὸς τὸ ἐὰν θέλης νὰ μηδὲν μηθύσης
πζ’ πρὸς ἀποφλεγματισμοὺς καθαίροντας τὴν κεφαλὴν
πη’ πρὸς τ. ἂν θέλης ἵνα καθαιρίσης τὸ κεφάλην σου ἀπὸ τὰ ὑγρά
πθ’ πρὸς τὸ ἐὰν θέλης ἵνα πτάρνης πολλά
ϙ’ πρὸς τὸ ἐὰν θέλης ἵνα μηδὲν πτάρνης
ϙα’ πρὸς τὸ ἐὰν ἀπὸ ψύξεως πονῇ τὴν κεφαλὴν αὐτοῦ
ϙβ ’ πρὸς τὸ ἐὰν πονῇ τὴν κεφαλὴν αὐτοῦ ἀπὸ τὸν ἥλιον
.γ’ πρὸς τὸ ὅταν ῥέει τὸ ὑγρὸν ἀπὸ τὸν οὐρανίσκον αὐτοῦ
.δ’ πρὸς τὸ ἐὰν ἀπὸ ψύξεως ῥέει τὸ ὑγρὸν ἀπὸ τὸν οὐρανίσκον
πρὸς τὸ ἐὰν πέσουν αἱ τῶν ὀφρύων τρίχες
πρὸς τὸ ἐὰν θέλης νὰ ποιήσεις τὰς τρίχας τῶν ὀφρίων μαῦρας
ϙζ’ πρὸς τρίχας ὅταν εἶναι διπλαὶ εἰς τὰ ὀφρύδια
ϙη’ πρὸς τὸ ὅταν τὸ βλέφαρον τοῦ ἀνθρώπου ἔχη κριθάριν
ϙθ’ πρὸς τὸ ὅταν τὸ ἄσπρον τοῦ ὀφθαλμοῦ γένηται κόκκινον πολλὰ καὶ ὡς ἂν φαίνεται ὅτι μεγαλότερον ἔνι τὸ ἄσπρον παρὰ τὴν κόρην
ρ’ πρὸς τὸ ὅταν τὸ βλέφαρον ὀγκοθῇ
ρα’ πρὸς σκληροφθαλμίαν
ρβ’ πρὸς ξηροφθαλμίαν
πρὸς τὸ ὅταν τὸ βλέφαρον τοῦ ἀνθρώπου ἐξηστραφῇ
πρὸς τὸ ὅταν εἰς τὸν μέγαν κανθὸν τ.. ὀφθαλμοῦ εἰς τὴν ἀπέσω μερέαν .... πλήσιον τῆς μύτης γένηται ἀ..στημα τὸ λέγουσι οἱ ἰατροὶ αἰ...λωπα
πρὸς τὸ ὅταν πέσ... αἱ τρίχες τῶν βλεφάρων
πρὸς τὸ ὅταν ἀπὸ τὸν μέγαν κανθὸν .εῖ τὸ τὸ δάκρυον παρὰ τὸ θέλημα.

4 πρὸς] L 114 20 τὰ] L 115 33 πρὸς] L 116

ὅπερ τὸ πάθος τοῦτο καλεῖται παρὰ τῶν ἰατρῶν ῥυὰς ἤτοι ἀπὸ τοῦ ῥέειν
πρὸς τὸ ὅταν ὁ ὀφθαλμὸς ἔχει ἄσπρον
ρϛ' πρὸς πτερύγιον ὀφθλαμοῦ
ρη' πρὸς ἀμβλυοπίαν καὶ ἀμαύρωσιν
ρθ' πρὸς ἀμβλυοπίαν
ρι' ὅταν ὁ ὀφθαλμὸς ἔχει ἄσπρον
ρια' ὅταν τὸ κάτω βλέφαρον γένηται μαῦρον
ριβ' ὅταν τὸ κάτω βλέφαρον γένηται κόκκινον
..γ' ὅταν γένηται ἀπέσω εἰς τὸ ὠτὶ πόνου διὰ ψύξιν ἀπὸ τῆς .οιάδας
ὅταν γένηται πόνος ἀπέσω ..ς τὸ ὠτίον διὰ ἥλιον πολύ
ὅταν .... θερμὴν δυσκρασίαν γένηται ... ὠτίον ἀπέσω πόνος
ὅταν γε..ται πόνος εἰς τὸ ὠτίον ἀπέσω ...ὰ παχεῖς χυμούς
πρὸς φλεγμονὴν .τίου
πρὸς αὔξησιν ὀδόντος παιδίου
ὅταν γένηται ὄγκος εἰς τὴν γλῶσσαν ὑποκάτω
πρὸς ἄφθας
ὅταν τὰ χείλη τῶν παιδίων ὅπου βιζάνουσι φαγένονται
ρκα' ὅταν φλεγμένουσιν οἱ σιαγῶνες ὅπου σαλεύουσιν ὅταν τρώγης
ρκβ' πρὸς βῆχας
δυναμίαν
ρπθ' ὅταν ὑπάγῃ πολλὰ ἡ κοιλία
πρὸς τὸ ἐὰν ἔχῃ ζέσιν ὁ στομαχός
πρὸς θεραπείαν σπληνός
πρὸς στρόφον ἀνθρώπου
πρὸς ἐὰν δάκῃ τινα ὁ σκορπίος
ὅταν συμβῇ εἰς πληγὴν καὶ γίνονται σκουλίκια
πρὸς τὸ ἐὰν δάκῃ τινὰ ἡ θαλασσία τρυγών
ρϙβ' πρὸς σπληναρίους καὶ πρισμένα
ρϙγ' πρὸς φλεγμονάς
ρϙδ' ὅταν κρατηθῇ ἡ φωνὴ τοῦ ἀνθρώπου
ρϙ.' πρὸς φθείρας ἐν τῇ κεφαλῇ
ρϙϛ' πρὸς χοιράδας τὰ λεγόμενα χελωνικά
ρϙζ' πρὸς ψώρας
ρϙη' πρὸς πόνον ὠτίου
ρϙθ' πρὸς τὸ ἐὰν θέλῃ νὰ μὴ διψᾷ τις
σ' πρὸς ὅταν ὁ ἄνθρωπος εὐγάνει ὡς ἂν φακὴν εἰς τὸ πρόσωπον
σα' ὅταν ἐκβάλῃ ὡς ἂν φακὴν ἄσπρην
σβ' πρὸς ἀλωπεκίαν ἤτοι λυκοκεφαλίαν
σγ' πρὸς παρωτίδας
σδ' ὅπου κατουρεῖ αἷμα

5 οπίαν] L 117 20 δυναμίαν] L 118 40 ρεῖ] L 119

πρὸς βήχας
πρὸς ὅταν ....μεῖ τὸ στόμα
πρὸς σῆψιν οὔλῶν
σϛ' πρὸς βουβῶνας ἤτοι ἀπορυφα....
σζ' πρὸς βιασμούς
ση' ὅταν καταπίῃ τις βδέλλας
σθ' πρὸς τὸ νὰ καταβασ... γυνὴ γάλα
σι' πρὸς τὸ ὅταν νὰ φαίνονται τὰ γόνατα κρύα ὡς ἂν χιώνην
σια' πρὸς δυσουρίαν
σιβ' πρὸς δυσεντερίαν
.ιγ' πρὸς τὰ λεγόμενα συνήθη γυναικῶν
.ιδ' πρὸς ἐσωχάδας
.ιε' ὅταν ὁ κόλος τοῦ ἄν.... εὐγένει
σιϛ' ὅταν ἔχῃ τις πλατέας ἑλμίνθας
σιζ' πρὸς ἐκβατὰ καὶ ἀπορυφὰς
σιη' ὅπου κατουρεῖ τὴν νύκτα
σιθ' ...ταν δάκῃ τινὰ ἔχιδνα
σκ' πρὸς ἡμίκρανον
σκα' ὅταν τις πονεῖ τὸ ἰσχίον του
σκβ' ὅπου ἔνι τὸ πρόσωπόν του κίτρινον
σκγ' πρὸς λέπρας
σκδ' πρὸς ψώραν
σκε' πρὸς λιχίνας
σκϛ' ὅταν θέλῃς νὰ νοήσῃς παρθένον
σκζ' πρὸς μὴ

BIBLIOGRAPHY

1. *Primary sources*

Aetius Amidenus. *Libri Medicinales.* Ed. A. Olivieri (Corpus Medicorum Graecorum VIII 1-2). 2 Vols. Lipsiae et Berolinae : Teubner et in aedibus Academiae Litterarum, 1935-1950.

Alexander von Tralles. *Original-Text und Übersetzung nebst einer einleitenden Abhandlung. Ein Beitrag zur Geschichte der Medicin.* Ed. Th. Puschmann. Wien: Wilhelm Braumüller, 1878.

The Chronicle of Morea, το χρονικον του Μορεως, *a History in Political Verse, Relating the Establishment of Feudalism in Greece by the Franks in the Thirteenth Century.* Ed. J. Schmitt. London: Methuen, 1904.

Claudius Galenus. *Opera Omnia.* Ed. C.G. Kühn. 20 vols. Lipsiae: Knoblochii, 1821-1833.

Hunain ibn Ishaq. *Le livre des Questions sur l'* . Ed. P. Sbath, M. Meyerhof (Mémoires présentés a l'Institut d'Égypte vol. 36). Kairo: Imprimerie de l'Institut Français d'Archéologie Orientale, 1938.

Ιωαννου του Σταφιδου Ιατροσοφιον. *Extraits.* Ed. É. Legrand (Bibliothèque grecque vulgaire vol. 2). Paris: Maisoneuve 1881.

Paulus Aegineta. Ed. I.L. Heiberg (Corpus Medicorum Graecorum 9.1.2). Lipsiae et Berolinae: Teubner, 1921-1924.

Stephanus the Philosopher. *A commentary on the prognosticon of Hippocrates.* Ed. John M. Duffy (Corpus Medicorum Graecorum 11.1.2). Berlin: Akademie-Verlag, 1983.

Theophanis Nonni Epitome de curatione morborum graece et latine, ed. I.O. Steph. Bernard. Gothae: C.W. Ettinger, Amstelodami: J.St. von Esveldt Heltrop et soc. 1794.

2. *Dictionaries*

M. Contopoulos 1880-1882. *A New and Complete Modern Greek and English Dictionary.* London: D. Nutt.

E. Kriaras 1968-1997. Λεξικο της μεσαιωνικης Ελληνικης δημωδους γραμματειας, 1100-1669. Thessalonike.

Kriaras online (www.komvos.edu.gr/dictionaries/dictonline/DictOnLineKri.htm).

A. Kyriakides 1909. *Modern Greek-English dictionary with a Cypriote vocabulary.* Athens: A. Constantinides.

H.G. Liddell, R. Scott, H.S. Jones 1996. *A Greek English Lexicon, with a Revised Supplement.* Oxford: Clarendon Press.

E.A. Sophocles 1887. *Greek lexicon of the Roman and Byzantine periods (from B.C. 146 to A.D. 1100).* New York: Scribner.

E. Trapp 1994-2001. *Lexikon zur byzantinischen Gräzität besonders des 9.-12. Jahrhunderts.* Wien: Verlag der Östereichischen Akademie der Wissenschaften.

3. *Secondary literature*

A.M. Bandini 1770. *Catalogvs Codicvm Graecorvm Bibliothecae Lavrentianae.* Florentiae.

D. Bennett 2003. *Xenonika. Medical Texts Asssociated with* Xenones *in the Late Byzantine Period.* Unpublished PhD thesis. University of London.

L. Bliquez 1984. "Two Lists of Greek Surgical Instruments and the State of Surgery in Byzantine Times". *Dumbarton Oaks Papers.* 38: 185-204.

L. Bliquez 1999. "The Surgical Instrumentarium of Leon Iatrosophista" *Medicina nei Secoli.* 11: 291-322.

M. Capone Ciollaro 2006. "Un ricettario medico attribuito a Giovanni Archiatra". In V. Boudon, A. Garzya,

J. Jouanna, A. Roselli (Eds.) *Ecdotica a ricezione dei testi medici greci. Atti del V Convegno Internazionale Napoli, 1-2 ottobre 2004 (Collectanea 24).* Napoli: D'Auria. 213-230.

P. Codellas 1942. "The Pantocrator, the Imperial Byzantine Medical Centre of 12th Century A.D., in Contstantinople". *Bulletin of the History of Medicine.* 12: 392-410.

D. Constantelos 1968. *Byzantine Philanthropy and Social Welfare (Rutgers Byzantine series).* New Brunswick: Rutgers University Press.

D. Constantelos 1999. "Medicine and Social Welfare in the Byzantine Empire". *Medicina nei Secoli.* 11: 337-355.

C. Constantinides, R. Browning 1993. *Dated Manuscripts from Cyprus to the Year 1570 (Dumbarton Oaks Studies 30, Texts and Studies of the History of Cyprus, Cyprus Research Center 18).* Washington, D.C. : Dumbarton Oaks Research Library and Collection, Nicosia : Cyprus Research Centre.

Ch. Daremberg 1853. *Notices et extraits des manuscrits médicaux grecs, latins et français des principales bibliothèques de l'Europe. Manuscrits grecs d'Angleterre.* Paris: Baillère.

W.R. Dawson 1932. *Manuscripta medica. A Descriptive Catalogue of the Manuscripts in the Library of the Medical Society of London* London.

H. Diels 1906. "Die Handschriften der antiken Ärzte. II Die übrigen griechischen Ärzte ausser Hippokrates und Galenos (Abhandlungen der Preussischen Akademie der Wissenschaften, Philosophisch-Historische Klasse)". Berlin: Verlag der Königlichen Akademie der Wissenschaften.

J.A. Fabricius 1724. *Bibliotheca Graeca.* Hamburgi: Sumtu Theodori Christophori Felgineri.

P. Franchi de' Cavalieri, G. Muccio 1896. "Index codicum graecorum Bibliothecae Angelicae". *Studi Italiani di Filologia Classica.* 9: 7-184.

A. Garzya 2003. "Pour l'édition des Iatrosophia démotiques". In A. Garzya, J. Jouanna (Eds.) *Trasmissione e ecdotica dei testi medici greci. Atti del IV Convegno Internazionale Parigi 17.-19. maggio 2001* (Collectanea 21). Napoli: D'Auria.

I. Hardt 1806. *Catalogus codicum manuscriptorum bibliothecae regiae Bavaricae.* Monachii: J.E. Seidel.

I. Hardt 1812. *Catalogus codicum manuscriptorum bibliothecae regiae Bavaricae.* Monachii: J.E. Seidel.

P. Horden 2005a. "How Medicalized were Byzantine Hospitals?" *Medicina e Storia.* 10: 45-74.

P. Horden 2005b. "Memoria, Salvation and Other Motives of Byzantine Philanthropists". In M. Borgolte (Ed.) *Stiftungen in Christentum, Judentum und Islam vor der Moderne (Stiftungsgeschichten 4).* Berlin: Akademie. 137-146.

H. Hunger 1969. *Katalog der griechischen Handschriften der Österreichischen Nationalbibliothek (Museion n.F. 1).* Vienna: Verlag Brüder Hollinek.

H. Hunger, H. 1978. *Die hochsprachliche profane Literatur der Byzantiner (Handbuch der Altertumswissenschaft* 12.5) 2 vols. München.

H. Hunger 1992. *Katalog der griechischen Handschriften der Österreichischen Nationalbibliothek (Museion n.F. 1).* Vienna: Verlag Brüder Hollinek.

A. Jakovljevic 2002. *Catalogue of Greek Manuscripts in the Library of the Monastery of St. Neophytos (Cyprus).* Nicosia: Holy Basilike and Stauropegiake Monastery of St. Neophytos, Cyprus.

E. Jeanselme 1930. "Sur une aide-mémoire de thérapeutique byzantin contenu dans un manuscrit de la Bibliothèque Natinale de Paris (suppl. gr.

764)". In Ch. Diehl (Ed.) *Mélange Charles Diehl. Études sur l'histoire et sur l'art de Byzance.* Paris: E. Leroux. 147-170.

G. Karas 1994. Οι επιστημες στην τουρκοκρατια. Χειρογραφα και εντυπα. Οι επιστημες της ζωης. (Κεντρο Νεοελληνικων Ερευνων 47). Athens: Βιβλιοπωλειον της εστιας.

A. Kouzis 1929. Παρατηρησις επι τινων ανεκδοτων ιατρικων κωδικων των βιβλιωθηκων της Ευρωπης και καθορισμος ενιων τουτων. Επητερις εταιρειας Βψζαντινων σπουδων. 6:375-382.

M. Maniaci 2002. "Papier". In H. Cancik *Der Neue Pauly* 15/2: 62-65. Stuttgart/Weimar: Metzler.

A. McCabe 2006. *A Byzantine Encyclopaedia of Horse Medicine: the Sources, Compilation and Transmission of the Hippiatrica (Oxford Studies in Byzantium).* Oxford: Oxford University Press.

T. Miller 1984. "Byzantine Hospitals". *Dumbarton Oaks Papers.* 38: 53-63.

T. Miller 1985. *The Birth of the Hospital in the Byzantine Empire.* Baltimore: Johns Hopkins University Press.

T. Miller 1997. *The Birth of the Hospital in the Byzantine Empire.* Baltimore: Johns Hopkins University Press. 190-206.

V. Nutton 1986. Review of T. Miller 1985. *The Birth of the Hospital in the Byzantine Empire. Medical History.* 30: 218-221.

H. Omont 1888. *Inventaire sommaire des manuscrits grecs de la Bibliothèque Nationale, seconde partie, ancien fonds grec, droit - histoire - sciences.* Paris A. Picard.

J. Sonderkamp 1987. *Untersuchungen zur Überlieferung der Schriften des Theophanes Chrysobalantes (sog. Theophanes Nonnos)* (Ποικιλα Βυζαντινα 7) Bonn: Habelt.

O. Temkin 1932. "Geschichte des Hippokratismus im ausgehenden Altertum". *Kyklos* 4: 1-80.

A. Touwaide 2007. "Byzantine Hospital Manuals (Iatrosophia) as a Source for the Study of Therapeutics". In: B. Bowers (Ed.) *The Medieval Hospital and Medical Practice.* Aldershot: Ashgate. 147-173.

B. Zipser 2005. "Zu Aufbau und Quellen der σύνοψις ἰατρικῆς des Leo medicus". In: Th. Fögen (Ed.) *Antike Fachtexte / Ancient Technical Texts.* Berlin/New York: Walter de Gruyter. 107-115.

INDEX

Materia medica

Compounds

Instruments and tools

Synonyms

Doctors and Names